Acoustic Communication in Insects and Anurans

Acoustic Communication in Insects and Anurans

Common Problems and Diverse Solutions

H. Carl Gerhardt
and Franz Huber

The University of Chicago Press
Chicago and London

H. Carl Gerhardt is Curators' Professor of Biological Sciences at the University of Missouri, Columbia.

Franz Huber is professor emeritus and retired scientific member of the Max Planck Society and former director of the Division for Neuroethology at the Max-Planck-Institute for Behavioral Physiology, Seewiesen.

The University of Chicago Press, Chicago 60637
The University of Chicago Press, Ltd., London

Printed in the United States of America

11 10 09 08 07 06 05 04 03 02 1 2 3 4 5

ISBN: 0-226-28832-3 (cloth)
ISBN: 0-226-28833-1 (paper)

Library of Congress Cataloging-in-Publication Data
Gerhardt, H. Carl.
Acoustic communication in insects and anurans: common problems and diverse solutions / H. Carl Gerhardt and Franz Huber.
p. cm.
Includes bibliographical references (p.).
ISBN 0-226-28832-3 (cloth) — ISBN 0-226-28833-1 (paper)
1. Insects—Behavior. 2. Anura—Behavior. 3. Animal communication.
I. Huber, Franz, 1925- II. Title.
QL496 .G35 2002
595.7159'4—dc21

2001008256

∞ The paper used in this publication meets the minimum requirements of the American National Standard for Information Sciences—Permanence of Paper for Printed Library Materials, ANSI Z39.48-1992.

Dedicated to the memory of Lore Huber

Contents

Preface

Animal communication is not only fascinating, but its study has also led to significant progress in our general understanding of motor and sensory systems, evolution, and speciation. A major appeal of studying communication is that a researcher can quantify how biologically important information might be coded in particular physical properties of a signal and then experimentally determine if the animals themselves use this information. For some systems, the selective consequences of behavioral decisions based on such information can also be quantified. Acoustic communication is especially well suited for this kind of approach because of the relative ease with which sounds can be recorded, analyzed, synthesized, and played back with high fidelity.

Studies of acoustic communication in insects and frogs are especially powerful because individuals of many species respond so reliably to playbacks in both the laboratory and in the field. For many species, large numbers of individuals can be tested in order to provide a quantitative analysis of receiver selectivity for information-bearing properties of signals. Such data have provided a framework for studying the mechanisms underlying signal recognition and sound localization as well as for generating predictions about patterns of sexual selection and speciation. Moreover, unlike many birds and mammals, the development of species-typical signals and their recognition does not require learning from other conspecific individuals. Some insect systems are amenable to quantitative genetic studies, and external fertilization in frogs allows the design of rigorous experiments that can reveal direct or genetic benefits that females might obtain by choosing males on the basis of specific acoustic properties. Studies of insects and anurans have provided us with important insights concerning many aspects of neural systems and animal behavior and will undoubtedly continue to do so.

This book began as a joint project during HCG's six-month visit, supported by a Senior Humboldt fellowship, at the Max-Planck-Institute for Behavioral Physiology at Seewiesen, Germany. However, the idea of considering acoustic communication in insects and frogs followed naturally from our mutual appreciation of the parallels between crickets and frogs that was already evident during our very first discussions—also at Seewiesen—in 1979. As we show in this book, these very different animals face a similar set of problems that have led to both convergent and divergent solutions.

It was also clear from the beginning that we share a philosophy about how communication should be studied. Namely, we believe that the greatest progress comes from a multidisciplinary approach that combines quantitative studies of the natural, adaptive behavior of the animal with studies that seek to discover the underlying mechanisms. We also appreciate the power of the comparative approach, which is necessary to discover generalizations and which can also provide insights about the evolutionary history of communication systems. These ideas are not new, and we acknowledge the inspiration of our mentors and several colleagues: the late W. Frank Blair (HCG), the late Karl von Frisch (FH), the late Werner Jacobs (FH), the late Kenneth D. Roeder (FH), Robert R. Capranica (FH and HCG), Mark Konishi (FH and HCG), M. J. Littlejohn (HCG), and Peter Marler (HCG).

Our wives—the late Lore Huber, to whom this book is dedicated, and Dayna Glanz—not only encouraged and supported us but facilitated the best part of writing this book: our personal time together in both Germany and the United States. HCG would also like to thank his parents—Howard and Elizabeth—who always encouraged his passion for biology.

We owe a great deal of gratitude to our colleagues, who gave us encouragement, critical feedback, and access to unpublished material. Friedrich Barth, Gerhard Neuweiler, Dale Roberts, and Heiner Römer read the entire manuscript, and the following people commented on one or more chapters: Winston Bailey, Henry Bennet-Clark, Rex Cocroft, Norbert Elsner, Albert Feng, Ulmar Grafe, Michael Greenfield, Berthold Hedwig, Otto and Dagmar von Helversen, Matthias Henning, Mark Konishi, Murray Littlejohn, Axel Michelsen, Mohammed Noor, Gerald Pollack, Christopher Murphy, Peter Narins, Mike Ritchie, Bernd Ronacher, Michael Ryan, Johannes Schul, Kerry Shaw, Leigh Simmons, Andreas Stumpner, Bill Wagner, Wolfgang Walkowiak, and Kentwood Wells. Finally, the book is much improved because of the comments and questions raised by students and postdocs at the University of Missouri: Mark Bee, Sarah Bush, Robert Dudley, Gerlinde Höbel, Sarah Humfeld, Michael Keller, Carlos Martinez, Vincent Marshall, Karthik Ramaswamy, Rafael Rodríguez, Joshua Schwartz, and Allison Welch.

HCG is especially grateful to Johannes Schul and Winston Bailey for teaching him so much about insects and to Rex Cocroft, Mike Ritchie, Johannes Schul, and Henry Bennet-Clark for in-depth discussions of controversial or technically difficult issues. We have not followed all of the advice we received and take responsibility for the inevitable mistakes that remain.

We also thank Theo Weber and Heidrun Bamberg, Max-Planck-Institute for Behavioral Physiology, for preparing the figures. HCG also thanks John D. David, director of the Division of Biological Sciences, for his tremendous support of all aspects of scholarship. Christie Henry at the University of Chicago Press provided enthusiastic encouragement and help at all stages, and Erin DeWitt did a fine job of copyediting. Susan Danzi Hernandez prepared the index.

1

Introduction

IMMENSE AGGREGATIONS of sound-producing insects and anurans (frogs and toads) are among the most impressive of biological phenomena. In North America, for example, periodical cicadas (genus *Magicicada*) emerge in populations of millions of individuals per hectare and produce painfully loud sounds. In many parts of the world, choruses of frogs produce a continuous din of sound extending for kilometers along the shoreline of lakes and rivers. In both insects and anurans, mixed-species choruses of more than a dozen species are common. Whether or not individuals signal from aggregations, each signal has relatively invariant features that can be as useful as morphological traits for identifying the species of the signaler. At the same time, other features of the signal vary within and among individuals in ways that provide information affecting the responses of rivals and prospective mates. This book considers the information content of acoustic signals, how signals are produced, recognized, and localized, their current communicative functions, and the evolutionary history of signal diversification and recognition mechanisms.

We focus on orthopteran insects (crickets, grasshoppers, and katydids), cicadas, and anurans for three main reasons. First, their communication behavior develops normally in the absence of even the opportunity to learn from other members of the same species, thus ensuring that much of the diversity within and between species is not attributable to individual differences in acoustic experience. Second, these animals face a set of common problems in the context of communication. The diversity of their solutions to these problems is impressive, and yet insects and anurans often adopt remarkably convergent solutions. Third, in comparison to other insect groups and most vertebrates, we know a great deal about many different attributes of their acoustic communication systems. Although we sometimes consider other insect groups, readers interested in broad taxonomic surveys should consult Ewing (1989), Bailey (1991), and Greenfield (2002). Greenfield (2002) also considers communication by chemical, visual, and tactile signals and points out that long-distance communication by sound is uncommon among insects other than orthopterans and cicadas.

Our main concern is intraspecific communication, which critically affects reproductive success in both senders and receivers; moreover, divergence in communication systems is a hallmark of speciation. Our definition is operational (appendix 1; Wilson 1975). That is, *communication*[1] occurs if after a signal is produced, there is a reduction of uncertainty about a signaler's identity or subsequent behavior or a change in the behavior of the receiver. This definition emphasizes observable events, such as signaling and *phonotaxis*, which is orientation relative to the location of a sound source. Such observations, in turn, raise fundamental questions about mechanisms. What mechanisms of sound production ensure that information about the signaler is reliably encoded in the acoustic waveform and transmitted to a distant receiver? How do receivers process acoustic signals and extract this information, especially in the noisy environments in which insects and anurans typically communicate?

Other questions about communication focus on the selective consequences of these behaviors, which are expected, on average, to benefit both signalers and receivers (Wiley 1983). Why are receivers highly selective with respect to variation in some aspects of a signal and unselective with respect to discriminable differences in other properties? What benefits, if any, do receivers obtain from mate choice based on particular acoustic criteria? What are the causes and consequences of signaling in aggregations?

Still other questions concern evolutionary history. How is evolutionary change in signals coordinated in time and space with that in receiver capabilities and preferences? How does the acoustic environment, which includes both physical and biotic components, influence signal design? What role does acoustic communication play in speciation?

We believe that the greatest progress in understanding acoustic communication results from efforts to answer as many of these different kinds of questions as possible in the same system. For example, the selectivity of receivers for variation in a particular acoustic property is likely to depend on its reliability in encoding information that currently benefits receivers or has done so in the past. At the same time, the nervous system's capacity to process differences in relevant acoustic properties and to localize the source of the signal sets limits on the selectivity of receivers.

Common Problems

Common problems in the realm of acoustic communication arise from similarities between insects and anurans in their physical and physiological attributes and natural history (table 1.1). The similarities and differences in the

1. Special terms will be italicized the first time they are defined, or, for some italicized anatomical terms, the reader will be referred to a figure.

Table 1.1. Common Problems and Diverse Solutions in Acoustic Communication by Insects and Anurans

Attribute	Problems/Constraints	Some Solutions	Chapters
Physical and Physiological Traits			
Small size	High-frequency signals: limited range because of excess attenuation, scatter	Baffles, burrows, elevated call sites Shift to short-range communication, use of other modalities Add mass to oscillator to lower frequency	2, 9, 10, 11
	Energetically inefficient and costly signaling	Limit signaling to optimal times for propagation or mate attraction Signal on or near food source Switch to alternative mate-acquisition tactics	2, 8, 9
	Closely spaced ears: small external binaural cues	Pressure-difference sound localization or use of very high-frequency signals	5, 7
Ectothermy	Temporal changes in signals and temporal, spectral shifts in auditory selectivity	Temperature coupling Wide acceptance range of receiver Restrict signaling to narrow temperature range	2, 4, 5, 6
Natural History			
Time constraints	Limited opportunity for learning or signal assessment	Developmental canalization of signal structure and auditory selectivity	2, 3, 8, 9
Signaling in aggregations	Masking interference and competition for resources	Temporal and spatial partitioning of the breeding habitat (inter- and intraspecific) Antiphonal signaling (alternation) Directional hearing (release from masking) Graded aggressive signaling, neighbor recognition Switch to alternative mate-acquisition tactics	5, 7, 8, 9, 11
Acoustically orienting predators/parasites	Increased signaling reduces survivorship	Reduction in signaling period, signal duration Switch to alternative mate-acquisition tactics	8, 9, 11

solutions to these problems within these two taxonomic groups provide insights about the mechanisms, function, and evolution of acoustic communication in general.

Physical and Physiological Attributes

Small size. The relatively small size of most insects and many species of anurans poses three major problems for communication by sound. First, because effective production of low-frequency sounds requires sound radiators with large dimensions, small size generally results in the production of relatively high-frequency signals (chapter 2; Michelsen 1983; Bennet-Clark 1998). High-frequency signals are more subject to attenuation and degradation than are low-frequency signals and hence limit communication range (appendix 3 and chapter 11; Wiley and Richards 1978; Michelsen 1983). Some small species of insects and frogs increase their broadcast range by behavioral tactics, such as using plants as acoustic baffles (e.g., Forrest 1991), by calling from elevated positions (Arak and Eíriksson 1992), or by emitting signals from burrows that function like horns (Bennet-Clark 1989). Other exceptional species (bladder grasshoppers, some cicadas and anurans) or groups, such as crickets, have structural modifications that allow them to produce much lower frequencies than other species of comparable size (chapter 2). These exceptions are especially important to study because of the generally strong correlation between body size and *carrier* or *dominant frequency*, which is the frequency component with the greatest acoustic energy. Such a relationship potentially provides information to receivers about the signaler's size, but different selection pressures on the two traits—size and frequency—will often be confounded and sometimes run counter to each other.

The second problem imposed by small size is that sound production will generally be more inefficient and energetically more costly than in large animals (chapter 9; Prestwich 1994; Wells 2001). In some species of anurans, individuals participating in many choruses over a prolonged breeding season suffer remarkably large losses of body mass. Energetic limitation is probably the main reason that most individuals attend relatively few breeding aggregations (Murphy 1994b), and such constraints might also play a role in promoting alternative mate-acquisition tactics (Lucas et al. 1996). Although a subject of controversy, energetic limitations are also likely to be important for some kinds of insects (Wagner and Hoback 1999).

A third problem of small size is that the narrow separation of insect and anuran ears reduces the magnitude of external differences between the two ears in sound intensity and time of arrival (chapter 7). Such external binaural cues are sufficient for sound localization in large animals, which have larger head dimensions, but in many insects and small frogs that do not produce very high-frequency signals, the available external cues are inadequate. Some insects and frogs have therefore adopted a different mechanism—that of a pressure-difference system—to generate binaural differences (review: Michelsen 1998a). Pressure-difference receivers can be highly dependent on sound frequency, a dependency that has other consequences for acoustic communication, such as limiting the range of frequencies that can be effectively used in advertisement calls. Finally, mechanisms of sound localization can interact with mechanisms involved in pattern recognition (Helversen 1984; Stabel et al. 1989), sometimes improving and sometimes interfering with the latter process (chapter 7).

Ectothermy. Some temporal properties and, to a lesser extent, spectral properties of insect and anuran signals are influenced by temperature (chapter 2). In some species, females shift their preferences in a roughly parallel fashion so that at any particular temperature females prefer the signals of a male calling at about the same temperature (chapter 4; Gerhardt 1978c; Doherty 1985b). In other species, receivers have a broad range of acceptance that covers all of the conspecific signals likely to be heard over the normal range of breeding temperatures (Helversen and Helversen 1983). In still other species, however, a shift in temperature can cause a shift in frequency sensitivity to spectral bands that are not emphasized in conspecific calls (chapters 3–5; Gerhardt and Mudry 1980). This is a clear example of how hearing mechanisms can impose constraints on communication or, seen another way, bias the selectivity of the animal toward signals that are not usually produced by conspecific individuals (e.g., Ryan 1990; chapter 11).

Natural History

Time constraints. Insects and anurans tend to be short-lived, have relatively brief episodes of breeding, or both. Parental care is also rare in the insects and anurans we consider in this monograph, and for many species an individual usually first hears the signals of conspecific individuals after it reaches sexual maturity. These factors probably contribute to the fact that, unlike many kinds of birds and mammals, no species of insect or anuran is known to learn any of its signals from parents or other conspecific individuals (Doherty and Hoy 1985; Gerhardt 1994a). Although learning to distinguish between the signals of different neighbors may occur in some territorial anurans (Bee and Gerhardt 2001a), limited time and opportunity for learning make it unlikely that acoustic experience generally affects the selectivity of receivers. The lack of learning and short breeding periods also create selective pressure for mating and other decisions to be made without extensive assessments and to be based on relatively simple acoustic criteria.

Signaling in aggregations. Choruses, whether dense or sparse, are focal points for competition for mates, which, in turn, require the resources (oviposition sites, food) that usually determine where and when choruses occur (chapter 8). This competition usually takes the form of the defense of resources through signaling and fighting, spacing of signalers, or some combination of these tactics. Spacing of signalers and signal-timing interactions between neighbors affect the extent of *masking interference*, whereby transmission to receivers is reduced by virtue of overlaps between signals of different individuals. Researchers have found a diversity of signaling interactions, ranging from alternation of mate-attracting signals to attempts to actively jam the signals of their rivals (Greenfield 1994a; 2002). These interactions between signalers are, in turn, influenced by the preferences of prospective mates, such as the commonly observed preference for leading calls (Greenfield 1994a; but see Klump and Gerhardt 1992; Grafe 1999). High levels of background noise generated by both conspecific and other signalers can also affect signaling behavior (Schwartz and Wells 1983a,b; Greenfield 1988) and the ability of receivers to detect and locate signalers (Gerhardt and Klump 1988a).

Masking interference among conspecific individuals might also be a factor promoting alternative mating tactics, such as *satellite behavior* (chapter 9). Here a male does not signal but situates himself near a calling male and tries to intercept females (e.g., Perrill et al. 1978). There are many possible factors that influence the adoption of alternative tactics, including the inability of some individuals to compete effectively for territories or calling sites and acoustically orienting predators or parasites. In particularly dense choruses, some males might not be able to produce calls with sufficient energy to be detected above the chorus background (Gerhardt and Klump 1988a).

Acoustically orienting predators and parasites. Predators and parasites as well as conspecific rivals and mates can use signals to identify and locate the signaler (review: Zuk and Kolluru 1998). The exploitation of communication signals undoubtedly affects the behavior of signalers and can perhaps promote the evolution of other kinds of signals, such as substratum- or plant-borne vibrations (e.g., Belwood and Morris 1987).

Many insects have ultrasonic frequency sensitivity, even if their signals lack such frequencies or even if they do not communicate by sound at all. This sensitivity is used to detect and evade echolocating bats (Hoy 1992), which have probably played a major role in the evolution of hearing in some groups and serve as a significant constraint on acoustic communication in other groups (e.g., Bailey 1991).

Advantages of Studying Acoustic Communication in Insects and Frogs

With few exceptions, insects and anurans have small repertoires of signals, and, as mentioned above, the structure of these signals is unlikely to be influenced by experience during their entire lifetime. The physical properties of signals provide hints about mechanisms of sound production, and statistical analysis of variation in these properties estimates their potential for encoding biologically important information. This potential can be directly assessed by playback experiments. Indeed, the greatest advantage of studying these animals is the reliability of playbacks in eliciting signaling and phonotaxis both in the field and in controlled laboratory situations. With rare exceptions, the selectivity of these responses appears to be unaffected by experience. For many insects and frogs, playback experiments using synthetic sounds, in which one or more properties can be varied systematically, have identified at least some of the information-bearing properties of communication signals and have estimated minimum changes in these properties that result in selective responses to different signals (reviews: Gerhardt 1988; Greenfield 2002). Playback experiments—including interactive designs in which computers vary the sounds played to subjects depending on the signals produced by those subjects (e.g., Schwartz 1994)—have also helped to discover "rules" governing signaling interactions (Greenfield 1994a; Grafe 1999). Finally, quantitative behavioral studies of phonotaxis have assessed the kinds and magnitudes of binaural cues required for sound localization (e.g., Helversen and Rheinlaender 1988; Jørgensen and Gerhardt 1991).

Insects and anurans are also good subjects for studies of the mechanisms underlying signal production, recognition, and localization. In insects, interneurons that trigger sound-producing circuits have been characterized both anatomically and physiologically. Moreover, in both insects and frogs, the orchestrated activity of motoneurons, muscles, or both that pattern acoustic signals has been described in detail (reviews: Gans 1973; Elsner 1994; Hedwig 2000a). Recordings from isolated brain preparations have begun to determine if and how pattern generators can be fine-tuned by peripheral feedback (e.g., Kogo et al. 1997). Insects and frogs are also suitable for studying sensory processes. As in specialized vertebrates such as electric fish (Heiligenberg 1994), barn owls (Konishi 1994), and bats (Suga 1989; Neuweiler 1999), single auditory neurons or populations of neurons within well-defined auditory nuclei respond selectively to sounds in ways that mirror the behavior of the whole animal (review: Feng and Schellart 1999). Many cells have been characterized physiologically, and connections among the relevant auditory nuclei are well-known. Orthopteran insects offer an additional advantage. All of the biologically relevant information available in the acoustic waveform is conveyed to the brain by a handful of ascending auditory neurons that are individually identifiable (Huber et al. 1989; Pollack 1998). Moreover, the responses of some of these neurons can even be recorded in behaving animals.

In the realm of evolutionary processes, playback experiments have generated estimates of the patterns and intensity of selection by female choice on male signals. Moreover, the potential benefits of mate choice based on particular acoustic properties can sometimes be identified and quantified (e.g., Gwynne and Simmons 1990; Robertson 1990; Brown et al. 1996; Welch et al. 1998). Playback experiments have also revealed hidden preferences for signals that are currently not present in the population or species (Ryan 1990). Such preferences may reflect evolutionary history and can bias the direction of future changes in communication systems. Some insects are also suitable for studying the genetic bases of acoustic structure and selective responsiveness (Ewing 1989; Shaw 1996b; Ritchie 2000). Other insects have been the subjects of artificial selection experiments that can estimate additive genetic variation and covariation in signal structure and receiver selectivity, which is assumed to arise from assortative mating (Butlin 1994: Bakker and Pomiankowski 1995).

In both insects and anurans, there are species complexes whose members show significant geographical variation in signals, or hybridize in narrow zones (review: Harrison 1990). Studies of these phenomena can address questions about the roles of the habitat acoustics, acoustic interference among species, and hybridization in population-level differentiation and, ultimately, in speciation (chapter 11). Traditional studies have recently been complemented by independent, phylogenetic information derived from molecular and morphological traits at the population, species, and higher taxonomic levels (e.g., Gleason and Ritchie 1998; Ryan et al. 1996). These data can sometimes throw light on the evolutionary order of appearance of new signals and auditory preferences or, less controversially, help to select appropriate populations or species for comparative studies that can test ideas about environmental factors that can lead to divergence or convergence in communication systems. The causal effect of environmental differences on signal divergence is, for example, more confidently inferred if the populations or species to be compared are closely related.

Acoustic communication in insects and anurans presents a diverse and fascinating set of opportunities for research by biologists interested in both proximate mechanisms and evolutionary processes. In addition to advocating an integration of the knowledge gained by these approaches, we highlight some especially attractive areas of current and future research on communication in these two groups of animals.

2

Acoustic Signals: Description and Peripheral Mechanisms

ACOUSTIC SIGNALS provide windows into the mechanisms and evolution of communication. In terms of mechanisms, the physical structure of a signal must reflect both the biophysics of the peripheral sound-producing structures and the activity of the neural circuits and muscles controlling these structures. In terms of function, all of the biologically important information transmitted from sender to receiver must be encoded in the physical structure of a signal. Different kinds of information can be carried by different and independently varying acoustic properties. Finally, comparisons of signal structure across populations and species can provide insights about speciation and the evolutionary history of communication systems. Understanding the mechanisms, function, and evolutionary history of acoustic signaling in insects and anurans also requires knowledge about other systems, such as locomotion and respiration, which usually share both peripheral structures and circuits that control their movements (Gans 1973; Ewing 1989). We first discuss the analysis and description of signals, emphasizing the advantages of statistical descriptions of variation at the individual, population, and species levels. We then describe the structure and function of peripheral structures that generate sounds. Here we emphasize the influence of these structures on the spectral (frequency content) and fine-scale (< 100 ms) temporal properties of signals. We also discuss how accessory structures augment and modify sound radiation.

Analysis and Description of Signals

Signals are analyzed and described in two fundamental ways. An acoustic analysis generates an inventory of the physical properties of a signal, such as the form (e.g., changes in amplitude or frequency with time) of repeating acoustic units, patterns of repetition of recurring units, carrier frequency or frequencies, and bandwidth. Appendix 2 presents some basic acoustic concepts. Box 2.1 shows common ways of displaying the amplitude versus time

Box 2.1. Description of Signals and Terminology

Classifying Acoustic Units

Some species of insects and anurans produce advertisement signals lasting a few milliseconds at intervals of seconds or minutes, others produce continuous trills for minutes or even hours, and still others organize pulses into complex sequences. Some examples are shown in fig. 2.1. These kinds of extremes make it difficult to define the duration of a signal or even to say what constitutes an episode, or bout, of signaling. Thus, many authors identify recurring acoustic units and then describe how they are temporally organized. In many species there are two or more higher-order patterns that repeat (examples in figs. 1, 2).

Acoustic elements in insect songs have also been defined in terms of mechanisms of sound production. For example, syllables in insect songs can be related to the sounds produced by one to-and-fro movement of the stridulatory structures (Broughton; Elsner 1974b; Ragge and Reynolds 1998) or by one closing movement of the forewings in field crickets, which produce sounds only during wing closing (e.g., Kutsch and Huber 1989). In a parallel fashion,

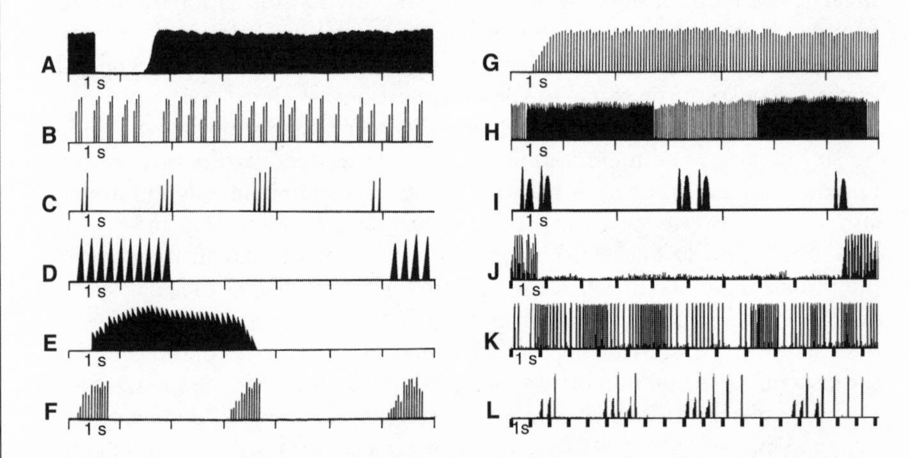

Box 2.1, Figure 1. Diagrams showing the gross temporal patterns (change in sound amplitude with time) of the calling songs of some gryllids (crickets and mole crickets) (A–C, *Gryllotalpa gryllotalpa*, *Gryllus campestris*, and *Acheta domesticus*, respectively), grasshoppers (D–F, *Chorthippus montanus*, *Gomphocerus rufus*, and *Chrysochraon dispar*, respectively), katydids (G–I, *Tettigonia cantans*, *Conocephalus dorsalis*, and *Ephippiger ephippiger*, respectively), and anurans (J–L, *Litoria ewingii*, *Hyla femoralis*, and *Litoria phyllocroa*, respectively). Each vertical line represents a pulse; solid areas are trains of pulses produced at such high rates that individual pulses are not resolved with the time base used. Whereas some species organize pulses into discrete trains that have a characteristic duration (e.g., B, C, D, F, J), others produce continuous or nearly continuous trains of pulses, which may be uniform in rate (A, G), alternate in rate (H), or have an irregular rate (K). A–I from Bellman 1985, pp. 63, 67 of English edition; J–L, original figures from the authors.

(box 2.1 continued)

McLister et al. (1995) define notes in frog vocalizations as the sound units produced by single expiratory events. For example, the complex pulse of the pine woods treefrog (fig. 2C) is produced by a single contraction of the body wall, which causes a single expiratory event. The subpulses are produced by passive amplitude modulation (see the text).

Whatever the rationale for identifying acoustic elements, attempts to establish a universal terminology are almost certainly doomed by traditional usage and the sheer diversity of signal structures (reviews: Thompson et al. 1994; Gerhardt 1998; Ragge and Reynolds 1998). Terms such as *pulse*, *impulse*, *syllable*, *note*, *echeme*, *chirp*, *element*, and *call* have all sometimes been used to label the very same kind of basic acoustic unit, even in the same taxa. Oscillograms and sonograms should thus be labeled with the terms chosen by the author in order to avoid the confusion that occurs because different researchers so often use different terminology. Ragge and Reynolds (1998) provide further discussion and examples of such labeling; they also propose a reasonable classification scheme for higher-order temporal structure in orthopterans (see also Alexander 1962).

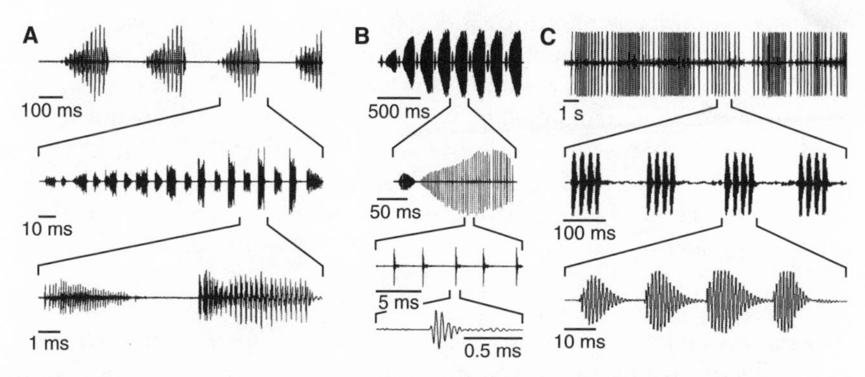

Box 2.1, Figure 2. Oscillograms of some acoustic signals. (A) *Top trace,* four verses (= echemes, phrases, notes, long pulses) of the katydid *Gampsocleis gratiosa; second trace,* one verse, showing pulses; *third trace,* two pulses (made during one opening and closing movement of the wings). Modified from Jatho 1995, fig. 3.30. (B) *Top trace,* one verse (= echeme, phrase, or long pulse) of the katydid *Ephippiger perforatus; second trace,* two pulses; *third trace,* five impulses from the pulse produced during wing closing; *fourth trace,* one impulse. Pulses are defined as the sound units made during one movement of stridulatory structures. Modified from Jatho 1995, fig. 3.11. (C) *Top trace,* pulses produced during approximately 10 s of calling by a treefrog *Hyla femoralis;* males of this species do not organize their complex pulses into discrete trains and vary pulse rate over a fairly wide range during a bout of calling; *second trace,* four pulses, each with four subpulses; *third trace,* one pulse with four subpulses. From a descriptive perspective, it would also be logical to consider each subpulse as a pulse (shortest recurring unit); the decision to describe these as subpulses is influenced by the fact that one body-wall contraction results in a complex pulse with subpulses. Modified from Gerhardt 1998, fig. 9.

waveform of insect and anuran signals and considers some of the challenges of acoustic terminology. A statistical analysis describes measures of central tendency (mean, median) and patterns of variability (range, standard deviation, coefficient of variation) in these physical properties. Together, physical and statistical analyses can be used to characterize the signals in the repertoire of a species and thus to provide a starting point for studying mechanisms of signal production.

Statistical analyses also provide hints about what biologically significant information might be available in each signal. Such information can include the signaler's (1) identity (individual, sex, population, and species); (2) size; (3) physical condition; (4) genetic fitness; and (5) motivational state, such as its readiness to mate or fight. Insects and anurans must often extract such information from arrays of signals that are emitted by two or more individuals of the same or different species. Thus, statistical estimates at different levels of analysis (within and among individuals, populations, species) can be used to assess the variation in signal properties that confront an individual. Such data are also used to assess variation among individuals that can be subject to selection, to characterize patterns of geographical variation, and even to help define species boundaries.

Although receivers are likely to attend to variation in signal properties that provide them with reliable information, studies of receiver selectivity are necessary to verify the communicative relevance of any acoustic property—regardless of its pattern of variability at any level of analysis. For example, as we show in chapter 4, even highly stereotyped properties of signals are not always used to identify conspecific signals. In subsequent chapters we examine the extent to which the potential information in signals can be decoded by the auditory system and used by prospective rivals and mates to make decisions about how to respond to the signaler.

Signal Repertoires

The most commonly heard insect and anuran signals (*calling songs*, *advertisement calls*) are used in long-range attraction of females by males; these same signals often affect signaling and aggressive interactions between males (Wells 1988; Loher and Dambach 1989). Many species also produce distinctly different signals during aggressive encounters with other signalers (*aggressive signals*, *encounter calls*, *rivalry songs*) (see chapter 9); and among anurans, both males and unreceptive females produce *release calls* when clasped by a male. Males and sometimes females also produce short-range *courtship* signals immediately prior to mating (Wells 1988; Ewing 1989; Schlaepfer and Figeroa-Sandí 1998; chapter 10). Finally, insects and anurans may produce distinctive signals when they are handled or caught by predators (*distress* or *protest* signals) (Alexander and Moore 1958; Dumortier 1963a; Simmons and Young 1978; Duellman and Trueb 1986; Ewing 1989). Although most species of insects and anurans have a limited repertoire consisting of a subset of these signals, several cicadas and

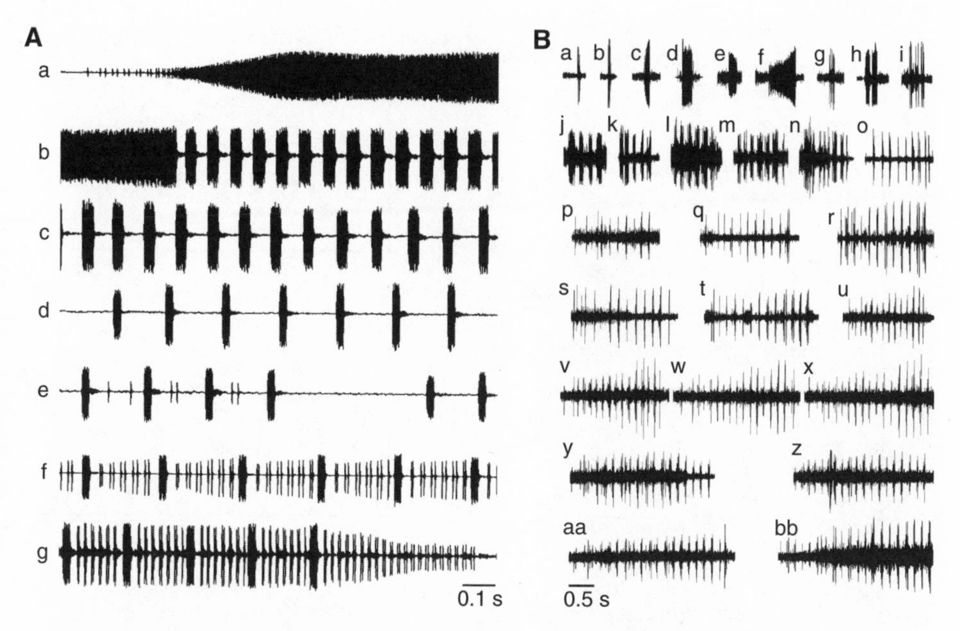

Figure 2.1. (A) Oscillograms of a song sequence by a cicada (*Purana* aff. *tigrina*) from southern Thailand. From Gogala 1995, fig. 5. (a) Beginning of sequence; (b) transition to a distinctly pulsed part of the song; (c) and (d) progressive decrease in pulse rate and duration; (e) first appearance of clicks between pulses; (f) double clicks plus pulses; (g) end of sequence, where the amplitude of double clicks equals that of pulses before the latter cease. The diversity of acoustic structures in this one sequence from a single individual is representative of the diversity that can be seen among different species. (B) Oscillograms of 24 of the 28 note types described by Narins and Lewis (1996) for the rhacophorid treefrog *Boophis madagascariensis* from Madagascar. The signals shown were not all produced by a single male. The first notes (a–g) differ in their spectral structure, temporal structure or both; notes h–bb (and the four types not shown) differ in number of clicks and were termed *iambic notes.* From Narins and Lewis 1996, fig. 2.

frogs produce a bewildering array of sounds whose diversity rivals that of a small community (e.g., Gogala 1995; Narins et al. 2000). We show two examples in figure 2.1. The functional significance of these different acoustic elements (or combinations thereof) is, however, still unknown.

This chapter is mainly concerned with long-range (advertisement) signals because they have been more thoroughly studied than other signals. Other signals in the repertoire nevertheless provide hints about the mechanisms and evolution of sound production. For example, as shown below, carrier frequency and pulse rate vary relatively little from one signal to the next in the advertisement signals of a typical individual insect or frog. However, these properties can have completely different values, or even show graded variation, in the aggressive or courtship signals of the same individual (chapters 9 and 10). Large differences in carrier frequency and pulse rate can also occur in the different elements of complex advertisement calls (e.g., figs. 2.1 and 2.2; box 2.1, fig. 1). This apparent flexibility suggests that constraints imposed by

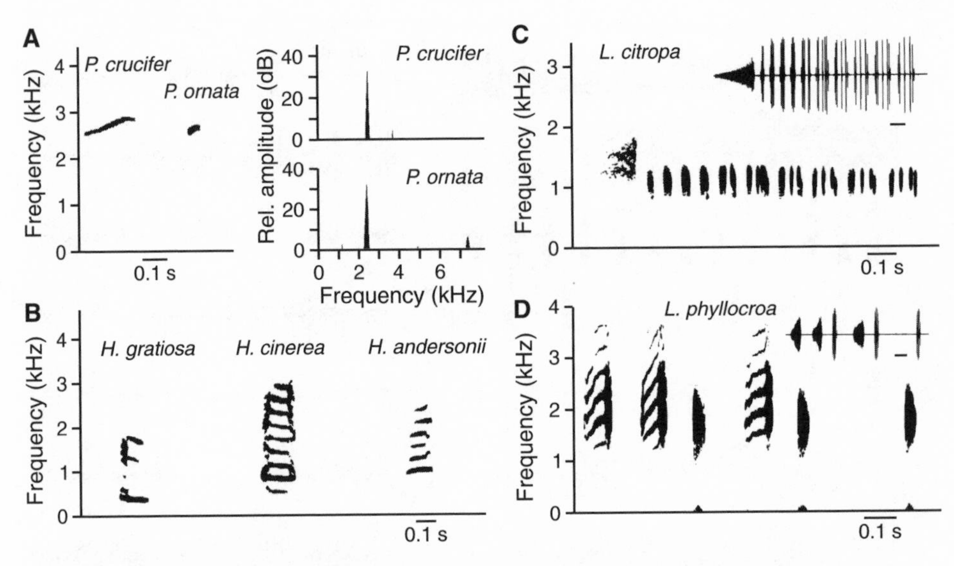

Figure 2.2. Qualitative differences in call properties that differ in spectral and temporal complexity. (A) Sonograms (*left*) and power spectra (*right*) of the relatively narrow-band (< 400 Hz) advertisement calls of the spring peeper *Pseudacris crucifer* and ornate chorus frogs *P. ornata*, which often share the same breeding sites in the southeastern United States. Both signals are nearly pure-tone bursts that show slight frequency modulation from beginning to end. (B) Sonograms showing the relatively broadband (> 1500 Hz) advertisement calls of the barking treefrog *Hyla gratiosa*, green treefrog *H. cinerea*, and pine barrens treefrog *H. andersonii*, which are often found in the same ponds in the southeastern United States. Frequency components in the last two species consist of harmonics and sidebands arising from amplitude modulation. From Gerhardt 1998. (C) Sonogram (*inset*, oscillogram) of the two-part advertisement call of the Blue Mountains treefrog *Litoria citropa* of Australia. The first part of the call is less variable than the second part, whose duration is determined by the number of repeating elements. (D) Sonogram (*inset*, oscillogram) of the two-part calls of the congeneric leaf-green treefrog *L. phyllocroa*, which is often found in the same streams as *L. citropa*. Notice that both the spectral and temporal properties of the call elements differ. In *L. citropa* the two-part calls are repeated at relatively long intervals. In *L. phyllocroa* the long and short elements are not produced in a fixed order, and the short elements are commonly produced in long sequences.

mechanisms of signal production are unlikely to be the sole explanation for the stereotypy of acoustic properties of any one kind of signal at least over evolutionary time.

Variation in Advertisement Signals

Animal signals are multidimensional. Each of the acoustic properties of a signal typically shows a characteristic range of variation at different levels of analysis: within and among individuals, among populations, and among species. At one extreme, some basic structural properties of the advertisement signals produced by every individual of a particular species are nearly invariant at all levels of analysis. A dichotomous categorization of these *qualitative properties* is straightforward: (1) pulsed versus nonpulsed (fig. 2.2A vs. 2.2C); (2) organization of pulses into discrete trains (*echemes, chirps, notes*) versus continuous trains (*trills*) (examples of both types in box 2.1, fig. 1); (3) narrow versus broadband spectra (fig. 2.2A vs. 2.2B). These kinds of categories are used in keys for species identification by humans (e.g., Bellman 1985; Ragge and Reynolds 1998), and qualitative properties are especially well suited for phylogenetic analyses at the species and higher taxonomic levels (e.g., Alexander 1962; Cocroft and Ryan 1995).

Within-individual variability: static and dynamic properties. Quantitatively variable properties such as carrier frequency and pulse rate[1] typically show relatively little variation from signal to signal within individuals. By contrast, other properties, such as song or call duration and call rate, often vary considerably during a single episode of signaling. The extent of variation in such gross temporal properties can be influenced strongly by the social context, density of other signalers, and even the time of day (Wells and Taigen 1986; Ragge and Reynolds 1998). Gerhardt (1991) computed the coefficient of variation (CV) for properties of advertisement calls produced by individuals of three species of treefrogs. Extending this kind of analysis to a much larger sample of species (fig. 2.3) confirms that in comparisons of the mean CVs of some acoustic properties, there is little overlap in distributions; hence these properties could be reasonably classified as static or dynamic. For example, the mean CV for carrier frequency and pulse rate is usually < 4% , whereas the mean CV for signaling (call) rate is usually > 10%. In a survey of eleven species of crickets and katydids, the mean CVs were as follows: carrier frequency ≤ 3%; pulse rate 1–7%, *usually* < 4%; call rate 7–35% (Gerhardt, unpubl. data). By comparison, signal duration shows an enormous and continuous range of within-male variability, from about 1% in some species to about 30% in others. There is a trend for relatively short signals (< 0.5 s) to show low variability (mean CV usually ≤ 4%) and for longer signals (> 0.5 s) to show higher variability (mean CV usually > 7%). The same basic pattern was found in the sample of orthopterans: short signals < 5%; long signals as high as 40% (Gerhardt, unpubl. data). Nevertheless, the considerable overlap of values in the range of 4–8% precludes classifying short signals as static and long ones as dynamic. On the one hand, the fact that variability in all properties is continuous means that categories must always be defined by some arbitrary cutoff. One solution is to consider, for a given species, the relative variability of each property of interest (e.g., Shaw and Herlihy 1998). On the other hand, the strong trends shown in figure 2.3 raise the question of how and why variation in some kinds of

1. This is especially true if the pulse rate over all or part of a signal is averaged, as we assume that receivers do to some extent. In some species, however, a consistent change in pulse rate, which could be important for signal identification, occurs from the beginning to end of each call.

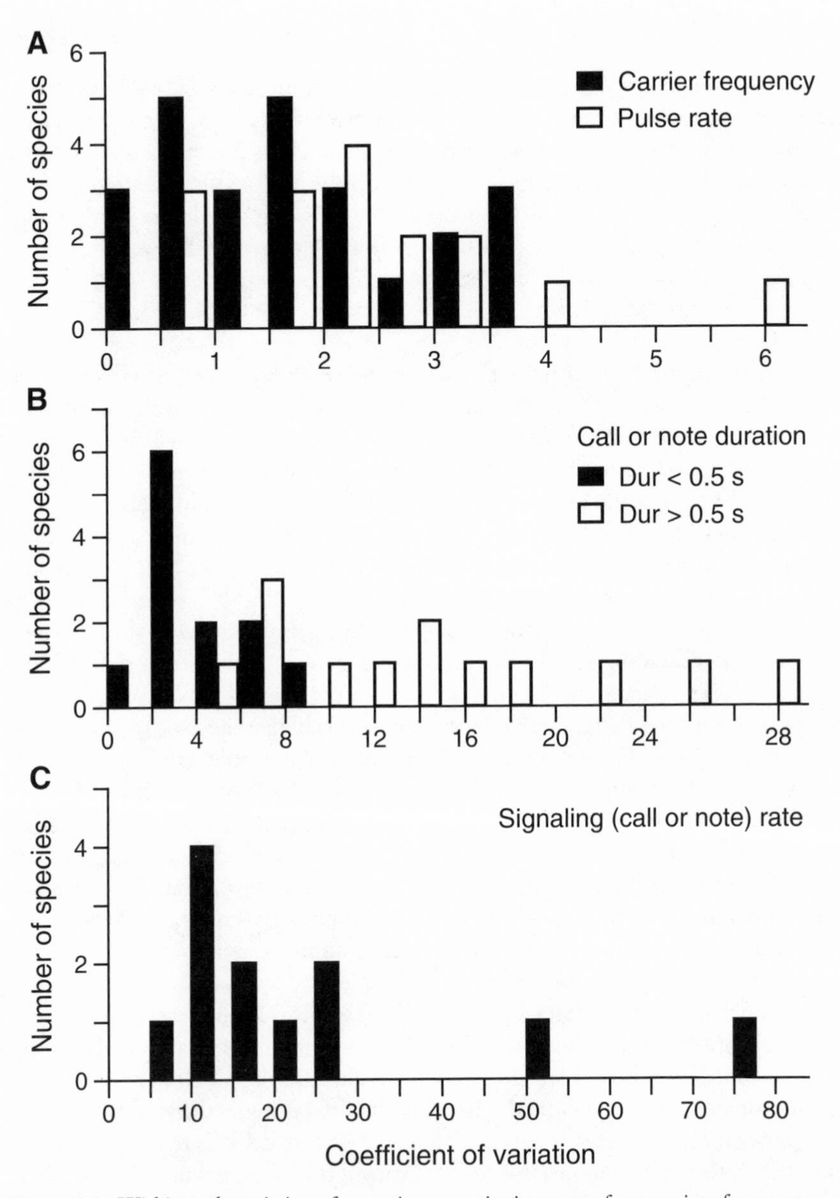

Figure 2.3. Within-male variation of acoustic properties in twenty-four species of anurans. Histograms show mean within-male coefficients of variation (CVs) for (A) carrier frequency and pulse rate, which are usually relatively invariant "static" properties; (C) call rate, which is usually a highly variable "dynamic" property; and (B) call duration, whose broad continuous range of variability defies categorization. Notice, however, that mean CV is usually lower in short signals than in long signals. The figure is based on published and unpublished data derived from measurements of at least eight individuals per species. Not all properties were analyzed for each species because of sample-size limitations; for example, recordings were not always long enough to estimate variability in call rate. A fully documented data set is available by request from H. C. Gerhardt.

properties is seemingly so narrowly constrained within individuals and others are seemingly free to vary.

With some interesting exceptions, the physical dimensions and properties of peripheral structures involved in sound production explain a significant part of the low variability in the spectral properties of insect and anuran signals. Interpreting mechanistic constraints on fine-scale temporal properties is more problematic. As mentioned above, individuals showing low variability in pulse rate in their advertisement calls may still produce a wide range of rates in other signals in their repertoire. Although this observation suggests that these individuals might, in principle, be able to produce a wider range of pulse rates than observed in their advertisement signals, the pattern generator (see chapter 3) controlling this kind of signal might well constrain variability. In turn, the stabilizing selection on pulse rate by receivers in the past and present would be expected to confer the pattern generator with this characteristic. What about pulse rate in the more flexible signals and the high variability of dynamic properties? It seems unlikely that their high variability reflects some limit on how precisely the nervous system can regulate signaling rates or the duration of relatively long signals. Rather, much of the variability of dynamic properties at least correlates with temporary or permanent attributes (e.g., physical condition, fitness, aggressive motivation). But how do the underlying pattern generator or generators work? What kind of selective history (and current selection) has allowed or perhaps even promoted this flexibility?

Variation in acoustic properties among individuals in the same population. Intuitively, we might suppose that only static properties would be useful for distinguishing among individuals in a population. In fact, the potential for any property to do so depends on the extent of among-individual variation relative to within-individual variation. For example, if a property has a very limited range of variation among signalers, then it is no more useful for distinguishing among individuals than is a qualitative property. Moreover, despite their relatively high within-individual variability, some dynamic properties are useful for distinguishing among individuals over various biologically meaningful time scales (e.g., Hedrick 1988; Runkle et al. 1994; Gerhardt et al. 1996). The extent of among-individual variation is often quantified in terms of *repeatability*, which sets the upper bound on *heritability* (see chapter 10).

Because the diversity of biological systems can be counted on to defy universal generalizations, it is hardly surprising that highly variable signals might even be reliable indicators of species identity. For example, Bennet-Clark and Leroy (1978) describe a species of drosophilid fly (*Zaprionus mascariensis*) in which all males produce signals with highly variable pulse rates, all confined, however, to the same overall range of values. The same is true for males of the North American pine woods treefrog *Hyla femoralis* (box 2.1). Whether receivers in these two species use such variability for signal identification is, however, still untested.

Variation in acoustic properties among populations. Most studies of geographical variation in the advertisement signals of insects and anurans consider only static properties such as carrier frequency and pulse rate (reviews: Ewing 1989; Gerhardt and Schwartz 1995; exceptions: Ralin 1977; Ryan and Wilczynski 1991; Ryan et al. 1996). Differences in pulse rate between widely separated populations can be as great as those between species (e.g., Littlejohn 1965), and in the *Litoria ewingii* complex of Australian treefrogs, significant geographical differences occur in the variability per se of some temporal properties (Gerhardt and Davis 1988; Littlejohn et al. 1993). By contrast, the magnitude of among-population variation in other properties can be limited and even comparable to that of evolutionarily conservative, morphological traits (e.g., carrier frequency in the cricket frog *Acris crepitans blanchardii*, Ryan and Wilczynski 1991). Just as variation within males does not always correlate directly with variation among males, among-male variation does not reliably predict variation among populations.

For example, in an analysis of variation in call properties in twenty populations of the gray treefrog *Hyla chrysoscelis*, among-population variation in pulse rate was nearly double that in carrier frequency even though both properties show very low within-male and between-male variation within single populations (Gerhardt, unpubl. data). By contrast, among-population variation in call duration and call rate was not much greater than that observed within and between individual males in typical populations (see also Castellano and Giacoma 1998). The latter pattern is explained by the fact that the range of variation in these dynamic properties observed in some individual males comes close to encompassing the entire range of variation in the species.

The existence of geographic variation and species differences in static properties that vary little within and among males raises a general question: How do these differences in signal structure arise in the face of the generally conservative influences that might be expected of the underlying mechanisms of signal production and the usual pattern of stabilizing selection by receivers? As a provisional answer to the question of mechanistic constraints, we show below that some striking differences in signal structure can be traced to seemingly trivial changes in the peripheral structures of closely related species. We address questions about how receivers influence population-level and species differences in signals in chapters 4, 10, and 11.

Peripheral Mechanisms of Sound Production

In this part of the chapter, we focus on the structures that influence static properties of insect and anuran signals. Even though the mechanisms of sound production used by insects and anurans are profoundly different, these and other animals have adopted the same basic solution to the production of long-distance signals: *frequency multiplier mechanisms.* That is, relatively slow muscle-contraction cycles, patterned by the central nervous system, drive peripheral structures that have much higher natural frequencies of vibration (Michelsen 1983; Bennet-Clark 1998, 1999a). For insects and small anurans, these frequencies are, in general, relatively high because of basic physical laws governing sound production. For example, effective coupling of a vibrating structure to the air occurs when the physical dimensions of the oscillator are no less than about one-fourth of the wavelength of the carrier frequency of the sound (Bennet-Clark 1971). The peripheral vibrating structures and secondary radiators are like a musical instrument in that they determine the basic spectral properties (carrier frequency or frequencies, bandwidth) and some of the fine-temporal qualities (pulse or impulse structure) of the acoustic signal. The central nervous system plays the instrument.

Insects produce signals by a variety of mechanisms (reviews: Dumortier 1963b; Bailey 1991; Ewing 1989; Greenfield 2001). The dominant mode of orthopteran sound production is *stridulation*. Crickets and katydids rub a scraper (*plectrum*) on the upper side of one wing against a file (*elytron*) with stridulatory teeth on the lower side of the other wing. The files and scrapers are usually well developed on both wings in crickets, but males typically use the scraper on the left wing to excite the teeth on the file of the right wing (chapter 3 discusses the mechanisms of wing-position preferences). In male katydids the opposite pattern is enforced because the right wing has a scraper, and the left wing, a well-developed file (fig. 2.4A; Dumortier 1963b; Bailey 1970; Elsner and Popov 1978); if females sing, their morphology may differ (fig. 2.4B). Whereas crickets produce sounds during the closing stroke of the wings, katydids can produce sound during the closing stroke, opening stroke, or both (fig. 2.4C; many examples in Heller 1988).

Most grasshoppers produce sounds by rubbing surfaces of the legs against a raised edge of the forewings (fig. 2.5A,B). In some groups (subfamily Gomphocerinae), the file is on the leg and the scraper is on the wing, and in others (subfamily Oedipodinae), the locations of the file and scraper are reversed (Elsner and Popov 1978). In some species the two legs move with different temporal patterns. Even in species in which the pattern is similar, there is typically a phase difference that causes the temporal pattern of the sound produced by the file and scraper on one side to be obscured by the sound produced by the file and scraper on the other side (fig. 2.5C). The behavioral significance of this "masking," which differs in degree in the male and female songs of the acridid grasshopper, has been well studied in the grasshopper *Chorthippus biguttulus* (e.g., Helversen and Helversen 1997; chapter 4). Other species of grasshoppers have stridulatory files on the abdomen that are rubbed by scrapers on the legs or wings (e.g., fig. 2.12A; van Staaden and Römer 1997), and still others produce acoustic signals by beating hardened parts of the wings together (Elsner 1974a). Many North American and some European grasshoppers also produce sound by *crepitation*, which is a poorly understood

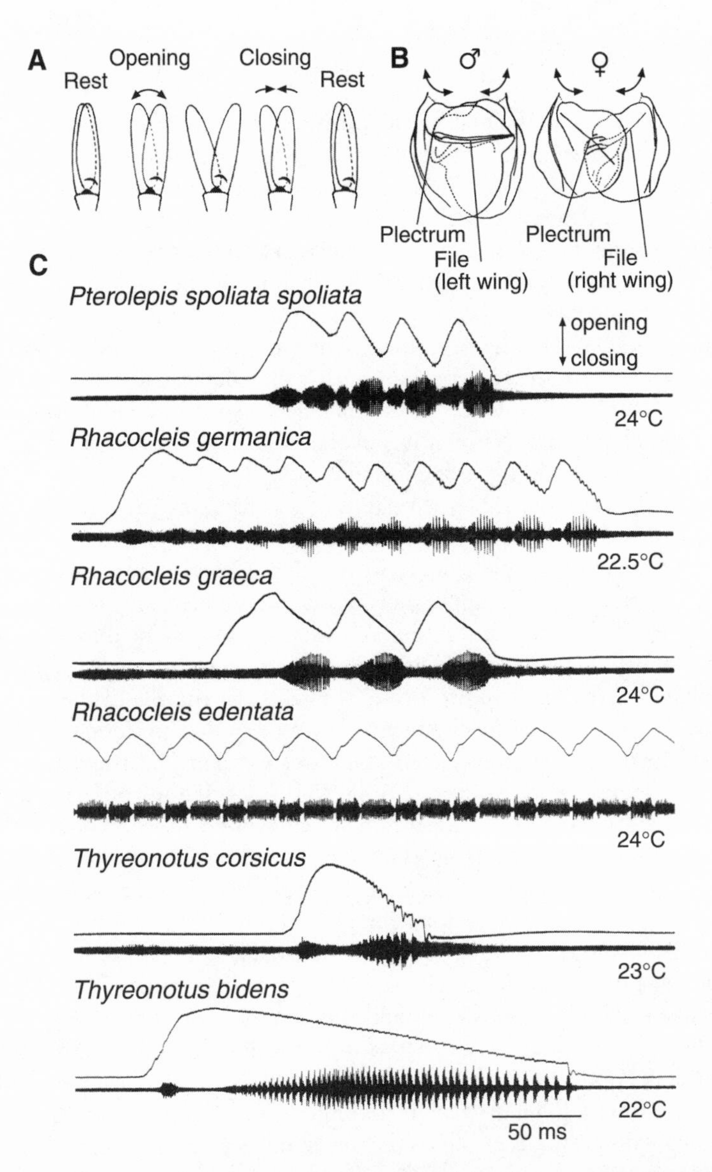

Figure 2.4. (A) Diagram of wing movements during stridulation by a generalized katydid. The dotted line represents the scraper (plectrum), and the solid curved line near the base of the wing to the right in the figure, the file. (B) Sexual dimorphism in stridulatory structures in phaneropterine katydids (*Poecilimon sanctipauli*). Notice the reversal of the locations of the scraper and files in the two sexes. (C) Oscillograms of calling songs of some European species of katydids. The lines above the oscillograms indicate the time patterns of wing openings (upward deflections) and wing closing (downward deflections). Notice that the opening and closing movements correspond well to acoustic units (trains of pulses) in some species, but that these trains produced during opening movements may have very different pulse rates, durations, and rise-fall characteristics than the pulse trains produced during closing movements. Furthermore, pulse rate within trains can change even when the velocity of the closing movement appears to be rather smooth and steady (e.g., *Thyreonotus corsicus* and *T. bidens*). These changes evidently depend on changes in the spacing or physical properties of the stridulatory teeth from one end of the file to the other. From Dumortier 1963c, fig. 220; Bailey 1991, fig. 5.9; and Heller 1988, fig. 70.

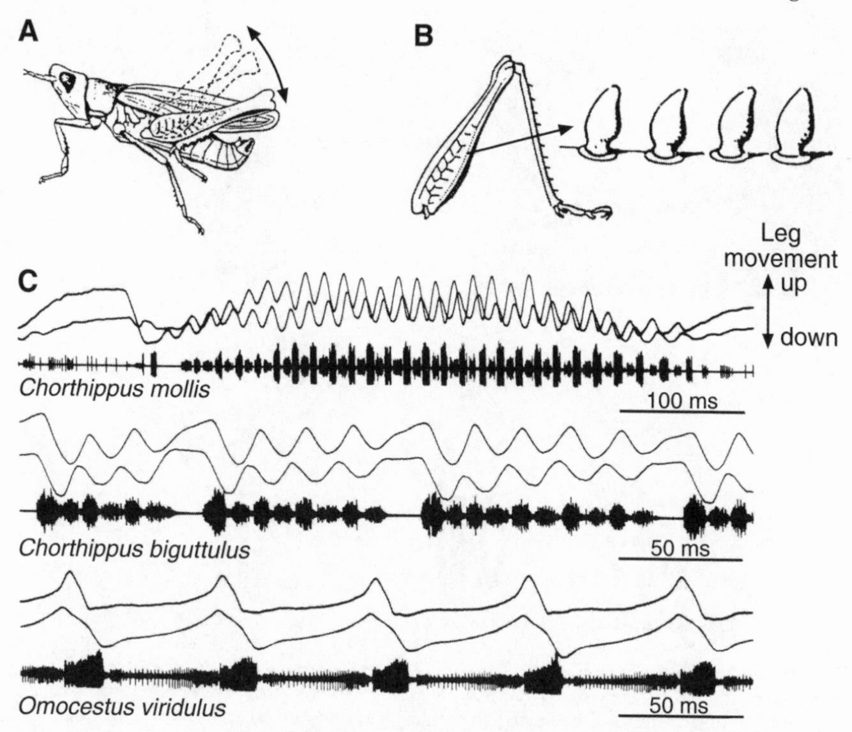

Figure 2.5. (A) Diagram showing the movement of the leg during stridulation in a grasshopper. (B) Details of a right leg, showing the location of the stridulatory file on its inner surface and detailed diagrams of some stridulatory pegs on the file. (C) Courtship songs or parts thereof in three species of European acridid grasshoppers. The lines above the oscillograms show the movements of the right and left stridulatory apparatuses, which can be seen to be out of phase. The oscillograms were made from animals in which only the right leg was allowed to stridulate in order to show the temporal patterns that are obscured when both legs participate. From Dumortier 1963c, fig. 221; Elsner 1983a, figs. 1, 3.

mechanism that usually occurs during flight and sometimes in the context of elaborate flight displays (Otte 1974; Ragge and Reynolds 1998). Crepitation also occurs in a few species of cicadas (Sanborn and Phillips 1999).

Cicadas generate sound with structures called *timbals*, which are paired organs on the abdomen (fig. 2.6). One or more pulses of sound are produced when a series of timbal ribs buckle inward; in some species, loud sounds are also emitted when timbal ribs buckle outward to their original positions (Fonseca 1994). Cicadas offer excellent examples of the diversity of sound patterns that can be produced by the same basic mechanism and anatomically similar structures (figs. 2.1A, 2.13). Timbals and associated muscles appear to be more specialized for sound production than the structures and muscles used in stridulation. Timbals are not located on appendages required for locomotion. Rather, these structures and their associated muscles might be homologous to those of bugs and hoppers in the same superfamily (Cicadoidea)

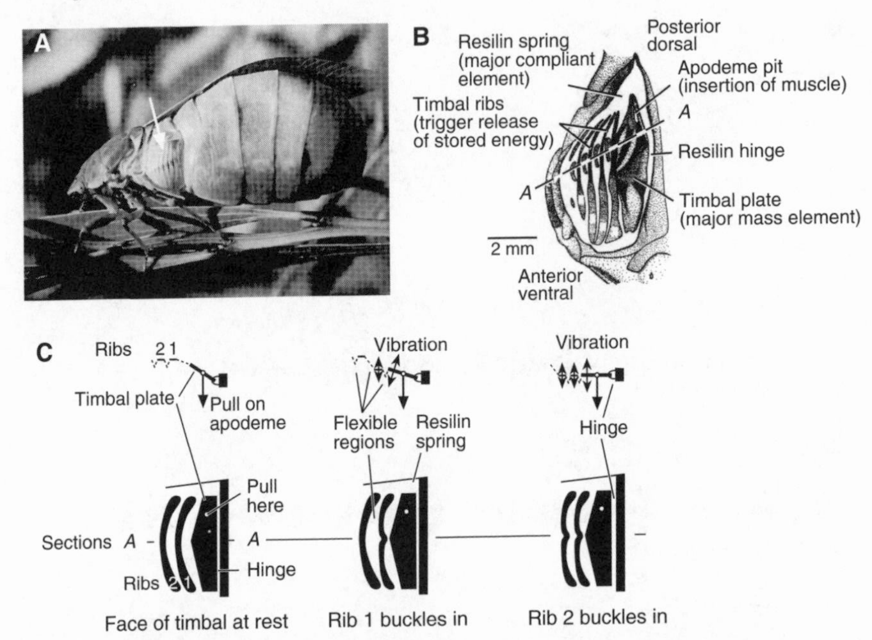

Figure 2.6. (A) Photograph of a bladder cicada (*Cystosoma saundersii*), with the left wings removed to show the location of the timbal (*indicated by arrow*). From Bailey 1991, fig. 5.10. (B) Diagram of the timbal of an Australian cicada (*Cyclochila australasiae*), showing the main sound-producing elements, which are also labeled in terms of their functions as a sound resonator. Bennet-Clark (1999a) provides details and comparisons with physical analogs. (C) Diagrams of the mode of excitation of vibration in cicada timbals. *Left*, inward movement of the timbal plate is brought about by contraction of the timbal muscle pulling on the apodeme. At first the movement is constrained by the timbal ribs (see part B). *Right and center*, as each timbal rib buckles inward along the line shown in part B, previously stored energy in the resilin spring is released, allowing the timbal plate to move rapidly inward and then to vibrate. From Bennet-Clark 1999a, fig. 5.

that communicate with vibratory signals (Ossiannilsson 1949; Ewing 1989). All of the timbal muscles that have been critically examined show a *neurogenic* pattern of response, that is, one muscle twitch per neural spike (Young and Josephson 1983a).

With a few exceptions in the family Pipidae (Rabb 1960; Yager 1992), anurans produce vocalizations by modifying breathing mechanisms. Two chambers, the lungs and buccal cavity, are involved in respiration. Each chamber has a valvular closure; the larynx occludes the glottis, and the nostrils (nares) close the buccal cavity. In most species, vocalization involves these and additional chambers, termed *vocal sacs*, and associated valves (Gans 1973; fig. 2.7A). The force of pressurized air from the lungs or buccal cavity on the vocal cords is greatest when these structures block the laryngeal opening maximally and is minimized when the laryngeal opening is completely dilated (Martin 1971). Sound production usually begins during expiration of air from

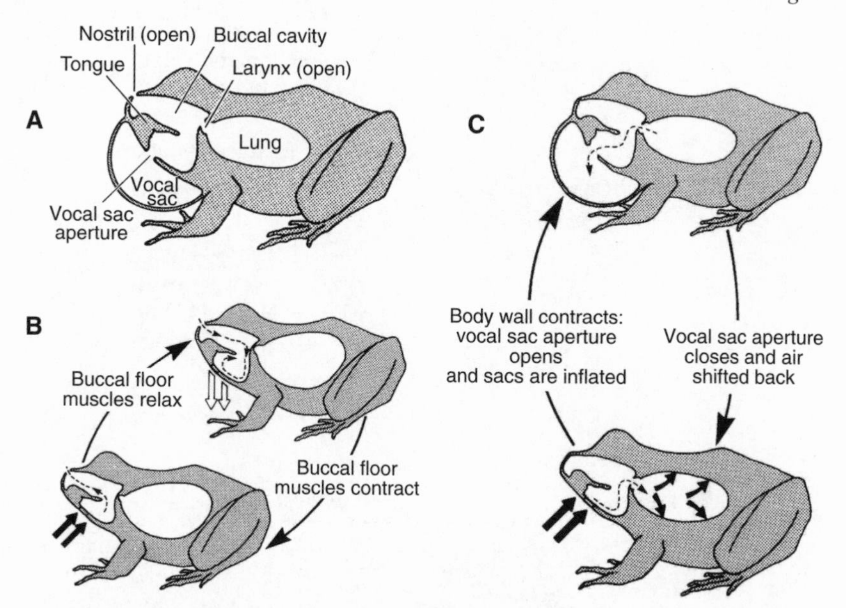

Figure 2.7. Anuran structures and their movements during respiration and vocalization. (A) Diagram showing the main structures used in vocalization by a typical anuran. (B) Diagram of the movements of muscles that occur during normal cyclical ventilation. (C) Diagram of the movements and airflow that occurs during prolonged vocalization, which occurs in many species of toads (genus *Bufo*); the prolonged call duration is achieved by shunting air back and forth from the vocal sac and lungs. From Duellman and Trueb 1986, fig. 4.9, and modified from fig. 4.10.

the lungs, helping to force open the larynx (fig. 2.7B). The air rushing through this constricted space excites the vocal cords into vibration and subsequently inflates the vocal sac. Males in the genus *Bombina* (fire-bellied toads), which lack vocal sacs, vocalize when the buccal cavity constricts and forces air through the larynx into the lungs, whereas *Discoglossus pictus* produces sounds both during inspiration and expiration (review: Schneider 1988). We present diagrams showing the locations and main structures of a typical anuran larynx in figure 2.8, but the morphology of these structures varies even among so-called advanced anurans (Schneider 1988 and below).

Files and stridulatory teeth, cricket harps, timbal ribs and plates, and vocal cords can be considered *primary oscillators* or *primary resonators.* Primary oscillators are not always the main radiators of acoustic signals, and they often drive a secondary resonator. In some systems the primary resonator drives the secondary resonator at frequencies close to its natural frequency of vibration (as in most crickets, some katydids and cicadas, and some anurans), resulting in the increased efficiency of radiation of loud sounds whose bandwidth is relatively narrow. In other systems the primary resonator produces a broadband sound, whose spectrum can be significantly modified by the secondary resonator (typical grasshoppers, katydids, and many anurans). Secondary resonators

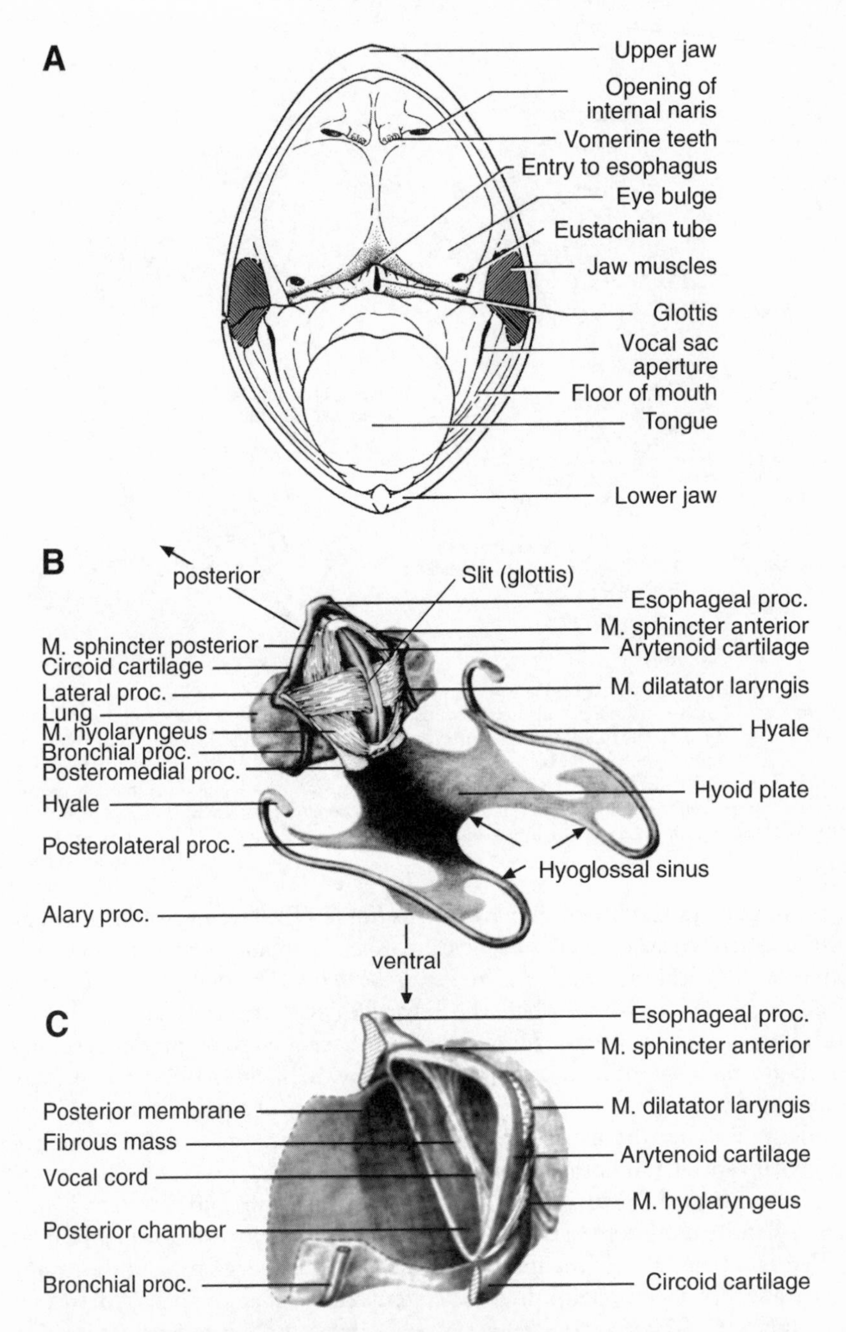

Figure 2.8. Location and morphology of the sound-producing structures of a generalized anuran. (A) The location of the glottis, which is located anteriorally to the larynx. (B) Laryngeal apparatus relative to the lung and supporting cartilaginous processes (e.g., hyoid plate). The vertically oriented slit between the arytenoid cartilages in the upper part of the diagram is the glottis. (C) Cross-sectional diagram of the larynx showing internal structures, including the main sound-producing structure, the vocal cord. The fibrous mass is typical of some anurans that produce lower-than-expected carrier frequencies based on their body size (see the text for further explanation). From Duellman and Trueb 1986, figs. 4.3, 4.4, 4.5.

include some wing structures (some orthopterans), tympanal membranes (some cicadas and frogs), air sacs (some cicadas and some grasshoppers), buccal cavities and vocal sacs (anurans), and even burrows (mole crickets and some anurans) (Penna and Solís 1996; Purgue 1997; Bennet-Clark 1999a). Secondary resonators, which typically have larger dimensions than primary ones, can also be considered *impedance matching devices* in the sense that they absorb vibrational energy more effectively from primary resonators than does the air, and then radiate more vibrational energy to the air than do the primary resonators (Martin 1971; Bennet-Clark 1989, 1999a; Fonseca and Popov 1994; Bradbury and Vehrencamp 1998). In some systems, secondary resonators, or parts of primary ones, can feed back on and affect the vibratory patterns of the primary resonators (Bennet-Clark 1999a; see below). Not surprisingly, therefore, the distinction between primary and secondary resonators can be arbitrary and subject to debate.

Some stridulatory oscillators approximate to *dipole sources*, which radiate sound mainly in two directions along the main axis of vibration (Nocke 1971; Michelsen 1983; Bennet-Clark 1989). The amplitude of sounds produced by small dipole sources is subject to a significant limitation: *acoustic short-circuiting*, which is a partial cancellation of a *compression* produced on one side of a dipole oscillator by the *rarefaction* produced on the opposite side (appendix 2). The problem is severe when small dipole sources produce relatively low frequencies, whose wavelengths are about the same as the size of the dipole. Behavioral solutions that can ameliorate this problem include raising and holding the wings together to form a baffle, using a leaf as a baffle, singing from a small depression in the ground, or calling from burrows (see chapter 11). In anurans, katydids, and some cicadas, the effective sound-radiating structures (vocal sacs, bladders, subalar spaces) may approximate to *monopole sources* having more or less omnidirectional patterns of radiation (Gerhardt 1975; Prestwich et al. 1989; Daws 1996; Michelsen and Fonseca 2000). Signaling from depressions or burrows can, however, make the acoustic output of such sources more directional, and the sound field around grasshoppers on level ground can be complex (Michelsen and Elsner 1999). As Michelsen (1983) has emphasized, insects and anurans are not idealized sound sources—dipoles or monopoles—located in free-field situations where sound waves are neither redirected nor scattered. Thus, the models and equations of physical acoustics cannot provide precise predictions about patterns of radiation from these animals.

In most insects and anurans, there is strong sexual dimorphism in the peripheral structures that produce sounds (Schneider 1988; Ewing 1989; Bailey 1991; Wells 2001). The peripheral structures are strongly developed in males, which typically produce loud advertisement signals, and weakly developed or absent in females, which, in most species, do not produce acoustic signals. In species in which female acoustic responses are well developed, the stridulatory apparatus is likely to differ from that of the male (Dumortier 1963b). In phaneropterine katydids, for example, the morphology of the file and scraper differs from that of the male and their locations on the wings are reversed (fig. 2.4B; Heller and Helversen 1986; see also van Staaden and Römer 1997; fig. 2.12). In the tick-tock cicada *Cicadetta quadricincta*, females answer male songs, which are produced by timbals, with clicks that are produced by wing flicking (Gwynne 1987). These observations indicate that sound production in insects must often have independent evolutionary origins in the two sexes.

Peripheral Determinants of Bandwidth and Modulation

Bandwidth is traditionally quantified as the range of frequencies that have amplitudes within a certain value (usually 3 dB or half peak power) of the amplitude of the carrier frequency (Bennet-Clark 1999b). Dividing the carrier frequency by the bandwidth results in a dimensionless number that is termed the *quality factor*, or *Q*. Narrow-band signals have relatively high Q-values (e.g., > 10), and broadband signals, low Q-values (e.g., < 4). Q-values are used in physical acoustics and bioacoustics to describe the sharpness of tuning of resonating structures and filters; such values can also be computed by assessing the rate and form of decay of pulses of sound (see appendix 2; Bennet-Clark 1999b). In this chapter all values of Q refer to Q_{3dB}.

Three main determinants of signal bandwidth are (1) the mode of vibration of the main oscillating structure; (2) the Q-values of primary and secondary resonators; and (3) the degree to which the frequencies produced by primary oscillators are matched or coupled to the resonant frequency or frequencies of the secondary resonators. Even if the carrier frequency has a nearly sinusoidal waveform, however, its frequency, amplitude, or both can vary with time. Sweeps in carrier frequency (one form of *frequency modulation;* see appendix 2) from low to high or vice versa are found in the signals of some species of insect and many anurans; rapid changes in amplitude (*amplitude modulation;* see appendix 2) are very common in the signals of both groups. Both forms of modulation effectively increase the bandwidth of signals, especially those with discrete carrier frequencies (see appendix 2).

Crickets. Among orthopterans, crickets are exceptional in that most species produce narrow-band advertisement signals without strong harmonics or frequency modulation. Some exceptions include the calling songs of species in the genera *Tartarogryllus* from central Asia (Elsner and Popov 1978), *Gryllodes sigillatus* in North America (Alexander 1960), and two tropical species to be discussed below.

Mechanisms of sound production have been particularly well studied in European field crickets (*Gryllus bimaculatus* and *G. campestris*), which produce nearly pure-tone (sinusoidal) signals; the amplitudes of the second and third harmonics are usually 20–30 dB less than that of the carrier frequency of about 5 kHz. These signals have a tonal quality because the vibration caused by a given tooth impact does not decay before the next tooth strike but rather is reinforced and maintained (Nocke 1971; fig. 2.9A,B). Furthermore, the song

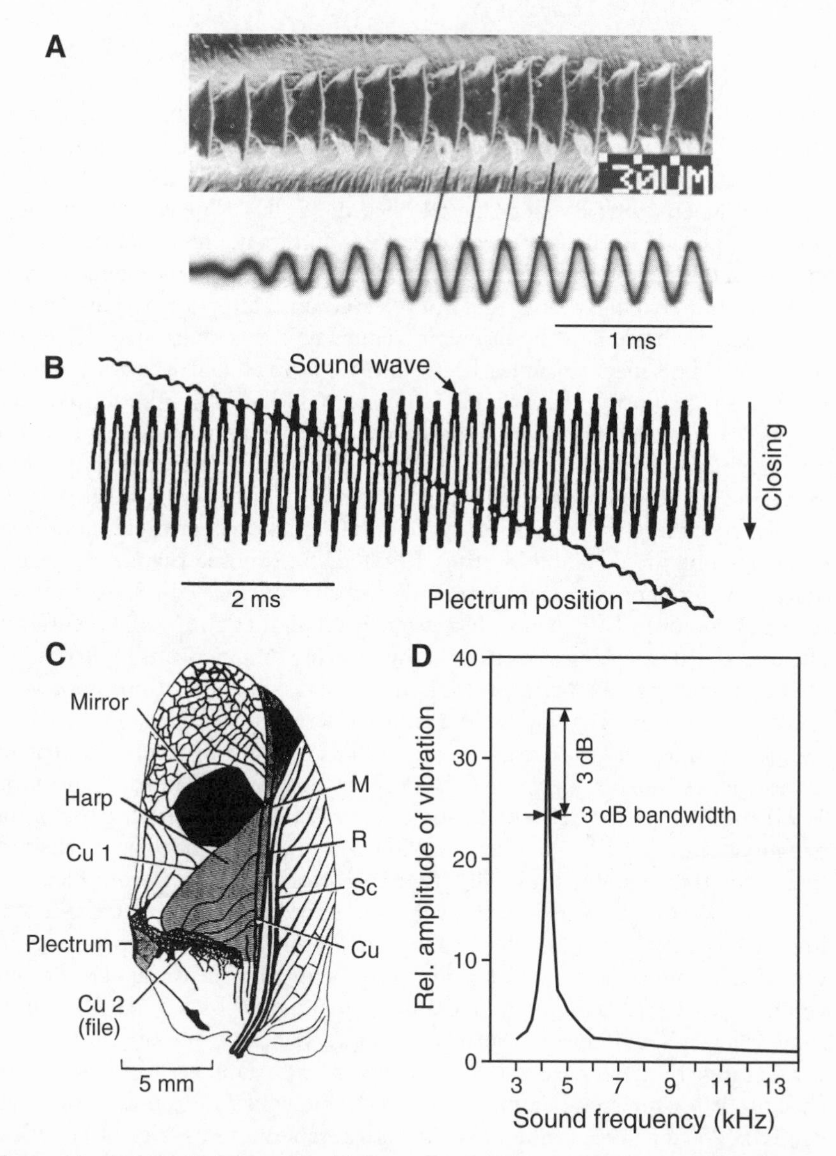

Figure 2.9. (A) Magnified view of a stridulatory file of a field cricket (*above*) and an oscillogram of part of a pulse (*below*), which shows reinforcement of the successive cycles of the waveform. Notice that the teeth are slanted, a feature that results in a momentary catchment of the scraper, which is freed after the resonator, the harp, undergoes a cycle of vibration. From Elsner and Popov 1978, fig. 1. This mechanism matches the tooth-impact rate to the resonant frequency of the harp and results in production of a coherent sinusoidal wavetrain, which is also shown in (B), with the scraper (plectrum) position superimposed on part of a pulse to represent the correspondence in time of cycles of a carrier wave with tooth impacts. Modified from Koch et al. 1988, from Huber 1990, fig. 1. (C) Tegmen of a male of *Gryllus campestris* seen from below with structures important for sound radiation. The dorsal field is separated from the thick veins (Sc, R, M) of the lateral field by a band of flexible cuticle and contains the harp (*hatched*), which is the primary resonator in most Gryllinae. The veins Cu 1 and Cu 2 (the file) and the mirror (*black*) enclose the harp. (D) Resonance curve for the harp of *Gryllus campestris*, the Q-value (3 dB) is approximately 28; see the text and appendix 2 for an explanation. From Bennet-Clark 1989, fig. 8.6A.

frequency is mainly determined by a sharply resonant (highly tuned) part of the wing called the harp (fig. 2.9C; review: Dumortier 1963c). Estimates of the Q-value of the harp of *G. campestris* range from 16 to 28 (Nocke 1971; Elliott and Koch 1985; Bennet-Clark 1999a; fig. 2.9D). The reinforcement of sound amplitude in the beginning of a pulse and its maintenance during the middle of the pulse occur because the scraper is caught and momentarily held in place after striking a tooth and not released until after the harp undergoes one cycle of vibration at the resonant frequency; the scraper then strikes the next tooth. This mechanism has been compared to a clockwork with an escape mechanism (Elliott and Koch 1985; Bennet-Clark 1989). The harp's resonance, like the frequency of oscillation of a clock's pendulum, determines the rate at which each tooth is struck. If the harp is artificially loaded to decrease its resonant frequency, then the tooth-impact rate (rate of wing closure) is decreased accordingly (Koch et al. 1988). The "clockwork cricket" model was challenged by Stephen and Hartley (1995), who proposed that the space under the wings acts as a cavity resonator and that crickets could use auditory feedback to control frequency. However, Prestwich et al. (2000) recently reaffirmed the classical (clockwork) view by showing that increases in carrier frequency occurring when crickets sang in a light-gas (heliox) atmosphere were much less than expected if a cavity resonator were involved.

Species differences also exist among gryllids in the wing structures that act as resonators. For example, in *Cycloptiloides canariensis* (Dambach and Gras 1995) and *Oecanthus nigricornis* (Sismondo 1979), the harp, other parts of the wing (e.g., mirror cells), or even the wing as a whole appear to act as the primary radiators of sound under different conditions (e.g., temperature).

The resonant properties of the wing sound radiators influence the tooth-impact rate in crickets. Yet changes in temperature will also, in principle, be expected to alter the speed of wing movements, and in these ectothermic animals, pulse rate is highly temperature-dependent (box 2.2). How then do some crickets maintain the tooth-impact rate at nearly the same value over a wide range of temperature? In *Gryllus rubens*, for example, carrier frequency changes little over a temperature range over which pulse rate doubled (box 2.2, fig. 1B). Walker (1962) found that at high temperatures, males of *G. rubens* decreased the number of tooth strikes per pulse and increased the proportion of time in the wing-stroke cycle occupied by wing closing, which is the phase during which sound is produced. The first change allows the wings to move a shorter distance and hence make more round trips per unit time, thus increasing the pulse rate, without a proportional increase in the speed at which the scraper moves across the file to produce a single pulse. The second change results in a much higher increase in the rate of wing opening, when no sound is produced, compared with the rate of wing closing. In *G. rubens*, for example, 65 tooth strikes per pulse were measured at 21°C and 55 per second at 31°C; however, the wing-closing rate increased by only 6.5% and that of wing opening, by 129% (Walker 1962). As shown in box 2.2, figure 1B, however,

Box 2.2. Temperature Effects on Acoustic Signals

Because most insects and anurans are ectotherms, temperature influences those attributes of acoustic signals that are controlled by the neuromuscular system. Some exceptions are some katydids and cicadas, in which singing itself can generate heat in the stridulatory muscles that raises the animal's body temperature above ambient temperature (e.g., Josephson 1984; Heller 1988; Sanborn 2001). In any event, comparisons of differences in these acoustic properties across individuals and populations require some form of statistical adjustment of values for the temperatures at which the animals are recorded. Aside from studies of the katydids just mentioned, however, to our knowledge no study has specifically addressed the question of how temperature affects neuromuscular mechanisms in the context of sound production. Hence, we provide only a brief overview of temperature effects, which we reconsider in other parts of the book that deal with the temperature-dependence of receiver selectivity. Ragge and Reynolds (1998) provide an extensive review of the orthopteran literature.

Pulse rate is highly temperature-dependent (e.g., insects: Walker 1957, 1962, 1975a,b; Gray et al. 2000; anurans: Zweifel 1968; Schneider 1977). Furthermore, as noted by Walker (1975b), the wing-stroke rate (= pulse rate in crickets; pulse rate during buzzing parts of katydid songs) is best described as a linear function of temperature (fig. 1A). The same is true for most species of anurans (fig. 1C), although, as might be expected, nonlinearities sometimes occur at high temperatures, often well above normal breeding temperatures (Schneider 1977). Pulse duration tends to decrease with increasing temperature in many anurans, so that the pulse-duty cycle (ratio of pulse duration to pulse period) remains nearly constant (e.g., Gayou 1984). In two species (*Bufo viridis* and *Hyla arborea*), however, interesting differences occur among individuals (Schneider 1977). In some males, pulse rate increases linearly throughout the temperature range, but the decrease in pulse duration becomes nonlinear at high temperatures. In others, pulse duration decreases linearly over the whole range, and pulse rate becomes nonlinear at high temperatures. The mechanisms and potential behavioral significance of these individual differences remain unexplored.

Other, gross temporal properties, such as call (phrase) duration and repetition rate are also temperature-dependent, but the correlation coefficients are seldom as high as those for pulse rate. In some species of insects, carrier frequency also increases with temperature, but again the rate of change is usually far below that for pulse rate. This can be easily seen by plotting carrier frequency against pulse rate (fig. 1B). See the text for a description of some of the compensatory mechanisms that tend to stabilize carrier frequency in the

(box 2.2 continued)

face of temperature change. In anurans, carrier frequency also varies much less, if at all, with shifts in temperature in comparison with pulse rate (fig. 1D; data and review in Gerhardt and Mudry 1980).

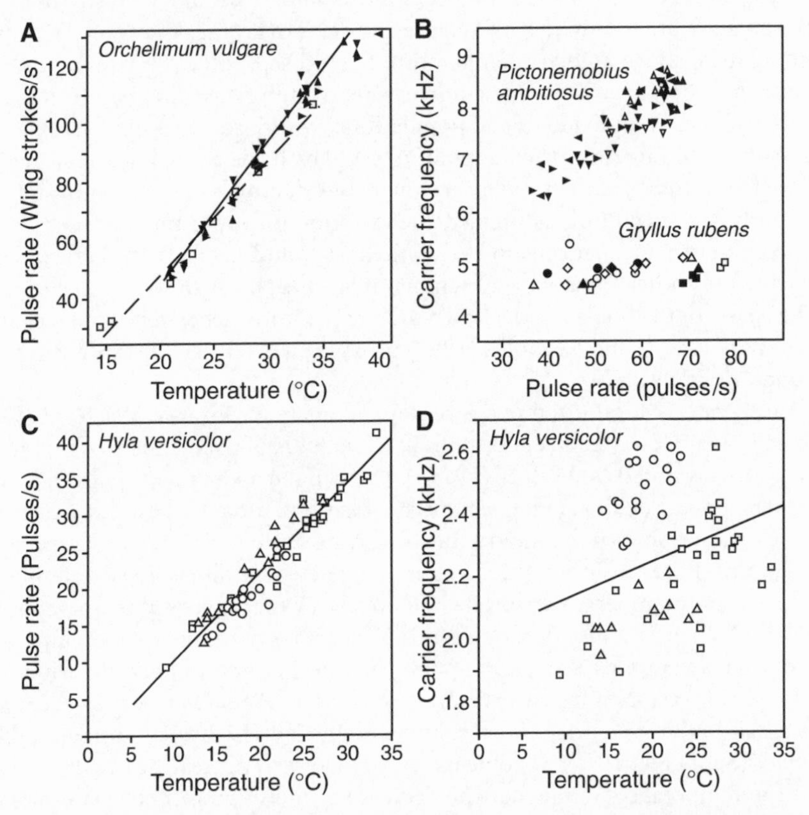

Box 2.2, Figure 1. Temperature effects on pulse rate and carrier frequency. (A) Wing-stroke rate (= pulse rate) during the buzz part of the calling song of the katydid *Orchelimum vulgare* as a function of temperature. Notice that the relationship is linear over a range of at least 20°C. Data are based on the songs of individual insects (see symbols) that were recorded multiple times but at different temperatures. From Walker 1975b, fig. 1. (B) Plots of carrier frequency as a function of pulse rate in two species of crickets. Notice virtually no change occurs in carrier frequency in *Gryllus rubens*, but that a substantial change (> 2 kHz) occurs in *Pictonemobius* (= *Nemobius*) *ambitiosus*. From Walker 1962, fig. 10. (C) Pulse rate versus temperature in the advertisement call of the gray treefrog *Hyla versicolor*. Data are shown for the calls of one individual recorded at different temperatures (*triangles*) as well as for samples of individual frogs each recorded at a single temperature in a semi-anechoic chamber (*squares*) or in the field (*circles*). From Gayou 1984, fig. 1. (D) Plot of carrier frequency versus temperature in *H. versicolor*, showing that very little of the variance in carrier frequency is explained by temperature. Symbols as in (C). From Gayou 1984, fig. 3.

carrier frequency does increase significantly with increasing temperature in another cricket (*Pictonemobious* [= *Nemobius*] *ambitiosus*). This observation raises questions about sound production and how receivers might cope with such changes. How can signal amplitude be maintained at different temperatures (frequencies) if a single resonator is used? Does the frequency tuning of receivers change in a parallel fashion with temperature?

In the field cricket *Gryllus bimaculatus*, intraspecific variation exists in the degree of frequency modulation within the pulses making up chirps. Simmons and Ritchie (1996), showed that the carrier frequency of the calling song is negatively correlated with the area of the harps and also discovered that the extent of frequency modulation is positively correlated with the degree of asymmetry in the areas of the two harps (fig. 2.10). Significant within-pulse frequency modulation occurs when the area of the right harp, which emits the second part of each pulse, is greater than the area of the left harp, which emits the first part. Bennet-Clark (1999a) reviews additional data and discusses the mechanisms that are probably responsible for the general observation that frequency of cricket pulses tends to drop from the beginning to end.

A much greater extent of frequency modulation than observed in European field crickets occurs in two tropical species of crickets (deSutter-Grandcolas 1998), but here the mechanisms generating modulation are more straightforward. In one species (*Lerneca fuscipennis*), the spacing of the teeth differs

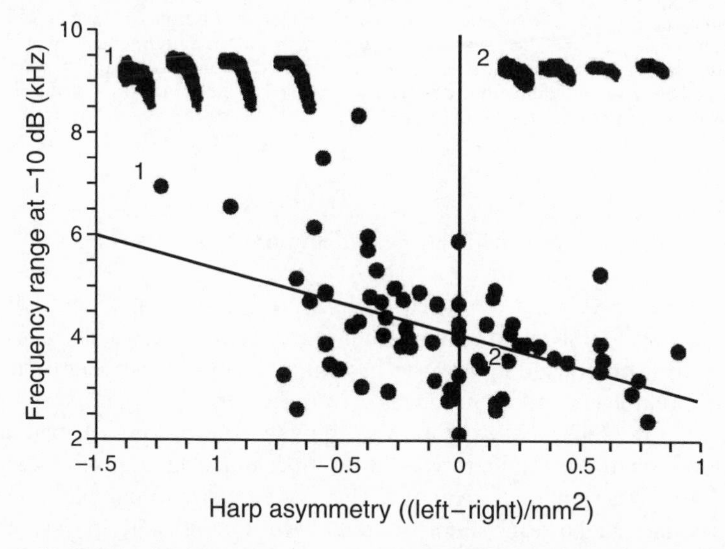

Figure 2.10. Relationship between asymmetry in the size of the two harps (within individuals) and the degree of frequency modulation of pulses in the calling song of *Gryllus bimaculatus*. The insets are sonograms of a single chirp from (1) a male with relatively asymmetrical harps and (2) a male with relatively symmetrical harps; the datum from each of these males is identified by the same number in the scatter plot. From Simmons and Ritchie 1996, fig. 1B.

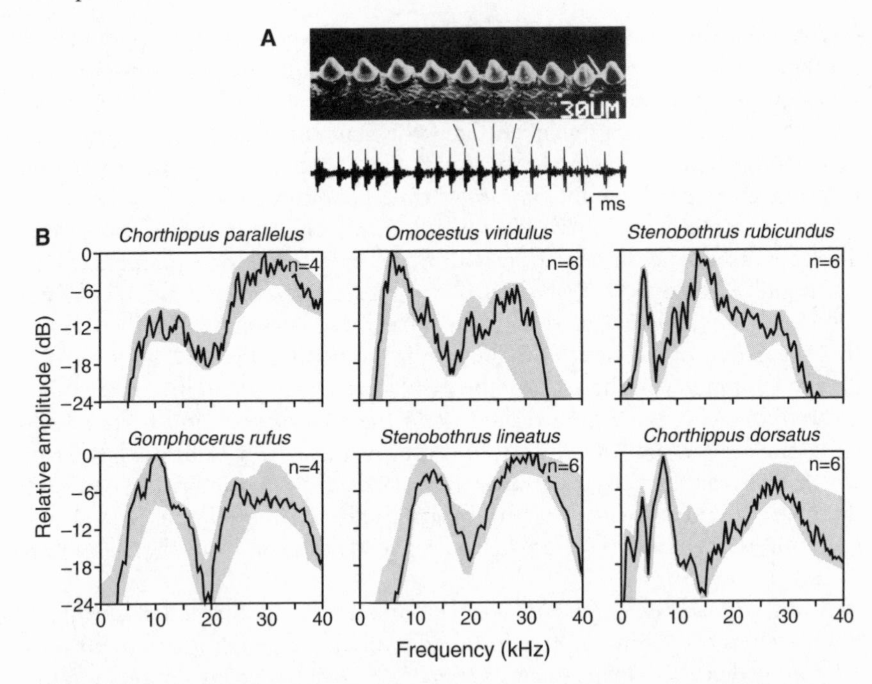

Figure 2.11. (A) Magnified view of a stridulatory file of an acridid grasshopper (*above*) and an oscillogram of part of its calling song (*below*). As indicated by the lines, each vibration of a peg results in a transient, highly damped pulse or impulse. From Elsner and Popov 1978, fig. 1. (B) Sample of the spectra of the signals produced by six species of European grasshoppers. Solid lines are based on one individual; gray areas show the range of variation among all individuals that were sampled. Notice that the spectra are usually bimodal and that the bandwidths of the two spectral peaks often differ within a species; n = number of individuals. Modified from Meyer and Elsner 1996, fig. 1.

significantly in different parts of the file, and in the other (*Eneoptera guyanensis*), half of the teeth are divided by a median furrow into two crests.

Katydids and grasshoppers. Stridulation in typical grasshoppers and katydids produces a train of damped pulses rather than a continuous tone (fig. 2.11A). The resulting broadband signal often has a bimodal spectrum because the wing resonators emphasize different ranges of frequency (fig. 2.11B; many examples for katydids in Heller 1988). Moreover, the rates of wing and leg stridulation are often high and variable, thus actively superimposing amplitude modulation on the fine-time-scale structure, that is, the series of damped pulses or impulses generated by each single scraping of the file (fig. 2.4). In the meadow katydid *Metrioptera sphagnorum* of Canada, the male generates different tooth-impact rates during an episode of signaling and thereby alternates between two distinct song types with different frequency spectra (Morris 1970).

Some exceptional species of katydids and grasshoppers produce songs with nearly sinusoidal waveforms, in which the sound waves produced by successive tooth strikes coalesce to form relatively long pulses (review: Sales and Pye 1974; see also Heller 1988; van Staaden and Römer 1997). For example, in one species of katydid from Trinidad, *Drepanoxiphus modestus*, the tooth impulses generated by a single scraping of the file fuse to produce each of three 6–10 ms pulses with a nearly sinusoidal waveform (Suga 1966). Experiments by Bailey (1970) and Bailey and Broughton (1970) showed that in another species of "pure-tone" katydid, *Ruspolia nitidula*, the carrier frequency corresponded to the tooth-impact rate or was a multiple thereof. Whereas the harp of field crickets is a relatively thick membrane with a light frame, the secondary resonator (radiator) in *R. nitidula* is probably the mirror frame, which behaves as a tuning fork (Bailey and Broughton 1970; Bennet-Clark 1989). Grasshoppers in the family Pneumoridae of southern Africa also produce nearly pure-tone signals. The mechanism involves coupling stridulation to a large, resonating air sac. In *Bullacris membracioides* the male produces a relatively long-duration, high-amplitude (90 dB SPL [sound pressure level *re* 20 m Pa; see appendix 2] at 1 m), narrow-band pulse with a carrier frequency of about 1.7 kHz using abdominal-leg (femoral) stridulation. Females produce softer (60 dB SPL at 1 m), broader-band signals with a peak at about 6 kHz using abdominal-wing stridulation (fig. 2.12; van Staaden and Römer 1997). Selection appears to have maximized the communication range of the male song, but the range of courtship duets is obviously limited by the much lower amplitude of the female's song.

Cicadas. The timbal buckling of most cicadas produces short, damped pulses. The Q-values of the secondary resonators are also quite low ($Q \approx 6$), resulting in the production of broad-spectrum songs (e.g., fig. 2.6C). In some species the timbals, air sac, and tympanal membranes produce sounds with different frequencies, and changes in the spectrum can be accomplished by extension of the abdominal air sac or by variation in the rate of contraction of tensor muscles (see below).

Rapid buckling of the timbal ribs can also result in the production of a series of coherent sound pulses with a narrow-band spectrum, whose frequency is similar to the resonant frequency of the abdominal air sac. The air sac can act as a cavity resonator, as in *Cyclochila australasiae*, the loudest known species (nearly 150 dB peak SPL just outside the tympana), which produces narrow-band signals ($Q = 15$) of about 4.3 kHz (Young and Bennet-Clark 1995; Bennet-Clark and Young 1992; Bennet-Clark 1997). Alternatively, the bladder cicada *Cystosoma saundersii* of eastern Australia produces a narrow-band ($Q = 15$) carrier frequency of about 850 Hz, and sound radiates directly from the thin walls of its large, hollow abdomen (Simmons and Young 1978; Daws 1996; Bennet-Clark and Young 1998).

Other species, such as the North American periodical cicada *Magicicada cassini*, have air sacs that act as cavity resonators but produce very broadband

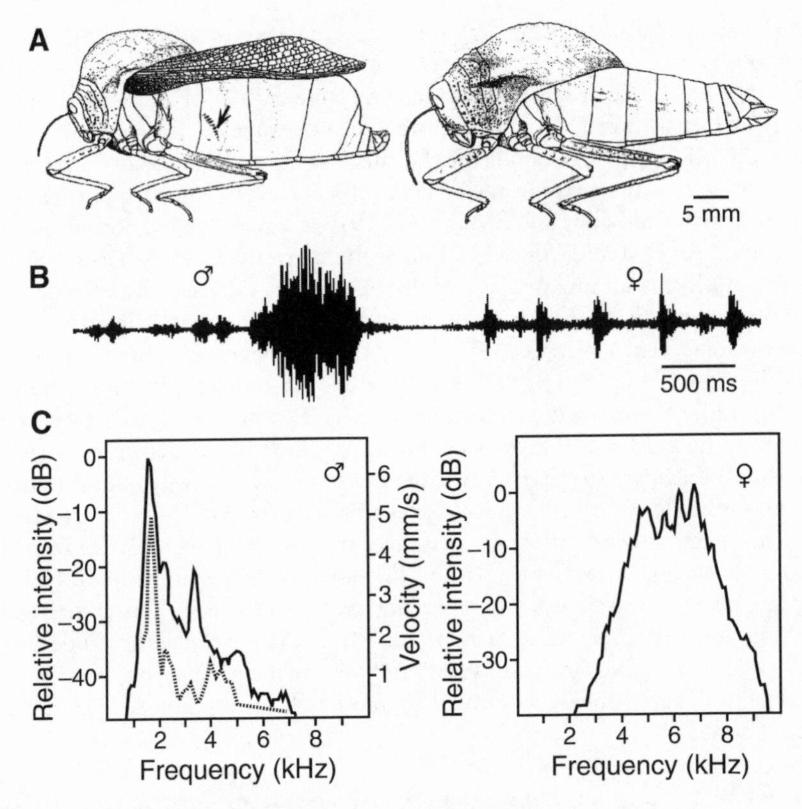

Figure 2.12. (A) Diagrams of the bladder grasshopper *Bullacris membracioides.* Large males (*left,* a typical winged morph) use abdominal-femoral stridulation; the location of the abdominal file is shown by the arrow. Large females (*right*) use abdominal-wing stridulation. (B) Oscillograms of the song of a male followed by the song of a female. (C) *Left,* male power spectrum (*solid line*) and vibration velocity (*dotted line*) of the inflated abdomen in response to acoustic stimulation with pure tones of 86 dB SPL. Notice that the maximal vibration velocity corresponds with the carrier frequency of the male calls. (C) *Right,* female power spectrum. From van Staaden and Römer 1997, figs. 1, 2.

signals (Q = 5) of about 6 kHz. This species provides an important contrast to its sibling species, *M. septendecim,* which is only slightly larger and has timbals with a similar structure. However, this species produces a pure-tone (Q = 25) calling song of about 1.3 kHz (fig. 2.13; Young and Josephson 1983b). We will return to these two species in a later section to examine the morphological differences that might be responsible for the dramatic differences in their songs.

Anurans. In vocalization the compression of pulmonary air by the trunk muscles is converted to periodically varying pressure waves (sound) by excitation of their thin, membranous vocal cords. Many species of anurans produce

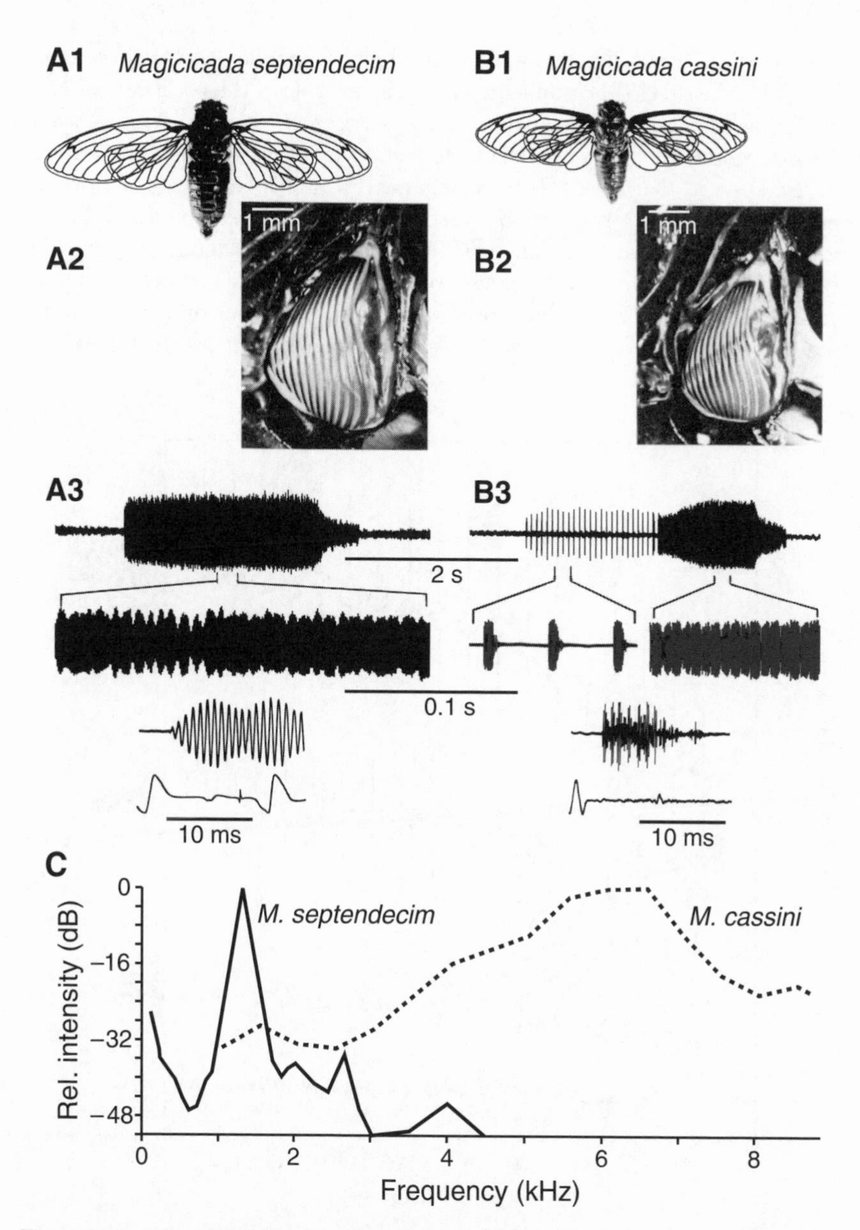

Figure 2.13. (A1) and (B1) Photographs of the periodical cicadas *Magicicada septendecim* and *M. cassini.* (A2) and (B2) Detailed views of their timbals. (A3) and (B3) Oscillograms with different time bases, showing the gross and fine-temporal differences in calling songs of the two species. The traces below the finest-scale oscillograms show the timing of timbal-muscle action potentials relative to sound production. Notice that one cycle of timbal activation produces a pulse with strong amplitude modulation resulting in 8–9 discrete cycles in *M. cassini*, whereas successive activation of timbals in *M. septendecim* produces slight amplitude modulation of a coherent train of sinusoids. (C) Diagrams showing the spectral differences in calling songs in the two species. From Young and Josephson 1983b; modified from figs. 1, 2, 3, 5.

narrow-band, tonal signals, with only slight frequency modulation (e.g., fig. 2.2A). Martin (1971) established the primary role of the vocal cords in determining the carrier frequency of the advertisement call of various species of toads (genus *Bufo*). By artificially activating larynges with pressurized air, Martin showed that their vibration was nearly sinusoidal or had one or two weak harmonics. In seventeen species the fundamental frequencies of the vocal cords, which were artificially activated at pressures of about 180 mm Hg, approximated the carrier frequencies of field-recorded calls. Three examples are shown in figure 2.14, which also indicates that at lower pressures, carrier frequency is slightly lower. This observation is consistent with the fact that in

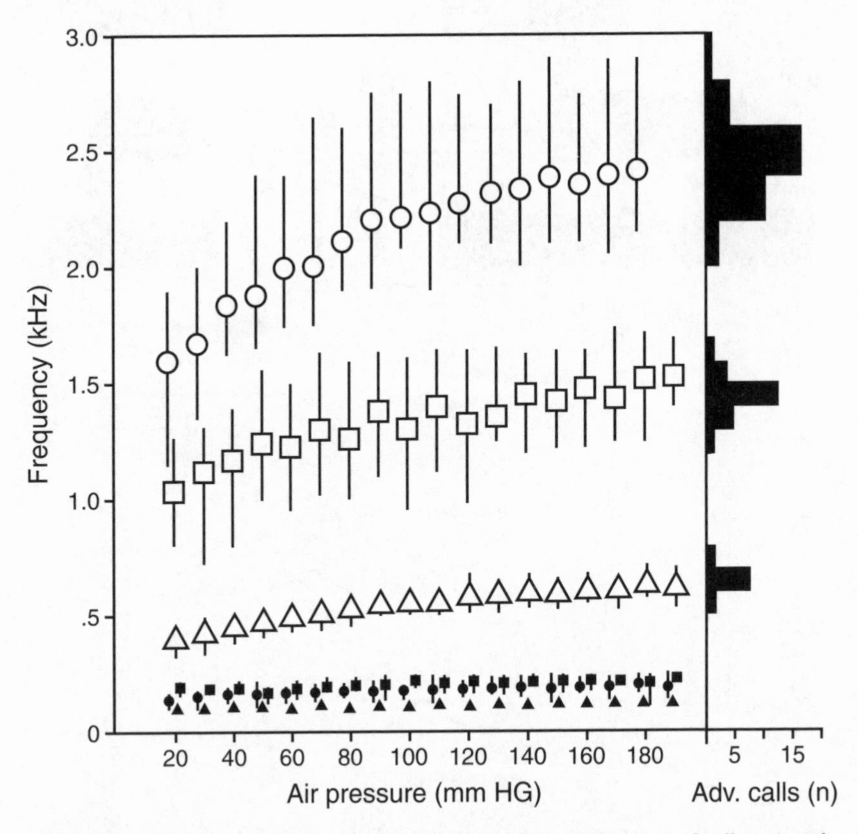

Figure 2.14. Examples of variation in the carrier frequency produced by artificially activated larynges of three species of toads with increasing activation pressure (mm Hg) (*circles, B. cognatus; squares, B. valliceps; triangles, B. marinus*). The fundamental frequencies of the vibration of the arytenoid cartilages are shown by the small solid symbols (all at or below 200 Hz; same key as for carrier frequency) and closely match mean pulse rates in advertisement calls. There was a slight-to-moderate increase in frequency with activation pressure, and at 180 mm Hg the carrier frequency matched that of the advertisement calls of all four species (histograms on the right side of the figure; n = number of advertisement [adv.] calls). From Martin 1971, fig. 4.

the long trills produced by most species of toads, the carrier frequency is slightly lower in the very first part of the trill when the pressure from the lungs is building up and the vocal sac is still not fully inflated. In other anurans, which have not yet been studied by the methods developed by Martin, nonsinusoidal vibration of vocal cords is the most likely source of signals rich in harmonics (examples in figs. 2.2 and 2.15A).

The pulses of most species of toads show only modest within-pulse frequency modulation, but dramatic sweeps in carrier frequency have been found in other families of anurans (fig. 2.15). Martin (1972) found that in a leptodactylid species, *Odontophrynes occidentalis*, the carrier frequency of artificially activated larynges was much more pressure-sensitive than those of bufonids. Moreover, there are also several pairs of laryngeal muscles, which can directly or indirectly vary the tension of the vocal cords and hence change carrier frequency. These muscles can also affect the positioning of the larynx relative to the glottis and can thereby vary the way in which the vocal cords are excited by the airflow from the lungs (Martin 1972).

Unlike in most vertebrates, most signals produced by the vocal cords of a typical anuran are not transmitted directly to the air surrounding the animal through an open mouth. Instead, the airflow generated by the lungs and exciting the vocal cords into vibration is contained within the oral cavity and vocal sacs (fig. 2.7), which are thought to be responsible for coupling internal

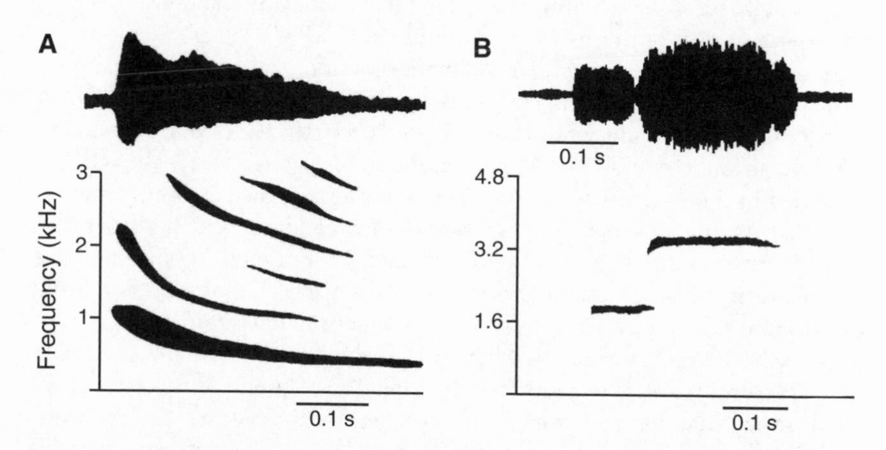

Figure 2.15. (A) Oscillogram and sonogram of the "whine" of the túngara frog *Physalaemus pustulosus* of Central and South America. Notice the down-sweeping harmonic series, which is produced as the vocal sac inflates. (B) Oscillogram and sonogram of the advertisement call of the boat-whistle frog *Eleutherodactylus johnstonei* of the Caribbean and northern South America. Its vocal sac also inflates during sound production, but here the frequency suddenly jumps by nearly an octave. Carrier frequency is thought to be mainly determined by tension on the vocal cords, their position in the air stream, and the pressure of the driving force of air from the lungs, with the vocal sac having a minor filtering effect (see the text). From Dudley and Rand 1991, fig. 3; and Watkins et al. 1970, fig. 1.

sound waves to the outside air. In most anurans these structures expand to a remarkable degree during vocalization and thus might store energy in a fashion analogous to a simple spring (Dudley and Rand 1991).

Whereas a cavity resonator has rigid walls that are assumed to reflect most sound energy back inside the cavity in order to generate standing waves, the vocal sac is an elastic membrane that potentially vibrates with considerable amplitude. Experimental manipulations, such as puncturing the vocal sac to reduce its volume (Martin 1972) or inducing a frog to call in a low-density gas (helium) atmosphere (Capranica and Moffat 1983; Dudley and Rand 1991), did not result in upward shifts in carrier frequency as expected of a cavity resonator. Furthermore, if the vocal sac were to function in this way, we might expect a decrease in frequency as it inflated and the volume of vibrating air increased. Although the fundamental frequency and harmonics in the calls of *Physalaemus pustulosus* decrease as predicted by this mechanism, the opposite is true of *Eleutherodactylus johnstonei* (fig. 2.15). These observations provide only circumstantial evidence, however. Another variable, which could differ between species, is whether the vocal sac becomes stiffer as it expands (expected to increase carrier frequency) or floppier (expected to decrease carrier frequency). As suggested above, different patterns of frequency modulation could also result from different degrees of compensatory actions, such as contractions of muscles to increase tension on the vocal cords (raising carrier frequency). In *P. pustulosus* the elasticity of the vocal pouch and its muscular activity might speed lung reinflation after each vocalization and also decrease the energetic cost of this process (Jaramillo et al. 1997).

The general consensus is that the main function of the vocal sac is to enhance sound radiation; furthermore, the vocal sac can also modify the spectrum of the sound produced by the vocal cords (Martin 1971; Martin and Gans 1972; Ryan and Drewes 1990; Dudley and Rand 1991). These ideas have been extended by a new approach to anuran vocalization, which assumes from the start that standing waves (resonance) are not formed inside cavities but rather on the elastic membranes associated with cavities (Purgue 1995). In bullfrogs —*Rana catesbeiana*, for example—the vocal sac transfer function (amplitude of vibration as a function of the driving frequency) matches the low-frequency peak in the advertisement call, whereas the vibration of the tympanic membrane (eardrum) not only augments the radiation of sound in this part of the signal's spectrum, but also contributes nearly all of the acoustic energy found in the high-frequency peak in the call (Purgue 1997). The seasonally variable tympanic membrane of an African treefrog (*Petropedetes parkeri*) might function in the same way (Narins et al. 2001). In this respect, these anuran systems are reminiscent of some species of cicadas, which emit considerable acoustic energy from their tympanic membranes (Young 1990). A striking sexual dimorphism exists in the size of the tympanic membrane in bullfrogs and some other species of ranid frogs. Tympanic membranes are much smaller in both sexes of most other anurans, which usually also have well-developed vocal sacs.

Although the vocal sac is likely to radiate and filter the sound produced at the larynx, appreciable acoustic energy can also be radiated from the lateral body wall overlying the lungs (Bennet-Clark and Gerhardt, unpubl. data; Purgue 1995). This hypothesis is circumstantially supported by the generally omnidirectional or modestly directional radiation patterns described for anurans (Gerhardt 1975; Prestwich et al. 1989). As hypothesized by Martin and Gans (1973), the anuran vocal system might function as some form of *coupled resonant system* in which the structures that produce and modify sound interact in complex ways that we have yet to study in a quantitative way.

In some species of toads (genus *Bufo*), males pulse the airflow from the lungs to larynx (Martin 1971). Each pulse of sound, which consists of a train of sinusoidal waveforms, in the call corresponds to a separate contraction/relaxation cycle of the thoracic musculature that compresses the lungs (fig. 2.16A). Martin termed such a mechanism *active amplitude modulation;* the timing of occurrence of each sound pulse is clearly determined by the central nervous system. By contrast, other kinds of toads produce pulsed vocalizations even though their thoracic muscles contract slowly and smoothly, resulting in a steady (direct current) flow of air from the lungs (fig. 2.16B; Martin 1971, 1972). Here the initial buildup of pressure opposes the closure of the larynx and the elastic properties of the vocal cords themselves. This results in the (elastic) storage of energy, which is released when the larynx suddenly springs open. The vocal cords vibrate, and their vibration is modulated by another laryngeal oscillator, the paired arytenoid cartilages, the vibration of which depends, in turn, on the presence of fleshy structures inside the larynx called *arytenoid valves* (fig. 2.8B,C; fig. 2.16). Elastic forces close the larynx, and the cycle is rapidly repeated. In an analogy drawn from physics, Martin (1971) likened the action of these structures to a *relaxation oscillator*, and he termed this mode of pulsing *passive amplitude modulation*. In contrast to the pulses produced by active amplitude modulation, which have relatively slow rise-fall characteristics, the pulses produced by passive amplitude modulation have fast rise-times and exponential decays (compare fig. 2.16A and B). In other species the airflow to the larynx is pulsed, but the arytenoid valve/cartilage oscillator is also activated, resulting in sound pulses with intrapulse amplitude modulation (fig. 2.16C). Thus, the temporal pattern of this signal results from a combination of active and passive modulation.

Comparison of modulation processes in insects and anurans. Just as frequency-multiplier mechanisms convert relatively slow movements of muscles into high-frequency sounds, passive amplitude modulation can be considered a mechanism for multiplying the rate of amplitude modulation beyond that which could be directly achieved by repetitive cycles of muscle contraction.

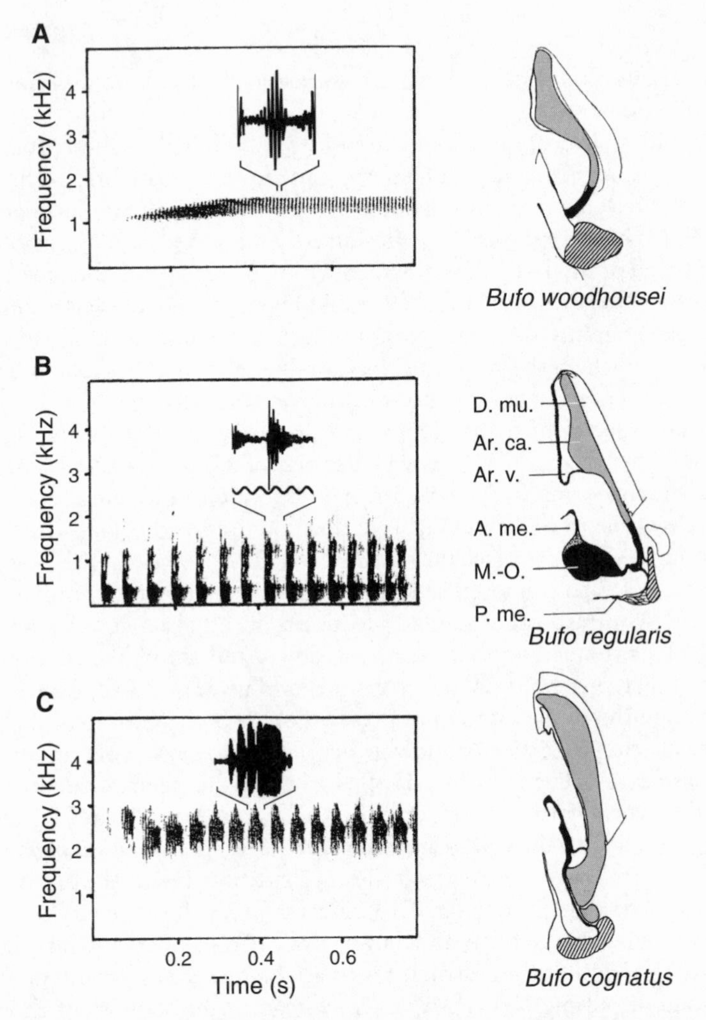

Figure 2.16. (A) *Left,* oscillogram showing a pulse from the advertisement call of a *Bufo woodhousei;* a sonogram with the beginning of the call is shown below. Note the slight increase in frequency at the beginning; the carrier remains constant, however, for the remainder of a long trill. *Right,* diagram of a frontal section of half of a larynx (see fig. 2.8C); airflow from the lungs comes from the bottom of each diagram. This species lacks an arytenoid valve and midrib ossicle, and its mode of amplitude modulation is purely active. The diagram to the right of (B) shows the locations of these and other laryngeal structures in a species that has all of them: D.mu. = dilator laryngis muscle; Ar.ca. = arytenoid cartilage; Ar.v. = arytenoid valve; A.me. = anterior membrane (vocal cord); M.-O = midrib-ossicle; P.me. = posterior membrane. (B) *Left,* oscillogram showing a pulse from the advertisement call of *Bufo regularis;* a sonogram showing part of an advertisement call is shown below. *Right,* this species has a well-developed arytenoid valve, which is responsible, in part, for the passive mode of amplitude modulation in this species. Moreover, its massive midrib-ossicle (= fibrous mass) results in the production of a much lower carrier frequency than in toads of comparable size that lack this structure. (C) *Left,* oscillogram showing a pulse from the advertisement call of *Bufo cognatus;* a sonogram is below. *Right,* this species has a well-developed arytenoid valve that is responsible for the passive amplitude modulation of each of the pulses, which are, however, produced by active modulation. See the text for further explanation. From Martin 1967, unpubl. M.A. thesis, figs. 5, 9, 13.

In stridulating insects, the central nervous system actively modulates the signal, determining the rate and pattern of opening and closing of the wings or of leg movements. The number, spacing, and acoustic properties of the file and stridulatory teeth also contribute to fine-scale patterns of amplitude modulation at any given tooth-impact rate. These structures can thus be considered functionally analogous to the structures that achieve passive amplitude modulation in toads and other anurans.

Passive modulation using the principle of a relaxation oscillator (sensu Martin 1971) is particularly well illustrated by some cicadas. The initial contraction of large and powerful timbal muscles causes the deformation of the timbal and the storage of potential energy, which is released when the further application of force by timbal muscle contraction finally causes a series of timbal ribs to buckle inward (Bennet-Clark 1997). Fast rise-time pulses of sound are nearly always produced when the ribs buckle inward, and in some species loud pulses are also produced as the ribs move outward to their original resting position (fig. 2.6; Fonseca 1994). These passive processes interact with the active modulation generated by the central nervous system, including different timing patterns within the same basic timbal cycle and differing phase relationships between the two timbals. Tensor muscles can also influence the shape, deformation, and stiffness of the timbals, thereby generating signals that show frequency modulation, amplitude modulation, or both (e.g., *Tympanistalna gastrica;* Fonseca 1994). In some cicadas, tensor muscle contraction causes an increase in pulse amplitude, and in others, a decrease (Fonseca 1994; Fonseca and Hennig 1996; further discussion in chapter 3).

Do Peripheral Mechanisms Constrain Acoustic Structure?

Comparative studies usually confirm the expected inverse relationship between the effective size of sound-producing structures and the carrier frequency of long-range signals (figs. 2.17 and 2.18; Bennet-Clark 1998). For anurans at least, the dimensions of sound-producing and radiating structures are usually also correlated with body size or mass; thus, in turn, reasonably strong correlations are usually found between body size and carrier frequency in both intraspecific (e.g., Zweifel 1968; Robertson 1990) and interspecific comparisons (e.g., Duellman and Pyles 1983; Zimmerman 1983; Ryan 1988). By contrast, a recent study of acridid grasshoppers failed to find statistically significant correlations between carrier frequency and body size or wing length (Meyer and Elsner 1996). One possibility is that grasshoppers, with few exceptions, do not have specialized resonating structures on their wings, as do katydids and crickets; moreover, estimating the carrier frequency of their typically broadband signals cannot be done precisely.

The coupling of carrier frequency and body size means that selection on either attribute could be countered or confounded by concomitant selection on the other trait. For example, strong selection by receivers to increase carrier

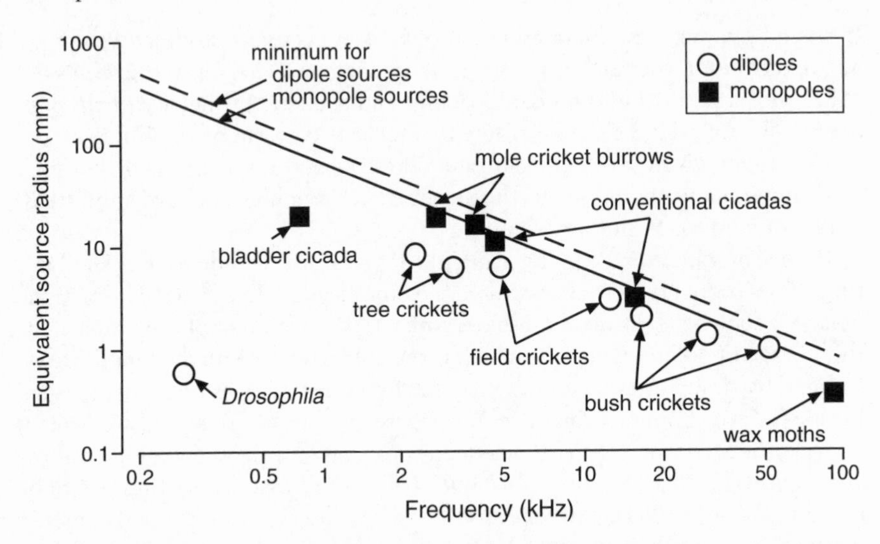

Figure 2.17. Effective source dimensions of peripheral sound-radiating structures plotted against carrier frequency. The solid line indicates the minimum source dimensions for optimal sound production for monopole sources, and the dotted line gives the same values for a dipole source. One interpretation of these results is that insects trade reduced efficiency in sound production for the reduced excess attenuation achieved by producing signals of lower frequency. From Bennet-Clark 1998.

frequency could be countered by selection for increased body size; selection causing changes in body size would result in a change in carrier frequency even if receivers do not have strong preferences. The correlation is not only a nuisance for biologists trying to understand the causes of evolutionary diversification of carrier frequency. If, for example, several species of about the same size signal at the same time and place, then the carrier frequencies of their signals are likely to overlap and thereby increase masking interference. Numerous examples have been reported in mixed-species choruses of anurans (Littlejohn 1977; Drewry and Rand 1983; Duellman and Pyles 1983).

These considerations make it especially important to understand how exceptional taxonomic groups and species have apparently broken the expected correlation between body size and carrier frequency. At a gross level of analysis, the crickets and their allies are one example. The evolutionary inventions of the clockwork mechanism, the harp and expanded wing areas, for example, are keys to the production of songs that are not only narrower in bandwidth but also much lower in carrier frequency than those of other orthopterans of comparable size.[2] As shown above, the same general solution occurs in some

2. A further reduction is achieved by an Australian cricket (genus *Rufocephalus*), which couples its wing resonance to that of its burrow, as in mole crickets. This species is about half the length of *Gryllus bimaculatus* and its carrier frequency is only 3.2 kHz (compared to about 5 kHz in the larger species) (Bailey et al. 2001).

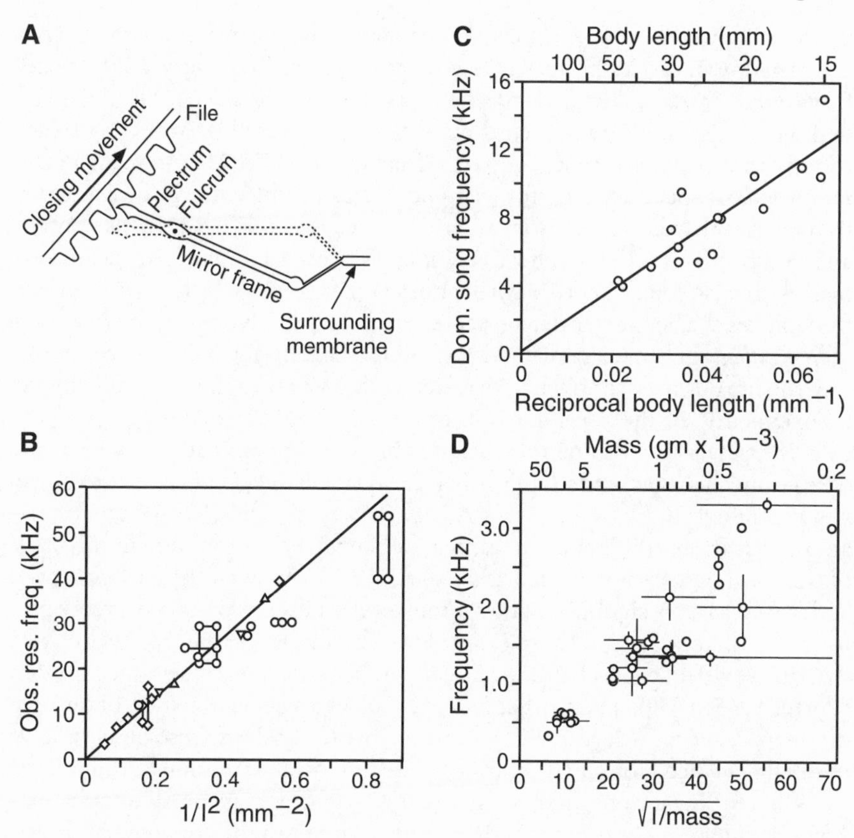

Figure 2.18. Scaling of song frequency. (A) Diagram of the mirror frame of a katydid and its hypothesized movements as the wings close and the scraper engages the file. (B) Scatter plot and fitted regression line for the lowest spectral peak (or peaks) in the calling song in nineteen species of katydids against the reciprocal of the squared length of the mirror frame (based on a physical-acoustics model of a cantilever system; Sales and Pye 1974 provide details). (C) Scatter plot and fitted regression line for the carrier frequency of the calling song of sixteen species of "typical" cicadas against the reciprocal of body length. Relative body length is a good approximation of the dimensions of abdominal air sac, which, in conjunction with the two tympana close to its ventral surface, can be modeled as a Helmholtz (cavity) resonator. (D) Scatter plot of mean carrier frequency of the advertisement calls of twenty-nine species of toads (genus *Bufo*) and four hybrid combinations against the square root of the inverse of the mass of the larynx, a metric assuming that the vocal cords vibrate as a circular membrane. Further details are given by Martin (1972), who provides a key to the species and hybrids as well as the sample sizes. The lines show the ranges of variation in carrier frequency and mass. From Sales and Pye 1974, fig. 5.9; Bennet-Clark and Young 1994, fig. 1; and Martin 1972, fig. 15.11.

exceptional species of grasshoppers and cicadas that use an air-sac resonator. An additional key for the gryllids is a second invention on the receiver side, which, as we show in chapter 7, allows these small insects to effectively localize their relatively low-frequency signals.

Some anurans produce calls with frequencies differing from those expected by their body size because (1) vocal cord mass, which explains a large part of the variation in carrier frequency (e.g., toads, Martin 1972; fig. 2.18D; cricket frogs *Acris crepitans:* McClelland et al. 1996), does not always correlate well with body size; and (2) vocal cord mass can be increased by accessory structures without an increase in body size (Martin 1972). In African toads in the *Bufo regularis* species group, for example, males produce much lower carrier frequencies than expected from their size because their vocal cords contain a large fibrous ossicle that effectively increases its mass (fig. 2.16B). Fibrous masses are also found in other small anurans that produce signals with carrier frequencies that are lower than expected from their body size. In the *Physalaemus pustulosus* group (example in fig. 2.15A), relatively small differences in the way the fibrous mass is attached may determine whether the mass can vibrate independently of the vocal cords in order to produce the complex two-part calls characteristic of some members of the group (Ryan and Drewes 1990).

The mechanisms underlying other exceptional species remain unknown and unstudied. In two species of toads (*Bufo cognatus* and *B. valliceps*), for example, significant differences in carrier frequency exist even though the mass of the vocal cords is nearly identical (Martin 1971). In two others (*Bufo coccifer* and *B. hemiophrys*), significant population-level differences in carrier frequency were just the opposite of those expected from population differences in body size: the smaller species had a lower carrier frequency than the larger species (Porter 1965, 1968). Thus, other factors, such as active modulation of tension on the vocal cords (Martin 1972), must be involved. All of these observations highlight the need for much additional research on anuran sound production.

Why are "inventions" such as increasing laryngeal mass or modifying tension on the vocal cords apparently so uncommon? Are there physical trade-offs that make them energetically uneconomical or acoustically inefficient for most species? Or does lowering carrier frequency incur environmental costs, such as increased predation or reduced mating success? For example, lowering carrier frequency might increase signal range but reduce the ability of receivers to localize the signal. Some hints are provided by three species (cricket frogs *Acris crepitans:* Wagner 1989a,b; American toads *Bufo americanus:* Howard and Young 1998; green frogs *Rana clamitans:* Bee and Perrill 1996) in which males lower the carrier frequency of advertisement calls during vocal interactions or in response to playbacks. Although the changes averaged only about 4–5% from the carrier frequency produced during solo calling, deviations of as much as 15% were recorded in a few male cricket frogs. Producing lower-than-average carrier frequencies results in a reduction in call amplitude (Bee and Perrill 1995; Howard, pers. comm.; Wagner, pers. comm.), possibly because the lower carrier frequency does not match the resonance of the secondary sound-radiating structures as well as does the unaltered carrier frequency. Because male anurans interact aggressively at close range, the drop in amplitude is probably inconsequential, but lowering carrier frequency is unlikely to be a good option for increasing the distance over which males attract females. Any gains in enhanced propagation must be offset by the reduced signal amplitude, energetic costs of less-efficient signal production, or both. However, one other species, in which males alter carrier frequency, deserves further study in this context. Males of the white-lipped frog of Puerto Rico not only lower but also raise carrier frequency in response to playbacks; moreover, a drop in frequency does not appear to reduce signal amplitude (Lopez et al. 1988).

We conclude this section by returning to a comparison of two closely related species that not only produce signals with different carrier frequencies but also very different bandwidths. As discussed above, the dimensions of the timbals, timbal muscles, and air sacs of the two sibling species periodical cicada *Magicicada septendecim* and *M. cassini* are comparable in size, and both species have eleven or twelve long timbal ribs (Young and Josephson 1983b; Bennet-Clark and Young 1992). Yet *M. septendecim* produces nearly pure-tone calling songs with a carrier frequency that is about two octaves lower than predicted by the Helmholtz equations for a cavity resonator with its air-sac dimensions (1.3 kHz vs. 4.4 kHz). Its sibling species, *M. cassini*, produces complex amplitude-modulated pulses and broadband signals with a peak frequency that is very close to the value predicted by physics (6.0 kHz vs. 6.7 kHz) (fig. 2.13). What has changed to cause the profound differences in the song structures of these two species? First, Young and Josephson (1983b) note that the timbal of *M. septendecim* is much less stiff than that of *M. cassini*, and they show that the timbal produces the carrier frequency of the calling song, albeit at a much lower amplitude, without the air sac. If, as in other cicadas with narrow-band signals, the ribs buckle in rapid succession within a timbal muscle cycle and the air-sac resonance matches the carrier frequency, there can be an interaction that serves to produce a long coherent wave. Second, Bennet-Clark and Young (1992) find that the tympanal membranes of *M. septendecim* are much thicker than those of *M. cassini*. They hypothesize that a thicker membrane mimics a reduction in the effective sound-radiating area of the air-sac resonator and hence lowers its resonant frequency and raises its Q-value. This example serves to emphasize the fact that relatively minor changes in peripheral structures can, in principle, result in significant differences in signal structure.

Summary and Suggestions for Future Studies

Signal structure and variability are constrained in two ways: (1) by the biophysics of the peripheral structures that generate and radiate sounds and the neuromuscular mechanisms that drive these structures; and (2) by the requirements of receivers that must detect the signal, extract the information encoded in the signal, and localize its source. The physical and statistical analysis of signals is thus a starting point for understanding how signals are produced and what information about the sender might be encoded in particular acoustic properties. In order to extract reliable information, receivers must often assess between-individual variation in an acoustic property as well as within-individual variation, neither of which necessarily predicts variability at the between-population level. Thus, statistical analyses at all of these levels of analysis can be useful. Static properties such as carrier frequency and pulse rate vary relatively little from signal to signal in an individual, whereas at the other end of the continuum, dynamic properties such as call rate can change dramatically, especially during acoustic interactions with other individuals. Within-individual variation in call duration is continuous, with a tendency for low variability in species with short signals and high variability in species with long signals.

Carrier frequency and pulse structure show limited variation because they are usually tied to the size of the individual and the mechanics and morphology of the peripheral sound-producing structures, respectively. Although pulse rate shows little variation in advertisement signals, it can vary significantly in other signals in the repertoire, probably because of different patterns of past and current selection on the different signals. Studies of mechanisms underlying graded signals, in which pulse rate varies widely from signal to signal, would be especially welcome.

Peripheral mechanisms in insects and anurans share with other animals the general principle of frequency multiplication. Primary oscillators or resonators vibrate at frequencies much higher than can be accomplished by repetitive cycles of muscle contraction and relaxation. In both insects and anurans, high pulse rates are also sometimes generated by analogous mechanisms—passive amplitude modulation—whereby single, relatively smooth muscle contractions are converted into a series of rapidly repeated sound pulses by peripheral structures. The sound produced by the primary oscillators is often coupled to other, secondary resonators that increase the efficiency of sound radiation. Crickets and their allies are special in that, in contrast to most other orthopterans and cicadas, they produce nearly pure-tone signals that are much lower in carrier frequency than other insects of comparable size. Two inventions—the clockwork coupling of the rate of stridulation to the resonant frequency of the harp or other sound-radiating wing structures and the ability to localize relatively low-frequency sounds—appear to be responsible. Interesting questions remain, however, about how this mechanism of sound production works in species in which carrier frequency varies significantly with temperature. And how do receivers whose auditory systems are tuned to the carrier frequency deal with such changes in frequency?

Correlations between carrier frequency and body size can confound the effects of selection on either trait. Thus, exceptional taxonomic groups, populations, and species, for which the correlation between body size and spectral properties is weak or nonexistent, present special opportunities to learn how selection might directly target and bring about evolutionary change in such generally conservative traits as carrier frequency and bandwidth. This work is well advanced in insects, where many estimates of the physical properties that determine acoustic ones are available for some species (cicadas and crickets, e.g., Koch et al. 1988; Bennet-Clark 1989, 1997, 1998). By contrast, there has been relatively little progress in understanding the mechanisms of sound production in anurans since the pioneering work of Martin (1971, 1972) with the bufonids, which do not show the great diversity of acoustic structures found in the anurans as a whole. For example, we also still know little about how laryngeal muscles and other structures in the larynx modify carrier frequency and generate different forms of modulation. Our understanding of sound radiation in anurans has improved because of new assumptions and methodological approaches (e.g., Purgue 1995, 1997).

3

Neural Control of Sound Production

Despite profound differences in their nervous systems and peripheral sound-producing structures, the functional organization of the neural and neuromuscular components of the sound-production system is similar in insects and anurans (fig. 3.1). In both taxa, *command systems* start and maintain sound production by activating *pattern-generating networks*. These networks, in turn, orchestrate the timing and strength of the neuromuscular output that drives the peripheral structures (chapter 2). Once triggered, pattern generators that are experimentally isolated from peripheral feedback can still produce the basic temporal pattern of an acoustic signal at least for a short time. In intact animals, however, the output of pattern generators is modified and fine-tuned by feedback mechanisms, which are also likely to stabilize the output over relatively long periods of time (Elsner 1994). In both taxa, elements of the sound-production system are shared with systems that also produce rhythmic movements, that is, locomotion in insects and respiration in anurans. The anatomy and function of these basic systems have undoubtedly placed some constraints on the evolution of signal production.

Diagrams of the cephalic and thoracic parts of the central nervous systems of crickets and grasshoppers and the approximate locations of command systems, pattern generators, stridulatory output, and acoustic input are shown in figure 3.2. The courses of fiber systems (descending and ascending) whose activity influences stridulation have been particularly well studied in grasshoppers (review: Elsner 1994). We know less about crickets and katydids, and we have only preliminary wiring diagrams of the vocal control system in anurans (fig. 3.3). This situation in frogs is likely to improve as new studies have delineated both the gross distribution of neurotransmitters and the anatomy and projections of single neurons in midbrain areas that are probably major components of the sensory-motor interface (see below; Luksch and Walkowiak 1998; Endepols et al. 2000; Endepols and Walkowiak 2001). Understanding

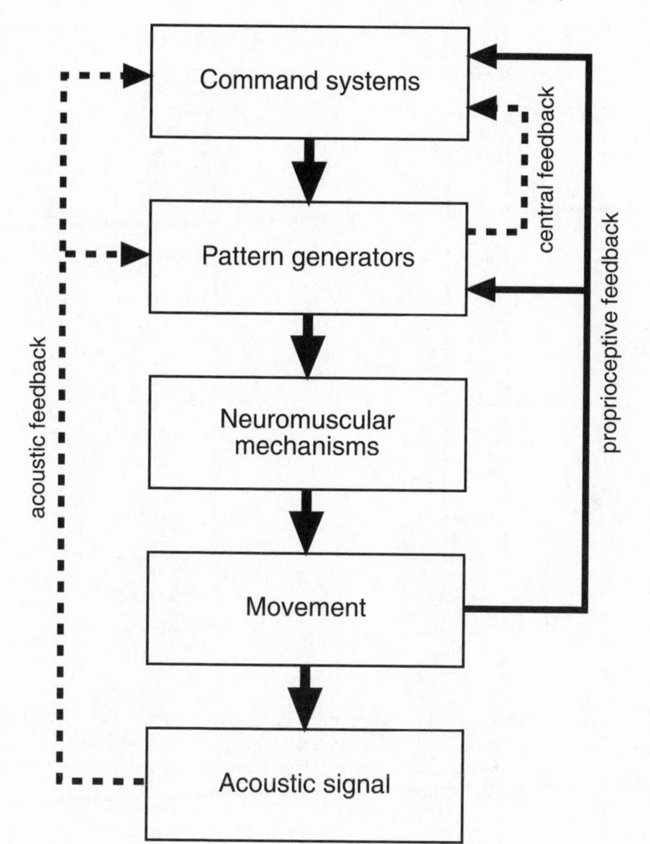

Figure 3.1. Subsystems involved in central and peripheral organization of sound production. Arrows indicate the flow of information. Solid lines indicate well-established functional and anatomical connections, and dotted lines, possibilities for feedback.

the neural control and patterning of signal production remains a major challenge for neurobiology. Such knowledge is vital for understanding both mechanistic constraints on signal evolution and the origins of new signals from rhythmic movements generated in other contexts.

Command Systems in Insects and Anurans: Overview

Signal Production Induced by Electrical Stimulation

Systems with command functions for stridulation were first discovered by Huber and his students, who used brain lesions and electrical stimulation to localize sites at which they could elicit one or more of the three types of songs produced by the field cricket *Gryllus campestris* (Huber 1955, 1960; Otto

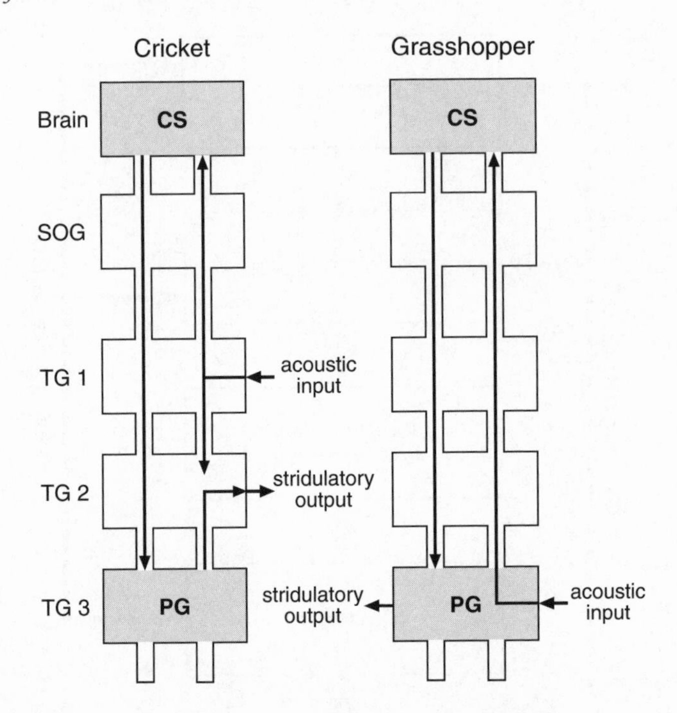

Figure 3.2. Diagrams of the cephalic and thoracic ganglia of a cricket and a grasshopper. The locations of command systems (CSs) in the brain and pattern generators (PGs) in the metathoracic ganglionic complex as well as stridulatory output and acoustic input are indicated. SOG = subesophageal ganglion; TG 1–TG 3 = thoracic ganglia (= pro-, meso- and metathoracic).

1971). Electrical pulses continuously applied to single brain sites in both tethered and freely moving animals either resulted in the production of one or more signals in the animal's repertoire or inhibited ongoing stridulation (fig. 3.4A,B). The same methods elicited calling and courtship songs in grasshoppers (Wadepuhl 1983; Hedwig 1986a). The sites are restricted to the protocerebrum. More specific locations for activating calling song are situated between the pedunculus and the dorsal surface of the a-lobe of the mushroom bodies. Courtship song activated by brain stimulation also requires intact nervous connections between sites near the spermatophore sac in the last abdominal ganglion and the thoracic ganglia, where the pattern generators are located (Huber 1955). The activation of sound production thus might also depend on monitoring by the central nervous system of the animal's reproductive state (Loher et al. 1993). Brain nuclei that concentrate reproductive hormones also play a major role in the control of vocalization in anurans (Kelley 1980; Endepols et al. 2000).

The temporal patterns of signals produced in response to electrical stimulation of the brain are independent of the temporal pattern of such stimulation.

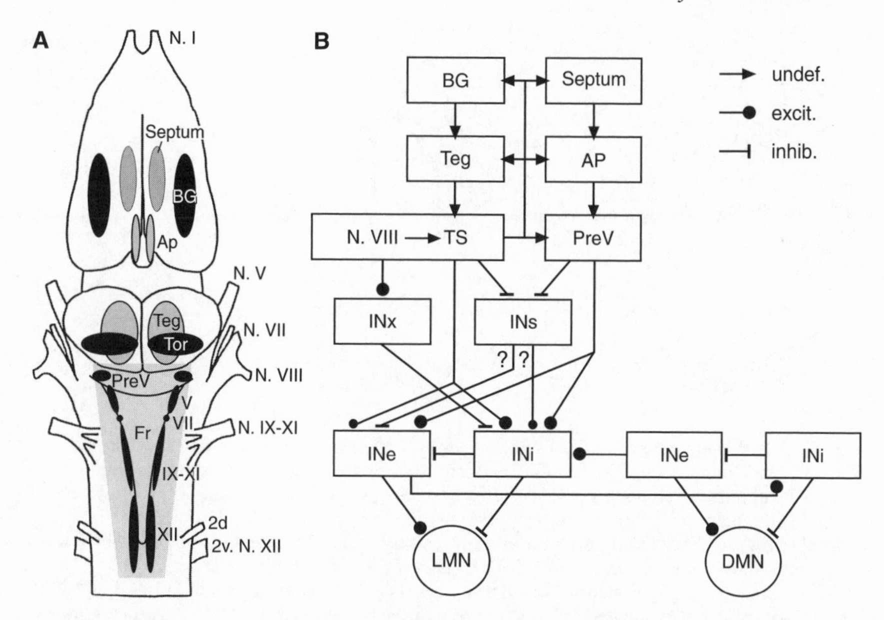

Figure 3.3. (A) Dorsal view of the brain of an anuran, showing the locations of nuclei in the medulla that are probably involved in triggering and patterning vocalization. Some of these and other nuclei are also involved in regulating respiration. As shown by Schmidt (1992), the main pattern-generating circuits for vocalization and respiration (the pretrigeminal nucleus [PreV] and pulmonary respiratory generators [anatomically and functionally coextensive with motor nuclei IX–X]) are located in the brain stem. (B) Tentative wiring diagram based on studies of frogs of the genus *Bombina.* Unlike most other species, these frogs produce advertisement calls during inspiration using antagonistic (depressor and levitator) muscles of the mouth cavity, and so the neuromuscular system is somewhat simpler than in most other anurans, which also use thoracic muscles to produce calls during expiration. The motor network apparently receives two parallel inputs: (1) a direct input from different levels of the auditory brain stem; and (2) an indirect input from the limbic system via septal nuclei, the preoptic area, and the pretrigeminal nucleus. The auditory brain-stem input may control reactions to the signals of neighbors that affect call timing (chapter 8), whereas the limbic input represents the main internal input to the respiratory and call-generating network. The proposed network thus includes classical auditory nuclei, such as the torus semicircularis, neuroendocrine-integrating centers such as the preoptic nucleus of the hypothalamus, and premotor and motor nuclei that directly control the muscles used in vocalization and respiration. Connections in the forebrain nuclei are established by anatomy, but their functional relationships are undefined, and in general we do not know how and over what time scales these neural components might interact to control vocalization. Auditory brain stem (N. VIII → TS); AP = preoptic area; BG = basal ganglia; DMN = buccal depressor motoneurons; IN(e–x) = different populations of interneurons (see text); LMN = buccal levitator motoneurons; N. VIII = auditory nerve; PreV = pretrigeminal nucleus; Teg = tegmentum; TS (Tor) = torus semicircularis.

Electrical pulses delivered at an arbitrary rate and even DC current elicited normal calling and courtship songs when applied to the cervical connectives (Otto 1967) or to descending neurons (Bentley 1977). Comparable results are available for grasshoppers (Hedwig 1986a). Moreover, a few male crickets, whose cervical connectives had been cut, produced normal singing, albeit

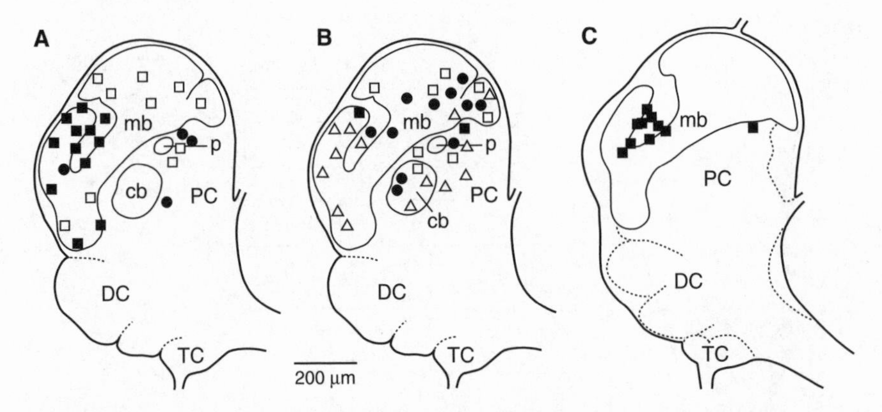

Figure 3.4. Schematic parasagittal sections through the brain of the cricket *Gryllus campestris*, showing sites for activation and inhibition of stridulation by focal electrical stimulation in tethered (A) and in freely moving animals (B). *Filled squares*, sites eliciting calling song often accompanied with courtship song; *open squares*, sites causing inhibition of stridulation; *filled circles*, sites eliciting aggressive song often followed by calling song; *open triangles*, sites eliciting courtship song followed by calling song or vice versa. (A) From Huber 1960; (B) from Otto 1971. (C) Schematic parasagittal section through the brain of *Gryllus bimaculatus* with the sites (*black squares*) for calling song elicited by injection of neuroactive substances. DC = deutocerebrum; TC = tritocerebrum; mb = mushroom body (corpus pedunculatum = PC); cb = central body; p = pons. From Wenzel et al. 1998.

only for a short time, indicating that thoracic pattern generators can work without the brain (Kutsch and Otto 1972).[1]

In parallel to the work of Huber and his colleagues, experiments with frogs show that reasonably normal advertisement and release calls are elicited by electrical stimulation of sites in several parts of the brain (Schmidt 1973; Knorr 1976; Wada and Gorbman 1977; review: Schneider 1988). Most of the effective sites that have so far been reported are in parts within the hypothalamus (diencephalon) and medulla that concentrate sex hormones (Kelley and Gorlick 1990). Implantation of testosterone in the preoptic area of the hypothalamus also increases calling activity in castrated males (Wada and Gorbman 1977). Higher centers in the telencephalon—the striatum and septum—are also likely to exert some control on the triggering and timing of sound production, and central control is almost certainly mediated by parallel descending pathways (e.g., Endepols and Walkowiak 1999, 2001; Walkowiak et al., 1999). Schmidt (1974a,b, 1976) showed that neural correlates of calling (fictive calling) survive complete denervation of the sensory input to the brain. As in crickets (Bentley 1969; Otto 1971; Hedwig 1996, 2000a), the temporal patterning of the signals observed during electrical stimulation are independent of stimulus frequency and persist for a short time after stimulation ceases

1. Albertus Magnus (1193–1280) was probably the first to observe stridulation in crickets after removal of the head (cited in Bodenheimer 1928 and quoted by Elsner 1994).

(Schmidt 1976). Recent research using isolated brain preparations has partially confirmed and extended Schmidt's work (see below).

Electrical stimulation of some sites in a treefrog brain not only elicits the production of advertisement calls in males and some females, but also the adoption of femalelike spawning positions by males (fig. 3.5; Knorr 1976). Thus, each sex apparently has the neural machinery to produce behaviors that are normally only observed in the opposite sex. The same is likely to be true of those grasshoppers in which females stridulate. Indeed, when prevented from mating for a prolonged period, virgin females show malelike courtship behavior (Loher and Huber 1965).

The *sensory-motor interface* consists of the anatomical and functional connections that make it possible for ascending auditory information to influence behavioral decisions that are executed by the motor system. Ironically, the greatest progress has been made by using isolated brain preparations, which allow researchers to trace the connections of single neurons and to obtain stable recordings from such neurons in response to electrical stimulation at various points in the brain or auditory nerve (Luksch et al. 1996). Such studies have identified neurons in certain nuclei in the midbrain (mainly the laminar nucleus of the torus semicircularis; see chapter 6) that both receive ascending auditory input and also project to lower and higher auditory nuclei and to premotor and motor regions (Luksch and Walkowiak 1998; Endepols and Walkowiak 2001). Electrical stimulation studies show that descending input from higher centers can modulate the responses of neurons in the laminar nucleus to excitation arriving from the auditory nerve (Endepols and Walkowiak 1999, 2001). Taken together, these studies show that auditory information is distributed widely throughout the anuran brain and that the torus semicircularis might serve as a center for integrating ascending auditory input and descending activity. The output of the torus (at least as inferred from its anatomical connections) might, among other things, control or modulate vocalization.

Stridulation Induced by Neuroactive Substances

Another approach, which is a first step in the analysis of the neurochemical specificity of control systems, has employed microinjection of neuroactive substances into the brain. In a preliminary study, Otto (1978) demonstrated that injection of acetylcholine (ACh) or eserine (an ACh-esterase inhibitor) can elicit calling, rivalry, and courtship songs in the cricket *Gryllus campestris*. His results were recently confirmed and extended by studies of other crickets and grasshoppers, in which inhibitory as well as excitatory mechanisms underlying stridulation have been examined (Heinrich 1995; Heinrich et al. 1997, 1998, 2001; Ocker et al. 1995; Wenzel and Hedwig 1999). In *Gryllus bimaculatus*, for example, all three types of species-specific song patterns and transitions among them can be elicited by pressure injection of ACh and cholinergic agonists into the protocerebrum (Wenzel and Hedwig 1999). Subsequent histological location of the micropipette tips showed that the sites at which

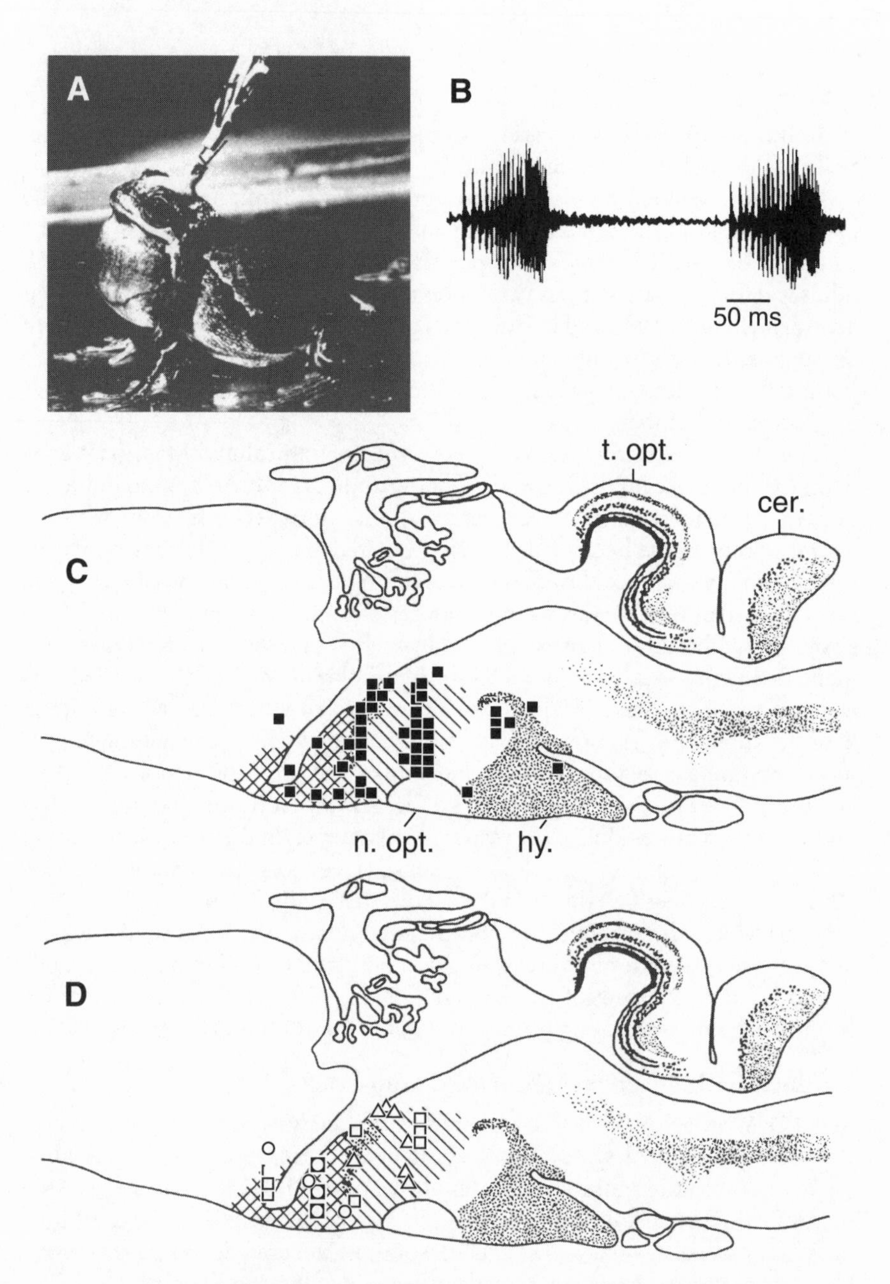

Figure 3.5. Brain stimulation of vocal and other reproductive behaviors in a treefrog. (A) Male of *Hyla arborea* in calling posture with an implanted electrode. (B) Oscillograms of two advertisement calls elicited by electrical stimulation. (C) and (D) Diagrams of longitudinal sections through the brain about 100 μm lateral to the midline. Sites where electrical stimulation elicited different behaviors are indicated by black squares (calling posture), white squares (advertisement calls), white circles (egg-laying posture); circles within squares (advertisement calls and egg-laying posture), and triangles (mucus production). Cross-hatching shows location of the preoptic nucleus; diagonal hatching shows the location of the posterior preoptic nucleus; and the stippled area corresponds to the approximate location of the magnocellular nucleus. cer. = cerebellum; hy. = hypothalamus; n. opt. = optic nucleus; t. opt. = optic tectum. From Knorr 1976.

these substances affect stridulation are the same as those at which stridulation is effectively elicited by electrical stimulation (fig. 3.4C). Moreover, the stimulation sites overlap with the arborizations of stridulatory command neurons in the protocerebrum (Hedwig 1996, 2000b).

In crickets, calling song generally starts several seconds after the application of a single pulse of ACh. At first the cricket produces small-amplitude wing movements and chirps of low intensity that contain only 2–3 pulses/chirp. The intensity of stridulation then gradually increases until the songs with the usual 4–5 pulses/chirp are produced at their normal intensity (fig. 3.6A). Such calling songs last up to two minutes. Microinjections of either cholinergic agonist— nicotine or muscarine—elicit calling songs, but with different time courses and durations. That is, singing activity is triggered more rapidly by nicotine than by muscarine, but the effects of muscarine last longer. As already reported by Otto (1978) and confirmed by Wenzel and Hedwig (1999), the chirp intervals of ACh-injected animals were shorter and pulse intervals somewhat longer than those of controls. Thus, pharmacological stimulation of the brain also apparently has some influence on the timing of chirp and pulse-rhythm generation.

Inhibition of ongoing stridulation can also be achieved pharmacologically (fig. 3.6C). Application of gamma-aminobutyric acid (GABA) immediately stops the stridulatory behavior elicited by a previous injection of ACh or muscarine to the same brain site. After GABA treatment, calling song then generally reappears at low intensity. Injection of picrotoxin, which is the most frequently used blocker of GABA-ergic action, leads to enhanced motor activity that incorporates the three different cricket song patterns (details in Wenzel and Hedwig 1999). The same effects have also been demonstrated in several species of grasshoppers, where long-lasting stridulation and complete courtship sequences are elicited by pressure injection of ACh or its agonist pilocarpine into the protocerebrum (fig. 3.6B). Injections of GABA or glycine inhibit stridulatory behavior (fig. 3.6D).

Command Neurons in Crickets and Grasshoppers

The facts discussed in the previous section predict the existence of descending brain neurons that trigger stridulation. Such neurons were first identified in the grasshopper *Omocestus viridulus* (Hedwig 1986a; see review: Elsner 1994) and later in the cricket *Gryllus bimaculatus* (Hedwig 1994, 1996, 2000a). These neurons are functionally equivalent despite the differences in their arborization patterns and the fact that their respective stridulatory neuropiles are located in different areas of the brain (Hedwig 2000b; fig. 3.7A,B). *Neuropiles* are concentrations of neuronal processes (axons, dendrites) that functionally connect populations of neurons; in insects they are usually found in central areas of ganglionic segments, while the cell bodies of the neurons are found mainly in the periphery.

Both neurons also partially satisfy two of the major criteria that have been

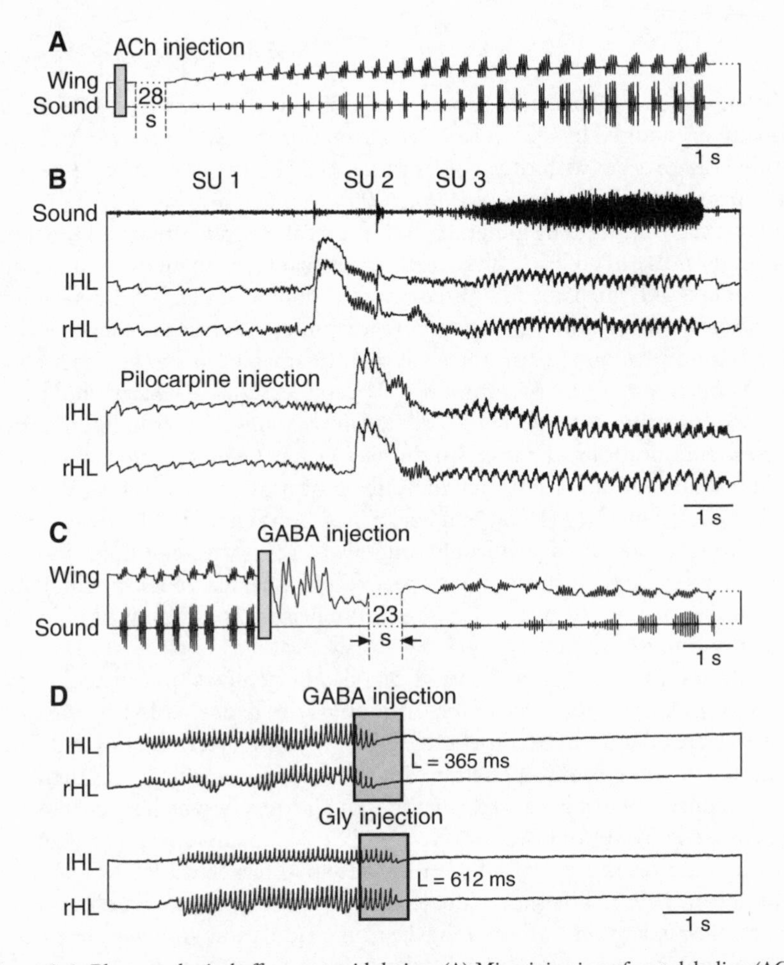

Figure 3.6. Pharmacological effects on stridulation. (A) Microinjection of acetylcholine (ACh) elicits calling song in *Gryllus bimaculatus*. *Upper trace*, movements of the right forewing; *lower trace*, calling song. Notice the delayed and slow buildup of calling song that followed ACh injection (*vertical bar*). From Wenzel et. al. 1998. (B) Activation of a complete courtship sequence of the male grasshopper *Gomphocerus rufus* by microinjection of pilocarpine into the brain. *Upper part*, sound produced during a single courtship unit. The unit is divided into three subunits (SU 1 = head shaking and small amplitude stridulation; SU 2 = jerking of the hind femora with sound pulses; SU 3 = continuous courtship song). *Middle part*, corresponding stridulatory movements of the left (lHL) and the right (rHL) hindleg in a natural courtship unit. *Bottom part*, stridulatory movements of both hindlegs elicited after injection of micromolar amount of pilocarpine. From Heinrich 1995; Heinrich et al. 2000. (C) Interruption of ongoing stridulation in *Gryllus bimaculatus* by microinjection of GABA (*vertical bar*). *Upper traces*, movements of right forewing; *lower traces*, calling song. Notice the beginning of the gradual recovery of song after a delay of more than 20 s. (D) Song sequences of the grasshopper *Omocestus viridulus* indicated by the up-and-down movements of the left and right hindleg (lHL, rHL) elicited by microinjection of pilocarpine (about 1nl, 10^{-4} M) into the brain. At one site within the protocerebrum, GABA (about 1nl, 10^{-4} M) terminates the song sequence, whereas glycine (Gly; about 1nl, 10^{-4} M) applied to the same site does not affect the ongoing sequence (see the first two traces). At another site of the same preparation (not shown), glycine is effective in terminating the sequence whereas GABA is not. This result suggests the existence of different membrane receptors for GABA and glycine in different brain neurons. *Horizontal bar* = pressure pulse duration; L = time difference between onset of the pressure pulse and termination of the sequence. From Heinrich et al. 1998.

used to define *command neurons* (review: Kupfermann and Weiss 1978): *sufficiency* and *necessity*. That is, the activity of these neurons is usually sufficient to trigger and maintain stridulation, and their continued activity is also necessary to maintain stridulation. Under normal circumstances, inhibition of these neurons using hyperpolarizing current causes the cessation of spontaneous stridulation (Hedwig 1994, 2000a). As shown in figure 3.8, both criteria are usually fulfilled in the grasshopper *Omocestus viridulus*.

Recent studies of the field cricket *Gryllus bimaculatus* have focused on an individually identified command neuron (review: Hedwig 2000a) in which changes in firing rate are correlated to some degree with the chirp rate (fig. 3.9A). In resting and nonstridulating crickets, intracellular depolarization of this brain neuron elicits a rapid tonic response (the neuron fires continuously as long as the stimulus is applied) and reliably evokes stridulation. As soon as the firing rate exceeds 60–80 spikes/s, the cricket raises its forewings and starts calling. Stridulation is maintained for as long as depolarization persists and then gradually vanishes following the end of increased neuron activity (fig. 3.9A). If stridulation is evoked repeatedly, however, crickets occasionally start to stridulate spontaneously, even if the command neuron discharges at a rate of just 30 spikes/s. If its spike rate in now increased by intracellular current injection, the chirp rate is transiently increased, the number of pulses per chirp is reduced, and the chirp rate appears to be reset by the command neuron activity. This last finding suggests that the effectiveness of command neurons depends on the activity state of the cricket and that command neurons can also modulate ongoing activity. We do not know, however, whether the slight alternation of firing rates in concert with the chirp rhythm in calling songs is inherent in the triggering circuitry of the brain or if it arises by peripheral or central feedback. As Hedwig (2000a) points out, the reset could be an epiphenomenon of modulating the driving input to the pattern generator or of some element of the generator.

The pattern of chirp and pulse generation can also be affected by wind stimulation of the receptors (filiform hairs) of *cerci*, which are structures found on the posterior part of the animal (Dambach 1989; Hedwig 2000a). Even though such stimulation can excite the command neuron, the wind-evoked silencing reactions of the animal are not reflected in the command neuron's discharge pattern. Indeed, stridulation can be suppressed by cercal inputs from the terminal ganglia even if the firing rate of the command neuron remains unchanged (Hedwig 2000a). The locus within the central nervous system at which this suppression acts is still unknown.

Intracellular stimulation of a stridulatory command neuron in the brain of a resting grasshopper (*Omocestus viridulus*) also causes a tonic discharge that declines slowly in firing rate and elicits a calling song when the neuron's spike rate exceeds 100 spikes/s (fig. 3.9B; review: Hedwig 2000b). The descending activity of the command neuron in grasshoppers has a modulatory effect on

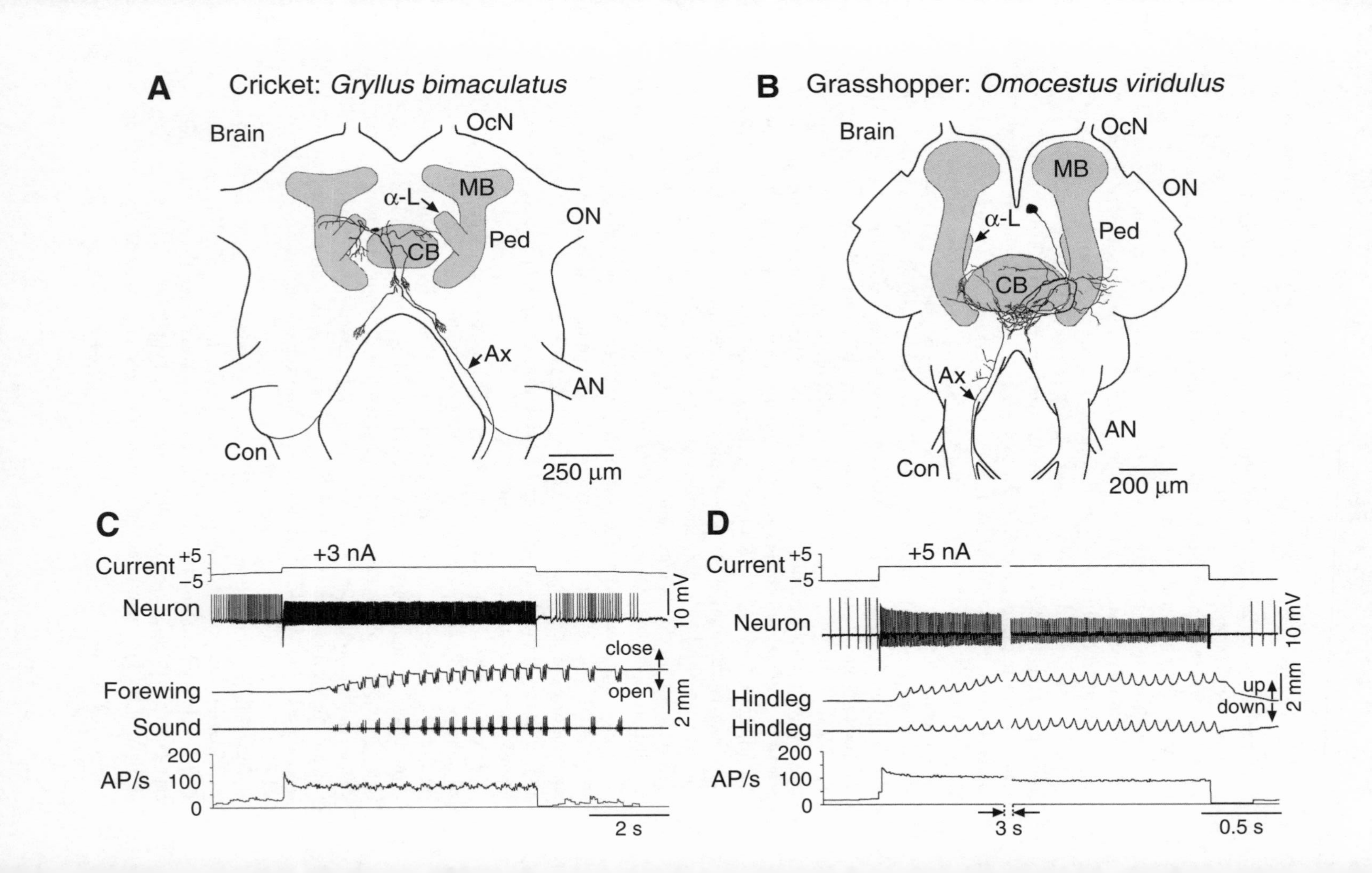
A
Cricket: Gryllus bimaculatus
Brain
OcN
MB
α-L
ON
CB
Ped
Ax
AN
Con
250 μm
B
Grasshopper: Omocestus viridulus
Brain
OcN
MB
ON
α-L
Ped
CB
Ax
AN
Con
200 μm
C
Current
+5
−5
+3 nA
Neuron
10 mV
close
open
Forewing
2 mm
Sound
200
AP/s
100
0
2 s
D
Current
+5
−5
+5 nA
Neuron
10 mV
Hindleg
up
down
2 mm
Hindleg
200
AP/s
100
0
3 s
0.5 s

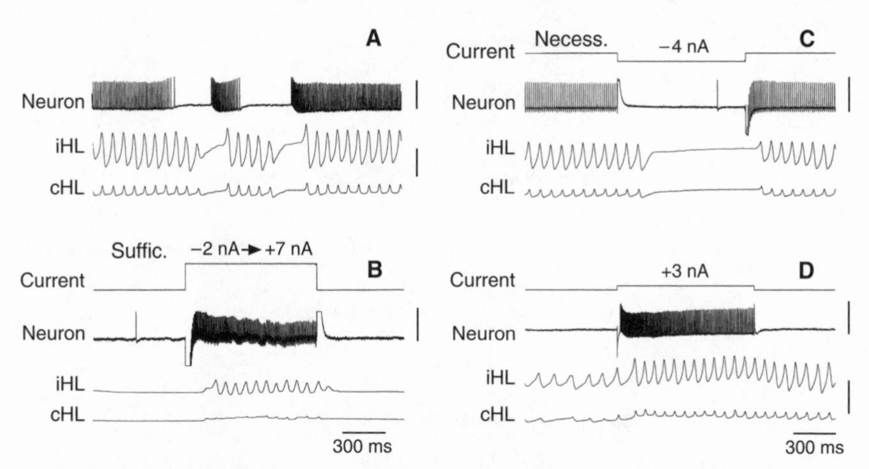

Figure 3.8. Partial demonstration of command criteria in a descending neuron in the grasshopper *Omocestus viridulus.* (A) During spontaneous bouts of stridulation, the command neuron fired at a rate of about 100 spikes/s, which is well above the rhythm of leg movements. The leg movement stopped as soon as the neuron's activity ceased. (B) The first criterion—sufficiency (Suffic.)—is fulfilled because nonsinging animals began to stridulate when the neuron was depolarized and began to fire at a rate greater than 100 spikes/s. (C) The second criterion—necessity (Necess.)—is fulfilled because when the application of hyperpolarizing current caused the neuron to stop firing, stridulation then ceases. (D) A few grasshoppers, however, sang spontaneously when this particular command neurons had been inhibited; as soon as the neuron was depolarized and began firing, the amplitude of stridulatory movements increased. Thus other neurons, possibly of the same type, appear to be involved in controlling stridulation. Traces below the oscillograms showing the activity of the command neuron show leg movements (iHL = ipsilateral hind leg; cHL = contralateral hind leg).

stridulation that also appears to be influenced by other, unidentified command neurons. That is, when the identified command neuron is inactive, some grasshoppers show stridulatory movements with a rather low repetition rate. When the command neuron is depolarized and increases its spike rate under these conditions, both the amplitude and repetition rate of leg movements increase.

(facing page)

Figure 3.7. Structures and arborization patterns of descending stridulatory command neurons in the brain of the cricket *Gryllus bimaculatus* (A) and the grasshopper *Omocestus viridulus* (B). Cricket brain is shown from a ventral view, grasshopper brain, from a dorsal view. AN = antennal nerve; Ax = axon; α-L = alpha-lobe; CB = central body; Con = cervical connective; MB = mushroom body; ON = optical nerve; OcN = ocellar nerve; Ped = pedunculus. (C) Release of calling-song stridulation in a resting cricket and (D) in a resting grasshopper. In each species, longer-lasting depolarization by intracellularly applied current of 3 or 5 nA (*first trace*) increased the instantaneous discharge rate to about 100 spikes/s (= AP [action potentials]/s), a rate far higher than the wing and leg rhythms (*second and fifth traces*). The depolarization also elicited forewing movements in crickets (*third trace in C*) and hindleg movements in grasshoppers (*third and fourth traces in D*) and sounds (*fourth trace in C*), which stopped soon after the end of depolarization. In crickets the slight modulation of the spiking rate corresponded with the chirp rhythm (*fifth trace*). Additional details in the text. From Hedwig 2000b, figs. 1, 3.

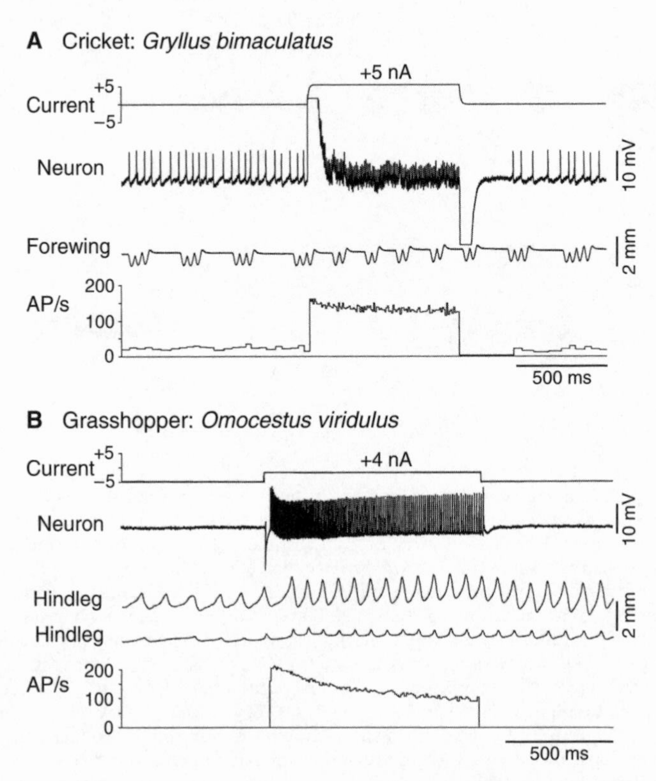

Figure 3.9. Modulation of the stridulatory motor output during spontaneous stridulation in a cricket *Gryllus bimaculatus* (A) and grasshopper *Omocestus viridulus* (B) after the activity of command neurons had been elevated by depolarizing current. In the cricket the chirp rate increased, the number of pulses/chirp was reduced, and the chirp pattern was reset during the burst of command-neuron activity (*third trace in A*). In the grasshopper the amplitude and repetition rate of stridulatory hindleg movements increased (*third and fourth traces in B*). AP/s = (action potentials [spikes])/s. See the text for further explanation. From Hedwig 2000b, fig. 4.

Local Brain Neurons and the Control of Stridulatory Movements

Ascending interneurons of crickets and grasshoppers that convey auditory information to the brain terminate outside the arborizations of the command neurons shown in figure 3.7A,B. Local neurons in the brain probably function as components of the sensory-motor interface by transferring information from the auditory neuropiles to stridulatory neuropiles. To date, only one local brain neuron has been discovered—in the grasshopper *Chorthippus biguttulus*—that has the structural prerequisite for this function. This neuron has arborizations that extend into the dorsal lateral neuropile of the protocerebrum, in which ascending auditory neurons terminate (fig. 3.10A). This cell thus satisfies the morphological requirement for information transfer between the ascending auditory pathway and the descending pathway for stridulation

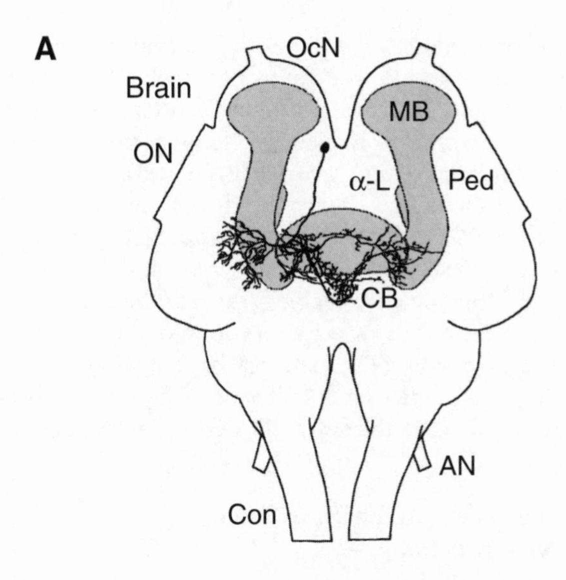

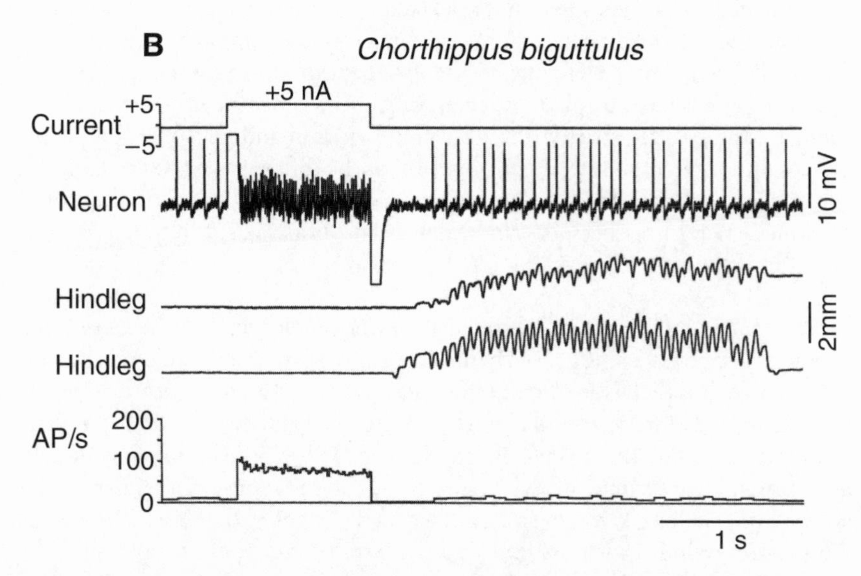

Figure 3.10. Transfer of auditory information to stridulatory neuropiles. (A) Structure and arborizations of a local neuron in the brain of *Chorthippus biguttulus* involved in the control of stridulatory behavior. See figure 3.7 legend for key to labels. (B) Depolarization (*first trace*) increases the firing rate to about 80 spikes/s (*second and fifth traces*). After a latency of around 1 s and after stimulation of the neuron ceased, both hindlegs start with stridulatory movements (*third and fourth traces*). AP/s = (action potentials [spikes])/s. From Hedwig 2000b, fig. 5.

(Hedwig 2000b). There is also strong physiological evidence for the involvement of this local brain neuron in the control of stridulation. If depolarized by intracellular current injection, the neuron increases its instantaneous spike rate to about 80 spikes/s; a previously silent grasshopper then starts to produce rhythmical stridulatory hindleg movements (fig. 3.10B). These movements occur after a latency of about 1.1 s, indicating that no direct activation of descending command neurons is involved because the direct stimulation of the command neuron would result in a much shorter latency. The movements of both hindlegs resemble the species-specific pattern of the calling song but are not as precise as those elicited by direct stimulation of descending command neurons. Although information about such local brain neurons is still very limited, this example constitutes the first evidence for an interface between the ascending auditory pathway and the descending command neurons triggering stridulation (Hedwig 2000b).

Command Neurons Associated with Different Song Types and Complex Songs

The different song types (calling, courtship, and aggressive songs) produced by some species of insects contain some rhythmic components in common. The addition of other elements or the exaggeration or modification of common elements serves to distinguish among these different song types. How, then, are these different rhythms activated by command neurons? This same general question applies to the production of the different phrases of complex songs, such as those of the calling songs of *Teleogryllus* and the courtship songs of some acridid grasshoppers, in which each phrase has different temporal patterns.

Control of different song types. In crickets (*Gryllus campestris, G. bimaculatus*) that produce distinctive calling, courtship, and aggressive songs, evidence is available that each type has its own command neuron (or a few neurons with the same function). The results of electrical (Otto 1971; Hedwig 2000a) and pharmacological stimulation (Wenzel and Hedwig 1999) of the brain of *G. bimaculatus* provide three lines of evidence. First, calling-song command neurons activate only calling song, even at induced spike rates as high as 160 spikes/s. This result excludes the possibility that different song types are controlled by different levels of activity of a single command neuron (Hedwig 2000a). Second, the transitions observed between song types (e.g., between calling and aggressive song or between calling and courtship song) during brain and connective stimulation (Huber 1960; Otto 1967, 1971) do not occur during any intracellular stimulation of the calling-song command neuron (Hedwig 2000a). Third, preliminary data indicate that a different type of descending neuron is involved in the control of courtship song (Hedwig, unpubl. data). These results support the assumption that during brain stimulation, spatially separated command neurons for the different types of songs might have been activated altogether (Hedwig 2000a). Single-cell activation, however, provides evidence that different song patterns in crickets are probably controlled by specific neurons, as in the complex courtship patterns of acridid grasshoppers, which we discuss below. This explanation probably does not apply to species, such as *Teleogryllus*, in which different song types have rhythmic elements in common (see below).

Control of different patterns within calling song and courtship sequences. The complex calling song of the field cricket *Teleogryllus oceanicus* is made up of chirps and trills with different pulse rates. Bentley (1977) provided evidence that a single command neuron (or a set of command neurons with the same function) activates this kind of signal, a result that is consistent with the finding that only one type of command neuron triggers the calling songs of *Gryllus bimaculatus* and *Omocestus viridulus* (Hedwig 2000b). The selection between chirps and trills is supposedly made within the thoracic pattern generator (Hennig 1992; and below).

By comparison, different command neurons trigger each of the three different rhythmic elements within the courtship sequence of grasshopper *Omocestus viridulus* (Hedwig 1995; Hedwig and Heinrich 1997). These different neurons are connected either to separate pattern generators, each producing one element of the song, or to separate parts of the same pattern generator. In the latter case, the difference from one element to another is probably caused by the recruitment of different interneurons within the pattern-generating network. These three types of descending brain neurons are similar in shape but vary in their cephalic branching pattern. Each type is probably activated at a certain time during courtship behavior (fig. 3.11). One type triggers normal stridulation, as exhibited in calling and courtship songs. The second type becomes active near the end of a whole courtship sequence and triggers the alternation of hindleg shaking. The third type triggers the rapid and synchronous precopulatory leg movements. The origin and mechanisms that control the sequence of activation of these neurons are still a mystery (Heinrich et al. 2001).

Pattern Generators

We have frequently referred to ensembles or neural networks called *pattern generators*, which are activated by descending input from command neurons and which control the movements of peripheral sound-producing structures. Pattern generators are not autonomous circuits that control all of the details of repetitive motor programs, nor can they maintain the stability of signals over relatively long periods of time (weeks, months) (reviews: Elsner and Popov 1978; Elsner 1994). Nevertheless, once activated, pattern generators can produce, through the neuromuscular system and in the absence of peripheral feedback, the basic temporal patterns of acoustic signals that carry

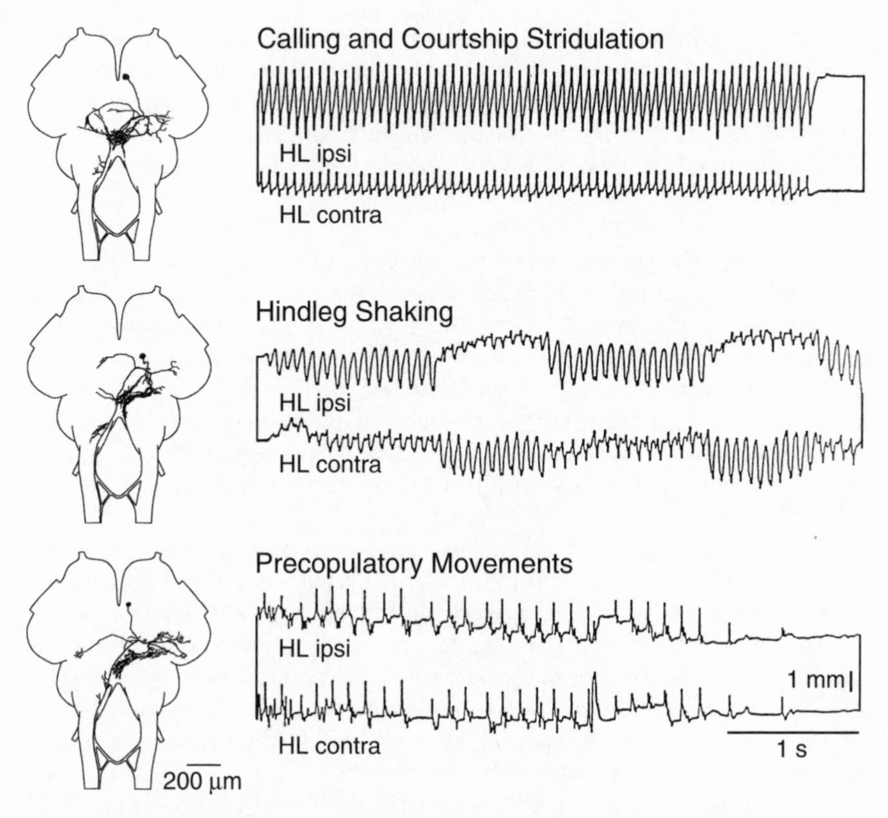

Figure 3.11. Influence of descending brain (command) neurons on different aspects of a complex courtship display in the grasshopper *Omocestus viridulus*. *Left*, frontal view of the brain showing the structure and arborization patterns of the three neurons. *Right*, hindleg movements activated by separate command neurons are associated with different parts of the complex courtship song. (HL ipsi [= ipsilateral, or same side], HL contra [= contralateral, or other side] with respect to the position of the descending axon). From Hedwig 1995.

biologically important information. A first, albeit indirect, step to understanding the action of pattern generators is to describe motor programs in terms of the patterned activity of the muscles that are involved in sound production.

Motor Programs Underlying Sound Production

Electromyograms (EMGs) are a major tool in the study of motor programs: the electrical activity of muscles preceding, during, and after contraction are recorded from thin wire electrodes implanted into single muscles. In insects the main individual muscles, and even *motor units* (i.e., the set of individual muscle fibers innervated by one motor neuron) within a given muscle, can be simultaneously monitored, even in unrestrained singing males (reviews: Elsner and Popov 1978; Kutsch and Huber 1989).

The muscles used for stridulation in orthopterans, and almost all the timbal muscles in cicadas, are fast-twitch and *synchronous* (neurogenic) muscles. A one-to-one relationship between motor neuron spikes and muscle contractions is even observed in some katydid and cicada species that produce songs with pulse rates up to 500 pulses/s (Josephson and Halverson 1971; Josephson and Young 1985). Motoneurons innervating the muscles used in stridulation are driven with different phase relationships by pattern-generating networks. For example, one set of muscles is responsible for closing and the other, for opening the wings (crickets and katydids) or, equivalently, for the up-and-down strokes of the hindlegs (grasshoppers) (fig. 3.12). Similar, more complex

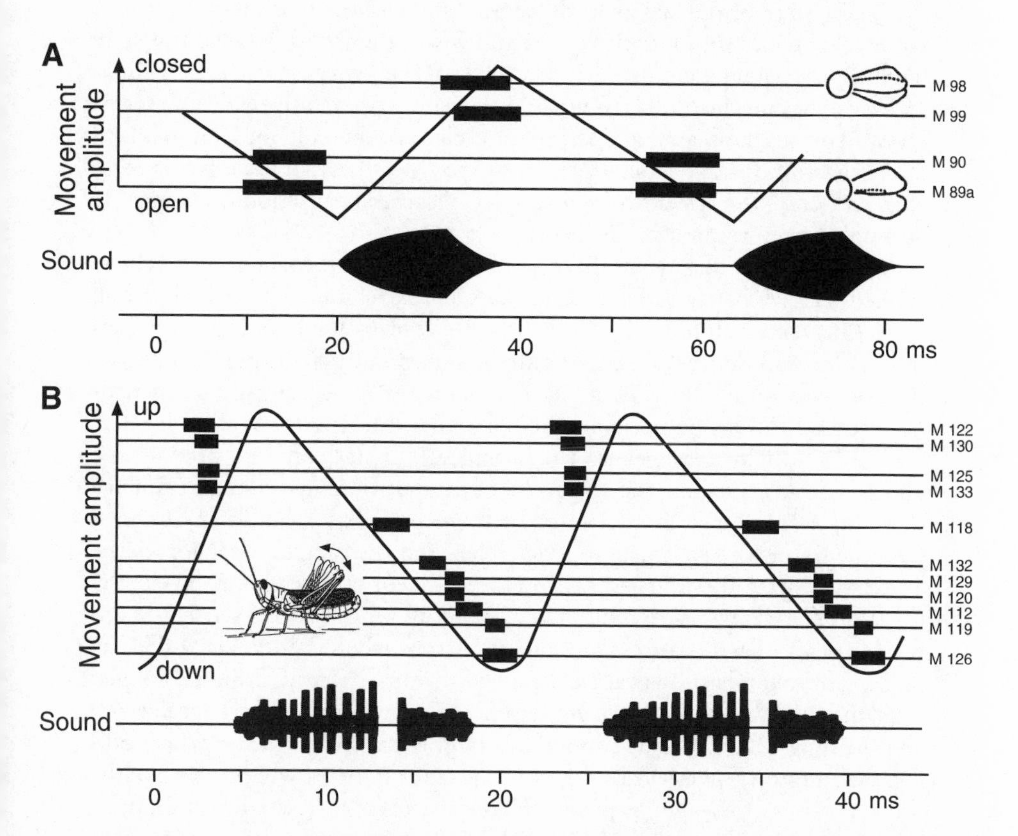

Figure 3.12. Antagonistic action of two sets of muscles in cricket and grasshopper sound production. (A) Cricket: diagram to show the timing of two synchronously active wing-closer (M 89a, 90) and two synchronously active wing-opener muscles (M 98, 99) of *Gryllus campestris* during two cycles of pulse production. From Huber 1975; data from Kutsch 1969. *Below*, pulses produced during the closing stroke of the wings. (B) Grasshopper: timing of nearly synchronously active downstroke muscles (M 122, 130, 125, 133) and sequentially active upstroke muscles (M 118, 132, 129, 120, 119, 126) during two cycles of the leg of subunit 3 during the courtship song of *Gomphocerus rufus*. From Elsner 1968. *Inset*, grasshopper showing the leg movements. From Helversen and Helversen 1981. *Below*, sound produced during a single up-and-down stroke.

phase relationships occur between the left and right hindlegs in grasshoppers and between the two timbal muscles of cicadas (e.g., Elsner 1975; Fonseca 1996). The force of muscle contractions preceding and during stridulation are controlled (1) by varying and synchronizing the number of motor units recruited; and (2) by varying the rate of excitation of single motor units (reviews: Elsner and Popov 1978; Kutsch and Huber 1989).

Most anurans use powerful trunk muscles to pressurize air in the lungs (Girgenrath and March 1977); as discussed in chapter 2, this air passes forcibly over and excites the vocal cords into vibration (Martin 1972). Vocalization also involves muscles controlling the dimensions of the mouth (buccal) cavity, the positioning of the larynx in the air stream, and the tension on the vocal cords (chapter 2). In modern representatives of the probable ancestral state (archaeobatrachian anurans, such as species in the genera *Bombina* and *Discoglossus*), males produce advertisement calls mainly or exclusively during inspiration. Hence, the buccal and tongue muscles are the only muscles involved in vocalization (Strake et al. 1994). In some frogs that call underwater (e.g., *Xenopus laevis*), the system is even simpler because vocalization is completely decoupled from respiration (Kelley 1999).

Muscle activity during vocalization in different species of anurans has been studied by EMGs (e.g., Martin and Gans 1972; Schmidt 1976; Strake et al. 1994; Girgenrath and Marsh 1997). The motor program underlying vocalization in the toad *Bufo valliceps*, for example, was monitored during the production of release calls by EMGs and simultaneous measurements of changes in air pressure in the buccal and pulmonary chambers (Martin and Gans 1972; Gans 1973). Before each pulse the laryngeal muscles contract, holding the glottis (paired arytenoid cartilages) closed and positioning it so that resistance to the increasing pressure generated by lung compression builds up. A sound pulse begins as the laryngeal muscles relax, and the pressured air forces the arytenoids apart: these structures and the vocal cords now vibrate, producing the subpulse (via passive amplitude modulation) typical of this species (chapter 2; see also fig. 11.14). Sound emission stops when the laryngeal dilators pull the arytenoids and vocal cords out of the air stream. During prolonged calls, the toad shunts (periodically reverses) airflow between the buccal cavity and the lungs. Each of the sound pulses of gray treefrog (*Hyla versicolor*) calls corresponds to a biphasic EMG spike in the trunk muscles, whereas each of the shorter pulses of its sibling species (*H. chrysoscelis*) is correlated with a single, simple spike (Girgeurath and Marsh 1997). Intrinsic properties of the calling muscles also differ: peak tension occurs after 25 ms in *H. versicolor* and after 15 ms in *H. chrysoscelis* (Marsh 1999).

Pattern Generators in Anurans

Anuran pattern generators were first studied in detail in the leopard frog *Rana pipiens*, by Schmidt (1974a,b, 1976). He developed an isolated brain preparation, stimulated vocalization triggering circuits in the preoptic region of the

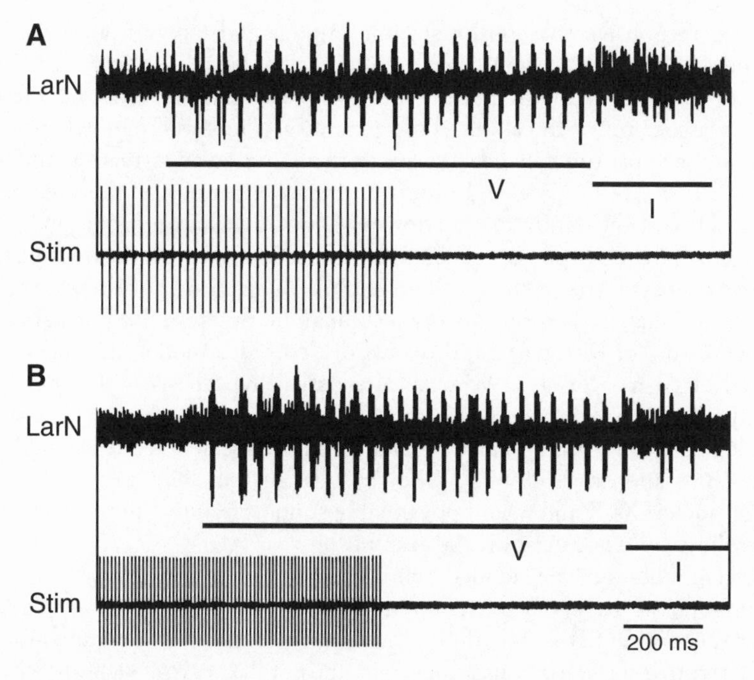

Figure 3.13. Fictive calling in the leopard frog *Rana pipiens* is represented by the activity of the laryngeal nerve (LarN) that innervates the glottis. In this preparation this activity was triggered in an isolated brain by stimulation (stim) with pulses of 50 Hz (A) or 100 Hz (B) within the area of the preoptic nucleus (part of the command system) of the hypothalamus. The vocal phase is indicated by the horizontal lines labeled V. After the vocal phase, there is rapid burst of activity that causes inspiration; this inspiratory phase is indicated by the horizontal lines labeled I. Note that the laryngeal timing patterns, which would correspond to production of sound pulses during the vocal phase in an intact animal, were the same whether the stimulus frequency was 50 Hz or 100 Hz. Modified from Schmidt 1974a.

brain, and then monitored *fictive calling* (patterns of nerve or muscle activity that are observed during calling by intact animals) by recording EMGs from the laryngeal muscles (fig. 3.13). The patterns of laryngeal muscle activity were essentially the same as those observed in intact, vocalizing frogs. Based on an extensive series of stimulation and lesion studies, Schmidt (1992) proposed that the vocal system of the leopard frog *Rana pipiens* consists of two semi-independent generators. One generator, mainly consisting of the pretrigeminal nucleus, which is located in the dorsal tegmental part of the medulla (fig. 3.3), is specialized for calling. Schmidt claimed that cells in this nucleus are not spontaneously active but generate rhythmic (vocal) activity after stimulation of other parts of the brain or of the dorsal medulla itself. At the same time, Schmidt (1992) considered the pretrigeminal nucleus to be homologous with the parabrachial complex of higher vertebrates, in which rhythm-generating cells are spontaneously active and influence respiration as well as vocalization (review: Kelley 2001). Some neurons in the pretrigeminal nucleus are also responsive to acoustic stimuli (Aitken and Capranica 1984). The second (pulmonary) generator of anurans has a slow, spontaneously active rhythm, and this structure must be intact in order to observe the fully pulsed activity associated with calling. This generator is less well defined anatomically and is approximately coextensive with the reticular formation and with motor nuclei IX–X (fig. 3.3). Lesions in the posterior part of this area of the brain stem disrupt respiration but not vocal control, whereas lesions in the anterior part have the opposite effect. Schmidt (1974b) also cut the midline connectives between the putative call generators on each side of the animal and established that each half of the brain stem alone can generate normal fictive calling. Some of the anatomical details of Schmidt's model have been confirmed by Wetzel et al. (1985). An alternative hypothesis is that rhythmic activity of all kinds are driven by specialized interneurons, which originate embryologically from *rhombomeres*[2] 7 and 8 and are located in the caudal hindbrain (Bass and Baker 1997). These cells are anatomically coextensive with motor nuclei IX–X and would presumably supply rhythmic inputs to circuits controlling both vocalization and respiration.

Recent studies have extended Schmidt's pioneering research with isolated brain preparations (e.g., Kogo and Remmers 1994; Kogo et al. 1994). In addition to recording fictive breathing, these researchers have simultaneously recorded from interneurons, motoneurons, and cranial nerves. Such studies have confirmed the existence of paired networks and that a single such network in one half of the brain can independently generate the rhythmic patterns of respiration and vocalization. The synchronization of these networks is mediated by numerous midline connections rather than any special set of crossing fibers (McLean et al. 1995). With respect to respiration, potential sources of peripheral feedback have also been identified (see below; Kogo et al. 1997).

Pattern Generators in Insects

In insects, rhythmically active networks capable of generating the temporal patterns of stridulation by driving the appropriate motoneurons are located in the thoracic ganglia (Kutsch and Otto 1972). Elsner (1994) provides a particularly valuable review. As in anurans, pattern generators in orthopteran insects exist in pairs, one on each side of the metathoracic ganglion. Even in crickets, where stridulatory output involves motoneurons in the mesothoracic ganglion, the metathoracic ganglion contains the pattern generators for forewing stridulation (fig. 3.14). Either of these so-called hemiganglionic pattern generators is capable of producing—via thoracic motoneurons—the timing observed in stridulation of each forewing in crickets (Hennig and Otto 1995/96) or hindleg in grasshoppers (Ronacher 1989, 1990, 1991).

Bentley (1969) was the first to record intracellularly from mesothoracic

2. Segmented compartments in the embryonic hindbrain that give rise to distinct complements of motoneurons and populations of interneurons.

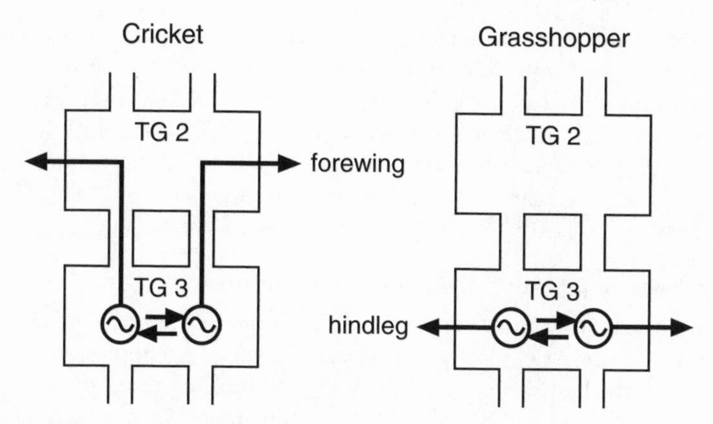

Figure 3.14. Location of thoracic pattern generators in crickets and grasshoppers. TG 2 = mesothoracic ganglion containing motoneurons that drive stridulatory muscles of the forewings; TG 3 = metathoracic ganglion with connections (*long arrows*) to the mesothoracic ganglion and outputs to the forewings in crickets and the hindlegs in grasshoppers. Both hemiganglionic pattern generators influence each other via intraganglionic connections (*short arrows*) Modified and simplified for crickets from Hennig and Otto 1995/96; original figure for grasshoppers, based on data provided by Ronacher 1989, 1990, 1991.

motoneurons and a few interneurons in crickets during singing that was elicited by brain lesions. He could associate the activities of these neurons with specific temporal properties of the songs, such as pulse and chirp rate. Because muscle contractions and peripheral thoracic and abdominal sensory feedback were eliminated in his preparations, Bentley concluded that sound patterns depend mainly on patterned activity that is generated within the thoracic ganglia.

The mechanisms underlying the connection and coordination of these hemiganglionic pattern generators have been compared in grasshopper species exhibiting slow and fast rhythms during stridulation (Fries and Elsner 1996; Heinrich and Elsner 1997). The metathoracic ganglion was split longitudinally to separate the two hemiganglionic pattern generators, and the stridulatory patterns were compared before and after splitting. In both groups the rather fixed phase relationships between the two legs observed before splitting disappeared. In species with slow rhythms, however, the movements of the two legs became synchronized, while in species with fast rhythms, the patterns were no longer fixed but drifted in and out of phase. These authors suggest that the synchrony of the legs in species with slow rhythms is controlled by descending input from higher centers either directly or, more likely, via ascending feedback. In species with fast rhythms, the coordination between the legs is exclusively achieved by intraganglionic connections.

Since the pioneering work of Bentley (1969), some progress has occurred in identifying elements of pattern generators at the cellular level, especially in grasshoppers. These elements occur both within each half and between the two halves of the metathoracic ganglion (review: Elsner 1994). Some interneurons are active only during stridulation and show a division of labor, whereas others shift and possibly reset the rhythm. These latter neurons are thought to be components of ganglionic pattern generators. Still other neurons probably coordinate the movements of the two hindlegs, and thus the two hemiganglionic pattern generators (Gramoll and Elsner 1987; Gramoll 1988; Hedwig 1992a,b).

Pattern Switching within One Song Type in Insects

We have already shown that crickets such as *Teleogryllus oceanicus* produce calling songs that consist of chirps and trills, organized into phrases, and similar changes in patterns are seen in the courtship and calling songs of some grasshoppers (e.g., *Omocestus viridulus*). The question arises, then, are these different temporal patterns within one song type or between song types controlled by separate pattern generators (e.g., different networks within the ganglion), or by a different recruitment of subsystems within a common pattern generator?

In *Teleogryllus*, changes in the patterns within the calling song type are fast, requiring as few as two or three cycles to shift from one stable period to the next (Hennig 1992), and Bentley's (1977) observations mentioned above rule out the existence of separate command neurons for chirps and trills as well as changes in the activity of a single command neuron. Thus cellular properties and connections within the thoracic pattern-generating network must be responsible for the production of chirps and trills. Hennig (1992) also recorded from a metathoracic interneuron, which was not directly connected to mesothoracic motoneurons. During the chirp and trill production, the same transitions in spike timing occurred as those observed in the motoneurons (fig. 3.15); similar changes in the membrane potential of the interneuron also occurred. Because the pulse rate within chirps is common to calling, aggressive, and courtship songs in *Teleogryllus*, a common pattern generator could, in fact, be involved in the control of all three signals in this species (Hennig 1992).

Insect Pattern Generators Involved in Stridulation and Flight

In orthopterans, several thoracic muscles move both legs and wings, and motoneurons that drive these bifunctional muscles are involved in both stridulation and flight. In crickets the pattern of activation and the recruitment of motor units within the same set of bifunctional muscles are somewhat similar during these two behaviors, causing similar rates of movements of the wings (Kutsch 1969; Kutsch and Huber 1989). In the grasshoppers *Gomphocerus rufus* and *Stenobothrus rubicundus*, the rate of activation of bifunctional muscles is also similar during leg stridulation and flight, but the phase relationships of firing in these muscles differ depending on whether the insect is stridulating or flying (Elsner 1974a, 1983; fig. 3.16). In other species, like *Omocestus viridulus* and *Stenobothrus lineatus*, the motoneuron discharge rates are often species-

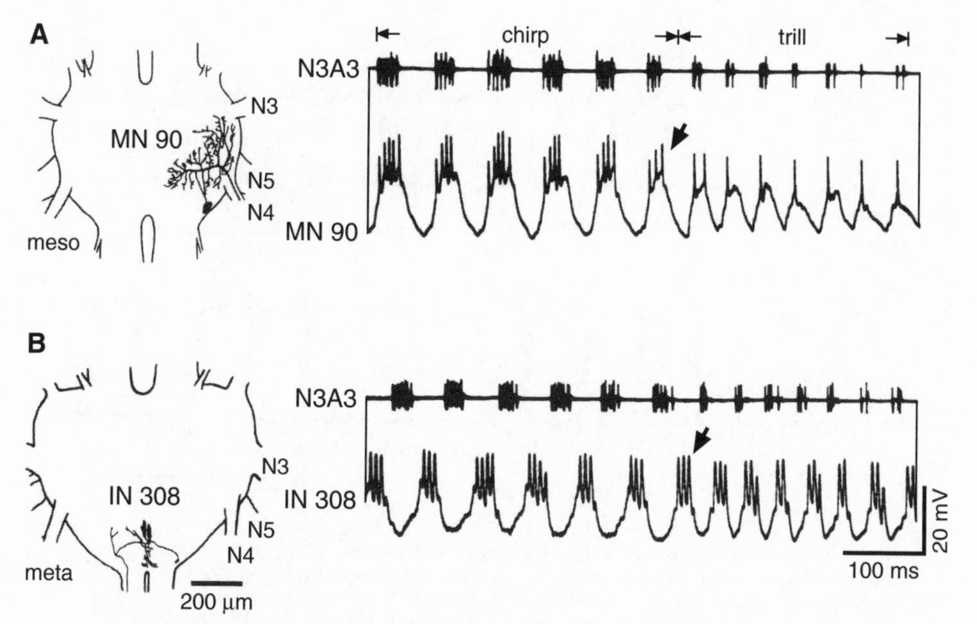

Figure 3.15. Neural activity recorded during the production of chirps and trills by the cricket *Teleogryllus oceanicus*. *Left*, morphology of one mesothoracic wing-closer motoneuron (A) and one metathoracic interneuron (B), which has no direct connection with the mesothoracic motoneurons. *Right*, motoneuron and interneuron activities are shown during production of chirps and trills. (A) The mesothoracic closer motoneuron MN 90 shows the switch from chirp to trill and the change in burst shape (*arrow*). (B) The same change appears in the metathoracic interneuron IN 308 showing different burst durations for chirps and trills and intermediate burst forms at the transition (*arrow*). Upper traces in (A) and (B) are recordings from nerve N3A3, which innervates the closer muscles. Modified from Hennig 1992, fig. 7.3.

specific and may differ during song and flight (Hedwig, pers. comm.). Are stridulation and flight controlled by separate pattern generators or by a single pattern generator using different subsets of neurons?

Some answers to this question are provided by studies of crickets, in which the same sets of muscles, controlled by the same motoneurons, move the wings, albeit in different ways, during stridulation and flight. Using a tethered preparation, Hennig (1990a,b) elicited stridulation in *Teleogryllus commodus* by brain stimulation, and flight, by puffs of air to the cerci. At the same time, he recorded the activity of mesothoracic motoneurons that opened and closed the wings during stridulation and depressed or elevated the wings during flight. He also recorded from two metathoracic interneurons with ascending axons. One interneuron was active during stridulation but suppressed during flight, and the other interneuron exhibited the opposite pattern (fig. 3.17). Thus, even though several interneurons are elements in both networks, their activity patterns differ during flight and stridulation, and they have different effects on motoneurons. Whether there are also separate interneurons controlling these two motor patterns is an open question.

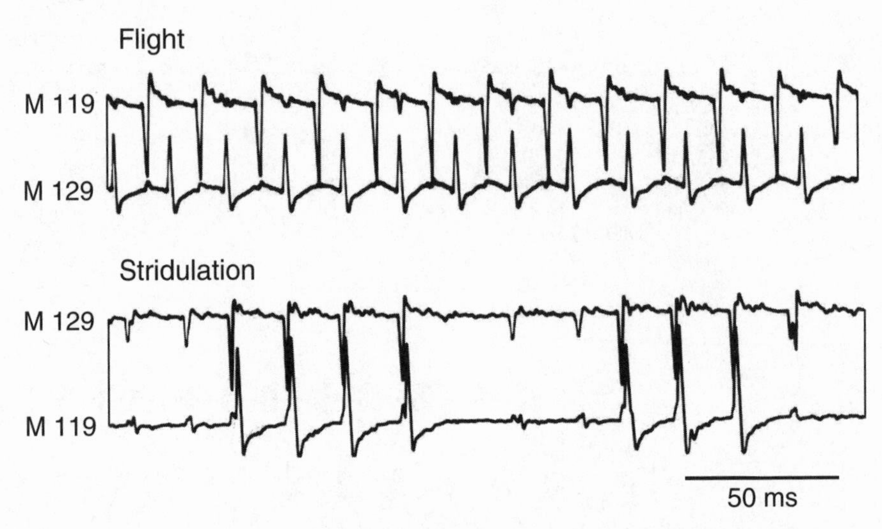

Figure 3.16. Temporal recruitment of two bifunctional muscles (M 119, 129) on the same side within the metathoracic segment for flight (*upper two traces*) and stridulation (*lower two traces*) in the grasshopper *Gomphocerus rufus.* The two muscles fire with the same rate, however, in an alternating fashion during flight, and nearly synchronously during leg stridulation. Modified from Elsner 1968, 1983b.

Modification of Signals by Intrinsic Properties of Muscles

Just as peripheral structures can impose passive amplitude modulation on temporal rhythms supplied by pattern generators, the relative contributions of muscles with different contraction/relaxation properties (e.g., "fast-twitch" and "slow-twitch" muscles) can in principle influence the temporal patterns of signals.

As previously mentioned, muscles used for stridulation in orthopterans and almost all the timbal muscles in cicadas are fast-twitch and synchronous muscles. In exceptional examples involving the cicadas, the intrinsic properties of the timbal muscles, together with mechanical properties of the timbal, account for a patterned output that differs from the firing pattern of the motoneurons. These muscles are therefore termed *asynchronous* because there is no clear-cut one-to-one correlation between motoneuron spikes and muscle contractions (Josephson and Young 1981). Whereas synchronous fast-twitch muscles lead to fast-twitch contractions, each producing a single timbal action with associated sounds, asynchronous muscles have a lower fusion frequency and slow-twitch duration and produce several timbal actions with associated sounds per contraction.

Intrinsic properties of muscles also influence the temporal properties of anuran signals. In most species the muscles and laryngeal structures that are specialized for sustained calling are highly sexually dimorphic, even in species

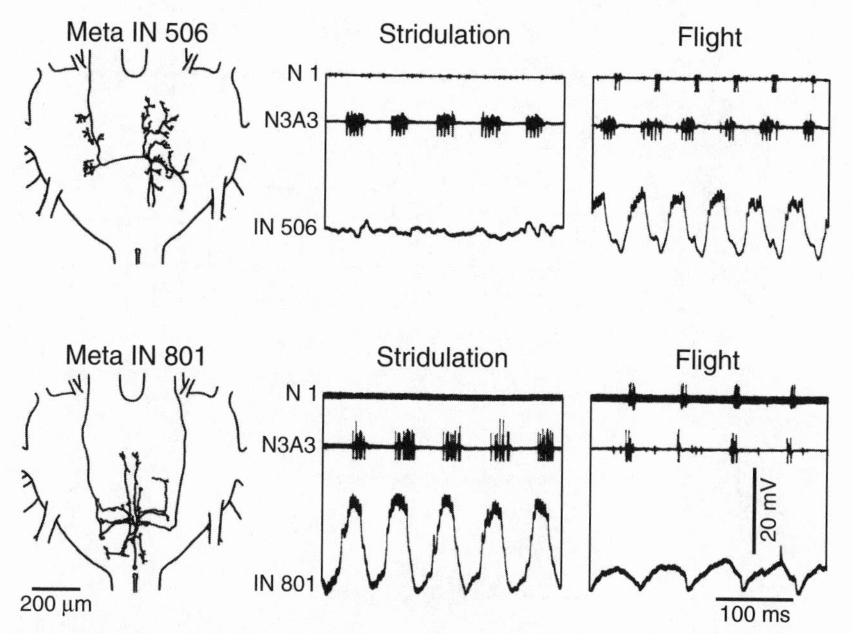

Figure 3.17. Different activation of interneurons during sound production and flight in crickets. *Left,* morphology of two metathoracic interneurons (IN 506, 801) with ascending axons. Activity recorded in two types of metathoracic interneurons and motor nerves (N1 and N3A3) during sound production (*left columns*) and during flight (*right columns*) of *Teleogryllus commodus.* During stridulation, IN 506 receives no rhythmic excitatory input in phase with the bursting indicating stridulatory output in the nerve N3A3. During flight, IN 506 shows rhythmical bursts in phase with the flight rhythm. By contrast, during stridulation, IN 801 shows strong oscillations in membrane potential approximately in phase with the bursts recorded in nerve N3A3. During flight, IN 801 exhibits strong inhibitory postsynaptic potentials approximately at the time of activation of the hindwing depressor muscle as indicated by the activity recorded in N1. From Hennig 1992.

in which females frequently vocalize (fig. 3.18; reviews: Kelley 1999; Wells 2001). This situation provides an opportunity to learn how intrinsic properties of muscles and laryngeal structures might contribute to differences in the signals of the two sexes.

Kelley and her colleagues have conducted extensive studies of sexual differences in the muscles of the call-production system of clawed frogs in the genus *Xenopus* (Kelley 1999). The calls of these anurans consist of clicks that are generated by contraction of laryngeal muscles. These sounds are produced underwater and do not depend on concomitant inspiration or expiration of air (Yager 1992). In *Xenopus laevis,* anatomical differences between the two sexes were found in the larynx and in the associated muscles. The laryngeal muscles in males are all fast-twitch in contrast to the predominance of slow-twitch muscles in females, and additional differences in muscle physiology lead to sexual differences in how laryngeal muscles respond to repetitive stimulation

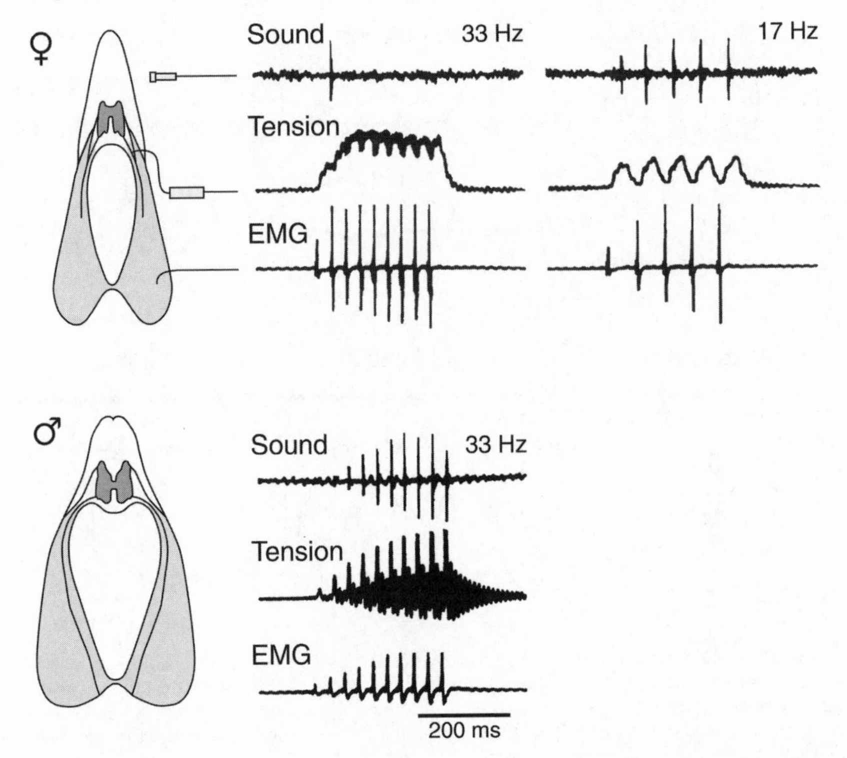

Figure 3.18. Vocal correlates of muscle-contraction properties in clawed frogs *Xenopus laevis*. *Left*, diagrams showing the general configuration of the larynges (*dark shading*) and associated muscles (*light shading*) of females and males; sites where electromyograms and measures of muscle tension were recorded are indicated in the top part of the figure. *Right*, the top traces show oscillograms of clicks, produced during stimulation of the laryngeal nerve at 33 Hz. The middle traces show changes in muscle tension (contractions and relaxations), and the bottom traces show electromyograms (EMGs). Notice that in the male, each EMG spike elicits a contraction of the laryngeal muscle, which, in turn, produces a click, whereas in females, the laryngeal muscle undergoes a sustained contraction (tetany), which produces only a single click at the onset of contraction. When stimulated with a rate as high as 17 Hz (*right trace*), the muscle of the female can respond to each EMG with a click. From Kelley 1999.

of the laryngeal nerve. As shown in figure 3.18, males produce calls with a pulse rate of greater than 30 pulses/s, with each sound pulse being triggered by a separate muscle contraction, which, in turn, corresponds in a one-to-one fashion with each EMG spike. Normal female calls, which consist of "ticking" and "rapping" sounds, have a lower pulse rate (maximum of about 10 pulses/s). In the production of such signals, female muscles respond in a one-to-one fashion to EMG spikes as in males and continue to do so in response to artificial stimulation at rates as high as 17 Hz. But at stimulation rates of 30 Hz, these muscles show only a sustained contraction (tetany) and produce only a single sound pulse (fig. 3.18). Recordings from the laryngeal nerve during vocalization show that the temporal patterning (pulse rate) typical of the different signals originates in higher centers (Yamaguchi and Kelley 2000). Nevertheless, the properties of the female neuromuscular system serve as a constraint on the production of signals with pulse rates as high as those of males.

Sexual differences also exist in the *strength of synapses* (= amount of transmitter released) between laryngeal motoneurons and laryngeal muscle fibers (Tobias and Kelley 1995). In females, periodic increases in circulating estrogen, which are associated with gonadotrophin-induced ovulation, can weaken synaptic strength and promote changes in the types of calls produced. Indeed, rapping, which is a signal of female receptivity, was produced only when synaptic strength was attenuated (Wu et al. 2001).

Feedback Mechanisms in Sound Production

Pattern generators do not control all of the details of acoustic signals, and temporal patterning is not the only important factor to consider. When switching from calling song to courtship or aggressive song, for example, some cricket species change not only the timing of stridulatory movements, but also their body position and associated movements. Males alter the static positioning (elevation) of the wings, which, in turn, may influence duration and strength of the contact between file and scraper required for effective sound output. Similar changes in body position and movements occur in grasshoppers when they switch from calling to courtship song.

Although peripheral influences on sound production in insects have been particularly well studied (reviews: Elsner and Popov 1978; Kutsch and Huber 1989), hardly any formal analysis of feedback is available for anuran vocalization. In *Bombina orientalis* the buccal depressor muscles contract in a variable fashion during the production of a series of calls that have nearly identical temporal properties (Strake et al. 1994). These variable contractions have thus been interpreted as stabilizing adjustments that might represent responses to variation in the state of the vocal tract, which is monitored by sensory neurons. Kogo et al. (1997) speculate that glottal closing during breathing is controlled in part by reflexes activated by pulmonary receptors that monitor the state of lung inflation and by laryngeal receptors that might be part of an airway protective reflex.

Acoustic Feedback?

The current view is that acoustic feedback received by the sender from its own signals seems to be unnecessary for modifying (shaping) the song pattern in insects and anurans. This conclusion stems from studies of preparations in which acoustic input was abolished, or from studies with "silently singing" animals (e.g., Kutsch and Huber 1970; Loher and Huber 1965; Möss 1971). Moreover, recent studies have found the crickets have normal mechanisms that tend to suppress auditory responses to their own songs (Poulet and Hedwig 2001). The same effect is achieved mechanically in some species of cicadas and anurans (e.g., Young 1990; Narins 1992b; see chapter 5).

Acoustic feedback in mole crickets and some burrowing frogs could contribute to optimal acoustic efficiency. Mole crickets build burrows that enhance certain frequencies in their songs, so that their songs have nearly pure-tone carrier frequencies. Not only must the burrow have a specific shape and the proper dimensions to emphasize a particular frequency, but also the insect must position itself at a point in the burrow that optimally increases the amplitude and purity of the song (Bennet-Clark 1987, 1989; Daws 1996). Models of burrows, acoustic analyses of vocalizations, and observations of calling frogs led Bailey and Roberts (1981) to suggest a similar strategy for frogs of the genus *Heleioporus* in Australia. Whether these animals find the proper position in the burrow on the basis of acoustic feedback or tactile cues is an open question.

Proprioceptive Feedback

In crickets the sound produced by wing stridulation has a near field component (see appendix 3) that stimulates cercal filiform sensilla (Dambach et al. 1983). These receptors transmit this information to a group of ascending interneurons that directly or indirectly have access to the song pattern generator (reviewed in Dambach 1989). When the filiform hairs of the anal cerci are stimulated with single weak air puffs within the pause between the chirps, the rhythm is altered. Depending on the phase of the stimulus, the onset of the following chirp is shifted in time. Because there is no compensation for this shift in the subsequent chirp intervals, the generator for chirps thus appears to be reset by external stimuli and phase-shifted with respect to the ongoing rhythm. Coupling and resetting occur only within a small time frame, that is, only when the difference between the externally applied rhythm and the chirp rhythm does not exceed 10%. Because phase-dependent coupling can also occur when the animal produces cercal stimulation by its own wing movements, this feedback mechanism could help to stabilize its own chirp frequency.

Short-term proprioceptive control is also provided by mechanosensory systems located on the cricket wing. These systems control the typical right-over-left positioning of the wings and the strength of the contact between file and scraper necessary to emit loud sounds (reviews: Huber 1990; Kutsch and Huber 1989). Wing-position preference in field crickets (e.g., file under the right wing positioned over the scraper on the top of the left wing) is controlled by two plates, covered with about fifty mechanoreceptive sensory hairs, which are found only in males. The plates occur on each wing, adjacent to the scraper and the file (Elliott et al. 1982). When the two hair plates situated between the wings (the lower one on the right and the upper one on the left) are removed, stridulatory movements are drastically impaired and the preference in wing position is reduced. Many pulses are also shorter than usual, and the sound pressure level of the sound is reduced by about one-tenth (Elliott and Koch 1983). These sensory plates are much less developed in mole crickets, which have no preference in wing position during singing (Kavanagh and Young 1989).

Finally, the proprioceptive feedback system for efficient calling involves yet another, even more powerful mechanosensory system. The sensory part of the system consists of a field of about twenty-five *campaniform sensilla*, which are located on each wing of the male along the cubital vein near the wing joint that controls the contact between file and scraper (Schäffner 1984; Schäffner and Koch 1987a,b; fig. 3.19). Removal of these inputs by cutting the nerve proximally to the sense cells results in songs with missing or shortened pulses and drastically changed pulse patterns. Recently, Hustert et al. (1999) reported that the pegs on the stridulatory files of grasshoppers contain two sensory cells. The fibers from these cells respond to deflection and pressure during stridulation, suggesting that stridulatory files are both sound-producing and proprioceptive organs.

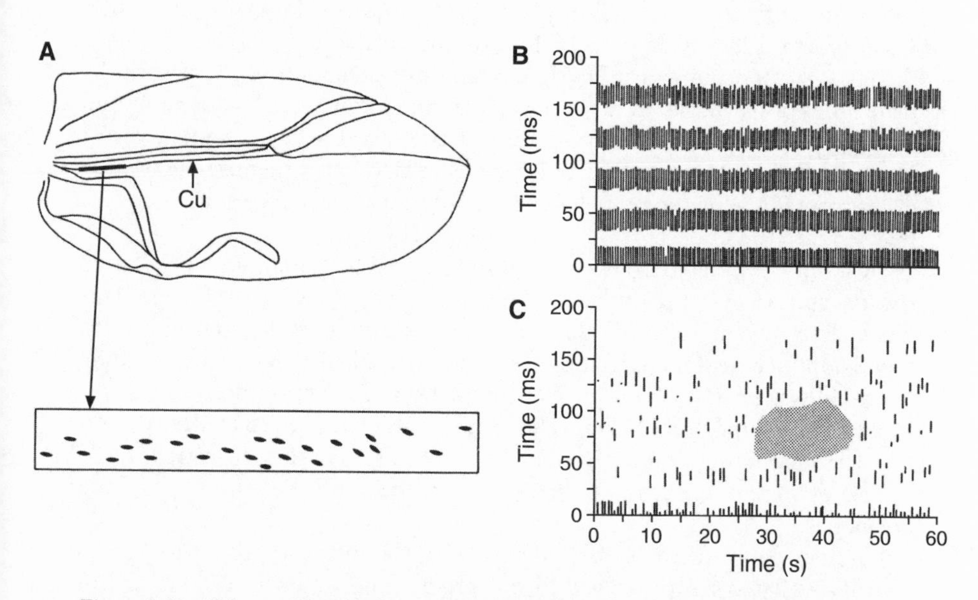

Figure 3.19. Male-specific cubital campaniform sensilla and their effects on song production in *Gryllus bimaculatus*. (A) Ventral view of the right wing. Cu = cubital vein; bar gives the location of the campaniform sensilla and, downward-pointing arrow, their distribution at larger scale. (B) Five-pulse calling chirps of a control male (ordinate: timing of pulses and pulse duration marked by bars; abscissa: time in milliseconds). (C) Songs of the same male after severing both nerves proximally to the campaniform sensilla showing severe changes in the pulse pattern. From Huber 1990; modified from Schäffner 1984.

Development of Neuromuscular and Neural Mechanisms for Signal Production

In neither insects nor anurans is there evidence that developing individuals must hear and learn the signals of adults in order to produce the spectral and temporal properties typical of conspecific signals (reviews: Ewing 1989; Gerhardt 1994a). In hemimetabolous insects with stepwise development, such as in orthopterans, the basic Bauplan (body plan) of the central nervous system is nearly finished at the end of embryogenesis, and the peripheral innervation pattern, except for the final muscle size and the final number of sensory cells, is completed during postembryonic development (Kutsch 1989). In crickets (*Teleogryllus commodus*), the development of stridulatory motor patterns must occur in the late postembryonic period, because motor output related to wing stridulation already exists in later larval crickets, despite the underdevelopment of the wings (Bentley and Hoy 1970; fig. 3.20).

In grasshoppers, larval instars stridulate with their hindlegs, although silently because they lack the forewings as counterparts of the file, whose pegs grow in size during postembryonic development (Reis 1996). Stridulatory movements are already produced by third- and fourth-instar nymphs in response to songs of adult males or playbacks (fig. 3.20B; Halfmann and Elsner 1978; Elsner 1981; Reis 1996). The existence of leg stridulation in larvae, the adultlike coordination of both legs, and the elicitation of the pattern by male songs or playbacks all support the hypothesis that pattern generators and command systems are already present in late larval instars. A more direct proof for functioning of command systems in larval grasshoppers is given by experiments in which leg movements occur after neuroactive substances are injected into the brain of larval insects (fig. 3.20C).

In studies of isolated brain stems of the leopard frog, Schmidt (1978) used the first appearance of pulmonary respiration as a reference point for aging individuals that were undergoing metamorphosis from tadpoles to froglets. He found adultlike neural correlates of electrically evoked advertisement calling in juveniles as early as four days after completion of metamorphosis (fig. 3.21). Moreover, Schmidt could elicit release-call movements of intact animals several days before the completion of metamorphosis. In both preparations these correlates were only seen in animals that had been treated with appropriate hormones.

The development of sexual differences in the intrinsic properties of the neuromuscular system involved in call production has been well studied in South African clawed frogs (Kelley 1999). Sexual differentiation of muscle type (slow- vs. fast-twitch), numbers, and the strength of neuromuscular synapses depend on secretion of gonadal hormones at specific times early in development. Muscle changes, which have also been studied at the level of the genes regulating laryngeal myosin, depend on male androgens, whereas the change from weak to strong synapses requires exposure to female estrogens.

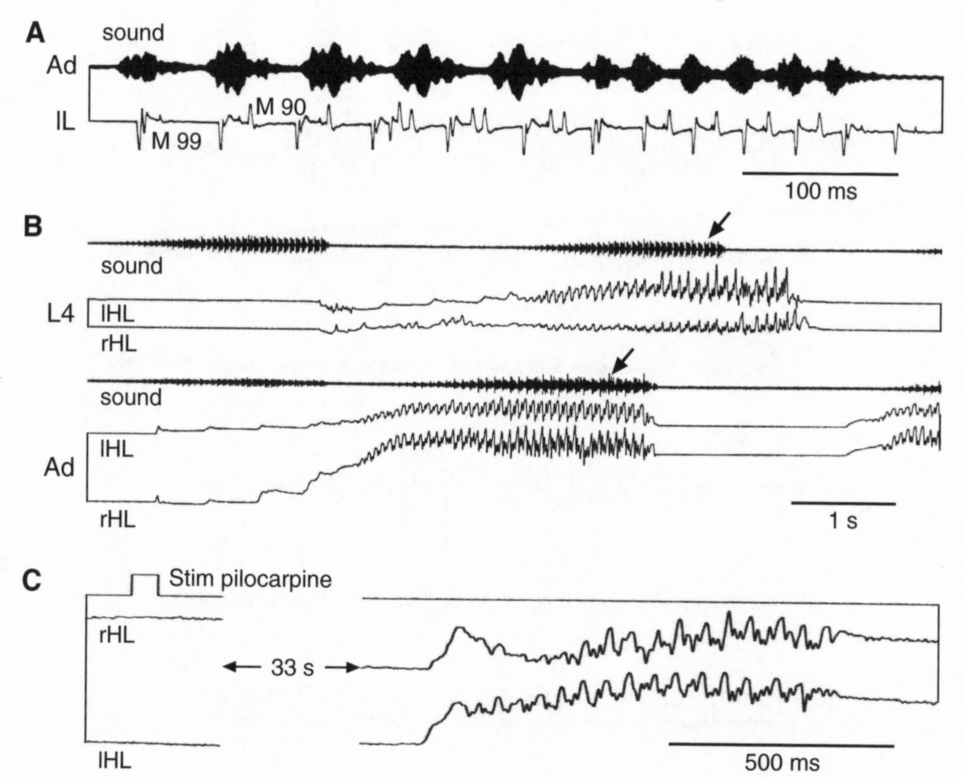

Figure 3.20. Evidence for early development of stridulatory circuits in orthopterans. (A) One phrase of the calling song of a *Teleogryllus commodus* adult male compared with muscle action potentials recorded from a last-instar nymph. Ad = song from adult; IL = simultaneous recording from the mesothoracic M 99, a wing-opener muscle (spikes initially downward), and from the mesothoracic M 90, a wing-closer muscle (spikes initially upward). The soundless larval muscle pattern, released by brain lesions, coincides well with the initial chirp part and the consecutive trill part of the song. Adapted and modified from Bentley and Hoy 1970. (B) Stridulatory hindleg movements in a fourth (ultimate) instar male larvae (L4) of *Chorthippus biguttulus*, elicited with playback of adult male songs (*upper trace*), compared with the patterns of adults (Ad). Arrows indicate the playbacks of the male song in L4 and the songs produced by the adult male in Ad. Modified from Reis 1996, fig. 34. (C) Functioning of the command system in larval grasshoppers indicated by eliciting species-specific stridulatory patterns after injection of neuroactive substances into the brain of *C. biguttulus*. A short pulse of pilocarpine elicited stridulatory sequences after a latency of about 33 s and for about 2 min in a fourth instar male larva. Each sequence lasted 0.8 to 2 s and showed the species-specific pattern of up-and-down strokes. Abbreviations: lHL, rHL = left and right hindleg; Stim = duration of pilocarpine injection. Modified from Reis 1996.

Summary and Suggestions for Future Studies

The role of the central nervous system in sound production in insects and anurans involves two main functions: activating and patterning of acoustic signals. In insects, single command neurons, whose activity is necessary and sufficient to activate and maintain different types of stridulation, have been

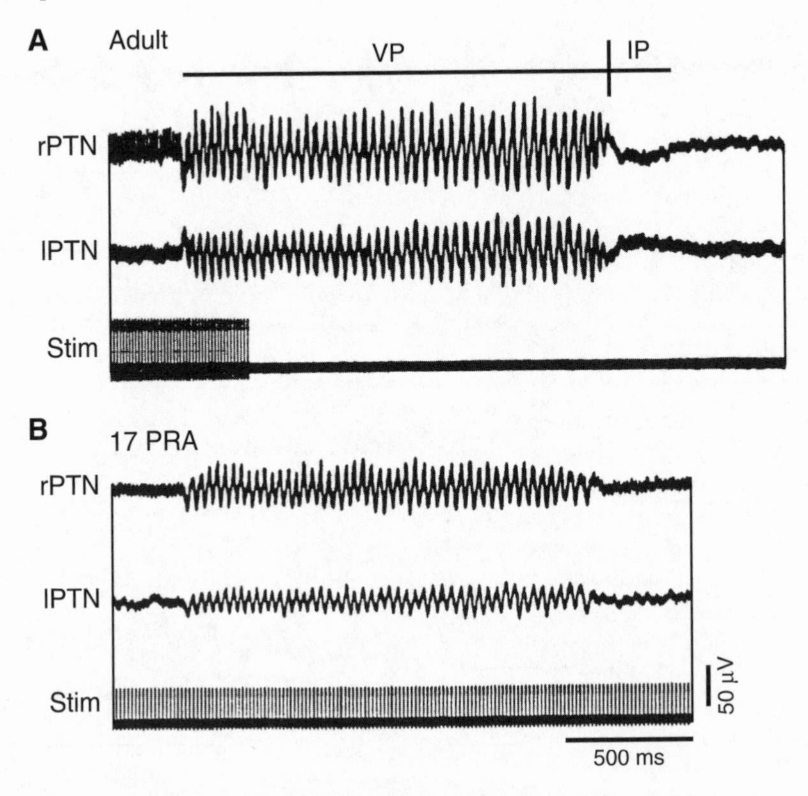

Figure 3.21. Neural correlates of calling evoked in the isolated brain stem by electrical stimulation (stim) in adult male anurans (*Rana pipiens*) (A) and in juveniles at pulmonary respiration age (PRA) of 17 (B). VP = vocal phase; IP = inspiratory phase. rPTN and lPTN = right and left pretrigeminal nucleus, respectively. From Schmidt 1978, fig. 2A,C.

identified in the head ganglia. The descending output of command neurons does not show temporal patterning that corresponds to the temporal patterns of acoustic signals. In anurans, signal production is controlled by parallel descending pathways; neurons that trigger vocalization in a fashion analogous to command neurons in insects probably originate in parts of the hypothalamus and medulla since nonspecific electrical stimulation of these areas elicits vocalizations with species-specific patterns. By contrast, much of the signal's temporal fine structure is present in the output of pattern generators. These circuits orchestrate patterns of firing in motoneurons, which, in turn, regulate the contractions of the muscles that move the peripheral sound-producing structures. The basic temporal properties of pattern-generator output and acoustic signals can be generated in the absence of peripheral feedback, at least over the short term. Although our conceptual view is well established, we still have only tentative wiring diagrams of the signal-production systems in insects and anurans. Indeed, we know few details about the anatomical and functional delineation of command systems and pattern generators, the cellular elements of pattern generators, the networks that coordinate activity in different parts of the signal production system, and circuits that modify ongoing pattern-generator activity.

We know much more about insects than anurans, but many important questions remain open: (1) How are the cells that form the command systems in each half of the brain coordinated? (2) How do they connect to the bilaterally occurring thoracic pattern generators and coordinate each of the pattern generators in the hemiganglia? (3) How do different command neurons that control different song types or different elements within a song act on the pattern generators? (4) How do single or multiple pattern generators produce temporally complex songs? (5) What is the significance of the large-scale representation of the stridulatory pattern throughout the insect central nervous system? That is, in addition to those neurons thought to be elements of command and pattern-generator systems, the activity of neurons in other parts of the nervous systems such as the subesophygeal ganglion is also coupled to sound production (review: Elsner 1994). The anuran vocal control system has been studied by only a handful of brain-stimulation and lesion studies, and in contrast to insects, we know almost nothing about proprioceptive feedback. The situation is improving, however, with the study of isolated brain preparations, where the anatomy and connections of single neurons can be resolved and where experiments using electrical stimulation can provide some hints about functional relationships between different parts of the brain.

For both groups, a major challenge is to describe the functional and anatomical relationship between the auditory networks that process signals and those that elicit acoustic and other responses (the so-called sensory-motor interface). How does sensory processing result in the activation of specific descending inputs (command systems) to the pattern generators? In our view, real progress in understanding this and many of the questions listed above will require the development of preparations in which recordings from multiple cells or populations thereof can be made during signal production that occurs in response to acoustic input. Moreover, studies of behaving animals will almost certainly be needed to discover the role, if any, of acoustic feedback. This approach might also help us learn more about the constraints imposed by the dual function of elements involved in sound production in locomotion (insects) and breathing (anurans).

Finally, the available evidence suggests that acoustic experience during development is not required for the production of normal signals. However, this dogma rests on a few experimental studies and stems largely from the observation that juveniles of many species develop in environments in which they will hear only inappropriate sounds. Experiments with species with prolonged breeding seasons in which developing individuals do predictably associate with signaling adults could provide compelling confirmation or perhaps some surprises.

4

Acoustic Criteria for Signal Recognition and Preferences

Field observations long ago suggested that acoustic signals produced by male insects serve to attract females (Rösel v. Rosenhof 1749), and the telephone playback experiments of Regen (1913) proved that sound alone was effective in attracting female crickets. By contrast, the mate-attracting function of anuran calls was still an issue until the playback experiments of Martof (1961). Playback experiments are required because other kinds of signals (visual, olfactory) can also attract mates, influence mating decisions, repel rivals, and so forth (Otte 1977; Wells 1977; Loher and Dambach 1989). The use of synthetic sounds, whose physical properties can be varied systematically, has made it possible to discover the key features (and their optimal values) that receivers require to initiate appropriate behavioral responses and to locate signalers (reviews: Gerhardt 1988, 1994a; Ewing 1989; Helversen 1997).

Determining the acoustic criteria used to evaluate signals helps to define the operating principles whereby the nervous system extracts biologically significant information from communication signals. Ultimately such behavioral data can also be used to judge the merits of hypotheses about neural mechanisms. Behavioral data also bear on questions about the contemporary function and evolution of communication. Comparing preferences based on variation in key acoustic properties with variation in these properties in signals within natural populations is a starting point for predicting patterns of intraspecific mate choice (chapter 10). Comparing preference criteria and mechanisms across populations and species can provide insights about broad-scale patterns of evolution of communication systems (chapter 11).

We focus here on particularly well-studied species of insects and anurans, including acridid grasshoppers, European katydids, field crickets, and North American treefrogs. For most of these species, some information is also available about how the key properties of communication signals are encoded in the peripheral auditory system (chapter 5) and then further processed in the central auditory system (chapter 6).

Several generalizations emerge from the behavioral studies to be discussed in this and subsequent chapters. First, the acoustic criteria used for mate attraction and mate choice often differ significantly between closely related species and even between the sexes within the same species (tables 4.1 and 4.2). Thus, there are multiple ways of accomplishing the same task. Second, many insects and anurans primarily use one or two main frequency channels for communication. That is, acoustic energy in their long-range signals is usually concentrated in one or two frequency bands to which their auditory system is reasonably well tuned. Third, signal recognition and especially frequency preferences are often intensity-dependent. Not only are preferences often abolished or reversed by reducing the relative intensity of a preferred signal, but also the absolute level at which the animal experiences alternative signals can influence its selectivity. Preferences based on fine-scale temporal differences are usually more robust in this respect. In any event, communication distance, which is generally correlated inversely with intensity at the position of a receiver, must often affect receiver selectivity. Studies employing a single playback level thus can assess only a limited part of the recognition or preference space that is likely to be relevant in nature. Fourth, preferences are usually based on multiple acoustic criteria that interact to determine the overall attractiveness of a signal. Relatively few behavioral studies have simultaneously varied even two key properties in a systematic fashion, even though a full understanding of the mechanisms underlying signal recognition and female choice depends on how receivers deal with complex signals that convey different kinds of information.

Terminology and Some Operational Definitions

Positive responses to any acoustic signal, including heterospecific and synthetic signals, are often taken as evidence for "recognition." In the psychological literature, however, the term *recognition* tacitly assumes cognitive processing, and the territorial behavior of male bullfrogs (*Rana catesbeiana*) is a good example of a noncontroversial application of the term. A male forms a memory (a cognitive process) based on repeatedly hearing its neighbor's calls (Bee and Gerhardt 2001a) and later compares new signals with this memory to classify these signals as those of the neighbor or of a strange male (Davis 1987; chapter 9). By contrast, acoustic responses of some female katydids take place within milliseconds after the onset of the male song (Heller and Helversen 1986; details below). Such reflexlike responses probably do not require signal processing by higher centers in the "brain" (Robinson et al. 1986). We could nevertheless argue that selection has generated a "phylogenetic template" (Herrnstein 1990) in the nervous system of the katydid. Recognition thus occurs when the peripheral input effectively matches that template, wherever it is located in the nervous system and regardless of whether acoustic experience is necessary to form it.

Because terms such as *recognition* (and *species recognition*), *preference*, and *discrimination* mean different things to different people, we introduce some operational definitions. In practice, an investigator usually assesses the effectiveness of an acoustic stimulus and makes a judgment about its recognition in terms of one or more arbitrary criteria: the percentage of positive responses, accuracy and speed of phonotactic movements, and so forth. An effective signal presumably has one or more acoustic properties (key features) with values that stimulate recognition circuitry in the auditory system. Ineffective signals presumably lack these properties. The same behavioral tests can be used to map the recognition and preference space of a key feature. That is, the effectiveness of a signal is quantified as the values of its key features are varied. The range of values that elicits responses meeting some (arbitrary) minimum response criterion for signal effectiveness defines the recognition space; quantification of the effectiveness within the recognition space estimates the preference space. If, for example, some range of values of an acoustic property much more reliably elicits responses than some other range of values, we expect that in a choice test, a signal with a value falling in the more effective range will be preferred to a signal with a value falling in the less effective range. Choice experiments not only can test this assumption, but also can be used to estimate directly the preference space. Choice tests can also be more sensitive than single-stimulus designs in revealing preferences based on very small differences in a key feature. If signal assessments occur sequentially (rather than simultaneously, as in a chorus situation), however, then single-stimulus estimates of a preference space are probably more biologically realistic than are estimates based on choice tests (Wagner 1998; chapter 10).

Differences in the relative effectiveness of two signals are evidence for *discrimination*. The sensory system can obviously resolve the physical differences between the signals. In studying the unconditioned responses of insect and anurans, however, a lack of preference is not evidence for a lack of discrimination. Some differences in signals might be functionally equivalent even if the animal can easily distinguish between them (Nelson and Marler 1990). Furthermore, the presentation of some signals might result in no response or even suppression of a response precisely because the signals are recognized as being inappropriate or indicative of a dangerous situation. In terms of signal detection theory (appendix 1), such a failure to respond is called *correct rejection*. Evidence for correct rejection involves showing that in a highly motivated animal, making some acoustic property more detectable (e.g., increasing its duration, repetition rate, or intensity within biologically realistic limits) renders a signal less attractive, if not ineffective. We provide several examples in this chapter. Signal detection theory also provides a way of viewing response variability other than as a nuisance. For example, the probability of a response often reflects an animal's reproductive state, and its selectivity can vary depending on the time available for assessment, the possibility of multiple matings, predation risks, and so forth (chapters 9–11). These factors influence the costs and benefits of responding selectively (Wiley 1994).

Selectivity for acoustic signals contributes to *species recognition* or *species isolation.* The implication of the latter term is that preferences for acoustic signals evolve in order to prevent matings with other species. But selective responses to conspecific signals should be expected *whether or not other species with similar signals are present* (e.g., Paterson 1985). Discrimination against heterospecific signals must often, if not generally, be a *consequence* of the fact that heterospecific signals lack key acoustic properties, or if present, the values of those properties are less effective than those in conspecific signals (e.g., Littlejohn 1981). Testing the hypothesis that divergent signals or acoustic criteria arise from interactions between two species requires comparisons of their signals and selectivity in *sympatric* and *allopatric populations,*[1] or other kinds of comparative approaches (see chapter 11). Finally, playback experiments sometimes show that some signal values, including those of other species, can be even more effective than values typical of conspecific signals, thus revealing "hidden preferences" (Arak and Enquist 1993; Enquist and Arak 1993) or "pre-existing biases" (Ryan and Rand 1993; chapter 11).

Selectivity Based on Differences in Sound Pressure Level

Female insects and anurans usually show preferential phonotaxis toward the more intense of two sounds that do not differ in other ways. In most tests, differences of 2–3 dB are adequate to elicit a significant preference (Gerhardt 1988; Forrest and Raspet 1994; Schul and Fritsch 1999). This generalization has to be qualified in two respects.

First, preferences based on differences in intensity are usually weaker as the absolute amplitude of alternative stimuli is increased. For example, female treefrogs (*Hyla cinerea*) preferred a standard synthetic call played back at 73 dB SPL to the same signal played back at 70 dB SPL, but did not show a significant preference in a test of the standard call played back at 83 dB versus 80 dB SPL (Gerhardt 1987). Females did choose more intense signals when the difference was increased by another 3 dB (86 dB vs. 80 dB SPL). Comparable data are available for crickets (Stout and McGhee 1988; Farris et al. 1997), mole crickets (Walker and Forrest 1989), and katydids (Schul and Fritsch 1999). Forrest (1994) points out that this pattern in insects and anurans violates *Weber's law* (i.e., that the minimum difference between the values of two stimuli required for discrimination is a constant proportion of the value of the

1. *Sympatry* (*sympatric*) refers to the coexistence of two species in the same geographical area; *syntopy* refers to their occurrence in the same location within sympatry. *Allopatry* (*allopatric*) refers to the existence of a particular species in a geographical area lacking other species with which it might be expected to interact.

reference stimulus). Some neural correlates of this phenomenon have been found in katydids (Römer et al. 1998; Schul and Fritsch 1999). The opposite effect has been demonstrated in birds and mammals, in which discrimination improves as the absolute intensity increases (the so-called near miss to Weber's law).

Second, modeling and empirical measurements predict that female attraction to the more intense of two sounds will be affected by the distance between them (Forrest and Raspet 1994). In particular, selectivity is expected to increase as the separation of the two sources is decreased (supporting empirical data in Farris et al. 1997). This last result is not especially pertinent to our discussion in this chapter, which mainly emphasizes selectivity that will be related to neural mechanisms, but has important implications for male spacing (chapter 8) and female choice in nature (chapter 10).

Selectivity for Spectral Properties of Signals

As shown in chapter 2, some species of insects and anurans produce narrowband signals with a nearly pure-tone carrier of constant frequency. Other species produce broadband signals, consisting of one or more frequency modulated carriers, a harmonic series, or a rapidly amplitude-modulated signal with many sidebands. Other broadband signals are aperiodic (noiselike) or highly transient (short pulses or clicks). To what extent are these structural features important for mate attraction, and how do differences in quantitatively varying properties, such as carrier frequency, influence signal effectiveness? Here we focus on frequency selectivity within relatively narrow frequency bands and the "rules" for responding to stimulation by acoustic energy in different, widely spaced frequency bands.

Preferences Based on Carrier Frequency

Preferences of insects and anurans for signals with frequencies typical of conspecific signals usually reflect the tuning of the peripheral auditory system. Such tuning, which also reduces stimulation by most heterospecific signals, has historically been interpreted in terms of *matched filtering* (Capranica and Moffat 1983; chapter 5). Besides contributing in some cases to species recognition, matched filtering can improve the detectability of conspecific signals in noisy multispecies choruses. A finer-scale analysis usually reveals that matches between signals and receivers are imperfect, as we might expect because the different systems (sound production and auditory) are subject to different kinds of selective pressures and constraints (Gerhardt and Schwartz 2001; see chapter 11). Recall, too, that selection on carrier frequency can be confounded by selection on body size because these traits are usually correlated (chapter 2). An extensive survey of the auditory system is presented in chapters 5–7, and the role of frequency preferences in female choice is treated in chapter 10.

In this section we show that frequency preferences, whether basically stabilizing (preferences for mean values in conspecific populations) or directional (preferences for lower- or higher-than-average frequencies), are not particularly strong. Small differences in amplitude can abolish or reverse preferences based on small differences in frequency. A second, subtler result is that the amplitude to which alternative stimuli are equalized can affect frequency preferences (Gerhardt 1987).

Sexual dimorphism and temperature effects complicate matched filtering in some systems. First, in some insects, both sexes may produce signals, which differ in frequency or other spectral properties. In the katydid *Ancistrura nigrovittata*, for example, emphasized frequencies are at about 15 kHz in the male and 27 kHz in the female. Behavioral experiments show that each sex preferred signals with frequencies corresponding to those emphasized in the songs of the opposite sex (fig. 4.1A; Dobler et al. 1994b). Acridid grasshoppers also show sex-specific preferences based on subtle differences in spectral structure (see below). Second, in anurans the tuning of the inner ear organs, especially the basilar papilla, is size-dependent, and because the two sexes often differ in size, both sexes are unlikely to be optimally tuned to the same narrow-band signal (review: Zakon and Wilczynski 1988). In two species (the coqui treefrog *Eleutherodactylus coqui* and the spring peeper *Pseudacris crucifer*), females appeared to be better tuned to male calls (or the presumed mate-attracting "qui" component in *E. coqui;* see chapter 9) than the males themselves (Narins and Capranica 1976; Wilczynski et al. 1984). Third, temperature shifts can result in poorer tuning if senders and receivers do not compensate in the same way. In some species of insects, carrier frequency, in which shifts of 1–2 kHz occur over a biologically realistic range of temperatures (see chapter 2), may be more temperature-dependent than auditory tuning. In other species of insects and in some anurans, frequency tuning can be more temperature-dependent than carrier frequency (see below; Gerhardt and Mudry 1980; Oldfield 1988; Lewis and Narins 1999).

Field crickets and katydids. Positive phonotaxis to the calling song of the male by the female is the first step in mate choice in field crickets and katydids. After close-range assessments that may be based on additional (courtship) sounds as well as visual, olfactory, and tactile cues, a female makes her final decision by walking onto the back of the male or moving away from him (Loher and Dambach 1989). The two pairs of species considered in this section have been subject to extensive behavioral and neurophysiological studies that we will review throughout this monograph.

The sibling species of field crickets *Teleogryllus commodus* and *T. oceanicus*, which occur in the Asian-Pacific region and sympatrically in areas along the east coast of Australia, were subjects of an early study of species recognition (Hill et al. 1972). Experiments using synthetic calls showed that females of each species responded reliably at low playback levels to signals with carrier

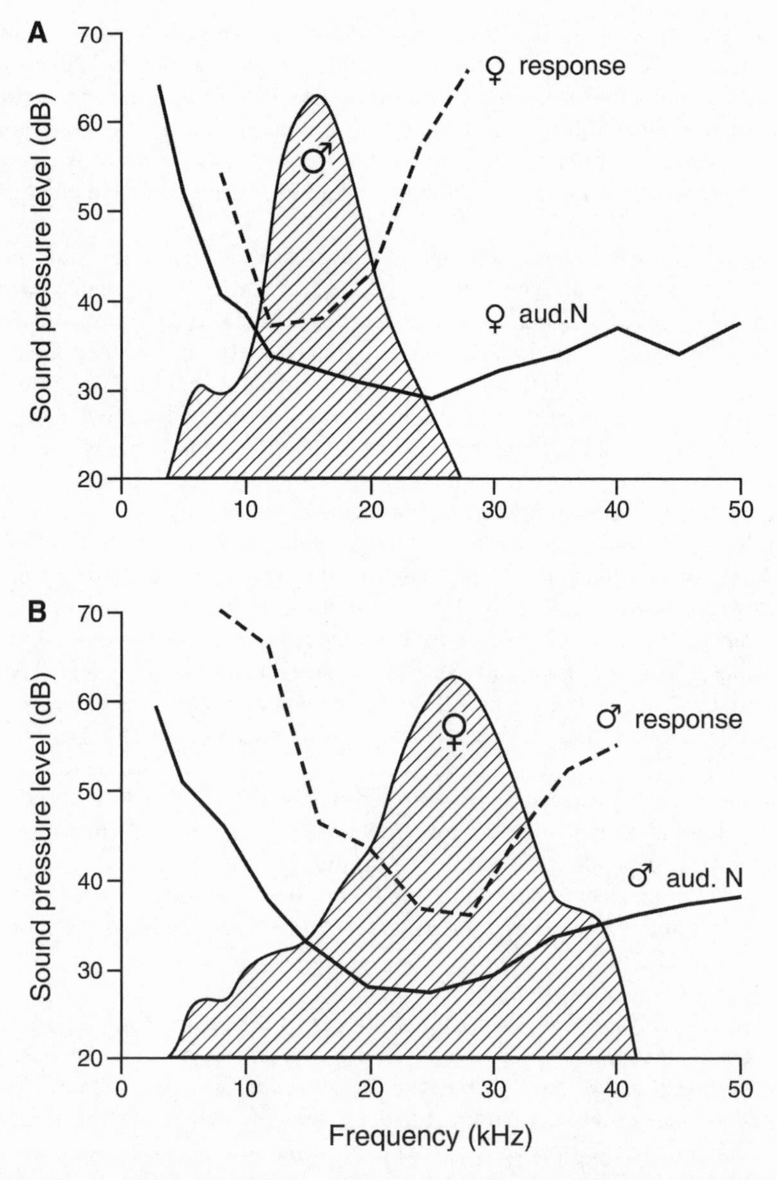

Figure 4.1. Sex discrimination based on differences in carrier frequency in duetting katydids. Hatched curves show the spectra of male (A) and female (B) acoustic signals in the duetting katydid *Ancistrura nigrovittata.* The dotted line in (A) shows the behavioral tuning curve of the female, which clearly matches the peak energy in the spectrum of the male song. The dotted line in (B) shows the behavioral tuning curve of the male, which clearly matches the peak energy in the female's acoustic response. Solid lines (A) and (B) show threshold curves, estimated from summed activity in the auditory nerve. In contrast to the behavioral data, the curves are nearly the same in both sexes, and later research involving ascending auditory interneurons has revealed the probable reason for the discrepancy (see chapter 6). From Dobler et al. 1994a, fig. 5.

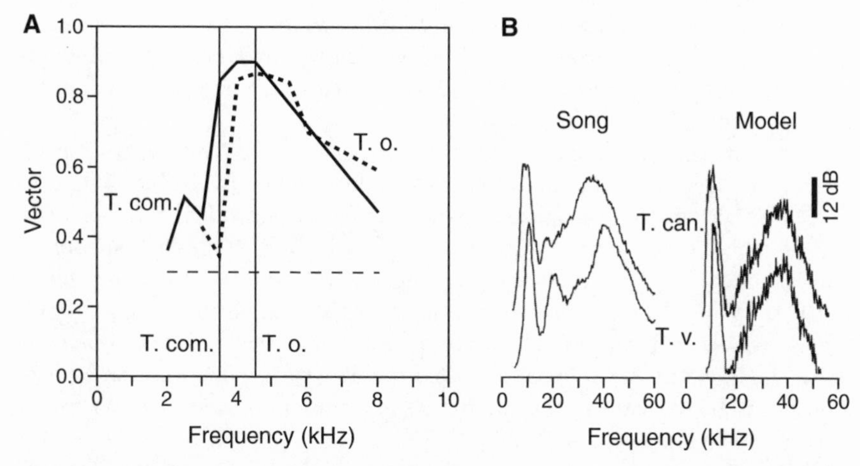

Figure 4.2. (A) Behavioral tuning curves for the field crickets *Teleogryllus commodus* (Tcom) (*solid curve*) and *T. oceanicus* (To) (*dotted curve*) plotted in relation to the mean carrier frequencies (*vertical lines*) of the songs of conspecific males (about 3.5 kHz in Tcom and 4.5 kHz in To). Notice that responses fall off much more sharply on the low-frequency sides of the curves than on the high side and that the carrier frequency of *T. oceanicus* better matches the behavioral tuning curve of conspecific females than does that of *T. commodus*. The vector score is a combined measure of phonotactic accuracy and speed, and values of about 0.3 (*dotted horizontal line*) or less are judged to represent random orientation. From Hennig and Weber 1997, fig. 7. (B) Spectra of natural ("song") and synthetic ("model") songs of the katydids *Tettigonia viridissima* (Tv) and *T. cantans* (Tcan). Notice that the spectra of the synthetic songs were adjusted to have the same shape and maximum values in the high-frequency region, whereas the species difference in the low-frequency region was maintained. The main difference is the slightly greater bandwidth of the low-frequency peak in the song of *T. cantans*, which overlaps that of *T. viridissima*. Females of *T. cantans* discriminate against synthetic calls with conspecific temporal properties and spectral properties typical of the songs of *T. viridissima;* females of *T. viridissima* do not discriminate against synthetic calls with conspecific temporal properties and spectral properties typical of *T. cantans*. From Schul et al. 1998, fig. 1D.

frequencies corresponding to the mean frequency in conspecific songs: 3.5 kHz in *T. commodus* and 4.5 kHz in *T. oceanicus* (Hill 1974; Hennig and Weber 1997). In both species, bands of equally effective frequencies are nearly 1 kHz wide, and behavioral effectiveness falls off more sharply on the low-frequency sides of these bands than on the high-frequency side (fig. 4.2A). The mean frequency in *T. oceanicus* songs falls close to the center of the most-effective band, and a signal with the much lower frequency typical of *T. commodus* songs was almost completely ineffective in eliciting phonotaxis by females of *T. oceanicus*, even if the temporal structure of the signal matched that in conspecific songs. In *T. commodus*, however, the mean frequency of the conspecific song falls at the low-frequency edge of the most-effective band. Because the upper end of this band includes the mean frequency of *T. oceanicus* songs, signals of this frequency with the appropriate temporal pattern were also attractive to females. Thus, the species difference in carrier frequency alone can contribute to discrimination against the songs of *T. commodus* by females of *T. oceanicus*, but not the other way around. As discussed below, however, temporal differences in calling songs are adequate for species-specific phonotaxis by females of both species.

Another example of a species difference in the usefulness of frequency preferences for discrimination between conspecific and heterospecific signals involves females of the European katydids *Tettigonia viridissima* and *T. cantans* (Schul et al. 1998). These two species have overlapping ranges of distribution and sometimes hybridize. As shown in figure 4.2B, the low-frequency peaks in the songs of these two species overlap, but the peak in the song of *T. cantans* is broader and extends to lower frequencies than that in *T. viridissima* signals. The high-frequency spectral peak is necessary for sound location but was adjusted to be the same in the synthetic signals used in playback experiments to be discussed here (fig. 4.2B).

In two-speaker experiments in which both alternatives had the conspecific temporal pattern, females of *T. cantans* preferred synthetic songs with a spectrum similar to that of conspecific songs to an alternative with a spectrum typical of the songs of *T. viridissima*. This preference was completely reversed, however, by decreasing the playback level of the conspecific stimulus by 12 dB. By contrast, females of *T. viridissima* were as reliably attracted to songs with a spectrum typical of *T. cantans* as they were to songs with the conspecific spectrum (Schul et al. 1998). Thus with the temporal pattern held constant, additional acoustic energy below about 10 kHz is relevant to females of *T. cantans* and has little or no effect on selective phonotaxis in females of *T. viridissima*. As discussed in chapter 5, this behavioral difference is correlated with a species difference in the degree to which auditory receptors are uniformly tuned to frequencies within this region of the spectrum (Schul 1998).

In summary, these two studies show that frequency preferences based on relatively small differences in frequency within a single region of the spectrum are relatively weak. Even when such preferences result in discrimination between sympatric species, they can usually be abolished by changes in relative intensity. Moreover, to the extent that frequency preferences alone might contribute to species identification, they operate in only one direction, thus highlighting the lack of generality of this kind of explanation.

North American treefrogs. Vocal communication in North American treefrogs has been studied extensively, and quantitative data about vocal structure, selective phonotaxis, auditory mechanisms, sound localization, and sexual selection are available for several species (review: Gerhardt 1994a). In these frogs, females approach and initiate mating (amplexus) with calling males and do not appear to use any sensory cues other than the long-range advertisement call for mate choice (courtship signals are absent). Thus, a phonotactic approach to a loudspeaker is equivalent to a final choice of a mate.

In the studies to be discussed in this section, the frequency selectivity of females was assessed by varying not only carrier frequency but also the relative and absolute playback levels of alternative signals. The results of these experiments allow us to relate behavioral results in a quantitative way to data from neurophysiological studies (e.g., Capranica and Moffat 1983; Schwartz and Gerhardt 1998; see box 5.3) and suggest how communication distance might influence receiver selectivity.

In the spring peeper *Pseudacris crucifer*, males from populations in central Missouri (U.S.A.) produce tonal advertisement calls with a carrier frequency ranging from about 2.7 to 3.3 kHz (mean of about 3.0 kHz) (Diekamp and Gerhardt 1992). In two-speaker tests, females showed little selectivity for synthetic alternatives that differed only in carrier frequency over this same range. Only one preference was observed at a playback level of 75 dB SPL (Doherty and Gerhardt 1984), and none at a level of 85 dB SPL (Schwartz and Gerhardt 1998). Moreover, when the playback level of a preferred stimulus with a carrier frequency near the mean was lowered by 6 dB, then alternatives with frequencies well outside the range of variation in male calls (1.8 kHz and 4.6 kHz, respectively) became equally attractive (Doherty and Gerhardt 1984). These results parallel those just discussed for field crickets. Although phonotaxis is elicited effectively by frequencies within the band typical of conspecific advertisement signals, signals well outside this range can also be nearly as effective at somewhat higher but still biologically realistic playback levels.

The green treefrog *Hyla cinerea* and its closely related congener, the barking treefrog *H. gratiosa*, showed frequency preferences that almost certainly contribute to species recognition in mixed-species choruses (fig. 4.3A). Females from sympatric and nearby localities continued to choose standard synthetic calls with the conspecific spectrum when its playback level was 12–21 dB less than that of alternatives with the heterospecific spectrum (Gerhardt 1982; Höbel and Gerhardt, unpubl. data). Notice, however, that the frequency of the low-frequency peaks in the calls of *H. cinerea* and *H. gratiosa* differs by a much greater percentage in comparison to the species-pairs of crickets and katydids discussed above.

In the green treefrog, the *absolute* sound pressure level to which alternatives were equalized also affected the selectivity of preferences based on differences in frequency within the two emphasized bands (Gerhardt 1987; fig. 4.4). Females were more selective for frequency differences within the low-frequency band at low-to-moderate sound pressure levels (65 to 75 dB SPL) than at high playback levels (85 dB SPL), at which females rejected higher-than-average frequencies but not lower-than-average ones. By contrast, selectivity for differences in frequency within the high-frequency band improved with increasing playback levels (Gerhardt 1987).

Taken together, these results support the hypothesis that insects and most anurans, unlike birds and mammals (Fay 1992), probably cannot resolve fine

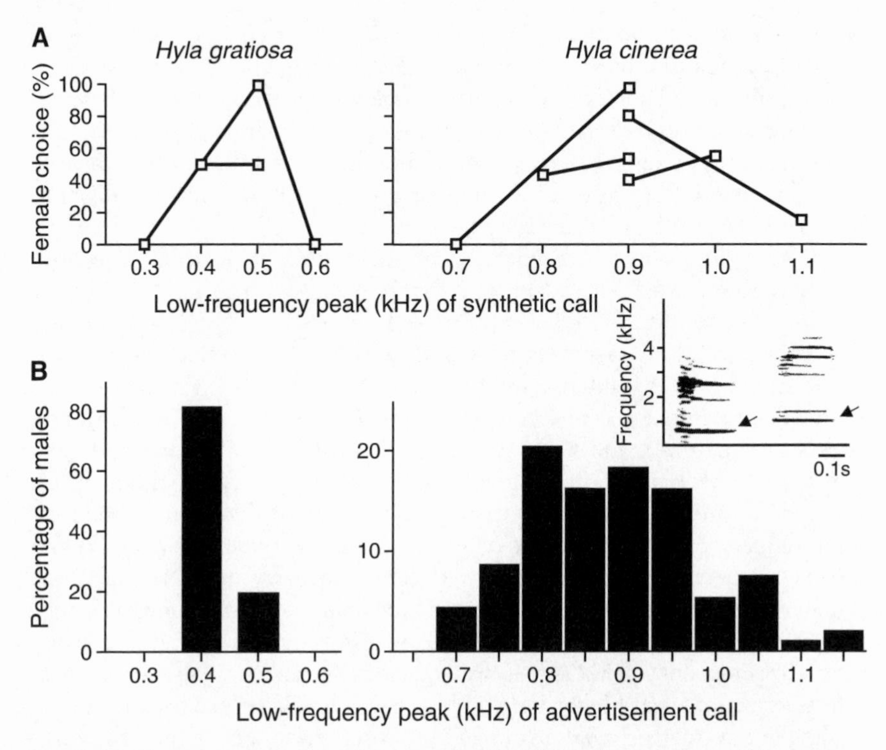

Figure 4.3. Frequency preferences of two species of North American treefrogs, *Hyla gratiosa* and *H. cinerea*. (A) Lines connecting squares show the percentage of females choosing each alternative of pairs of synthetic calls having a different low-frequency peak. For example, 100% of the females of *H. gratiosa* chose the call of 0.5 kHz to alternatives of 0.3 and 0.6 kHz. The sound pressure level of all pairs of stimuli was equalized at 75 dB. (B) Distributions of the frequency of the low-frequency peak in the advertisement calls (*inset*, sonograms of advertisement calls with arrows indicating the low-frequency peak; *left, H. gratiosa; right, H. cinerea*) of males from the same populations. The original data are in Gerhardt (1981a, 1987). From Gerhardt and Schwartz 1995, fig. 8.

differences (say, ≤ 5%) in frequencies in an intensity-independent fashion, at least when frequency is relatively high (say, ≥ 2 kHz). One exception might be the aquatic clawed frog *Xenopus laevis*, which in a conditioning paradigm was found to discriminate differences of just over 2% in the range of 1.6 to 2.5 kHz (Elephandt et al. 2000). While we are comfortable with the current dogma with respect to most species considered in this monograph (see also Pollack 1998), further research with cicadas and other, terrestrial anurans could uncover more exceptions. First, as pointed out earlier, the unconditioned natural responses used to assess frequency selectivity in insects and anurans do not necessarily reveal their abilities to distinguish between signals. Such abilities might be expressed, however, in contexts where frequency discrimination has significant selective consequences. For example, the most acute selectivity in unconditioned animals (discrimination based on 4–5%

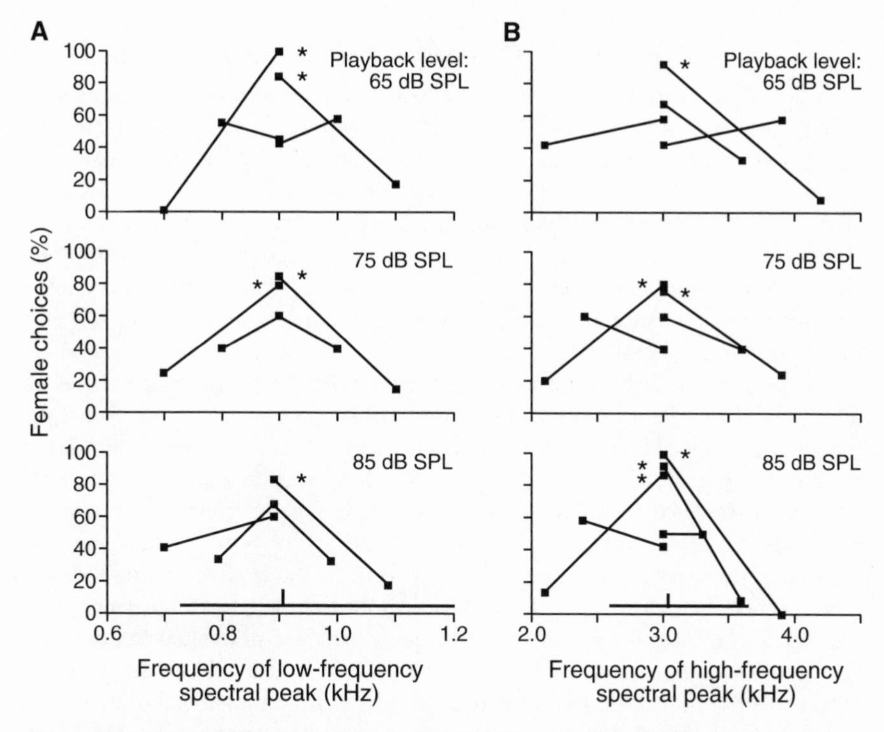

Figure 4.4. Preference functions based on frequency in green treefrogs *Hyla cinerea* as a function of playback level and based on (A) variation in the frequencies representative of the low-frequency spectral peak; (B) variation in the frequencies representative of the high-frequency peak. Points connected by lines show the proportions of females choosing each alternative (sample sizes ranged from 10–25 animals per test; Gerhardt 1987). *Statistically significant ($p < 0.05$, two-tailed binomial test) preferences. The horizontal and vertical lines along the bottom panels of (A) and (B) show the ranges of variation and mean frequencies, respectively, in the calls of males from the same populations from which females were tested.

differences in carrier frequency when playback levels are equalized) was found in an Australian species (*Uperoleia laevigata*), in which incorrect mate-choice decisions reduce fertilization success (Robertson 1990; chapter 10). Differences in frequency of this order of magnitude or less might also be discriminated in the context of territorial defense, in which males sometimes drop carrier frequency to a small extent (e.g., Wagner 1992; Bee and Perrill 1996; see chapters 2 and 9). Because these encounters are ultimately critical for mating success, these frogs are likely to express the ability to discriminate in this behavioral context if they have it. Second, anatomical and physiological studies suggest that that some species might have the capacity for fine-scale frequency discrimination in some parts of their hearing range (see chapters 5 and 6). For example, Fonseca et al. (2000) speculate that the "advanced" hearing system of some cicadas might be adapted to processing songs that are strongly frequency modulated.

Temperature-Dependence of Frequency Preferences in Anurans

Tuning of receptors in the inner ear organs of anurans is temperature-dependent (review: Lewis and Narins 1999; chapter 5), whereas the spectral properties of the signals of anurans change relatively little with temperature (Gerhardt and Mudry 1980 and references therein). The expectation that at some temperatures there are mismatches between frequency preferences and the frequencies emphasized in conspecific calls was confirmed by a study of the green treefrog *H. cinerea* (Gerhardt and Mudry 1980). Whereas the low-frequency peak in the calls of individual males recorded at 18°C and 27°C was nearly the same, females reversibly changed their frequency preference over the same general range of temperatures. At normal breeding temperatures of about 24–26°C, females preferred a standard call (close to the mean) with a frequency of 0.9 kHz to an alternative with a low-frequency peak of 0.6 kHz, which falls outside the range of variation of conspecific calls (see fig. 4.4). At temperatures of about 18–20°C, however, females preferred alternatives as low as 0.5 kHz to the 0.9 kHz standard call. Temperature had less effect on preferences based on frequency differences in the high-frequency range: a preference for an average (3.0 kHz) to a lower-than-average high-frequency peak (2.1 kHz) at normal breeding temperatures was abolished but not reversed at low temperatures. This difference in temperature-dependence probably reflects differences in the extent to which intrinsic, metabolically dependent mechanisms contribute to tuning within the anuran's two different hearing organs (chapter 5).

In terms of adaptive consequences, large male green treefrogs, which have lower-frequency calls than small males, might be expected to have a mating advantage early in the season, when temperatures are somewhat cooler than during the bulk of the breeding season. There might also be an increased risk of mismating under these conditions because the low-frequency peak in the calls of barking treefrogs is often as high as 0.5 kHz (fig. 4.3).

Selectivity Based on Frequency Modulation

Many insects and anurans produce signals that show some degree of frequency modulation, yet only a handful of studies have explored the potential relevance of this property for mate choice. As discussed in chapter 2, the pulses within the chirps of some species of field crickets are frequency modulated, and Simmons and Ritchie (1996) found that in *G. bimaculatus*, the extent of frequency modulation depends on the degree of asymmetry of the two wings. These authors also showed that frequency modulation can either increase or decrease the relative attractiveness of a signal. The simplest explanation is that depending on the starting frequency of each pulse, frequency modulation either increases or decreases the overall acoustic energy in the preferred range of 4.25–4.5 kHz. Thus frequency modulation per se is unlikely to be an acoustic criterion for signal choice. In two species of anurans (*Physalaemus pustulosus:* Wilczynski et al. 1995; *Kassina:* Grafe, pers. comm.), males produce signals in which the carrier frequency or frequencies are strongly modulated. Females required not only frequency-modulated signals but also that the direction of frequency change be the same as that in conspecific calls. In *P. pustulosus* the frequency sweep per se does not appear to be required because two constant-frequency tones of different frequency were also attractive when presented in the correct sequence (Wilczynski et al. 1995). Although gray treefrogs (*Hyla chrysoscelis* and *H. versicolor*) and spring peepers (*Pseudacris crucifer*) produce frequency-modulated signals, synthetic signals lacking frequency modulation were equally attractive (Doherty and Gerhardt 1984; Gerhardt 1978c, unpubl. data).

Selectivity Based on Spectral Structure

So far we have considered the selectivity of insects and anurans with respect to variation in carrier frequency within single, well-defined frequency bands. In this section we examine the behavioral significance of female preferences based on the presence of multiple frequency bands. In a later section, where we consider negative phonotaxis in both sexes of crickets, we show that simultaneous stimulation by widely separated frequencies can have unpredictable effects on behavioral responses.

Acridid grasshoppers. The grasshopper *Chorthippus biguttulus* has been the subject of extensive behavioral and neurophysiological experiments (reviews: Helversen and Helversen 1987, 1994, 1997). Males stridulate as they move around in the natural habitat. When a female responds with her own song, males rapidly turn toward her, a behavior that has been used to explore the sound localization capabilities of this species (Helversen and Rheinlaender 1988; Helversen 1997; chapter 7). Females remain more stationary than males, and in the laboratory, females duet with a computer-controlled system that automatically presents synthetic signals and records acoustic responses (Helversen and Helversen 1987).

The signals of the two sexes differ in their spectral characteristics (fig. 4.5A). The male song has a broad band of ultrasonic frequencies centered around 35 kHz, in addition to a strong peak at about 7 kHz. Females required both the low- and high-frequency components in order to respond reliably. Female songs have much less energy in the high-frequency band, and males showed robust turning movements when high-frequency energy was absent. In fact, male turning responses actually decreased when both high- and low-frequency bands were present. Thus, the difference in the spectral balance of energy in the two frequency bands is used by both sexes, but their criteria for attractiveness are just the opposite (table 4.1).

North American treefrogs. The green and barking treefrogs discussed above and the gray treefrogs *Hyla chrysoscelis* and *H. versicolor* all produce calls with a

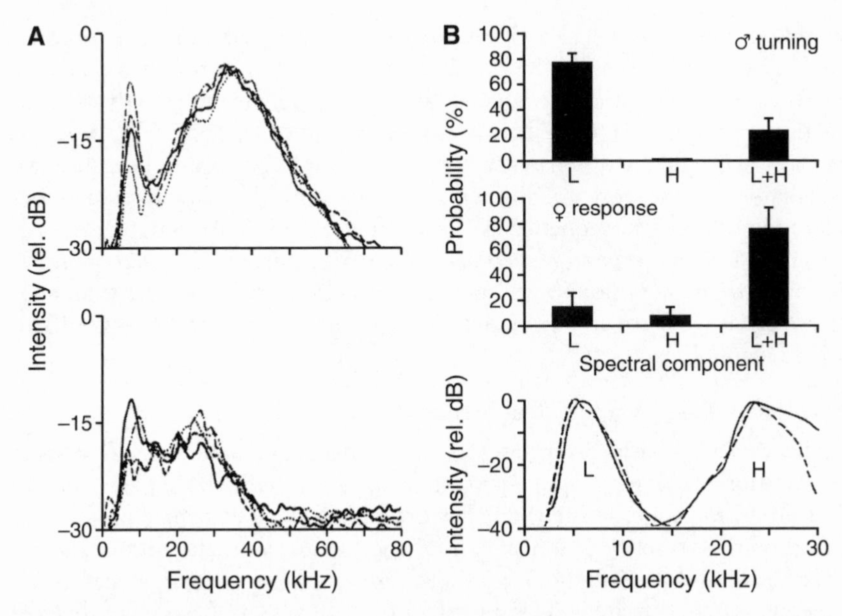

Figure 4.5. Sexual differences in selectivity for spectral structure in grasshoppers *Chorthippus biguttulus.* (A) Spectra of the songs of four males (*top*) and four females (*bottom*). (B) Effects of low- (L) and high- (H) frequency components presented alone or simultaneously. *Top*, mean turning probability and standard deviation for six males. L was played back at 58 ± 4 dB SPL, and H was tested at levels between 58 and 75 dB SPL. When L and H were presented simultaneously, the H component was increased to 75 ± 3 dB. Notice that fewer responses occurred to L+H than to L alone. *Middle*, mean response (acoustic answer) probability and standard deviation for five females. The L and H components were presented at 42–48 dB SPL and 60–66 dB SPL, respectively. Notice that the combination (L + H) was more effective than L or H alone. *Bottom*, frequency spectra of the synthetic calls played to males (*solid line*) and females (*dashed line*). From Helversen and Helversen 1997, figs. 2, 10.

bimodal spectrum. In single-speaker tests, females of all four species showed phonotaxis to synthetic calls having just one of the two spectral peaks. In choice tests, however, signals with bimodal spectra were preferred to signals with just one of the two peaks (Gerhardt 1981a,b; Gerhardt and Doherty 1988; see Schwartz 1987b for similar results from a Neotropical species).

Quantitative studies of preferences based on spectral structure have been conducted with *H. cinerea* (Gerhardt 1976, 1981b). The results indicate not only that female selectivity for spectral patterns depends on sound pressure level, but also serve to emphasize how single-stimulus and two-stimulus (choice) experiments could lead to very different conclusions. For example, in single-stimulus tests, there was no difference in the estimated behavioral threshold (about 48 dB SPL) for phonotaxis when females were offered a stimulus having just the low-frequency peak (0.9 kHz) or a stimulus with both spectral peaks (0.9 + 3.0 kHz). The estimated threshold for a stimulus of 3.0 kHz was 90 dB SPL. Although these single-speaker tests predict that

Table 4.1. Sexual Differences in Acoustic Signals and Criteria for Signal Recognition in the Grasshopper *Chorthippus biguttulus*

Acoustic Characteristics	Acoustic Criteria
Male Song	*Female Response*
Gaps in syllables are masked by phase difference in leg movements	Continuous syllables (> 30 ms) without gaps; detection of gaps reduces probability of response.
Large amplitude of first pulse in syllable	Large-amplitude pulse increases probability of response.
Broad-spectrum with two peaks (7 and 35 kHz)	Combination of low- and high-frequency peaks required for reliable response.
Female Song	*Male Turning Response*
Distinct sound pulses not masked phase differences in leg movements	Short syllables with gaps and pulses < 15 ms
Slow rise-time pulses	Slow rise-time pulses at moderate to low SPL
Broad-spectrum but with reduced energy in high-frequency peak	Low-frequency peak alone elicits response; addition of high-frequency peak inhibits

the high-frequency peak would be relatively unimportant in selective phonotaxis except at very high sound pressure levels, two-stimulus experiments provided a different picture (fig. 4.6B, *top panel*). That is, females offered a simultaneous choice between the single-peak (0.9 kHz) and bimodal (0.9 + 3.0 kHz) sounds reliably preferred the bimodal stimulus at playback levels of 54 dB SPL and higher (Gerhardt 1981b). Thus, the presence of high-frequency energy becomes important in eliciting *selective phonotaxis* at relatively low levels *when it is presented in combination with low-frequency energy.* The fact that females did not prefer the bimodal stimulus at a playback level of 48 dB SPL fits well with the difference in sensitivity of the two inner ear organs, each of which is tuned to one of the spectral peaks (see chapters 5 and 6; box 5.3).

The biological significance of this pattern of preference might be that the low-frequency spectral peak, which on average should propagate better than high-frequency components of the call, might initially attract and guide females to a calling male or chorus at a distance. The detection of high-frequency components would then be expected to influence preference as the female approaches the male more closely. At still closer distances (= higher playback levels) females of *H. cinerea* might also assess the relative amplitudes of the two spectral peaks. As shown in figure 4.6B (*middle panel*), they usually preferred signals in which the two peaks had about the same amplitude at playback levels of about 70 dB SPL and higher when the alternative had an attenuated high-frequency peak. Not surprisingly, however, attenuation of the low-frequency peak reduced the relative attractiveness of a synthetic call over the entire range of playback levels (fig. 4.6B, *lower panel*).

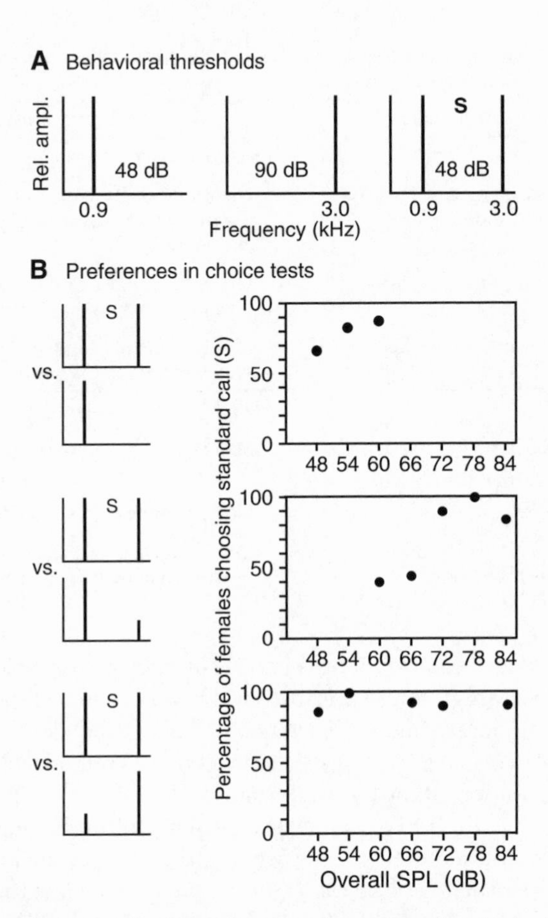

Figure 4.6. Spectral pattern selectivity in female green treefrogs *Hyla cinerea*. Synthetic calls differed in the number or relative amplitude of spectral peaks. The standard synthetic call (S) had components of 0.9 and 3.0 kHz of equal relative amplitude. (A)Behavioral thresholds for phonotaxis: minimum playback level (in decibels SPL) at which at least 50% of the females tested approached the speaker. Notice that the threshold for the standard call was the same as for a stimulus with only the low-frequency peak. This is consistent with the much higher threshold for synthetic calls with a single high-frequency peak. (B) *Top panel*, when the playback level of the two synthetic calls presented in a choice situation was adjusted to 48 dB SPL —the threshold for the standard call—then, as expected from the threshold estimates, a statistically significant proportion of females did not prefer the standard call to the call with the low-frequency peak alone; however, when the playback level was increased to 54 dB or 60 dB SPL, females strongly preferred the standard call, indicating that the high-frequency peak is important for selective phonotaxis when presented in combination with the low-frequency peak. *Middle panel*, females also preferred the standard call when the high-frequency peak of the alternative was attenuated—provided that the playback level was at least 72 dB SPL. *Lower panel*, females preferred the standard call to an alternative in which the low-frequency peak was attenuated over the entire range of playback levels. Data from Gerhardt 1976 and modified from Gerhardt 1981a.

Carrier Frequency and the Orientation of Phonotaxis

Communication typically involves movement toward the source of signals produced by the opposite sex, and such positive phonotaxis is often observed in aggressive interactions between members of the same sex (chapter 9). Studies of field crickets, in particular, have also provided important insights concerning how variation in carrier frequency can mediate qualitatively different phonotactic responses. In general, relatively low frequencies, typical of calling songs, usually elicit positive phonotaxis, whereas relatively high frequencies, typical of sounds produced by bat predators, can result in *negative phonotaxis* (movement away from the sound source) or other forms of evasive behavior (Pollack and Hoy 1989; Libersat and Hoy 1991). The situation is, in fact, more complicated because positive and negative responses to low and high frequencies are influenced by the context of the presentation, sound intensity, and perhaps by whether or not a species has courtship signals with frequencies that fall in the range of bat signals.

Negative phonotaxis in field crickets is most reliably elicited in "flying" preparations that are stimulated by signals with high frequencies that are almost certainly "interpreted" by the cricket as originating from bats (box 4.1; reviews: Pollack and Hoy 1989; Pollack 1998). Even at the lower carrier frequencies typical of calling songs, however, flying crickets may show negative phonotaxis, erratic turning, or no response when these sounds are presented at high playback levels (Doolan and Pollack 1985; Nolen and Hoy 1986a). Joint variation in carrier frequency and temporal pattern also affected orientation during phonotaxis in both walking and flying crickets (see below and chapter 7; Thorson et al. 1982; Pollack et al. 1984; Pollack and El-Feghaly 1993). Thus, the behavioral and neural systems underlying mate identification, predator evasion, and sound localization in crickets, and probably in other insects, are not entirely compartmentalized, but rather interact to determine the sign and selectivity of phonotaxis. Considerable progress has been made in understanding the neural bases of positive and negative phonotaxis (chapter 6).

Selectivity for Temporal Properties of Signals

Static fine-scale temporal properties such as pulse rate show little variation within episodes of signaling and among signalers within a population (chapter 2; reviews: Ewing 1989; Bailey 1991; Gerhardt 1991, 1994a; Pollack 1998). Preferences based on fine-temporal properties are usually stabilizing within populations in contrast to the highly directional preferences based on dynamic gross-temporal properties such as the rate of signaling (Gerhardt 1991; chapters 2 and 10). We concentrate here on preferences based on fine-temporal properties because they have been the focus of nearly all of the neurobiological studies that we examine in chapters 5 and 6. The results of many of the studies we consider in this section illustrate the danger of assuming that closely

Box 4.1. Walking and Flying Crickets

Walking crickets have been studied in arenas, where they can move freely in any direction from the release point (e.g., Stout et al. 1983); on locomotion compensators, where the cricket stays in the same place and its orientation and walking speed on a rotating sphere are recorded (e.g., Thorson et al. 1982); and on Y-mazes. Y-mazes are either structures on which crickets walk until they reach a right-left decision point (e.g., Popov and Shuvalov 1977) or three-dimensional lightweight structures that the cricket manipulates with its legs (virtual walking), even though it is tethered in one place (e.g., Hoy and Paul 1973). The phonotactic orientation of tethered "flying" is determined by monitoring the steering movements of the abdomen (fig. 1; Pollack and Hoy 1989). One advantage of tethered preparations and locomotion compensators is that the sound field is constant throughout an experiment, rather than being dependent on the cricket's position relative to the sound source(s). Pollack and Hoy (1981) discuss this and other possible explanations for the differences in selectivity they found in crickets tested with the same acoustic stimuli in walking and flying assays.

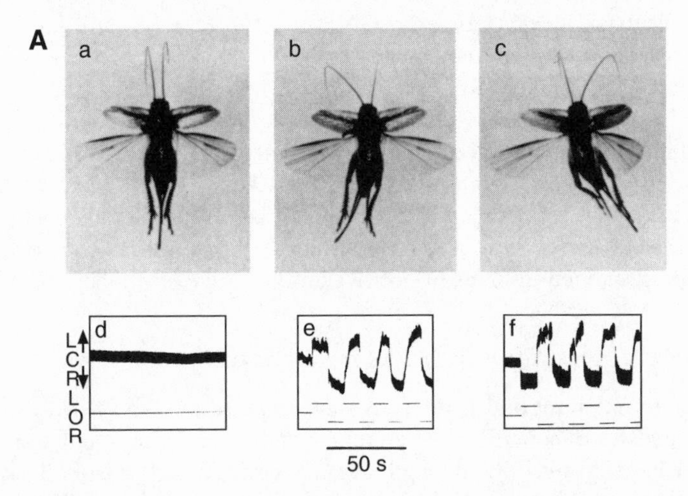

Box 4.1, Figure 1. (A) Photos of tethered "flying" crickets showing abdominal position changes and (*below*) oscillograms of electronic analogs of turning movements as sound is switched on and off from speakers located to the right and left of the preparation. (B) Percentage of positive and negative phonotactic responses in tethered *Teleogryllus* adult females (n = 12) to different frequencies. From Pollack and Hoy 1988, figs. 11.1, 11.2.

(box 4.1 continued)

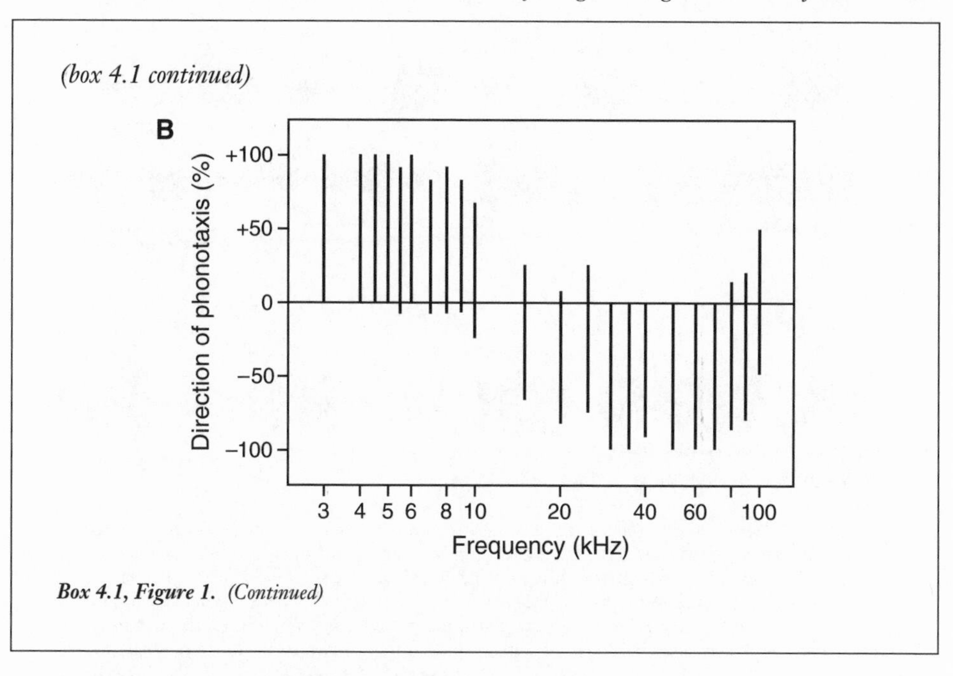

Box 4.1, Figure 1. *(Continued)*

related species, or even the two sexes in the same species, use the same acoustic criteria for signal identification.

Signal Recognition and Preferences Based on Sex-Specific Temporal Properties in Acridid Grasshoppers

The signals of the two sexes of the grasshopper *Chorthippus biguttulus* differ in both spectral and fine-temporal properties (figs. 4.5 and 4.7; Helversen and Helversen 1997). In males, short pulses (usually six), which make up a longer, noisy syllable, are produced by three short up-down movements of each hindleg. Because the movements of the two hindlegs are slightly out of phase, intervals between pulses produced by one hindleg are masked by the pulses produced by the other hindleg. Pauses between each syllable are also partially obscured by the sound produced by the lagging leg (fig. 4.7B, *left*). Indeed, pulses within syllables are clearly evident only in songs produced by males with one leg (fig. 4.7C). Syllables are, in turn, organized into phrases, which vary in duration (2–3 s) and number (usually three) during a song (fig. 4.7A).

Females also produce six-pulse syllables. Even though the movements of the two legs are also slightly out of phase as in males, the pulses in female song are not strongly masked; several pulses and gaps are usually evident within syllables (fig. 4.7B). In comparison with male song, the pulses of female song have noticeably slower rise-times (fig. 4.7C) and song phrases (verses) are shorter and less intense (fig. 4.7A).

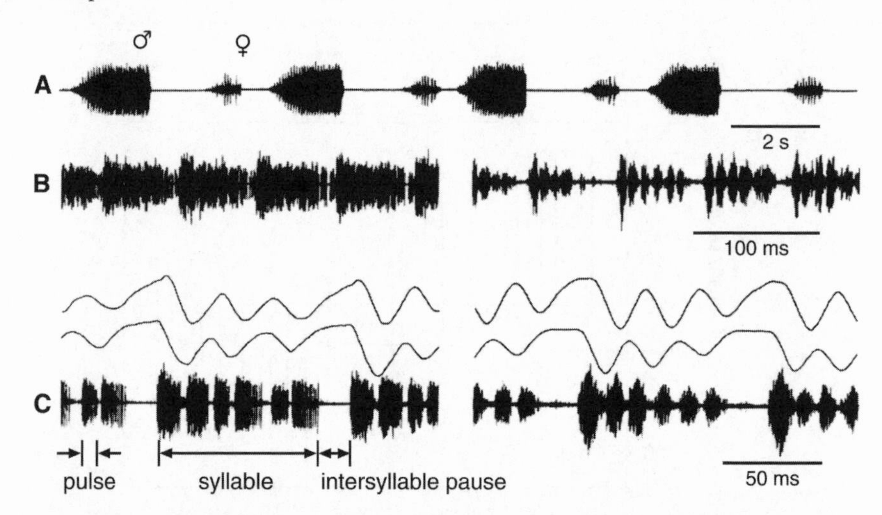

Figure 4.7. Temporal differences in the songs of male and females of the bow-winged grasshopper *Chorthippus biguttulus.* (A) Oscillograms of male songs and the less intense female reciprocation songs. (B) Oscillograms with an expanded time base showing a series of syllables in intact males (*left trace*) and females (*right trace*). Notice that pulses within syllables and, to a lesser extent, pauses between syllables are obscured in the male song, but not in the female song. Below are traces showing (*top*) the slightly out-of-phase stridulatory movements of both legs that result in masking of pulses within syllables in male songs in intact animals. (C) Syllables shown with an expanded time base were produced by a male with one leg so that the pulses of male songs can be seen (*left trace*). Syllables produced by an intact female (*right trace*). Notice that the rise-time of male pulses is faster than that of females. From Helversen and Helversen 1997, fig. 2.

Just as in their selectivity with regard to spectral differences, the two sexes have different criteria for distinguishing between the fine-temporal differences of male and female signals. The main results of playback experiments are summarized in table 4.1. Males showed turning responses to synthetic signals with distinctive pulses with slow rise-times (see below). If the pause between syllables was varied, males responded over a broad range of values extending beyond that found in female songs (fig. 4.8A). Females were much more selective than males. The probability of a female acoustic response was influenced by variation in both the duration of synthetic male syllables and the duration of pauses between them, that is, the intersyllable pause. Females were somewhat more selective with respect to differences in the intersyllable pause than to differences in syllable duration (fig. 4.8B). Considerable between-female variation was found, however, and some females responded well to longer-than-average pauses, including values outside the range of variation in male song (fig. 4.8B,C; Helversen and Helversen 1994, 1997).

We might intuitively expect that the masking of the pulses within syllables would reduce the ability of females to recognize male song. In fact, the opposite is true. Females responded poorly, if at all, to songs, such as those of one-legged males, in which the pulse structure and syllable repetition rates are

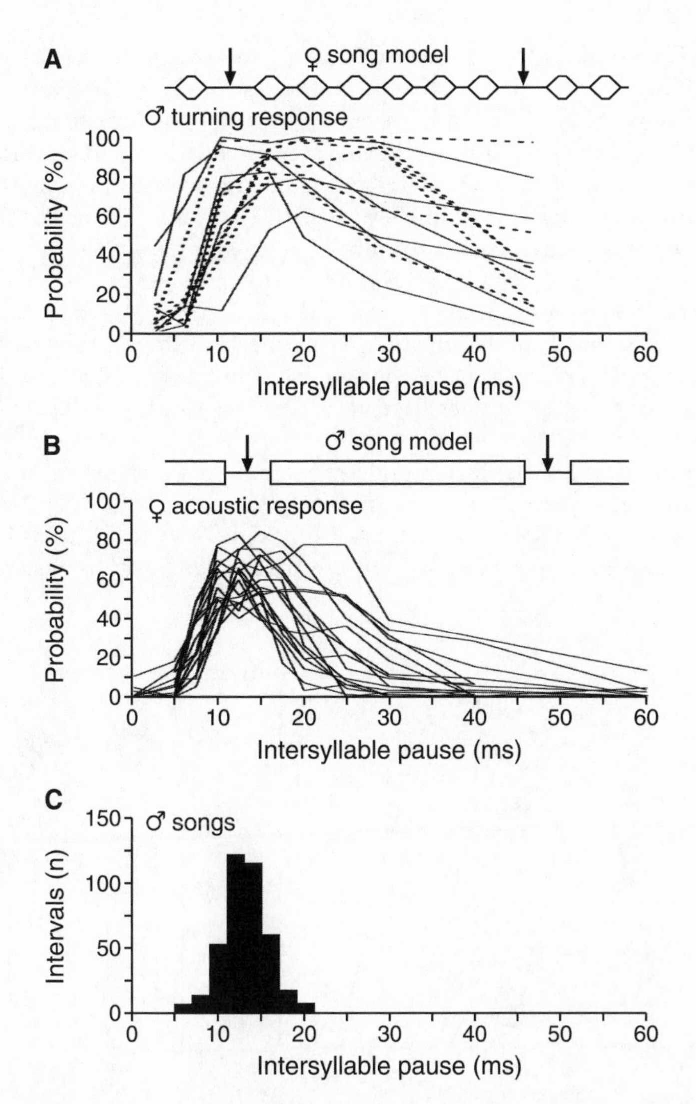

Figure 4.8. Fine-temporal criteria for song recognition in male and females of the grasshopper *Chorthippus biguttulus.* (A) Response functions (probability of a turning response) of individual males as the intersyllable pauses (*arrows* in diagram above) were varied between artificial syllables of synthetic female songs. Syllables with distinct pulses with slow ramps are optimal (see diagram above). (B) Response functions (probability of the production of a reciprocation song) of individual females as the pauses (*arrows* in diagram above) were varied between syllables of synthetic male songs. Syllables without gaps and having a fast onset were optimal for eliciting the response. Modified from Helversen and Helversen 1997, fig. 5. Although the overall range of effective values is much narrower in females than in males, some individual females had broad preference functions that included longer syllable intervals than produced by any male as shown in (C), which shows the distribution of syllable pauses in male song. Modified from Helversen and Helversen 1994, fig. 5.

clearly expressed. Indeed, at normal breeding temperatures, females stopped responding reliably when small gaps as short as 1.5–2 ms were inserted into synthetic syllables. There is, however, an intensity window of about 60 dB to 75 dB, below which females tolerated somewhat larger gaps (3–4 ms), and above which female responses usually declined regardless of the presence or absence of gaps (Helversen and Helversen 1997). Because a failure to respond occurs when gaps are detected within the normal intensity window, the behavior of female *Chorthippus* in this test situation provides an example of correct rejection (see also Helversen and Helversen 1998). Helversen and Helversen (1998) conclude that the underlying mechanism probably operates in the time domain but cannot be explained as a simple band-pass mechanism (see chapter 6). Why females have an upper intensity threshold is an open question.

In contrast to females, males preferred sounds in which pulses and pauses within syllables were distinct, as in female song (fig. 4.8). Males also responded best when pulses had slow rise-times (fig. 4.9A). The probability of response depended, however, on both rise-time and intensity, thus suggesting

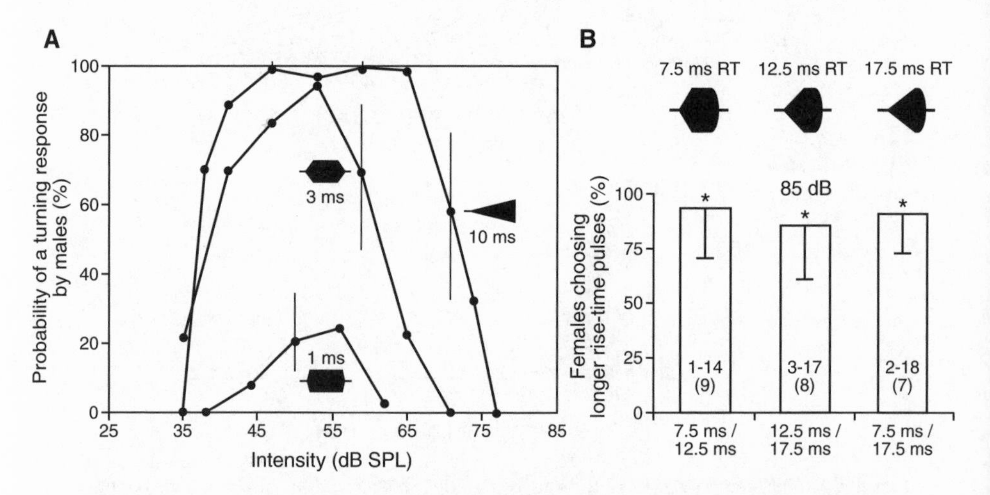

Figure 4.9. Rise-time discrimination in grasshoppers and treefrogs. (A) Probability of turning responses (*vertical lines:* ranges of variation) in males of *Chorthippus biguttulus* when presented with syllables containing pulses with different rise-fall characteristics (see the labeled diagrams of pulses). Other experiments (not shown) indicate that the different fall times did not influence the accuracy of the turning response and that the abnormally long pulse (needed to test the longer [10 ms] rise-time) was just as effective as pulses typical of female songs. From Helversen 1993, fig. 5a; see the text for further explanation. (B) *Top,* oscillograms of synthetic pulses used in playback tests with female gray treefrogs *Hyla versicolor.* Rise-time is given in ms. *Bottom,* percentage of females choosing the alternative with the longer rise-time, which is typical of conspecific calls. Error bars show lower 95% confidence limits, and the numbers of females choosing each alternative are shown within each bar. The number of females that did not respond to either stimulus is given in parentheses. *Statistically significant preference ($p < 0.05$) in two-tailed binomial test. Modified from Gerhardt and Schul 1999, fig. 1.

that the selectivity might be based on the rate of change in amplitude, or *steepness* (Helversen 1993; Helversen and Helversen 1997). Steepness, unlike *rise-time* (defined here as the time from the beginning of the pulse to the time of maximum amplitude), increases as a function of intensity. Helversen (1993) showed that the upper intensity limit for the turning response to synthetic pulses with the 3 ms rise-time typical of female song was about 60 dB SPL. By presenting pulses with abnormally long rise-times of 10 ms, the upper limit was extended to 65–70 dB SPL. In contrast, the selectivity of females of the gray treefrog *Hyla versicolor*, which also prefer signals with slow rise-time pulses, is not based on steepness and is robust over a 20 dB range of playback levels (Gerhardt and Schul 1999; see below).

Selectivity Based on Pulse Rate in Field Crickets

Selective phonotaxis in female crickets is based on relatively few temporal properties. Indeed, despite the wide variety of experimental designs and testing procedures, a near-universal conclusion is that pulse rate (or pulse or syllable period) is the most important temporal property used in signal selection by female crickets (e.g., tethered flight studies: Doolan and Pollack 1985; compensated walking: Thorson et al. 1982; Doherty 1985b; Doherty and Callos 1991; Doherty and Storz 1992; Y-maze walking: Popov and Shuvalov 1977; arena tests: Stout and McGhee 1988; Shaw and Herlihy 2000; see general commentary by Thorson et al. 1982). As might be expected, the most attractive signals usually have a pulse rate close to the mean values typical of conspecific calling songs (figs. 4.10 and 4.12). When tested, pulse-rate preferences were usually at least moderately intensity-independent. In two-stimulus playbacks, for example, Doherty (1985a) found that females of *G. bimaculatus* preferred synthetic songs with pulse rates near the species-typical mean pulse rate (22 pulses/s) over alternatives with lower and higher pulse rates, even when the sound pressure level of these alternatives was 5–10 dB greater (fig. 4.10A). Comparable selectivity occurred in a North American cricket, *Gryllus firmus*, in which a preference for a synthetic song with the mean conspecific pulse rate survived a 12 dB reduction in sound pressure level relative to an alternative with a higher (40%) pulse rate (fig. 4.10B; Doherty and Storz 1992).

With respect to preferences observed when the intensity of alternatives is equalized, the preference function in *G. firmus* appears to be asymmetrical (fig. 4.10B). Preference functions appear somewhat more symmetrical in *G. bimaculatus*, *G. rubens* (fig. 4.10A,C) and *G. campestris* (Thorson et al. 1982). However, the apparent symmetry of a preference function depends on how the periodicity is expressed and on the scaling of the X-axis. For example, a plot of the pulse-rate preferences of *G. bimaculatus* appears highly asymmetrical if scaled linearly, but if the preferences are plotted in terms of pulse period on a linear scale, then the pattern looks highly symmetrical. The first method is typically used by North American researchers, and the second, by European workers. Researchers studying anurans measure and report fine-temporal

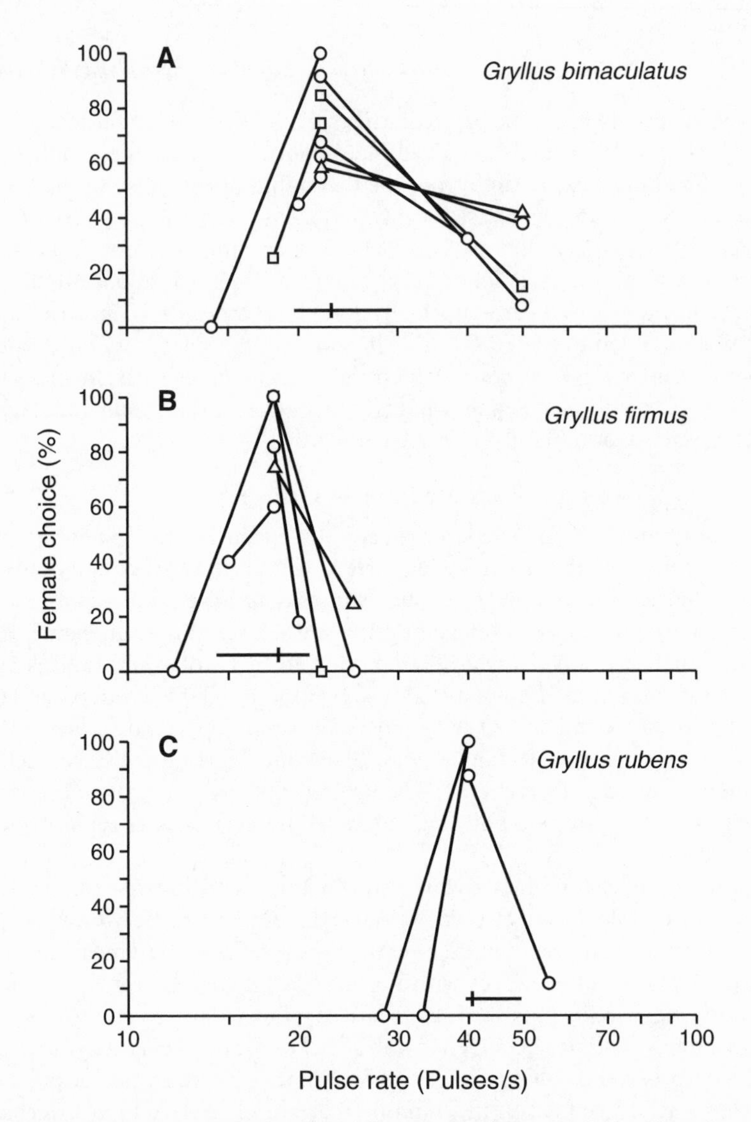

Figure 4.10. Selective phonotaxis based on differences in pulse rate in field crickets. Summary of experiments in which females of *Gryllus bimaculatus* (A) and *G. firmus* (B), which are chirping species, and females of *G. rubens* (C), which is a trilling species, were given choices between a standard synthetic calling song (= pulse rate typical of conspecific males at the test temperature) and alternatives with different pulse rates. Lines connect points showing the percentages of females that chose the standard pulse rate and an alternative with a different pulse rate. Open circles = the sound pressure levels of the alternative stimuli were equal; open squares = the sound pressure level of the standard stimulus was 5 dB (*G. bimaculatus*) or 6 dB (*G. firmus*) less than that of the alternative stimulus; open triangles = the sound pressure level of the standard stimulus was 15 dB (*G. bimaculatus*) or 12 dB (*G. firmus*) less than that of the alternative stimulus. The trills of the synthetic songs of *G. rubens* were equalized in duration, so that the number of pulses in the stimulus of 55 pulses/s was greater than in the standard stimulus. In all other paired stimuli, the number of pulses per trill was equalized. The vertical and horizontal lines at the bottom of each panel show the range of variation and mean, respectively, of pulse rate in male cricket calling songs recorded over about the same range of temperatures at which females were tested (20–23°C). Data replotted from Doherty 1985a, fig. 4; Doherty and Storz 1992, fig. 5; Doherty and Callos 1991, fig. 6.

patterns and preferences based on them in terms of pulse rate and use linear scaling. Regardless of the metric (pulse rate or period), the use of a logarithmic scale has the advantage of more closely approximating how sensory systems probably deal with variation in fine-scale temporal properties than does a linear scale.

In *Gryllus campestris* and *G. bimaculatus*, pulse-rate preferences were little, if at all, influenced by simultaneous variation in pulse shapes and pulse duration, which influences the *pulse-duty cycle* (= the ratio of pulse duration to pulse period) (Thorson et al. 1982; Doherty 1985b). Hence, selective phonotaxis in these species seems to be based on differences in pulse rate alone. Females of both species also tracked continuous trains of pulses with the correct rate, suggesting that the chirp structure of conspecific calls is not a key feature for recognition. However, other experiments challenge the conclusion, which could reasonably be made from these results, that the correct pulse rate is the *only* temporal feature required by females (Thorson et al. 1982). First, in no-choice tests, some females of *G. bimaculatus* responded phonotactically to continuous tones or to tone bursts (of equal duration to chirps) lacking the species-typical pulse structure (Popov et al. 1974; Tschuch 1976; Doherty 1985a,c; Wendler 1990). Second, as shown below, simultaneous variation of other temporal properties, such as chirp duration and chirp rate, can influence the range of phonotactically effective values of pulse rate.

In other species of field crickets with more complex songs, pulse-rate selectivity is less impressive than in *Gryllus*, and pulse duration can also sometimes affect phonotactic selectivity (Doolan and Pollack 1985). As shown in figure 4.11, males of *Teleogryllus commodus* and *T. oceanicus* produce calling songs with two kinds of pulse trains ("chirps" and "trills") that differ in duration and pulse rate. Are both of these song elements important, and, if so, is pulse-rate selectivity for each element the same?

In *Teleogryllus commodus*, females required the two different pulse rates typical of trills and chirps, respectively (Hennig and Weber 1997; see Zaretsky 1972 for comparable results in another species). Females nevertheless responded reliably to a much wider range of values than occur in conspecific song (fig. 4.12A). Indeed, Hennig (2001) reports that females of this species were actually more selective for pulse duration (best values: 18–35 ms) than for pulse rate. These results are consistent with effectiveness of "shuffled" song patterns (Pollack and Hoy 1979). In such signals, pulse intervals were randomly sequenced while statistically preserving their relative frequency of occurrence in conspecific songs; pulse duration was held constant.

In *T. oceanicus*, by contrast, walking and flying females preferred a narrow range of pulse rates typical of chirps over a wide range of pulse duration (≥ 20 ms) (Pollack 1982; Hennig 2001). Females also preferred songs composed of just chirps to songs made up only of trills (Pollack 1982; Pollack and Hoy 1981). In both flying and walking assays, females showed no selectivity for pulse rates typical of conspecific trills and also responded well over a large

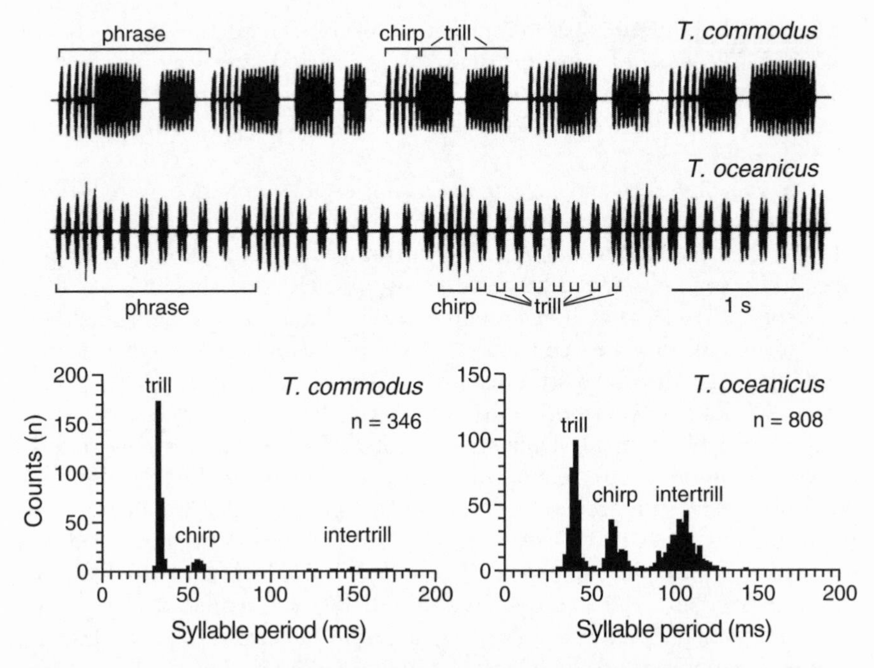

Figure 4.11. Oscillograms of typical calling songs of the Asian-Pacific crickets *Teleogryllus commodus* and *T. oceanicus*. The songs consist of repeating phrases with two kinds of pulse trains, termed *chirps* and *trills*. In *T. commodus*, chirps and trills are about the same duration but have different pulse rates (periods); in *T. oceanicus*, chirps are similar to those of *T. commodus* in duration and have slightly slower pulse rates, but trills consist of a series of six or more double pulses. *Bottom*, pulse-period histograms showing the distribution of all intervals between pulses in the calling song; recall that the pulse period is the reciprocal of the pulse rate.

range of intertrill intervals (fig. 4.12B; Pollack and Hoy 1981; Hennig and Weber 1997). Females did, however, prefer conspecific trills to the songs of *T. commodus* (Pollack and Hoy 1981). By contrast, flying males of *T. oceanicus* were less selective than females for the pulse rates typical chirps and even preferred trills to chirps (Pollack 1982).

One study of *Teleogryllus oceanicus* shows an interesting effect of absolute playback levels on temporal selectivity (Doolan and Pollack 1985). In flight assays, females were highly responsive to synthetic signals with the pulse duration and pulse rate typical of the chirps of conspecific males over a broad range of playback levels. However, variation in intensity strongly affected female responsiveness to signals having values of these two properties falling outside the range of conspecific variation (Doolan and Pollack 1985). More specifically, as intensity was increased, females stopped responding or showed steering movements interpretable as negative phonotaxis. Because increasing the intensity of marginally attractive signals might be expected to increase their attractiveness, these results probably represent another example of correct rejection. Perhaps at low intensities, the encoding of pulse rate and

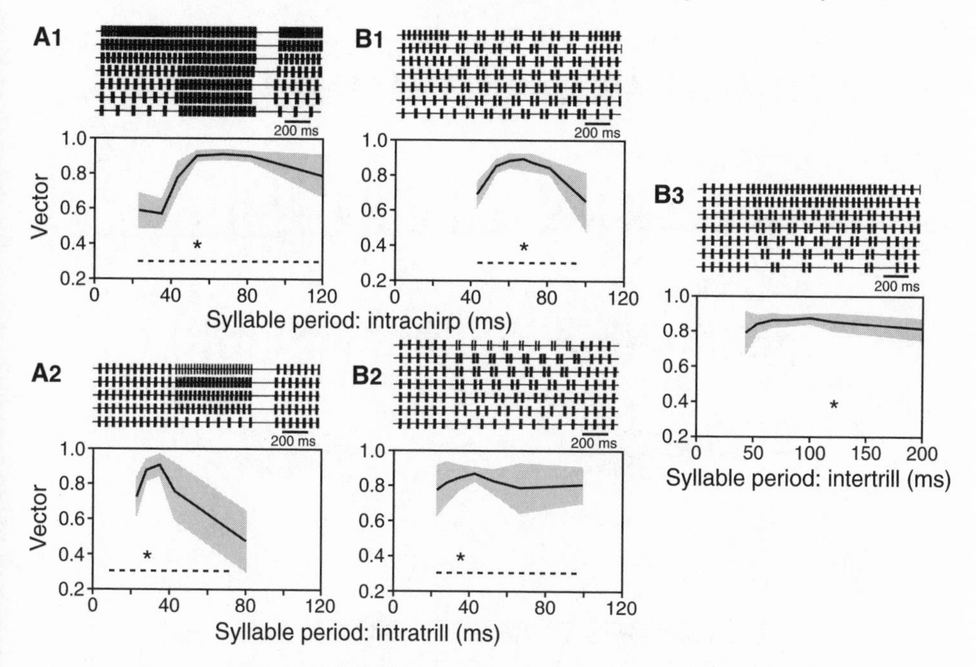

Figure 4.12. Phonotactic responses of female *Teleogryllus* that tracked synthetic songs on a locomotion compensator. The syllable (pulse) period within chirps or trills was systematically varied. (A) *T. commodus;* (B) *T. oceanicus,* which was also tested with songs in which the intertrill interval was varied. Solid lines indicate mean responses; gray areas indicate the range of individual variation. Dashed horizontal line at vector value of 0.3 indicates random orientation typically shown when the stimulus was unattractive; values of 0.6 or higher are considered reliable tracking. Diagrams of the test patterns are shown above each plot; * = approximate mean value in conspecific songs. From Hennig and Weber 1997, figs. 2, 3A,B, 5A,B.

duration is too poor for reliable discrimination of conspecific and heterospecific values, but improves as intensity and hence signal-to-noise ratios increase. Because both low intensity and degraded temporal structure would be expected in nature at some distance from the male; females might respond somewhat unselectively at distance and only begin to discriminate against heterospecific song as they move closer to its source (Pollack 1998). Bladder cicadas (*Cystosoma saundersii*) provide a parallel example of distance-dependent changes in temporal selectivity (see below; Doolan and Young 1989).

Selective Phonotaxis in European Katydids

The temporal properties used in song recognition in the katydids *Tettigonia cantans* and *T. viridissima* have been studied extensively with both behavioral and neurophysiological approaches (Jatho et al. 1994; Jatho 1995; Schul 1997, 1998, 1999; Schul et al. 1998). Some data are also available for a third species, *T. caudata.* Striking species differences exist in the temporal criteria used by females, and, consistent with a recurring theme in this chapter, some

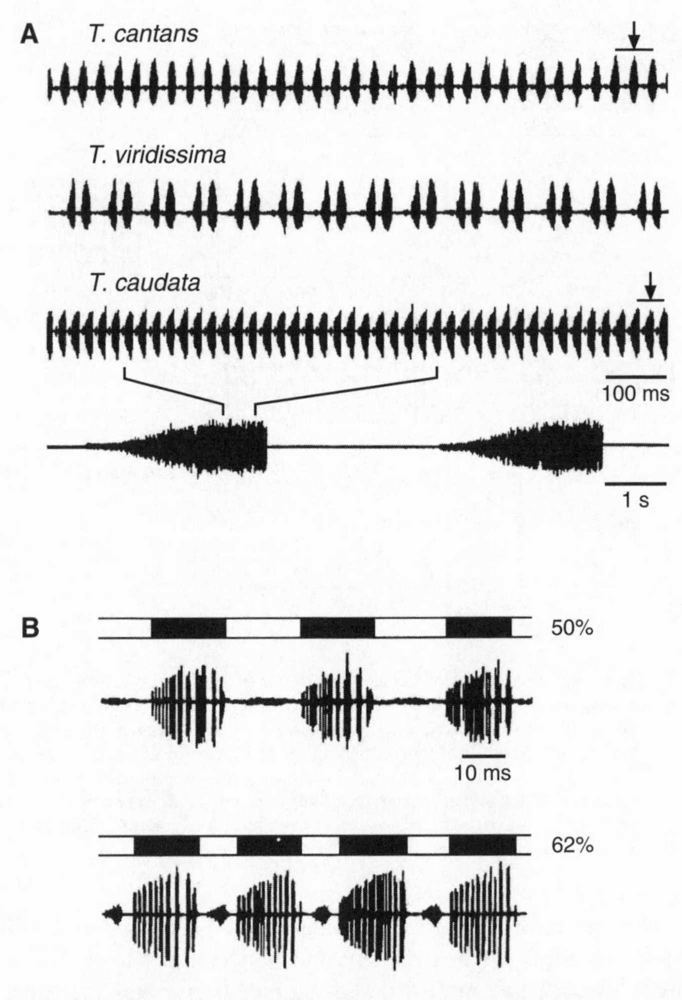

Figure 4.13. Calling songs in three European katydids. (A) Temporal patterns of calling songs. Oscillograms show parts of the trills, which normally last for many minutes, that make up the songs of *Tettigonia viridissima* and *T. cantans;* the pulses shown for *T. caudata* are from a verse, which lasts about 2 s and has a period of about 4 s. (B) Oscillograms of representative pulses, indicated by arrows in (A) of *T. cantans* (*top trace*) and *T. caudata* (*bottom trace*) showing the difference in pulse-duty cycle (*black-and-white bars above the pulses*). Modified from Schul 1998.

stereotyped (qualitative) properties of conspecific songs are irrelevant for song recognition.

As shown in figure 4.13A, males of *T. caudata* organize pulses into discrete verses, but the songs of the other two species consist of long trills. Songs of *T. cantans* and *T. caudata* have pulse periods corresponding to pulse rates of about 28 and 40 pulses/s, respectively; pulse-duty cycles are about 62% and

50%, respectively (fig. 4.13B). The trills of *T. viridissima* have a double-pulse structure, and hence there are two pulse periods, corresponding to rates of about 11/s (repetition of double pulses) and 30/s (reciprocal of period of the two pulses of each double pulse).

Females of *T. cantans* responded reliably to synthetic songs with the temporal pattern of *T. viridissima* songs in no-choice situations (fig. 4.14; Schul et al. 1998). In simultaneous playbacks of conspecific and heterospecific signals with the same spectral properties, the mean directional vector for females of *T. cantans* was deflected by about 18° away from the location of a source of conspecific signals and *toward* that of a synthetic model of *T. viridissima* songs (fig. 4.14A). This seemingly positive influence of the heterospecific signal might be expected from its attractiveness in no-choice playbacks. By contrast, females of *T. viridissima* and *T. caudata* were never attracted by playbacks of the synthetic songs modeled after those of the other species (Schul 1998). Even though females of *T. viridissima* did not respond to the songs of *T. cantans* in no-choice tests, the mean directional vector was deflected from the location of a source of conspecific signals in a direction *away from* the source of *T. cantans* songs (fig. 4.14C). As discussed in chapter 7, however, deflections toward or away from conspecific sound sources are probably best interpreted as an interaction between mechanisms underlying sound-pattern recognition and localization rather than as some form of positive or negative influence of these heterospecific stimuli on the attractiveness of conspecific signals.

The bases for such selective phonotaxis in *T. cantans* were studied in both arenas (Jatho 1995) and with a locomotion compensator (Plewka 1993). Females preferred the conspecific pulse rate to lower and higher values regardless of pulse duration (Plewka 1993), thus indicating that pulse rate alone is the main, if not sole, temporal criterion underlying preferences. More recently, behavioral testing on a locomotion compensator showed that there was a secondary peak of attractiveness for pulse rates that were about one-half of the value of the mean conspecific value (Schul and Bush 2000). This discovery has important implications for the underlying neural mechanisms and suggests that this species might use some kind of resonance filter or recurrent-inhibition mechanism rather than a band-pass mechanism based on a combination of low- and high-pass filters (see chapter 6).

Although we might anticipate that females of *T. viridissima* would prefer signals with a double-pulse structure to simple trills, and that females of *T. caudata* would prefer signals in which pulses with a rate of about 40 Hz are organized into verses with ramped beginnings, these expectations are not realized (Schul 1998). As shown in figure 4.15A, females of *T. viridissima* were attracted as effectively by a series of single long pulses (equal in duration to the two paired pulses of natural song plus the interval between them) as they were to songs with the double-pulse structure (control vs. test 6). In *T. caudata* (fig. 14.15B), females did not require the ramped onset of verses (control vs. test 3)

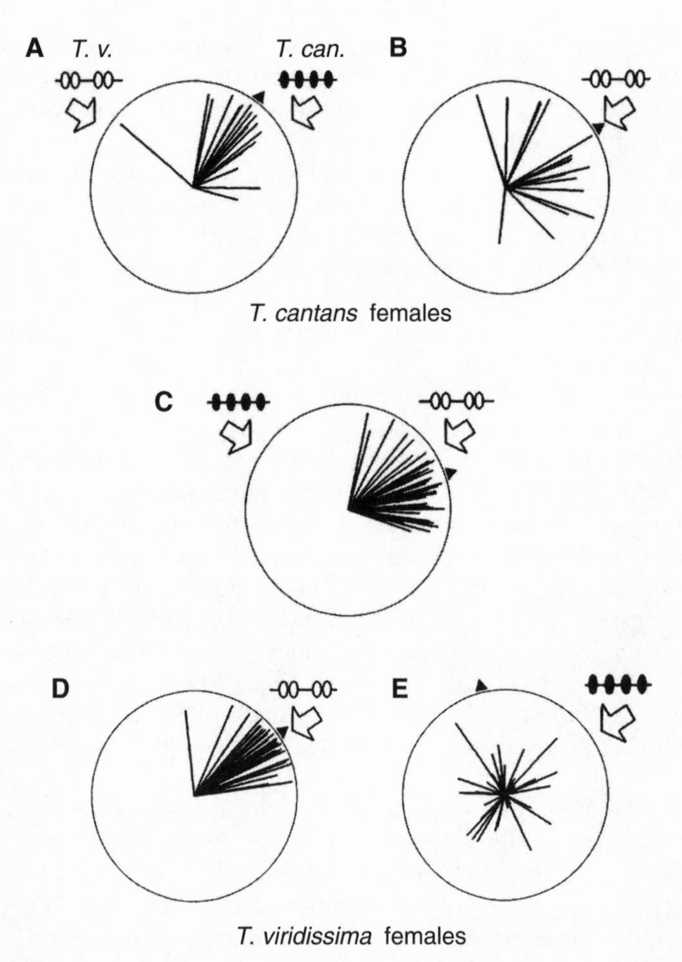

Figure 4.14. Polar diagrams showing the phonotactic responses of European katydids walking on a locomotion compensator. Speaker locations are shown by hollow arrows, and mean directional vectors, by filled triangles. (A) Two-speaker (choice) tests of *Tettigonia cantans* tested with synthetic calls modeled after conspecific songs (T. can.) and those of *T. viridissima* (T. v.). Notice that the mean vector was deflected from the location of the speaker playing back the conspecific song toward the location of the speaker playing back the song of *T. viridissima.* (B) No-choice tests of *T. cantans* with synthetic *T. viridissima* songs. The response magnitudes (*length of lines*) are expressed relative to a control stimulus with the pattern typical of conspecific songs; the heterospecific song was usually effective. (C) Two-speaker (choice) tests of females of *T. viridissima.* Notice that the mean vector was deflected from the direction of the speaker playing back the conspecific song away from the location of the speaker playing back the synthetic *T. cantans* song. (D) and (E) No-choice tests of females of *T. viridissima* with synthetic *T. viridissima* songs (D) or *T. cantans* songs (E). In (E) the response magnitudes are expressed relative to a control stimulus (*T. viridissima* synthetic song); the heterospecific song was usually ineffective. From Schul et al. 1998, figs. 3, 4.

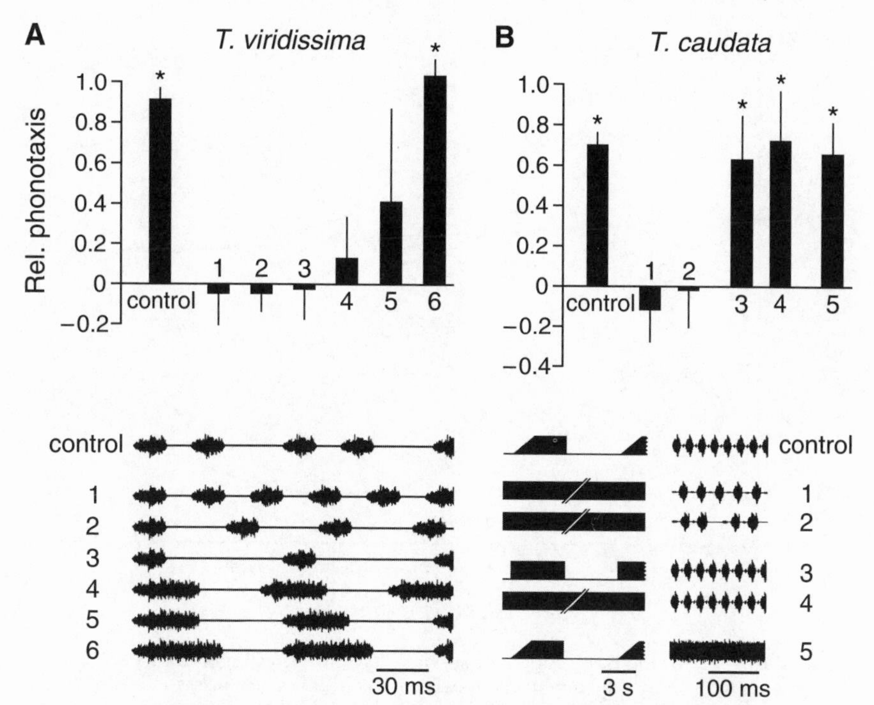

Figure 4.15. Summary of phonotaxis experiments with females of (A) *Tettigonia viridissima* and (B) *T. caudata. Top,* histograms show the relative effectiveness of the temporal patterns shown in the oscillograms below. In (A) notice that a single long pulse equal in duration to the double pulses and the interval between them was just as effective as the natural pattern (control). Changing either pulse duration or the interval between single pulses affected relative effectiveness (tests 1–5). (B) *Bottom,* notice that continuous trills with pulses having the temporal patterns of *T. viridissima* or *T. cantans* were completely ineffective (tests 1 and 2), but that continuous signals, lacking the conspecific verse structure (test 4) but having the conspecific pulse rate and duty cycle, were just as effective as the natural song (control). The effectiveness of noise bursts (test 5) shows that signals need have only some minimum pulse-duty cycle, given that this stimulus lacks pulses altogether and hence has a duty cycle of 100%. From Schul 1998, figs. 2, 3.

but responded just as well to continuous trills (control vs. test 4). Moreover, females of *T. caudata* also responded equally well to noise bursts (control vs. test 5), which lacked pulses altogether!

We might also expect that females would be most attracted to the values of key properties that fall within the ranges of conspecific values. In *T. viridissima,* however, intervals between pulses having a duration of about 45 ms or more could be varied over a wide range (10–60 ms), and within this range of pulse intervals, pulse duration could be increased to well over 100 ms (fig. 4.16). In *T. caudata,* signals with a duty cycle of 66% (typical of conspecific songs) were effective, but signals with lower values (e.g., the 50% duty cycle typical of the signals of *T. cantans*) were not (fig. 4.15B; control and tests 3 and 4 vs. test 1). However, since noise bursts, which effectively have a duty

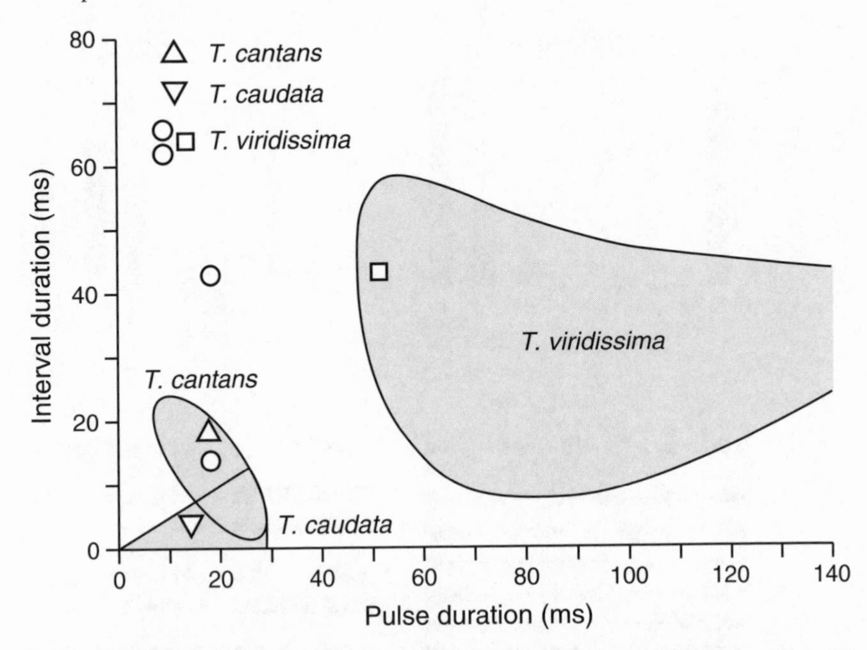

Figure 4.16. "Recognition" spaces of three species of closely related katydids (genus *Tettigonia*). The gray areas encompass the combinations of silent intervals between pulses and pulse durations that reliably elicited phonotaxis in females. The diagonal field for *T. cantans* is the expected pattern for a species using some form of pulse-rate or pulse-period criterion. The triangle indicates mean values of the interpulse interval and pulse duration for this species. The broad field for *T. viridissima* indicates that females are attracted by many combinations of pulse duration above about 55 ms and silent intervals above about 10 ms. The square indicates the mean duration of the double pulses (including the silent interval between them) and the silent interval between double pulses in a typical conspecific song. The circles indicate the mean duration of each pulse (of a double-pulse pair) and the two silent intervals (between each pulse in a pair and between each pair of double pulses). The field for *T. caudata*, which has not yet been fully tested, is consistent with the hypothesis that females are only attracted to songs in which the pulse-duty cycle (ratio of pulse duration to pulse period) exceeds about 66%. This criterion would allow females to reject songs of *T. cantans*, whose pulse rate would overlap that of *T. caudata* at high temperatures but whose duty cycle would be about 50% or less. The inverted triangle shows the mean values of pulsed duration and silent intervals for *T. caudata*. From Schul 1998.

cycle of 100%, were just as effective as verses with a pulsed structure (control and tests 3 and 4 vs. test 5), the simplest description of the main temporal criterion used by *T. caudata* is that a signal must have a minimum duty cycle of at least 66%.

In summary, comparative research with these three species of katydids shows striking differences in phonotactic selectivity for conspecific temporal patterns. First, females of *T. cantans* are less selective than females of the other two species, which do not respond to heterospecific signals even in no-choice situations. Second, the criteria underlying preferences for temporal patterns found in the conspecific song differ markedly (table 4.2), and only in

Table 4.2. Preferences Based on Fine-Scale Temporal Properties in Closely Related Species of Insects and Anurans

Species	Temporal Properties	Preferences
Asian-Pacific Field Crickets		
Teleogryllus oceanicus (To)	Calling song contains "chirps" (relatively lower pulse rate) and "trills" (relatively high pulse rate) of comparable duration	Females prefer chirps to trills but treat shuffled song as equivalent to models of normal song if shuffled song contains sufficient proportion of pulse intervals typical of those in To chirps. Pulse rate, pulse duration, and pulse-duty cycle all influence preferences, which are intensity-dependent. Males prefer trills to chirps.
T. commodus (Tcom)	Calling song contains chirps that are about twice the duration of the double-pulsed trills	Most effective songs contain both chirps and trills with pulse-rate typical of those in Tcom songs; pulse rates outside To values are also effective provided that pulse duration falls within the conspecific range.
European Katydids		
Tettigonia cantans (Tcan)	Simple, continuous trill (pulse rate 28 Hz)	Strong preference for Tcan pulse rate alone but responds to Tv songs in single-stimulus tests; its orientation is slightly away from Tcan speaker toward Tv speaker in simultaneous playbacks (locomotion compensator).
T. viridissima (Tv)	Double-pulse continuous trills (pulse rate of double pulses 17 pulses/s)	Preference mainly for pulse duration equal to that of the double pulses plus the short silent interval between them; the longer interval between double pulses also influences preferences. No response to Tcan songs in single-stimulus tests, but female orientation is slightly away from both Tv speaker and from location of Tcan speaker in simultaneous playbacks (locomotion compensator).
T. caudata (Tcd)	Pulses organized into verses (pulse rate within verses 40 pulses/s; pulse-duty cycle 66%)	Females do not require verse structure; prefer songs with minimum pulse duty > 60%, including 100% (= no pulses but rather a continuous noise burst).
North American Treefrogs		
Hyla chrysoscelis (Hch)*	Fast (inverse exponential) rise-time pulses Pulse rate 35–50 pulses/s at 20°C, depending on geographic location Pulse duration changes with pulse rate so that pulse-duty cycle stays nearly constant.	No preference for Hch pulse shape to Hv pulse shape; strong discrimination against pulse rates lower than population mean; weaker discrimination against higher pulse rates; pulse-rate preferences independent of pulse duration (duty cycle). Pulse-rate preferences temperature-dependent.
H. versicolor (Hv)**	Slow (nearly linear) rise-time pulses Pulse rate 20–25 Hz at 20°C, depending on location Pulse duration changes with pulse rate so that pulse-duty cycle stays nearly constant	Strong preference for Hv pulse shape to Hch pulse shape, based on rise-time; strong discrimination against pulse rates higher than population mean; weak discrimination against lower pulse rates; pulse-rate preferences dependent on pulse duration (duty cycle). Pulse-rate preferences temperature-dependent.

*diploid species; **tetraploid species, derived from *H. chrysoscelis* (Ptacek et al. 1994)

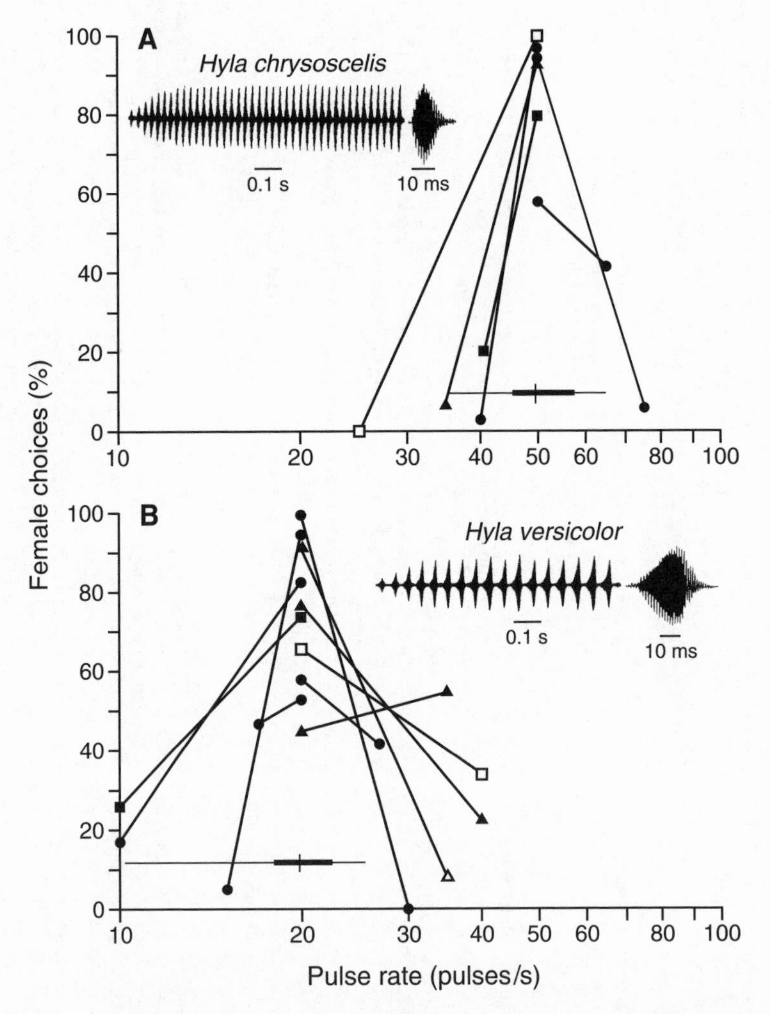

Figure 4.17. Diagrams showing the results of female-choice experiments with females of the gray treefrogs *Hyla chrysoscelis* (A) and *H. versicolor* (B). The lines connect points indicating the percentages of females choosing the standard call (50 pulses/s for *H. chrysoscelis* and 20 pulses/s for *H. versicolor*) and alternatives with different pulse rates. For example, the first two points (*open squares connected by a line*) in (A) show that 100% of the females of *H. chrysoscelis* chose the standard call of 50 pulses/s, and none chose the alternative of 25 pulses/s. Tests in which the two alternatives had the same SPL (85 dB) are plotted with circles; tests in which the SPL of the standard stimulus was reduced relative to that of the alternative are shown by solid squares (−6 dB), solid triangles (−12 dB), and open squares (−18 dB). The horizontal lines just above the X axes show the range of pulse rate in calls of males from the same populations at 20°C (*thick lines*) and at all temperatures (*thin lines*); the vertical lines indicate mean values. In (B), the open triangle indicates that the alternative with the higher pulse rate (35 pulses/s) had pulses with the pulse shape typical of *H. chrysoscelis*. Notice that females preferred the standard call (20 pulses/s, conspecific pulse shape) even though its SPL was 12 dB less than that of the alternative; in a test of these same two pulse rates, females did not prefer the standard call when both alternatives had pulse shapes typical of *H. versicolor* calls.

T. cantans is pulse rate alone the key feature underlying phonotactic selectivity in choice experiments.

Selective Phonotaxis Based on Fine-Temporal Patterns in Gray Treefrogs

The behavioral significance of fine-temporal properties has been extensively studied in cryptic species of gray treefrogs (table 4.2). *Hyla chrysoscelis* is a diploid species, and *H. versicolor* is the current designation for at least three tetraploid lineages (species) that have arisen independently from the diploid (Johnson 1959; Wasserman 1970; Ptacek et al. 1994). Males of both species produce trains of pulses (calls) that last, on average, just under one second (*insets*, fig. 4.17); the calls differ in pulse rate, pulse duration, and pulse shape. Pulses in the calls of *H. chrysoscelis* are shorter, have faster rise-times, and are repeated at higher rates than in *H. versicolor*; pulse-duty cycle is about 50% in both species (Gerhardt and Doherty 1988). Some females of both species show weak phonotactic responses to playbacks of the calls of the other species in no-choice situations (Gerhardt 1978c, 1994b; Gerhardt and Doherty 1988).

In figure 4.17 we summarize the results of experiments in which female gray treefrogs from Missouri, U.S.A., were offered choices between standard calls, with pulse rates and shapes close to the mean for the calls of males from the same populations, and alternative signals with different pulse rates (data from Gerhardt and Doherty 1988; Gerhardt, unpubl. data). In these experiments, which were conducted at 20°C, pulse-duty cycle was held constant at 50% by simultaneously varying pulse duration and pulse rate. Calls with conspecific values (adjusted for the test temperature) were strongly preferred to calls with heterospecific values.

Females of *H. chrysoscelis* discriminated strongly against synthetic calls with pulse rates that were just 20% lower in pulse rate (40 pulses/s) than those of the average conspecific male (50 pulses/s) at the same temperature (fig. 4.17, *upper panel*). When the alternative had a pulse rate 30% lower than the standard, more than 90% of the females still chose the standard even when its sound pressure level was reduced by 12 dB. Females did not, however, strongly prefer the standard call (50 pulses/s) to alternatives with higher pulse rates: females discriminated against an alternative of 75 pulses/s, but this preference was reduced by lowering the sound pressure level of the standard call by 6 dB. Thus, females of *H. chrysoscelis* discriminate much more strongly against alternatives with *lower-than-average pulse rates* than they do against alternatives with *higher-than-average rates*. This apparent asymmetry (with respect to the distribution of pulse rate) in preference strength could, in part, be attributable to the fact that a preference function generated by single-stimulus trials is symmetrical but peaks at a somewhat higher value than the mean in the population (Bush et al. 2001).

Females of *H. versicolor* discriminated against alternatives with pulse rates that were 25–50% lower (10–15 pulse/s) than the standard call (population mean) of about 20 pulses/s; the preference vis-à-vis the call of 10 pulses/s persisted when the sound pressure level of the standard call was reduced by 6 dB (fig. 4.17, *lower panel*). Most females also chose the standard call (at −6 dB) when an alternative had a pulse rate that was 100% higher (40 pulses/s), and about 60% of the animals continued to choose the standard call after a further 12 dB reduction of its intensity (fig. 4.17, *lower panel*). These results predict that females of *H. versicolor* might sometimes choose the calls of *H. chrysoscelis* if these signals are more intense than conspecific calls (see the results of the tests of the standard call against alternatives of 35 pulses/s and 40 pulses/s).

These conclusions about female selectivity in *H. versicolor* have to be qualified, however, because other fine-temporal properties that influence female preferences were held constant in the experiments just discussed. If, for example, the pulse shape of alternatives with higher pulse rates—close to or within the range of variation of *H. chrysoscelis*—were modeled after those of *H. chrysoscelis*, then preferences for the standard call with pulses having the shape typical of conspecific calls were robust in the face of decreases in its sound pressure level of 12–18 dB. Thus, differences in pulse rate and shape have a synergistic effect on female selectivity that virtually insures that mating mistakes by females in sympatric areas will be rare (Diekamp and Gerhardt 1995). Additional experiments indicate that females of *H. versicolor* probably base their preference for the conspecific pulse shape mainly on rise-time; females can discriminate between signals in which rise-time differs by as little as 5 ms (fig. 4.9B; Gerhardt and Schul 1999). These preferences were maintained at different playback levels, and hence females are apparently not using steepness as a criterion as in the grasshopper *Chorthippus biguttulus*. By contrast, females of *H. chrysoscelis* do not discriminate between synthetic calls that differ in pulse shape (Gerhardt 1994b).

Although the differences in pulse rate, pulse shape, or both found in natural calls are adequate for species discrimination by female gray treefrogs, the experiments so far discussed did not investigate the possible effect of varying pulse duration. Rather, pulse duration was changed along with pulse rate so that the pulse-duty cycle remained constant as in the calls of males of both species when, for example, males call at significantly different temperatures (Gayou 1984; Gerhardt, unpubl. data). When pulse duration (and hence pulse-duty cycle) was systematically varied, then, as in the katydid *T. cantans*, the pulse rate preferences of *H. chrysoscelis* were not influenced. These two species thus appear to use differences in pulse rate alone for signal identification. By contrast, the usual preference of females of *H. versicolor* for a standard call of 20 pulses/s for an alternative of 40 pulses/s could be abolished or even reversed by varying pulse duration so that the pulse-duty cycle was greater in the alternative call (Gerhardt and Schul, in prep.). This result suggests that fe-

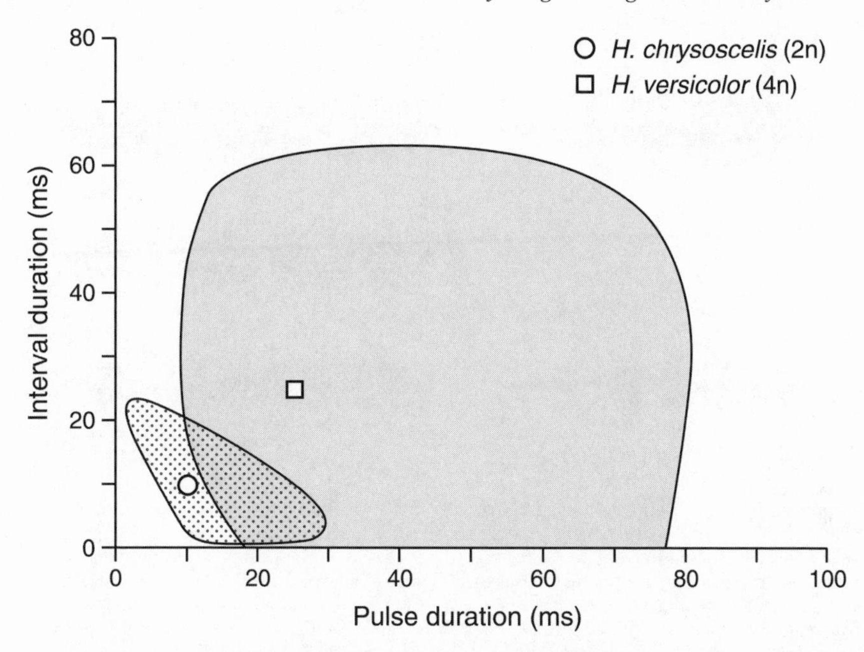

Figure 4.18. "Recognition" spaces of two species of gray treefrogs. The gray areas encompass the combinations of silent intervals between pulses and pulse durations that reliably elicited phonotaxis in females as judged by their response time relative to a control stimulus with conspecific values. The roughly diagonal field for *H. chrysoscelis* is the expected pattern for a species using some form of pulse-rate or pulse-period criterion. The broad field for *H. versicolor* indicates that females of this species, as in the katydid *Tettigonia viridissima,* are attracted by many combinations of silent intervals and pulse durations that are never found in conspecific signals. The circle and square indicate the mean values of silent intervals and pulse duration in *H. chrysoscelis* and *H. versicolor,* respectively.

males of this species have more complicated temporal criteria, and an analysis of phonotaxis in single-speaker tests shows that females of *H. versicolor* are similar to females of *Chorthippus biguttulus* and *Tettigonia viridissima* in that the recognition space is a complex function of variation in both pulse duration and interpulse interval (Schul and Bush, unpubl. data; fig. 4.18).

Fine-Temporal Discrimination and Correct Rejection in the Green Treefrog

The advertisement calls of the green treefrog *Hyla cinerea* are not obviously pulsed, except at the beginning (fig. 4.19A). Females preferred prerecorded advertisement calls to aggressive calls, which males produce by pulsing the advertisement call at a rate of about 50/s (fig. 4.19A; Oldham and Gerhardt 1975; Gerhardt 1978b,c). Females also preferred a synthetic call representative of the advertisement call to a similar call upon which 50 Hz amplitude modulation was superimposed, but only when the depth of modulation equaled or

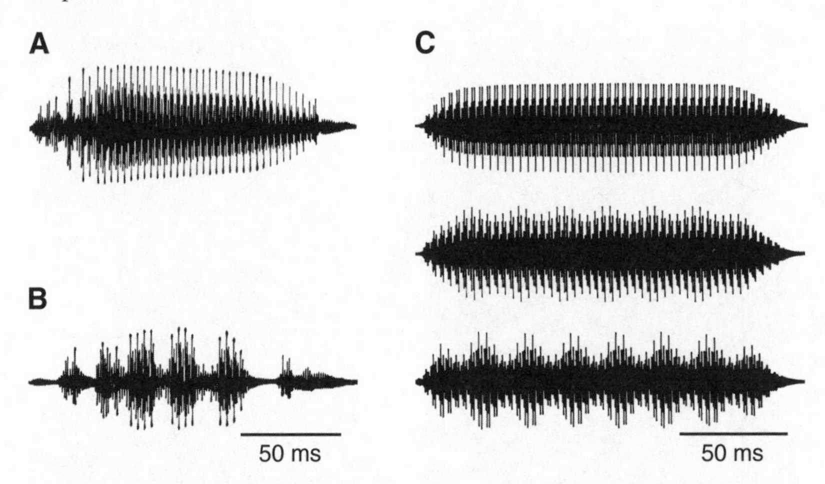

Figure 4.19. Advertisement and aggressive calls of the green treefrog *Hyla cinerea.* Oscillograms show (A) advertisement call; (B) aggressive call of the same male; and (C) oscillograms of synthetic calls used to test females for preferences based on the degree of pulsing (depth of amplitude modulation) at the rate (50 pulses/s) typical of aggressive calls. *Top,* unmodulated call, representative of advertisement calls; *middle,* call with a modulation depth of about 10%; *bottom,* call with a modulation depth of about 50%. Females did not show a preference for the unmodulated calls unless the modulation depth was at least 50%. Original figure.

exceeded 50% (fig. 4.19B; Gerhardt 1978c; see also Allan and Simmons 1994 for data from males). Thus, detection of modulation makes the signal less attractive, a result suggesting correct rejection. Females also showed remarkable selectivity with respect to the number of cycles of 50 Hz modulation, usually discriminating on the basis of a difference of one cycle and preferring signals with fewer cycles, that is, signals more similar to advertisement calls (Gerhardt 1978b). Avoiding aggressive calls might reduce the chances of an attack on the female by aggressively interacting males.

Duetting Katydids: Temporal Properties of Male Signals and the Timing of Female Responses

Among our final examples of the role of fine-temporal properties in intraspecific communication are three species in which females are rather unselective with respect to male signals. In *Leptophyes punctatissima* (Robinson et al. 1986), *Poecilimon ornatus* (Heller et al. 1997), and *Phaneroptera nana* (Tauber and Pener 2000), the male song is relatively simple and females respond with one or more short ultrasonic clicks. Females respond to a wide range of stimuli, including some sounds that differ considerably from those of conspecific males. The relatively poor selectivity of females for male songs, and the fact that female acoustic signals are also unspecific, potentially creates problems of mate identification. In these katydids, however, recognition does not depend merely on the structure of male and female signals. Rather, as in some firefly systems (review: Lloyd 1977), it is the male's selectivity for a species-specific time delay between his signal and the female's response that provides a reliable mechanism for promoting conspecific pairings (Heller and Helversen 1986). In *L. punctatissima*, for example, males do not respond phonotactically to a click simulating a female's response unless it occurs in a narrow time window between about 20 ms and 50 ms after the start of the male's song (Robinson et al. 1986; Zimmerman et al. 1989; fig. 4.20A). Thus, even if the female responds to an incorrect signal, produced, say, by a heterospecific male, no nearby conspecific male is likely to move toward her because the female's signal would not fall within the male's window of acceptance. The single, soft ultrasonic clicks of the female are also unlikely to help potential predators find her and, aside from their timing, contain little temporal information for the male.

What is remarkable about the female response is its very short latency (23 ms at 28°C in *L. punctatissima* and even shorter—15 ms—in *Ancistrura nuptialis*) (fig. 4.20A; Heller and O. Helversen 1986; Heller et al. 1997; Hartley and Robinson 1976). Because an auditory T-fiber in the mesothoracic ganglion requires 13 ms to fire, leaving only about 10 ms for the generation of a muscle contraction of the wing muscles via at least one motoneuron, Robinson et al. (1986) consider the female acoustic response to be a kind of acoustic motor reflex.

In another species of duetting katydid, *Ancistrura nigrovittata*, the male's song is more elaborate than in the species considered above. Females are selective for fine-scale temporal properties (pulse number and intervals between pulses) of the main part of the song (the syllable group). Females are also sensitive to the duration of the pause between the syllable group and a final single syllable, which acts as a trigger for each female response (Dobler et al. 1994a; fig. 4.20B). To reliably elicit female responses, the trigger pulse must occur within about 250 ms of the end of the syllable group. The female's response latency is only 25 ms from the end of the trigger pulse, but the time delay between the syllable group and trigger pulse is sufficiently long for higher centers in the female's auditory system to process the information in the main syllable group.

Duetting species of katydids provide a rich opportunity for comparative studies of phonoresponses (Heller 1990; Stumpner and Meyer 2001). The male's use of the timing delay for species identification is a diagnostic feature of the system, but species like *A. nigrovittata* deserve further study. Are there ecological conditions or differences in the mating system that can help to explain why the selectivity of females more closely resembles that of typical insect and anuran systems than other members of this subfamily? For example, could the assessment of males in this species result in greater direct benefits (e.g., nuptial gifts; Heller et al. 1997) or indirect benefits than in the less selective species? In a recent study of *Phaneroptera nana*, Tauber et al. (2001)

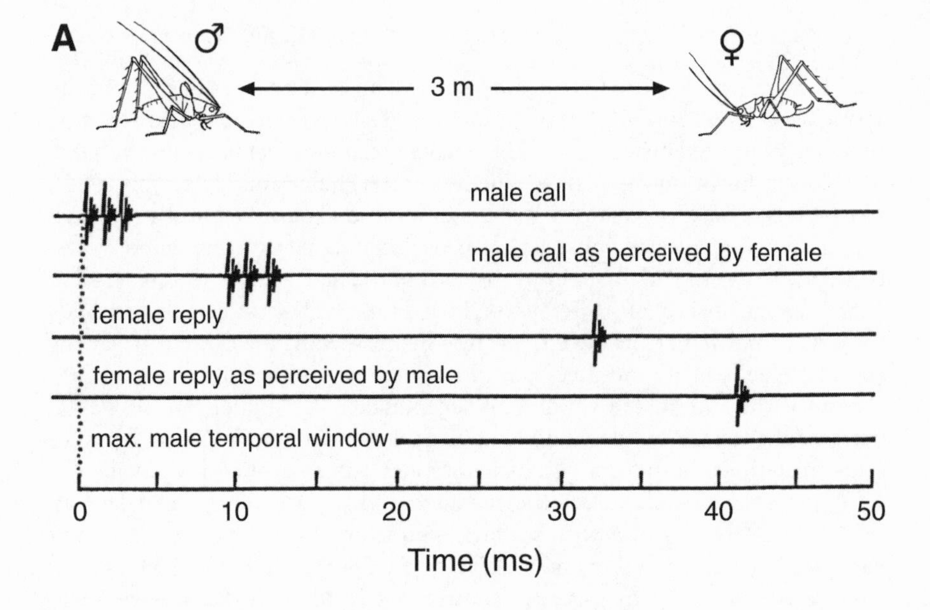

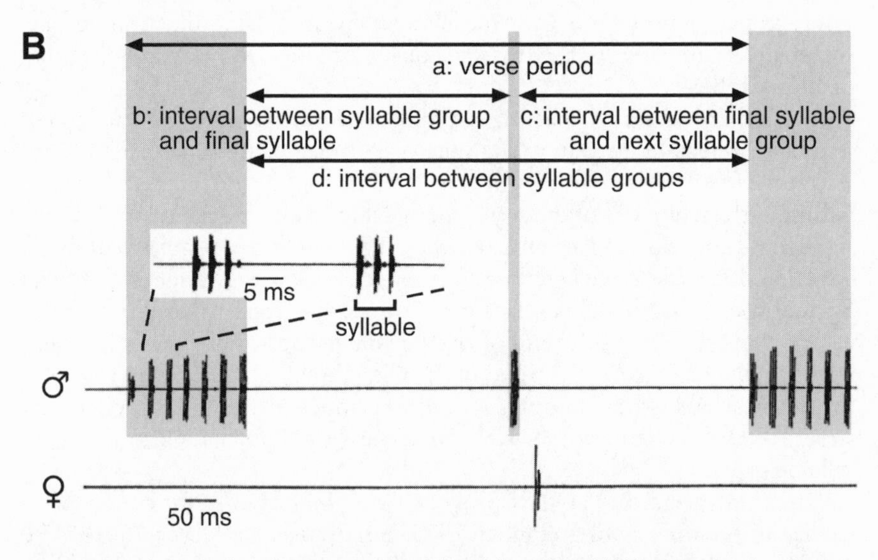

Figure 4.20. Acoustic duets between male and female katydids. (A) Diagrams showing the timing of signals in a hypothetical interaction between a pair of katydids (*Leptophyes punctatissima*) separated by a distance of 3 m, hence time delays exist between the end of the male call and his receipt of the female reply. The time window within which a male must hear the female's response in order to show phonotaxis is shown by the thick line at the bottom of the figure; this window is not adjusted by the intensity (a distance cue) of the female's response. Such an adjustment could, in principle, be used to compensate for different delays due solely to the time required for the signals to travel between males and females. From Robinson 1990, fig. 9. (B) Oscillograms of male signals (*top two traces*) and a female response (*bottom trace*) in the katydid *Ancistrura nigrovittata*. Notice the complex temporal structure of the syllable group (each containing three pulses). The single syllable that occurs after the group serves as a trigger pulse for the timing of the female response. From Dobler et al. 1994a, fig. 1.

found that females preferentially answered the longer chirps produced by large males, which might provide greater benefits than small males.

Temperature Effects on Selectivity for Temporal Properties

Changes in temperature affect the fine-scale temporal properties of insect and anuran signals, and pulse rate in particular, much more than spectral properties (chapter 2; Gerhardt and Mudry 1980; Walker 1962). Because temperature often varies within the breeding season, the female must have some strategy for dealing with such changes. Walker (1957) first demonstrated that the pulse rate preferred by female tree crickets (*Oecanthus fultoni*) depends on temperature. More specifically, females responded best when the pulse rate corresponded to that produced by a conspecific male at about the same temperature. This phenomenon—termed *temperature coupling* by Gerhardt (1978a)—has also been found in field crickets (*Gryllus bimaculatus* and *G. firmus*), some species of grasshoppers (*Chorthippus montanus* and *C. parallelus*), and the gray treefrogs (Helversen and Helversen 1981; Doherty 1985b; Gerhardt and Doherty 1988; Pires and Hoy 1992a). An example from *G. bimaculatus* is shown in figure 4.21. In at least one species of grasshopper, *Chorthippus biguttulus*, however, females solve the problem of temperature-dependent temporal properties in another way. They simply respond to any signal that has values of fine-temporal properties that are produced by conspecific males over the entire range of breeding temperatures (Helversen and Helversen 1981).

Temperature-dependent shifts in preference require relatively large differences in test temperature (minimum of 4°C in *H. versicolor:* Gerhardt and Doherty 1988; minimum of 7°C or more in *G. bimaculatus* and *G. firmus:* Doherty 1985b; Pires and Hoy 1992a). Moreover, in the treefrog, the temperature-dependent preferences break down at temperatures at the high end of the normal range of breeding temperatures (about 24°C; Gerhardt and Doherty 1988), whereas pulse rate increases with temperature in a linear fashion up to at least 30°C (Gayou 1984). If temperature coupling were based on the same neural oscillator as proposed by Alexander (1962), we have to postulate that other sensory components being driven by such an oscillator show a nonlinear behavior at relatively high temperatures. Experiments in the grasshoppers *Chorthippus montanus* and *C. parallelus* provide more direct evidence against a common oscillator. Differential heating of different parts of the head, where the "brain" resides, and the thorax, where pattern generators of song rhythm are located (chapter 3), can, in effect, create a mismatch between preferences for particular time intervals and the time intervals produced in the song of the same individual (Bauer and Helversen 1987). In the cricket *Gryllus firmus*, Pires and Hoy (1992b) found that heating of both the head and thorax was necessary to change female preferences for pulse rate. We return to this topic in chapter 10, where we discuss the issue of common gene control of sender and receiver mechanisms, which, in principle, can be served by entirely separate neural circuitry (Butlin and Ritchie 1989).

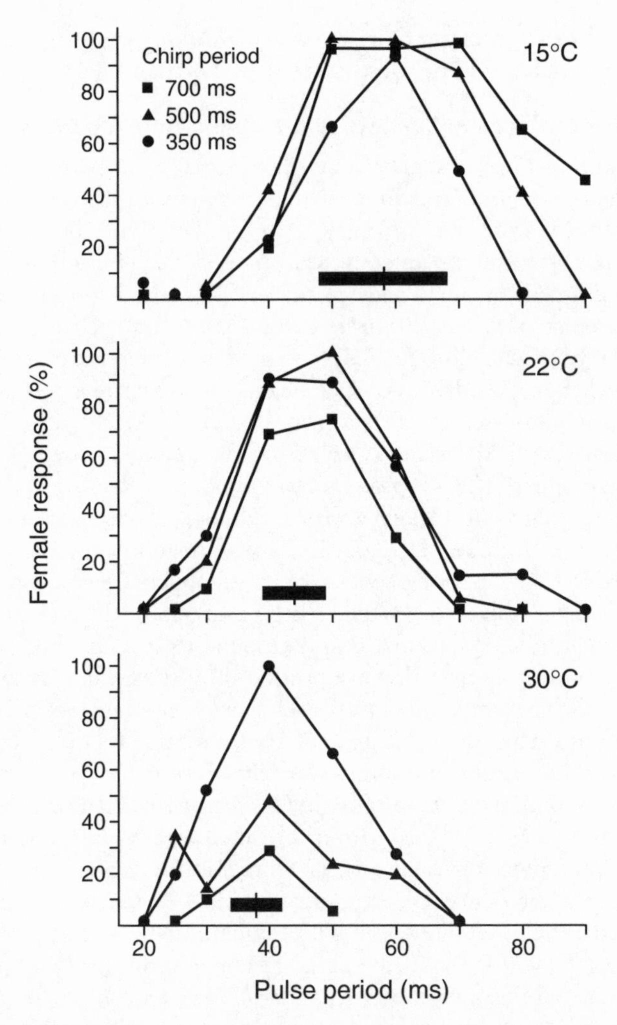

Figure 4.21. Temperature coupling between sender and receiver in the field cricket *Gryllus bimaculatus*. The percentage of responses of females is shown as a function of pulse period (reciprocal of pulse rate) and temperature. The mean and range of variation of syllable period is shown at the bottom of each panel. The results were, in general, independent of variation in chirp period (symbols show values of three periods used in the experiments). From Doherty 1985b, fig. 7.

Interaction of Preferences Based on Different Acoustic Properties

Two or more key properties of signals often vary independently and are likely to be conveying different kinds of biologically important information (see also chapter 10). This section reviews some of the few studies that have explored the effects of simultaneous variation in two properties.

Interaction of Preferences Based on Static and Dynamic Temporal Properties

Many studies have shown that preferences based on pulse rate in field crickets are modifiable by simultaneous variation in other temporal properties of the song (e.g., Doherty 1985c; Pollack and Hoy 1989; Popov and Shuvalov 1977; Stout et al. 1983). For example, Doherty (1985c) found that in *Gryllus bimaculatus* the range of pulse periods that elicit phonotaxis was sometimes expanded to values outside the normal range of variation when another variable chirp period had values within or slightly lower than the range of variation in calling songs (fig. 4.22A). When pulse period was held constant at average values, the ranges of effective values of chirp rate and duration became much broader relative to the natural range of variation in these traits (fig. 4.22B,C). Thus, pulse rate appears to influence the overall effectiveness of a signal more than chirp rate and duration.

Although female preferences have a stabilizing influence on pulse rate in *Hyla versicolor*, females show highly directional preferences based on call duration and call rate, preferring values that are even beyond the species range of variation (Klump and Gerhardt 1987; chapter 10). Increasing the call duration or call rate of signals with nonpreferred pulse rates can increase their attractiveness relative to signals with preferred pulse rates, but only up to a point. For example, when the difference in pulse rate was 50% (e.g., 20 pulses/s vs. 30 pulses/s), a few more females chose the alternative with the higher pulse rate when its duration was greater than that of the standard call than when the duration of the two stimuli was equal (Gerhardt and Doherty 1988). If the difference in pulse rate was 100% (e.g., 20 pulses/s vs. 40 pulses/s), females reliably chose the stimulus of 20 pulses/s, even if the higher pulse-rate alternative was much longer in duration or had a much higher call rate (Gerhardt 1983). In *H. chrysoscelis* the relative importance of pulse rate and call duration varies geographically and depends on the presence or absence of *H. versicolor* in the same area (chapter 11; Gerhardt 1994b).

Interaction of Preferences Based on Spectral and Temporal Properties

Relative importance of spectral and temporal properties in the bladder cicada. The bladder cicada *Cystosoma saundersii* is an Australian species in which females fly to singing males during a brief period of time just after sunset (Doolan and Young 1989; see also chapter 8). Tethered flying females (a preparation simulating relatively long-distance phonotactic behavior) were selective for the carrier frequency typical of the songs of conspecific males (0.85 kHz); alternatives with carrier frequencies of 0.5 kHz and 1.2 kHz were completely ineffective. The temporal properties of synthetic songs (pulse duration, pulse rate, and duty cycle), by contrast, could be varied over wide ranges, well beyond the values in calling songs, with no significant reduction in the attractiveness of a 0.85 kHz signal.

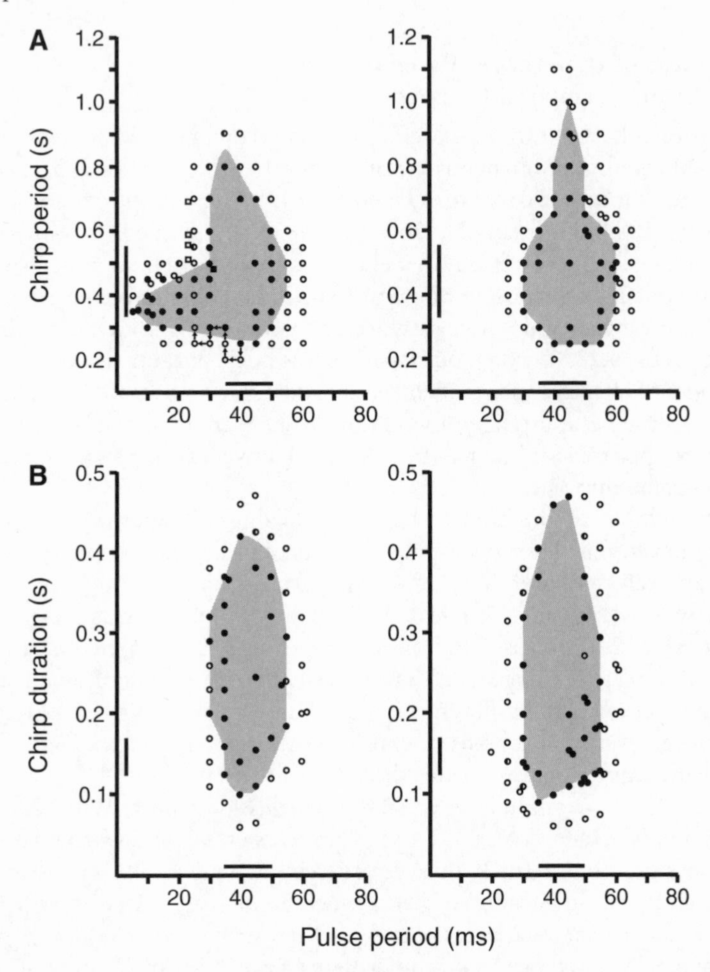

Figure 4.22. Interaction of chirp period (A) and chirp duration (B) with pulse period (pulse rate) in determining phonotactic effectiveness of synthetic calling songs in four females of *Gryllus bimaculatus.* The solid lines inside the axes of each figure show the ranges of variation of the temporal properties in conspecific calling songs. The shaded area estimates the space (combinations of the values of the two variables) in which a signal elicited reliable positive phonotaxis. Open circles = the combination was tested and reliable phonotaxis did not occur; closed circles = the combination elicited reliable phonotaxis. See the text for more details. From Doherty 1985c, figs. 8, 10.

At close range, female cicadas show responses that indicate their sexual receptivity, namely a wing-flicking behavior that disperses a pheromone. After this behavior begins, males switch from their loud calling songs to softer courtship songs. Using wing flicking as a response criterion, Doolan and Young (1989) found that caged females were much less selective with respect to carrier frequency than flying ones. Although they still did not react to a signal of

0.50 kHz, they now responded reliably to signals with carrier frequencies as high as 2.2 kHz. Moreover, unlike flying females, caged females preferred temporal properties with values similar to those in male courtship songs. Thus, female bladder cicadas apparently use only the carrier frequency for long-distance attraction and orientation to conspecific males, but require a close-range analysis of the temporal properties before making a final mating decision.

Effects of spectral and temporal properties on positive and negative phonotaxis in crickets. Using the tethered flying preparation, Pollack et al. (1984) first showed that the pulse rate of a stimulus influences the tracking errors of females of *T. oceanicus* when the carrier frequency is increased from the value typical of the calling song (5 kHz) up to 15 kHz. When the pulse rate matched that typical of the chirp part of the calling song (about 16 pulses/s), females still showed phonotaxis, but there was a deviation away from the sound source of about 60°; when the pulse rate was 2 pulses/s, females showed negative phonotaxis (mean direction of about 180° relative to the speaker), as they did when the carrier frequency was 30 kHz. Indeed, tethered flying females shift from positive to negative phonotaxis when the carrier frequency of signals with the correct pulse rate exceeds 10 kHz (Moiseff et al. 1978; box 4.1). Pollack and El-Feghaly (1993) found that phonotaxis in response to a stimulus with a 30 kHz carrier frequency shifted from negative to positive as the pulse rate of the stimulus was increased. However, the sustained positive orientation followed a transient negative phonotactic turn, suggesting that an initial avoidance response might be negated if a signal with an attractive temporal structure is repeated regularly. The echolocation signals of bat predators would probably be perceived as a few transient signals preceding an attack, whereas calling songs would be monotonously repeated. This topic deserves much more experimental study.

Summary and Suggestions for Future Research

In some insects, females respond with nearly reflexive acoustic responses to crude imitations of male signals, whereas other species of insects and anurans show remarkable intensity-independent selectivity based on small differences in the value of a key property. Failures of highly motivated individuals to respond to signals with particular attributes can sometimes be interpreted as correct rejection. Most often, however, ineffective stimuli probably lack key acoustic properties or have key properties with values that do not adequately stimulate the neural circuits used to identify appropriate signals. The costs and benefits of responding, which depend in turn on the reproductive ecology and physiology of a given species, are also likely to influence the probability of responding to suboptimal signals (chapter 10).

Most insects and anurans show biases for the carrier frequency or frequencies typical of conspecific signals, but the precision of the match varies among species, and the two sexes within a single species may differ in frequency selectivity. As shown in chapter 5, the hearing range of insects and anurans typically includes enhanced sensitivity in two or more frequency bands, which are often widely separated. Preferences based on relatively small differences in frequency *within emphasized bands* are usually intensity-dependent in that preferences are abolished or reversed by relatively small reductions in the intensity of the preferred stimulus, but conditioning experiments are needed to be sure that such differences cannot be discriminated. The absolute playback level to which two signals are adjusted can also affect female frequency preferences. In one species of anuran, a significant effect of temperature on frequency preferences has been demonstrated, and it will be interesting to learn if this is a more widespread phenomenon that might also apply to insects.

In some species, multiple (usually two) channels, whose relative excitation might provide biologically important information, are involved in processing broadband signals. The "rules" for reacting to simultaneous excitation of these channels can vary between the sexes and among species. In many insects, enhanced sensitivity in the high-frequency range can serve to detect both the signals of conspecific males and the echolocation sounds of bat predators. Although negative phonotaxis and other evasive behaviors are commonly observed in response to loud, high-frequency stimulation, behavioral responses to simultaneous stimulation of the low-frequency channel, usually tuned to calling songs, are unpredictable and can depend on temporal properties. Moreover, orientation of responses may depend on frequency, and mechanisms of sound localization and pattern recognition often interact (chapter 7).

In both insects and anurans, the most intensity-independent preferences are usually based on fine-temporal properties, such as pulse rate and pulse duration. Differences in pulse rate that typify those between conspecific and heterospecific signals can mediate preferences that survive two- to eightfold differences (6–18 dB) in sound pressure level favoring alternatives with heterospecific rates. In other species, however, females often respond to pulse rates well outside the conspecific range of variation.

Not all pulsatile signals are evaluated solely in terms of pulse rates, as exemplified by some grasshoppers (*Chorthippus biguttulus*), katydids (*Tettigonia viridissima*), and gray treefrogs (*Hyla versicolor*). In these species, particular combinations of the duration and intervals between fine-scale acoustic units (pulses or syllables) are the critical identifying features. Most studies have not conducted the appropriate experiments to show that pulse rate alone mediates selective phonotaxis. It is also important to learn if stimuli with values that are fractions or multiples of the most preferred (usually conspecific) value are effective; such tests can help to differentiate among various models that have been proposed for temporal-pattern discrimination (Helversen and Helversen 1998). The rise-time of pulses also affects the phonotactic selectivity of some grasshoppers and anurans; however, the basis for their selectivity is different. Differences in selectivity based on fine-temporal properties are likely to be strongly influenced by absolute intensity (distance), and we emphasize again the need to generalize preference tests over a wide range of biologically realistic levels.

Confronted with the same choices (conspecific vs. heterospecific calls or synthetic equivalents), species in groups of closely related species—*Tettigonia cantans*, *T. viridissima*, *T. caudata*, *Teleogryllus oceanicus* and *T. commodus*, and *Hyla chrysoscelis* and *H. versicolor*—use different criteria to choose among alternative signals. Sexual differences in temporal acoustic criteria also exist in some insects. These species and sexual differences serve to emphasize that evolution has a wide range of options for coordinating communication between senders and receivers, a view that is reinforced by species differences in how females deal with temperature-dependent changes in temporal properties.

Insects and anurans often evaluate more than a single independently variable acoustic property. Interactions between preferences based on these different properties determine the overall attractiveness of a signal. Much more research on multidimensional determinants of signal selection is needed in light of the complexity that has already been found in the rules governing the interaction of just two relevant properties at a time. Any complete model of how the nervous system mediates signal recognition and preferences will have to account for these interactions. As we argue in chapter 10, the relative importance of different properties for mate choice can probably be understood ultimately by the relative importance of the biological information they convey to receivers. Finally, although we have emphasized that female insects and anurans are generally selective for relatively narrow ranges of values of acoustic properties typical of conspecific signals, some variation among females is almost always observed. This variation must also be reflected in the neural mechanisms that we consider in the next three chapters as well as serving as the raw material for evolution (chapters 10–11).

5

Processing of Biologically Significant Acoustic Signals in the Auditory Periphery

THE MAIN role of the peripheral auditory system is to translate the information encoded in the acoustic signals into neural activity that will be decoded within the central auditory system. In insects the axons of auditory receptors project into the central nervous system. In anurans the axons of primary auditory neurons, which innervate the auditory receptors, serve the same function. The mechanical properties of the peripheral structures and the response properties of receptors and primary auditory neurons ultimately set the limits on an animal's ability to resolve differences in the spectral and temporal properties of acoustic signals (e.g., Ball et al. 1989; Römer and Tautz 1992; Lewis and Narins 1999). Although the information transmitted by any single receptor or neuron is further limited by its physiological properties, the central nervous system overcomes this problem by processing the input from populations of many such cells.

In insects with tympanal organs and in most anurans, the first step in hearing usually involves specialized peripheral structures, such as tympanic membranes or eardrums, which are driven into vibration by sound waves. In anurans additional pathways provide direct input of sonic and substrate vibrations to the inner ear organs (Narins et al. 1988). Insects also have subgenual organs, which are excited by low-frequency sound and substrate vibration, and several insect taxa possess cercal receptors, which are sensitive to air-particle oscillations and low-frequency sounds (chapter 3; reviews: Dambach 1989; Römer and Tautz 1992).

Hearing mechanisms in most orthopterans and cicadas share three basic features with anurans. First, tympanic membranes can be driven into vibration not only by sound waves arriving directly from the external environment but also by sound waves arriving via pathways inside the animal's body. The consequences of this dual (or of multiple) input to tympanic membranes on sound localization are examined in chapter 7. Second, sound reception is influenced by substrate vibration and by movements during signaling and respiration, which can affect tympanic membrane vibrations as well as inner ear receptors via extratympanic pathways. The net effect is usually a reduction in sensitivity to external acoustical signals (e.g., Schildberger et al. 1988; Narins 1992b; Hedwig and Meyer 1994; Christensen-Dalsgaard and Jørgensen 1996b; Poulet and Hedwig 2000; see box 5.1). Third, auditory receptors and primary auditory neurons in insects, anurans, and other animals have similar physiological characteristics and coding principles.

Auditory Organs and Sound Inputs

Insects

The hearing organs of insects evolved independently in different taxonomic groups and are located in different parts of the body (Fullard and Yack 1993; Hoy 1998; Yager 1999). Michelsen and Larsen (1985) provide a comprehensive description of the functional anatomy of the tympanal organs of orthopterans and cicadas, and here we summarize their main anatomical features.

Crickets and katydids bear the tympanal organs in the proximal parts of their forelegs. Tympanic membranes (usually two per organ) are located either on the outer surface of the leg or inside cavities, which are open to the outside by slits (fig. 5.1B,D). In grasshoppers and cicadas, the tympanal organs, each of which contains one tympanic membrane, lie ventrolaterally on the first and the second abdominal segment, respectively (fig. 5.1E).

In crickets the posterior tympanal membrane, the only one known to function in audition (Huber et al. 1984; see box 5.2), is exposed to the external environment. The other side of this membrane is backed up by a side branch of the main leg trachea to which the *auditory sensilla* (receptor cells plus accessory cells) are attached (fig. 5.1A). This leg trachea traverses the leg, and one of its branches opens on the lateral surface of the thorax via an acoustic spiracle, thus providing a pathway for sound to reach the inside of the tympanic membrane. A second branch of the leg trachea crosses the prothoracic segment and connects to the tracheal system of the ear in the other foreleg. The traversing leg trachea is separated by the central membrane (= *medial septum*) (fig. 5.1B). Thus, each tympanic membrane has four sound inputs: one from the ipsilateral acoustic spiracle, one from the contralateral spiracle, one through the contralateral tympanum, and the external sound impinging on the ipsilateral tympanum (Larsen and Michelsen 1978).

In most katydids the two tympanic membranes on the front and back of the proximal tibia are backed up by paired tracheae, which meet along the midline to form a secondary central membrane where the auditory sensilla are attached (fig. 5.1D). The final part of each acoustic spiracle widens to form the *hearing trumpet*, which is a horn-shaped tube (Lewis 1974). This structure controls the amount of sound input to the inner side of the tympanic membrane via the internal pathway. Because the two tracheal systems associated with hearing are separated in most katydids, each tympanic membrane has just

Box 5.1. Effects of Nonauditory Stimuli Generated by Movements on Auditory Function

One outstanding unsolved problem in acoustic communication is how animals deal with acoustic input during the execution of movements, which influence the transfer of acoustic information from the receptors to interneurons in the ascending pathway. In insects, for example, stridulation and other self-produced sounds, ventilation, and walking displace the tympanic membrane and thereby increase activity in the auditory nerve, which can mask the response of auditory receptors and modulate the activity of auditory interneurons to external acoustic signals (e.g., Wolf and Helversen 1986; Hedwig 1988, 1990; Schildberger et al. 1988; Hedwig and Meyer 1994; Lang and Elsner 1994; Poulet and Hedwig 2000).

In grasshoppers, vibrations of the tympanic membrane occur in phase with the stridulatory leg movements. These movements, in turn, create a large amount of background activity in auditory receptors, and only during a small window of reduced activity (at the lower reversal point of the stridulatory leg movement) can auditory interneurons detect sound signals with fidelity (fig. 1). Because this window is brief (about 10% of the stridulatory cycle), stridulating males are likely to miss a female's acoustic response. Indeed, in order to increase the chances of detection during the window, phonotactic responses of male grasshoppers usually cease when the male himself is stridulating (Hedwig and Meyer 1994).

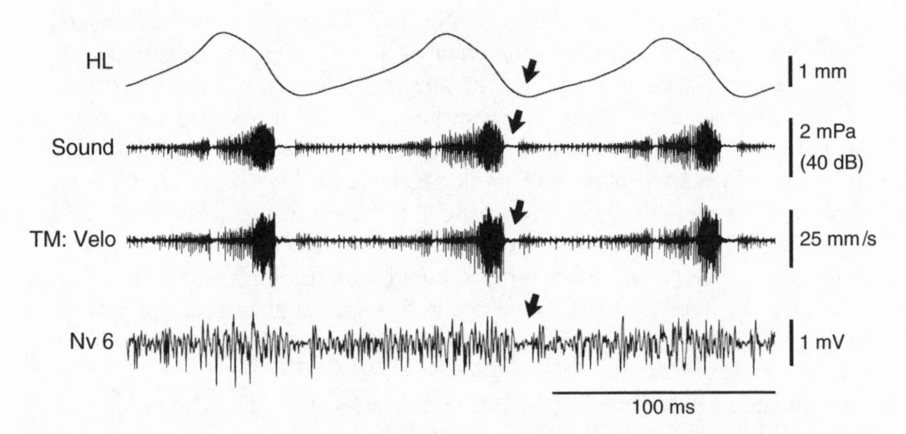

Box 5.1, Figure 1. Simultaneous recordings of stridulatory movements (*uppermost trace*) with up-and-down movements of one hindleg and acoustic output (*second trace*). Tympanic membrane velocity (*third trace*) and summed auditory activity (*bottom trace*). The lower reversal point of the leg movement and the associated gaps in the sound, the velocity of the tympanal membrane, and nerve activity are indicated by arrows for one stridulatory cycle. Modified from Hedwig and Meyer 1994.

(box 5.1 continued)

In crickets the acoustic spiracle is open during stridulation, and the tympanal membrane responds to self-produced sounds (Poulet and Hedwig 2000). During walking (phonotaxis in females), vibration can, depending on the substrate, excite receptors in the tympanal and subgenual organs as soon as the legs touch the ground. Indirect evidence suggests that, as in grasshoppers, responses to calling songs at low intensity are masked during particular phases of the leg-stepping cycle (Schildberger et al. 1988). Effects on central neurons (ON1 and AN2) to acoustic stimuli during self-produced chirps and silent stridulatory movements are discussed in chapter 6.

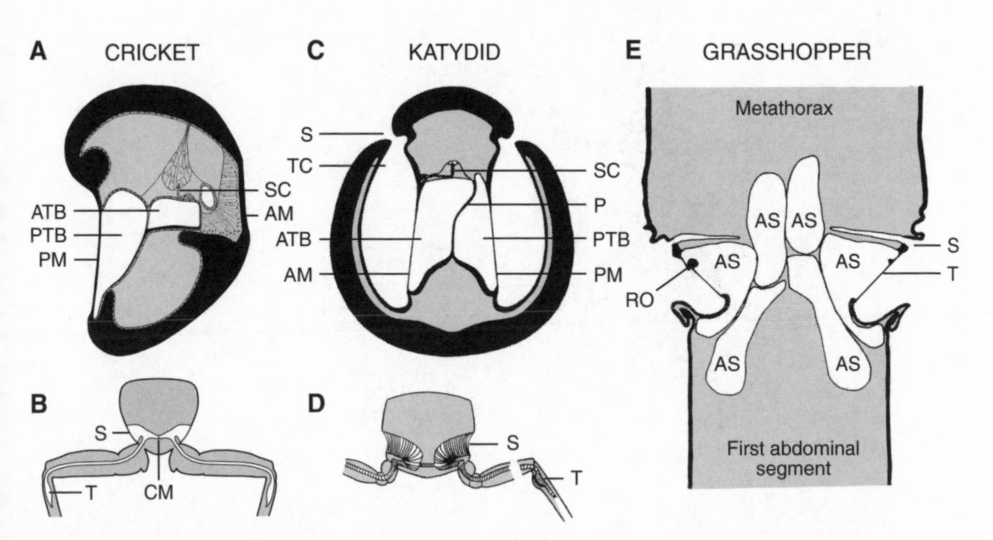

Figure 5.1. Tympanal organs of orthopterans. (A) Section through the proximal part of the foreleg tibia containing the ear in crickets; AM and PM = anterior and posterior tympanal membrane; PTB = posterior tracheal branch (main leg trachea); ATB = anterior tracheal branch (side branch of the main leg trachea) to which the sense cells (SC) are attached. From Larsen and Michelsen 1978. (B) Four-input system: tracheal tubes connect the internal surface of the tympanic membrane (T) of one ear with the other ear and with the ipsilateral and contralateral spiracle (S). The trachea connecting both sides is separated by the medial septum or central membrane (CM). From Michelsen 1998a,b. (C) Section through the ear in the region of the tibia of a katydid species with covered tympana. AM and PM = anterior and posterior tympanal membrane; ATB and PTB = anterior and posterior tracheal branch; P = central membrane separating ATB and PTB; S = slit; SC = sense cell; TC = tympanal cavity. From Michelsen and Larsen 1978. (D) Two-input system: sound reaches each ear (T = tympanic membrane) directly from outside and indirectly via the spiracle (hearing trumpet: S). From Michelsen 1998a. (E) Schematic horizontal section through the region of the ears in grasshoppers showing the tympanic membranes (T), associated air sacs (AS), the location of the receptor organ (RO = Müller's organ), and the spiracle (S). From Michelsen 1998a.

Box 5.2. Interference Experiment

In crickets there is definitive proof that vibration of the posterior tympanic membrane is essential for hearing. The interference experiment of Kleindienst et al. (1983) not only confirms the critical role of posterior tympanic membrane vibration, but also shows that sound is transmitted through the internal tracheal pathway. This experiment excludes an alternative hypothesis, which considered changes in sound pressure in the air sac underneath the larger (posterior) tympanum as the adequate stimulus (Fletcher and Thwaites 1979a).

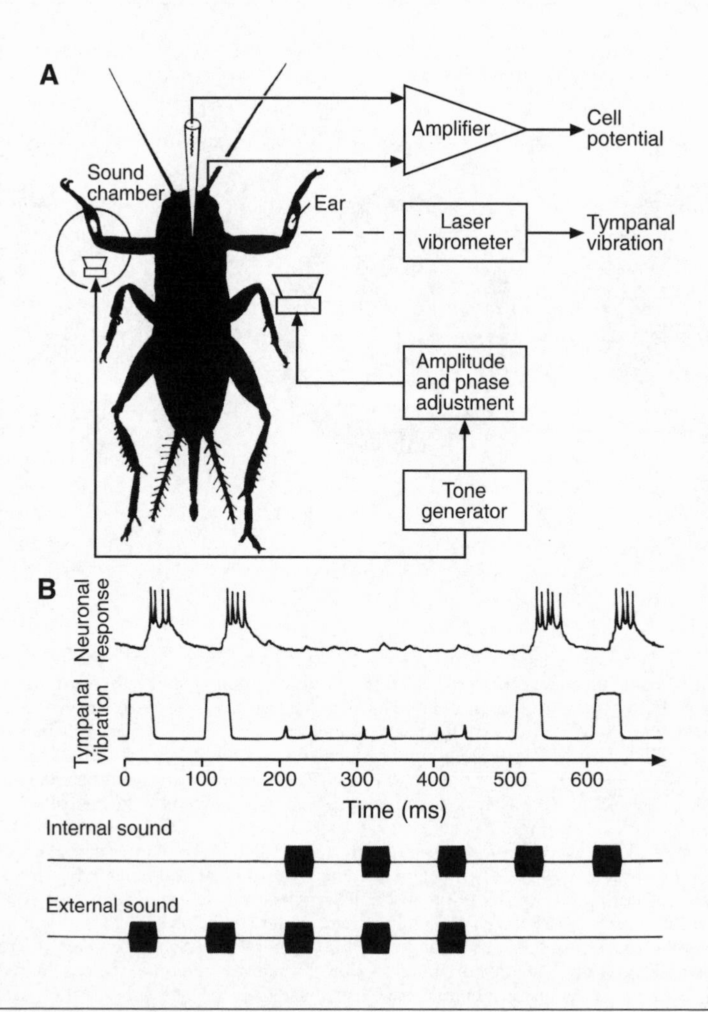

(box 5.2 continued)

(facing page)

Box 5.2, Figure 1. (A) Diagram of the experimental setup. Sound generated within the closed sound chamber is guided through the tracheal tube and elicits tympanal membrane vibration in the opposite ear. Its vibration can be canceled by an external sound directed to it. The phase and amplitude of the external sound are adjusted by means of a phase shifter and an attenuator until the tympanal vibration disappears. Tympanal vibration is measured with a laser vibrometer. Neural responses are recorded simultaneously from an acoustic interneuron, the omega cell, in the prothoracic ganglion, known to collect most of the auditory input. From Larsen et al. 1989, fig. 12.6; modified from Kleindienst et al. 1983, fig. 2. (B) Results of the interference experiment. Time pattern of external and internal 5 kHz sound pulses (*lower two traces*) and response of the omega cell in the prothoracic ganglion and tympanal membrane vibration (*upper two traces*). From Larsen et al. 1989, fig. 12.6; modified from Kleindienst et al. 1983, fig. 5. As long as only one sound source is active (*first two and last two pulses*), each sound pulse elicits both a tympanal membrane vibration and a neural response. With both sound sources activated simultaneously and adjusted (*middle three pulses*), both the neural responses and the tympanal membrane vibrations were almost canceled. Since both sound signals differ in their time of arrival at the active ear, the canceling effect is not complete and short-lasting oscillations of the tympanum can be seen at the beginning and end of each sound pulse, correlated with subthreshold synaptic activity in the neural response.

two ipsilateral inputs, one through the hearing trumpet and the other through the tympanic membrane (fig. 5.1D; review: Bailey 1990).

In grasshoppers and cicadas the tympanic membranes are backed up by one or more air sacs (fig. 5.1E). In both groups there appear to be only two inputs to each tympanic membrane, one from outside and the other through the contralateral tympanic membrane and the air sacs (Doolan and Young 1981; Fonseca 1994; reviews: Helversen 1997; Michelsen 1998a).

Anurans

Although anurans lack external ear canals and pinnae (external ears), the overall design of some of the peripheral structures that guide sound waves to the inner ear is usually similar to that in higher vertebrates. Vibration of the tympanic membranes is transmitted via the *columella* (stapes of higher vertebrates) to the oval window of the fluid-filled inner ear (otic) capsule via an air-filled middle ear (Wever 1985; Jørgensen and Kanneworff 1998; fig. 5.2). The *impedance mismatch* between air and fluid, which reduces transfer of sound between these two media, is partially compensated because the area of the tympanic membrane is much greater than that of the oval window. The lever action of the columella, which is slightly higher in the grass frog *Rana temporaria* than that in higher vertebrates, also increases the efficiency of sound transmission from the tympanic membrane to the otic capsule (Jørgensen and Kanneworff 1998).

Some anuran species, however, lack tympanic membranes, columella, or

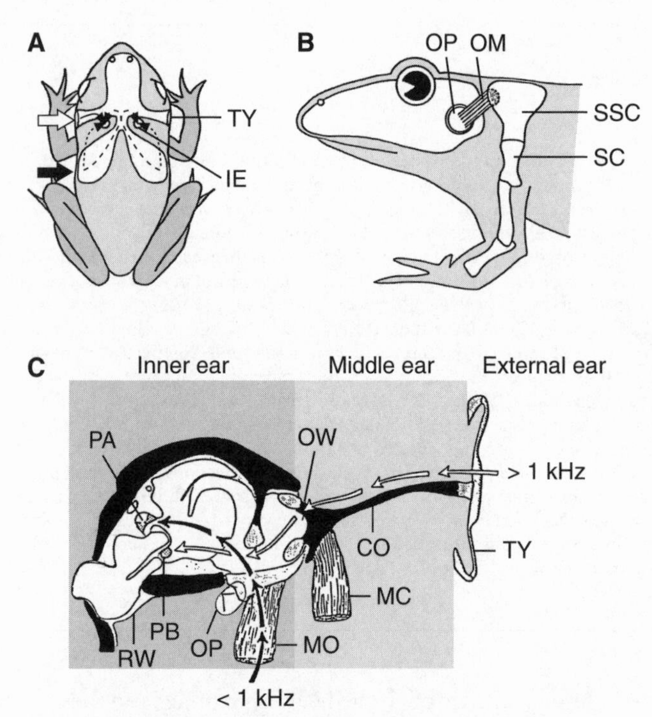

Figure 5.2. (A) Diagram of the dorsal view of a typical anuran; the white areas are internal pathways through which sound can be transmitted. The large, open arrow points to the tympanic membrane (TY), which is conventionally considered to be the main input to the inner ear (IE), via the columella (see C). The dark arrow points to the lateral body wall, which has been shown to be a second input for sound; such input can potentially affect the vibration of the tympanic membrane and might also stimulate the inner ears directly by as yet unknown pathways. (B) Diagram of the lateral view of a typical anuran, showing the opercularis system. The opercularis muscle (OM) originates on the shoulder skeleton and inserts on an operculum (OP) that rests in the oval window (OW; see C) of the inner ear. SC = scapula; SSC = suprascapula. From Hetherington 1992b, figs. 21.3, 21.2; after Narins et al. 1988. (C) Diagram showing a cross section of the structures of the right ear of a typical anuran with a well-developed middle ear. The columella (CO) connects the tympanic membrane (TY) to the oval window (OW) of the inner ear; contraction of the columellar muscle (MC) tends to immobilize the columella and reduce sound transmission. The oval window is also partially occupied by the operculum (OP) to which the opercularis muscle (MO) attaches (see B), and this is thought to be a second pathway for the conduction of sound and substrate vibrations to the inner ear. The locations of the two main auditory organs—the amphibian and basilar papillae (PA and PB) and the round window (RW)—are also shown. From Duellman and Trueb 1986, fig. 4.11.

even middle ear cavities (Jaslow et al. 1988) and nevertheless communicate effectively with airborne sounds (e.g., Lindquist and Hetherington 1996). One explanation is that the *opercularis system*, which is unique to amphibians and functions in the detection of substrate vibrations, can serve as an extratympanic pathway for low-frequency sounds (fig. 5.2B,C; Eggermont 1988; Hetherington 1992a,b; Christensen-Dalsgaard and Jørgensen 1996b; Lewis and Narins 1999). However, some frogs that lack tympanic membranes and middle ear structures have high-frequency sensitivity comparable to that of typical anurans (Jaslow and Lombard 1996). As discussed below, this sensitivity might be derived from sound-induced displacements of the body wall overlying the lung (Narins et al. 1988; Lindquist et al. 1998). Another possibility is that sound can still be transmitted effectively to the inner ears from the sides of the head where tympanic membranes are usually located, despite the absence of these and the other accessory structures.

In anurans with middle ear cavities, the tympanic membranes are coupled internally through the buccal cavity and permanently open eustachian tubes (Chung et al. 1978). Sound can reach the inner surface of the tympanic membrane from the contralateral tympanic membrane and possibly from the lateral body wall (via the lungs and buccal cavity) (Narins et al. 1988). The state of the inflation of the lung, which changes during breathing, affects thresholds and frequency responses of the tympanic membrane (Jørgensen 1991; Ehret et al. 1994). Larger pressures arising in the vocal sac and buccal cavity during vocalization impinge on the inside of the tympanic membrane and probably reduce self-stimulation (fig. 5.2A; Narins 1992b). Recall, too, that in the bullfrog *Rana catesbeiana*, as in some cicadas, the large tympanic membranes of the male are also significant radiators of sound (Young 1990; Bennet-Clark and Young 1992; see also chapter 2). Hearing during signal production is thus likely to be even more constrained in these animals than in other species.

In summary, the anatomy of the anuran periphery represents a multi-input system, both in terms of sonic inputs to the tympanic membrane and extratympanic inputs to the inner ear organs. More research is required to learn how the extratympanic inputs influence hearing and sound localization (Michelsen 1992; chapter 7).

Structure of Auditory Receptors

The auditory receptors of insects are bipolar sensory cells with a specifically modified ciliary structure in the outer dendritic segment (review: Field and Matheson 1998). Each receptor is associated with several accessory cells in a unit called the *scolopidium* (fig. 5.3A). There are fewer than a hundred scolopidia in crickets, katydids, and grasshoppers and as many as two thousand in cicadas (Fonseca et al. 2000). The peripheral parts of the scolopidia (attachment cells, cap cells, and dendrites) are mechanically deformed during tympanal membrane vibrations and cause excitation of the sensory cells (e.g., Bennet-Clark 1984; Römer and Tautz 1992).

The auditory receptors of anurans are *hair cells* that are located in two auditory papillae, the amphibian and the basilar papilla, and in the saccule, which is sensitive to low-frequency sound and vibration (fig. 5.3B; review: Smotherman and Narins 2000). Located in a fluid-filled inner ear capsule, hair cells get their name from the presence of cilia (multiple *stereocilia* plus a single

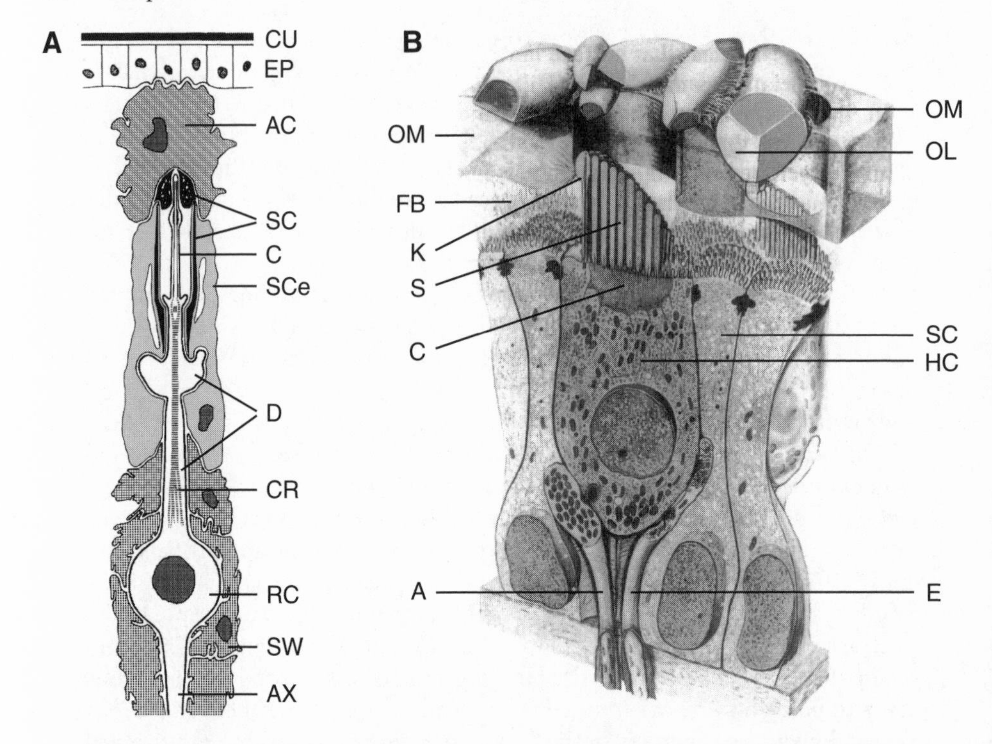

Figure 5.3. Insect and anuran receptors and accessory structures. (A) Structure of a scolopidium in the tympanal organ of a grasshopper (reconstructed after electron-microscopical data). The sensory cell (RC) is classified as bipolar because it has two processes (dendrite [D] plus cilium [C] and axon [AX]). Other structures: AC = attachment cell; CR = ciliary root; CU = cuticle; EP = epidermis linking the scolopidia with the tympanum; SC = scolopal rod; SCe = trichogen cell; SW = Schwann cell. Modified from Gray 1960. (B) Diagram of a generalized hair cell from the saccule of a bullfrog. The longest of the cilia, with a bulbed ending, is the kinocilium (K), and the other cilia, arranged in a bundle, are stereocilia (S). Deflection of the cilia in the direction of the kinocilium causes depolarization of the hair cell and increases the likelihood of firing in the afferent axon (A); deflection in the opposite direction causes hyperpolarization and decreases the likelihood of firing. Histological preparations of hair cells allow researchers to construct sensitivity maps (see fig. 5.5), which show the direction of deflection (toward K) that causes excitation. Other structures: C = dense cuticular plate; E = terminal of efferent neuron; FB = fibrous layer between the cell surfaces and the gelatinous membrane; HC = hair-cell body; OL = otoconial crystal; OM = gelatinous membrane at edge of a pore; SC = epidermal cells in close proximity to the hair cell. From Lewis and Narins 1999, fig. 4.8; modified from Hillman 1976.

kinocilium), some of which are embedded in an overlying *tectorial membrane*. Movements of the tectorial membrane in response to fluid-borne sound waves create shearing forces that bend the cilia. These mechanical forces are probably transduced to neuroelectrical signals by the intrinsic, metabolically powered mechanisms first proposed by Davis (1965). Lewis and Narins (1999) provide a detailed review, and Purgue and Narins (2000a,b) provide an analysis of acoustic energy flow in the inner ear of the bullfrog.

The two papillae, which are probably homologs of the macula neglecta and basilar papilla of fish (Lewis and Narins 1999), differ considerably in complexity. The amphibian papilla has five to twenty times as many hair cells as the basilar papilla (e.g., about 1500 in the amphibian papilla versus about 100 in the basilar papilla of the bullfrog; Lewis and Li 1975). Alfs and Schneider (1973) estimated that the basilar papillae of the European treefrog *Hyla arborea* has only 14 hair cells, whereas the basilar papillae of other hylids have 25–81 hair cells (Wever 1985).

Spatial Arrangement and Tonotopic Organization of Auditory Receptors

The close spatial arrangement of receptors (and auditory neurons throughout the nervous system) with the same frequency sensitivity is called *tonotopy*, which is usually confirmed by intracellular recording and staining of receptor cells. Tonotopy is maintained up to the level of the prothoracic ganglia of katydids, the metathoracic ganglion of grasshoppers, and within three auditory nuclei in the ascending pathway of anurans (see chapter 6).

Insects

In grasshoppers the tympanal membrane of the hearing organ (Müller's organ) varies in thickness and has different modes of attachment via different sclerites to which the dendrites of three groups of auditory sensilla project. The different groups of receptors cells are tuned to different frequencies (fig. 5.4A; Michelsen 1971a; Jacobs et al. 1999). In crickets and katydids, the scolopidia are arranged in rows along an inner membrane called the *crista acustica*. In general, the attachment cells and the length of the dendrites of the sensory cells decrease gradually in size proximally to distally (details in Michel 1974; Oldfield 1982; Rössler 1992a; Stölting and Stumpner 1998; fig. 5.4C). The location of auditory receptors along the crista acustica is correlated with the frequency to which they are most sensitive. Cells located proximally within the crista acustica are tuned to low frequencies, and those located more distally to high frequencies (Oldfield et al. 1986; Stumpner 1996).

In cicadas the auditory receptor cells are densely packed within an auditory capsule. On their outer side is the attachment horn and on their inner side, a rodlike structure called the *tympanal apodeme*, which, in turn, contacts the tympanic membrane (details in Doolan and Young 1981; Michel 1975; fig. 5.4D). Limited data indicate that auditory receptors terminate in different regions of the auditory neuropile, suggesting some kind of tonotopic organization (Münch 1999).

Anurans

In most frogs the hair cells within the amphibian papilla are found in two distinct patches (fig. 5.5A; Lewis and Li 1975). Several morphological types

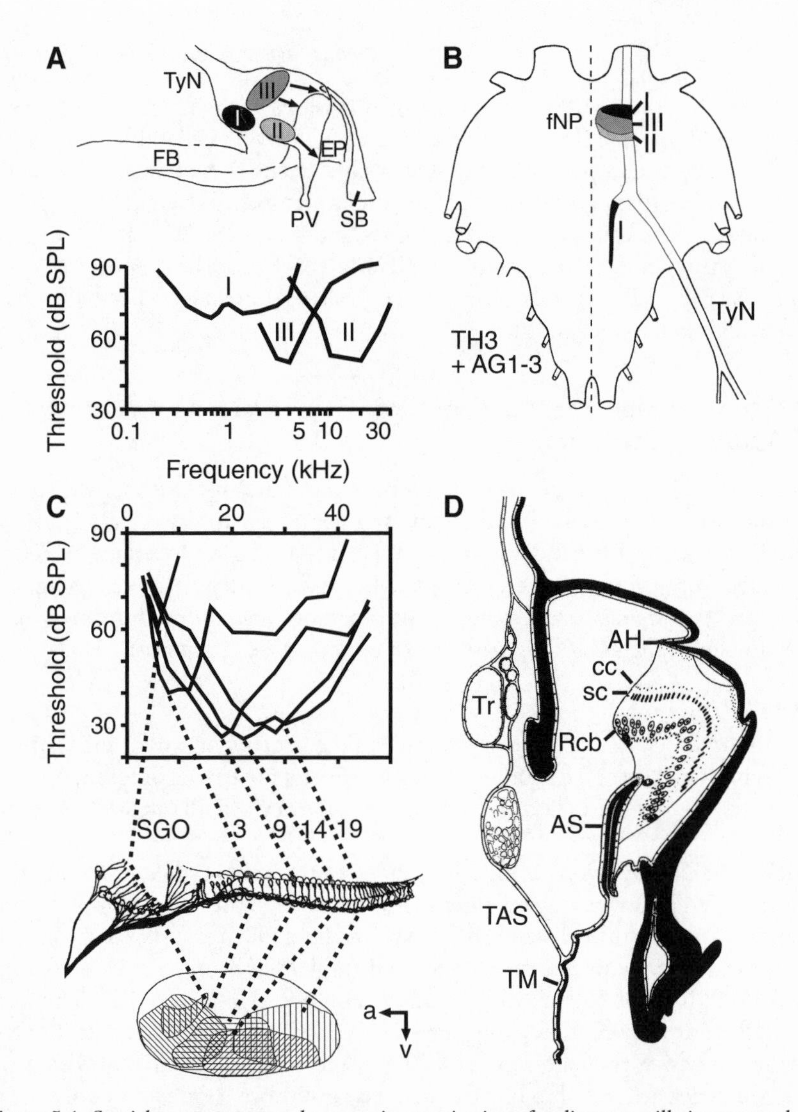

Figure 5.4. Spatial arrangement and tonotopic organization of auditory sensilla in tympanal organs of orthopterans and cicadas. (A) Grasshoppers: schematic diagram to show the positions of the cell body (soma) and dendritic attachment sites (*arrows*) of the three groups (I, II, III) of tympanal auditory sensilla within the Müller's organ of *Schistocerca gregaria.* EP = elevated process; FB = folded body; PV = pyriform vesicle; SB = styliform body. All receptor axons run within the tympanal nerve (TyN). *Below,* mean threshold curves of the three groups. From Jacobs et al. 1999. (B) Within the metathoracic complex (TH3 + AG1-3), the axons of the sensory cells terminate in a frontal auditory neuropile (fNP). Within the fNP they arborize tonotopically. From Jacobs et al. 1999. (C) Katydids: linear arrangement of auditory sensilla along the crista acustica in *Ancistrura nigrovittata. Top,* threshold curves for five auditory sensilla; *middle,* diagram shows the locations of auditory receptors within the crista acustica (3, 9, 14, 19) and one located within the subgenual organ (SGO); *bottom,* diagram of the auditory neuropile, showing the projection regions of receptors tuned to different frequency bands. From Stumpner 1996. (D) Cicadas: cross section through the right auditory capsule of *Cicada orni* showing the arrangement of the receptor cell bodies (Rcb), the scolopals (sc) and the cap cells (cc). AH = attachment horn; AS = tympanal apodeme to which the sense cells are attached; TAS = tracheal air space; TM = tympanal membrane; Tr = trachea. Modified from Michel 1975.

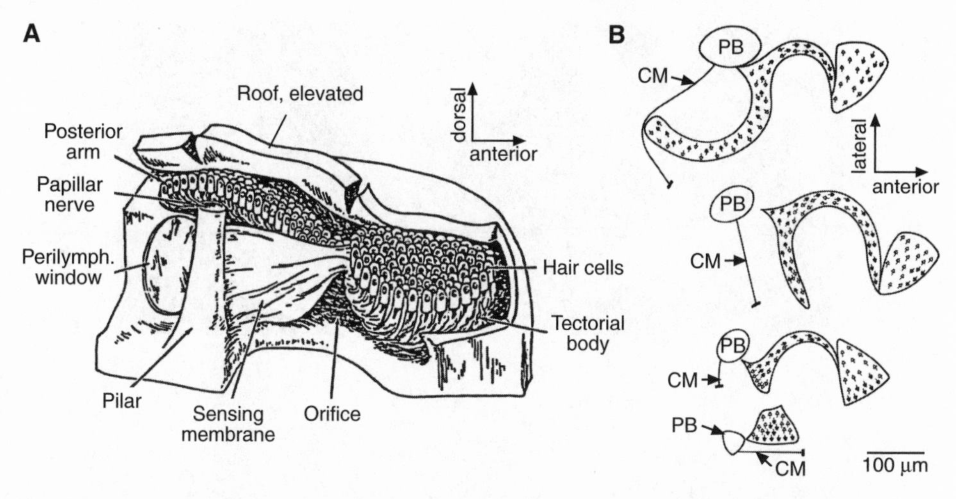

Figure 5.5. (A) Diagram of the amphibian papilla of a typical "advanced" anuran, showing the locations of the two patches of hair cells in relationship to other structures in the inner ear. The hair-cell bases are attached to the roof of the inner ear, and their cilia are embedded in the tectorial membrane (roof, elevated). From Wever 1985, fig. 3.62. (B) Outlines of the amphibian papillae of several species of anurans. Each arrow shows axis of excitation (location of kinocilium with respect to the stereocilia) of many hair cells (see the legend of fig. 5.3). CM = contact membrane (perilymphatic window of [A]); PB = papillar branch of the eighth nerve. The papillae are arranged from bottom to top from "primitive" to "advanced" based on other morphological criteria: *Ascaphus truei; Bombina orientalis; Scaphiopus couchi; Kassina senegalensis.* Only in the most advanced species—represented here by *Kassina* but including members of the families Hylidae, Rhacophoridae, Ranidae, and Microhylidae—is there a long caudal extension of the papilla that projects well down the posterior wall of the papillar chamber. From Lewis 1981, fig. 1.

exist, and even neighboring cells can have opposite polarities, as shown by the sensitivity maps of figure 5.5B. The anatomy of the amphibian papilla varies among species (review: Lewis and Narins 1999). In particular, a posterior extension, which extends the hearing range to relatively higher frequencies (up to about 1.6 kHz), is lacking in species considered to be primitive (fig. 5.5B; Lewis 1981). Differences also exist in the sensitivity maps of primitive and advanced anurans (Lewis 1978).

The tonotopic organization of the receptors in the amphibian papilla of anurans was established by dye tracing in the bullfrog *R. catesbeiana* (Lewis et al. 1982). Afferent fibers in the auditory nerve exhibiting characteristic frequencies between 100 and 300 Hz arise from the anterior patch; neurons tuned to about 200–1000 Hz arise from an S-shaped patch. Within this caudal extension, receptors tuned to about 200 Hz are at the rostral-most end, receptors tuned to about 1000 Hz, at the caudal-most end; afferents with characteristic frequencies of intermediate frequencies are tonotopically distributed between the two ends (fig. 5.6). As in higher vertebrates, there is evidence for traveling

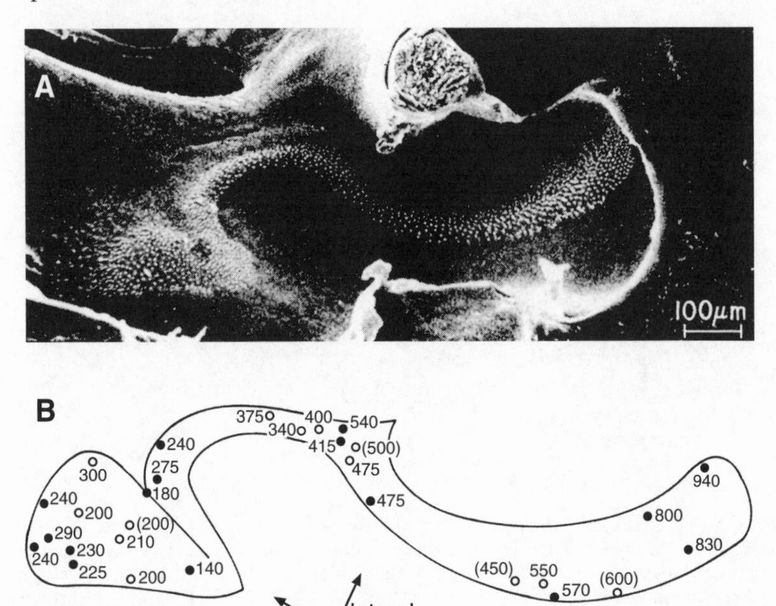

Figure 5.6. (A) Scanning electron micrograph of the amphibian papilla (tectorial membrane removed) of the bullfrog *Rana catesbeiana*, representing an "advanced" anuran. (B) Diagram showing the locations and characteristic frequencies (CF) of primary auditory neurons and demonstrating their tonotopic organization. CFs were determined with both open- (*unfilled circles*) and closed-field (*filled circles*) acoustic stimulation. From Lewis et al. 1982.

waves within the amphibian papilla (Hillery and Narins 1984), but hair-cell tuning also depends on intrinsic factors (see below).

The basilar papilla is anatomically simpler than the amphibian papilla. Although the polarity of the hair cells has been described as uniform, species differences, which are uncorrelated with phylogenetic relationships, exist in sensitivity maps (Lewis 1978; Wilczynski and Capranica 1984). In the bullfrog, afferent axons of primary neurons typically innervate one hair cell, but in the leopard frog *Rana pipiens*, primary fibers branch extensively within the basilar papilla and each innervates an average of five hair cells (Simmons et al. 1992). There is no evidence for tonotopic organization of hair cells within the basilar papilla, and most primary neurons in the same animal have similar tuning properties (Zakon and Wilczynski 1988).

Encoding of Acoustic Signals by Auditory Receptors and Primary Auditory Neurons

In this section we focus on two levels of encoding. First, we show that the information about acoustic properties can be represented in the firing patterns of individual receptors or primary auditory neurons. The range over which receptors can encode this information is, however, limited, and discrimination of two different properties, such as intensity and frequency, is confounded at the level of single cells. Second, we show how in principle the potential information about the signal can be extracted by considering the activity profiles of populations of receptors and primary neurons. Such population or *ensemble coding* overcomes some of the limitations of information encoding by single neurons, thereby increasing the dynamic and frequency range of the auditory system as a whole. The central nervous system assesses the activity of these populations over time, space (tonotopy = labeled lines), or both. At the same time, a recent study of crickets suggests that there are also distinctive populations of auditory receptors that can be recognized anatomically and on the basis of their tuning properties and amplitude responses (Imaizumi and Pollack 1999; Pollack 2001). These populations might have different functional roles in communication and predator detection.

Representation of Stimulus Intensity

As in other sensory systems, stimulus intensity is encoded as firing rate and the response *latency* (delay). Recordings from auditory receptors in insects and primary auditory neurons in anurans usually show a sigmoidal relationship between increasing stimulus amplitude and the number of nerve impulses per unit time (insects: e.g., Michelsen and Larsen 1985; anurans: e.g., Narins 1987; Zakon and Wilczynski 1988). Such intensity spike-rate curves may be linear over a range as small as 20 dB or as great as 60 dB (Capranica and Moffat 1983). At high-stimulus levels, when a further increase in intensity does not result in an increase in firing rate, the neuron is said to be *saturated.* The linear operating range of some of the more sensitive primary auditory neurons innervating the amphibian papilla shifts in the presence of background noise to a higher range of intensities (fig. 5.7A; Narins and Zelick 1988).

Groups of receptor cells tuned to similar frequencies often have different thresholds, so that the ensemble as a whole shows a higher dynamic range than any single receptor. Linear dynamic ranges of as much as 100 dB have been estimated for some insects (fig. 5.7B, Römer et al. 1998) and anurans (e.g., Capranica and Moffat 1983). This form of ensemble coding is called *range fractionation:* as intensity increases, new receptors are recruited.

Modeling of the population coding in populations of auditory receptors in the katydid *Requena verticalis* showed that the overall difference in firing by the two ears would be expected to decrease as absolute intensity increases (Römer et al. 1998). Recall that increasing absolute intensity also reduces the behavioral selectivity of insects and frogs offered choices between sounds that differ in sound pressure level (see chapter 4). Studies of another katydid (*Tettigonia cantans*), suggests that the reduction in amplitude discrimination occurs at the level of the ascending auditory interneurons (Schul and Fritsch 1999).

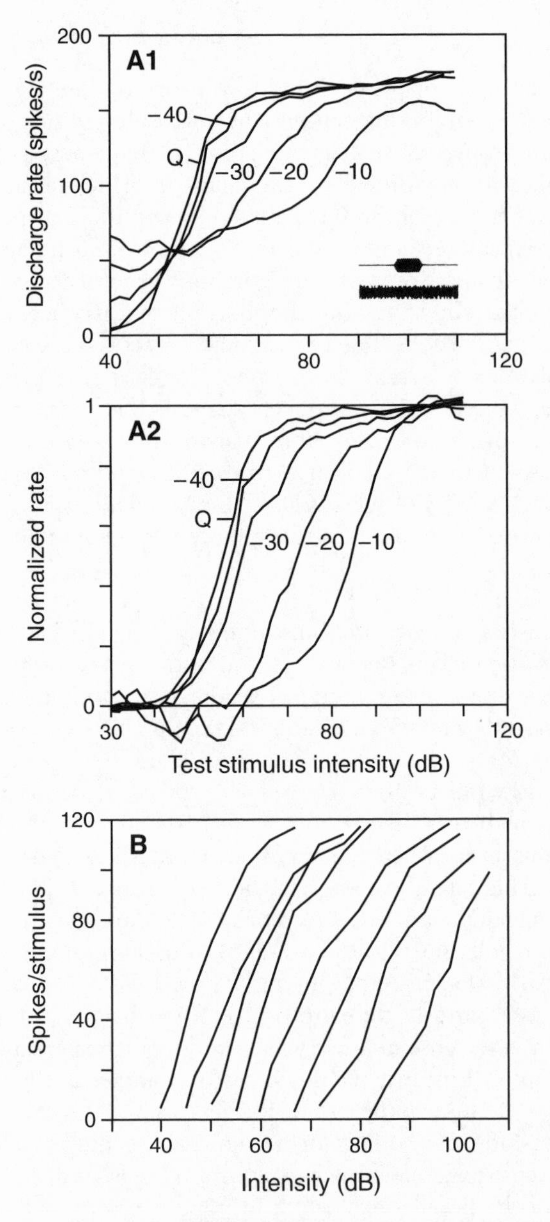

Figure 5.7. (A) Rate-intensity curves and shifts in the operating range of a primary auditory neuron in response to tone bursts presented against a continuous background noise (see inset) in the coqui treefrog *Eleutherodactylus coqui*. (A1) Absolute spike rates; (A2) normalized spike rate function (maximum rate = 1.0). The curves labeled Q show the spike-rate change with stimulus intensity without background noise. Increases of background noise are indicated by the labels −40, −30, −20, and −10, with the last value being the highest noise level. The shift in the linear operating range at the highest (−10 dB) continuous noise level was estimated to be 28.4 dB. From Narins and Zelick 1988. (B) Range fractionation shown by intensity response (IR) functions of eight different receptor fibers in the katydid *Requena verticalis* in response to conspecific songs. Note that the dynamic range of each receptor is limited to about 20 dB, but the ranges of the population of single receptors contribute to a much greater dynamic range, which is estimated to be 40–100 dB SPL. From Römer et al. 1998, fig. 3.

Representation of Stimulus Frequency

Stimulus frequency can be represented in the activity of auditory receptors and primary auditory neurons in two ways. First, at relatively low frequencies (up to about 0.9 kHz; see Narins and Hillery 1983), frequency can be represented in a *synchronization code*, where the timing of spikes correlates with a particular phase of the amplitude-time waveform of the signal. Second, and most commonly, the encoding of stimulus frequency is viewed from the perspective of tuning, the sharpness of which is often quantified in terms of *Q-values* (see appendix 2, part 2). In practice, stimulus frequency and intensity are varied, and the threshold (lowest intensity that elicits action potentials) of a neuron's response is plotted as a tuning curve (fig. 5.8A,B,D). The *characteristic frequency* corresponds to the lowest point of these typically V-shaped curves, or, in other words, it is the frequency to which the receptor or neuron is most sensitive.

Because many combinations of stimulus frequency and intensity within the linear response range can elicit a threshold response or any particular firing rate, the encoding of these two attributes of a signal is confounded. The representation of stimulus frequency and intensity is even more ambiguous at stimulus levels above the linear operating range of a neuron, which can be seen by comparing the tuning curves and iso-intensity functions of the same neuron (fig. 5.8C).

Iso-intensity functions show the rate of firing of a neuron as frequency is varied with intensity held constant. At low intensities the iso-intensity function mirrors the tuning curve, but at high intensities neither differences in stimulus intensity nor frequency are accurately represented by spike rate. In some neurons the firing rate of a neuron reaches saturation over a wide range of stimulus frequencies (fig. 5.8C). In other neurons some frequency selectivity is still seen in the iso-intensity function at high intensities, but a shift occurs in the stimulus frequency that elicits the maximum firing rate (Capranica 1992). As suggested above, tonotopy and population-wide processing in higher centers are two solutions to these limitations.

As in the cochlea of higher vertebrates, the responses of some primary auditory neurons in anurans indicate that strong nonlinear interactions occur at the level of the amphibian papilla at stimulus levels well below those that cause saturation. First, the response of many neurons with low characteristic frequencies can be suppressed by the addition of a second tone, a phenomenon termed *tone-on-tone suppression* (fig. 5.8D; Lewis and Narins 1999). Amphibian papilla neurons with relatively high characteristic frequencies (those in the caudal extension in advanced anurans) and basilar papilla neurons do not show tone-on-tone suppression. Second, some primary neurons are spontaneously active (fire in the absence of acoustic stimulation); their thresholds have to be estimated by assessing changes in firing rate in response to acoustic stimulation. Sometimes the spontaneous rate decreases when sounds are presented, a

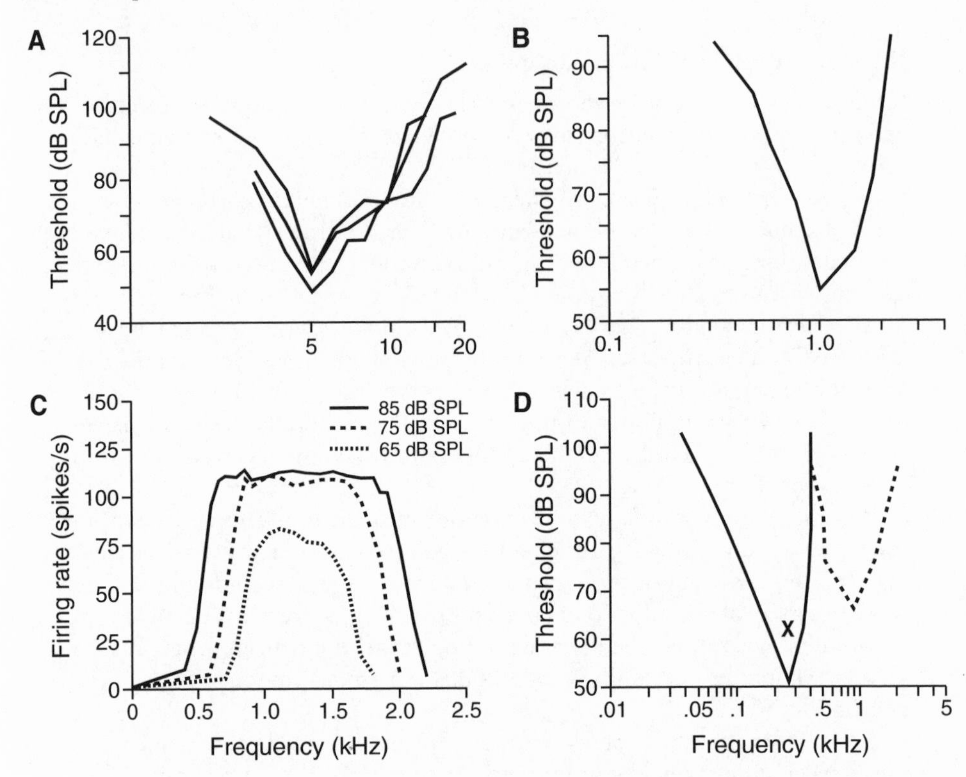

Figure 5.8. (A) Tuning curves of three auditory receptors of *Gryllus bimaculatus* with thresholds at the CF of 5 kHz of 45–55 dB. From Oldfield et al. 1986. (B) Tuning curve of a primary auditory neuron in a green treefrog *Hyla cinerea*, with a characteristic frequency (CF) of 1.02 kHz. (C) Iso-intensity functions for the same primary neuron, whose tuning curve is shown in (B). When stimulus intensity is held constant at a low level and frequency is varied, then the highest firing rate is elicited at the neuron's CF of 1.02 kHz. At higher levels the same (saturated) firing rate is observed over a frequency range of 1–2 octaves. From Capranica 1992. (D) Excitatory and inhibitory tuning curves for a low-frequency sensitive primary auditory neuron, innervating the amphibian papilla, in the green treefrog *Hyla cinerea*. The excitatory curve (*solid line*) has a characteristic frequency (CF) of 0.275 kHz and a threshold of 50 dB SPL. The same neuron showed tone-on-tone suppression. If excited at its CF and 10 dB above threshold (X in the figure), the addition of tones of a higher frequency and SPL completely suppressed the responses of the neuron. Dotted line indicates the tuning curve for suppression, which has a CF of 0.9 kHz. From Capranica and Moffat 1983.

phenomenon that has been termed *one-tone suppression* (Christensen-Dalsgaard and Jørgensen 1996a). Third, as in other vertebrates, some primary auditory neurons respond to *difference tones* that have frequencies above the neuron's characteristic frequency. For example, a neuron with a characteristic frequency of 150 Hz would not respond to tones of 650 and 500 Hz presented separately at a relatively low SPL. When these tones are presented at the same time, however, the neuron responds to the difference tone of 150 Hz (650 minus 500 Hz), even though there is no acoustic energy present at that frequency in the power spectrum. In insects tone-on-tone and one-tone suppression at the level of the population of the receptors have not been described.

Peripheral Determinants of Frequency Selectivity in Insects and Anurans

The mechanical properties of peripheral structures contribute to but do not explain the frequency selectivity of receptors and neurons at higher levels of the auditory system. Studies using laser vibrometry indicate, for example, that the displacement amplitude of the posterior tympanal membrane of *Gryllus bimaculatus* shows a band-pass property and is maximally stimulated by frequencies of 5 kHz, with two smaller peaks at higher frequencies (fig. 5.9A; Larsen et al. 1989). The sharpness of tuning of the posterior tympanal membrane (Q_{3dB} = 4.7) is, however, considerably less than that of single auditory

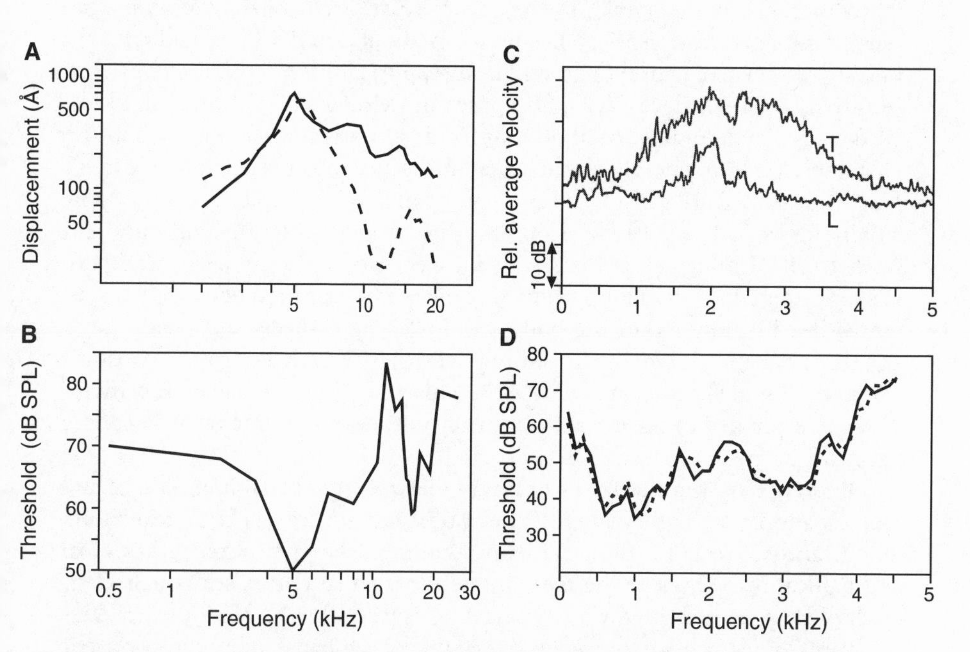

Figure 5.9. (A) Band-pass characteristic of the posterior tympanal membrane of a female cricket (*Gryllus bimaculatus*). Plots of the displacement amplitudes are shown as a function of stimulus frequency measured by laser vibrometry (SPL = 100 dB). Solid line = acoustic spiracle was open; dashed line = acoustic spiracle was closed with wax. The sound direction was always perpendicular to the membrane. From Larsen et al. 1989. (B) Audiograms based on recordings from the whole auditory nerve of *Gryllus campestris* match to some extent the band-pass characteristics of the posterior tympanal membrane, especially when the acoustic spiracle is closed. From Nocke 1971, 1972. (C) Frequency-response characteristics (vibrational amplitude measure by laser vibrometry) of the tympanic membrane (T) and lateral body wall (L, the lung input) of a green treefrog (*Hyla cinerea*). From Ehret et al. 1994. (D) Representative audiograms based on evoked potentials recorded in the midbrain (torus semicircularis) of *H. cinerea*. From Lombard and Straughan, fig. 3.

receptors (Q_{3dB} = 12) (Oldfield et al. 1986), and the same is true for *Teleogryllus oceanicus* (Q-values of 3 for the tympanic membrane and 5.8 for receptors) (Larsen et al. 1989). The sharpness of tuning of receptors in some cicadas (Q_{3db} = 30) not only exceeds the usual values in orthopterans but also that of anurans (Fonseca et al. 2000).

In anurans, laser vibrometry has been used extensively to estimate the frequency sensitivity of various external surfaces that vibrate readily in response to sounds (Narins et al. 1988; Jørgensen 1991; Jørgensen and Gerhardt 1991; Hetherington 1992a,b; Ehret et al. 1994; Lindquist et al. 1998). The tympanic membrane and the lateral body wall input to the lungs also show a general band-pass characteristic corresponding to the general hearing range (fig. 5.9C).

Variation among species exists in the degree to which maximum displacements of the tympanic membrane and body wall match the threshold minima in audiograms. For example, in the green treefrog *H. cinerea*, the tympanic membrane's greatest response lies between about 2 and 3 kHz, and the frequency response of lateral body wall (lung input) shows a region of enhanced sensitivity at 2 kHz (fig. 5.9C,D; Capranica and Moffat 1983; Ehret et al. 1994). By contrast, audiograms based on multi-unit activity in the midbrain and the characteristic frequencies of primary auditory neurons both show maximum frequency sensitivities at about 0.3, 1, and 3 kHz, with a distinct drop in sensitivity between 1 and 3 kHz (Lombard and Straughan 1974; Capranica and Moffat 1983; Ehret et al. 1994). In several species of harlequin frogs (*Atelopus*), the maximum displacement of the tympanic membranes and body wall in species with eardrums and of the body wall in "earless" species matched one of the three enhanced regions of auditory sensitivity estimated from recordings in the midbrain (Lindquist et al. 1998). This same high-frequency sensitivity also corresponded to the carrier frequency in the advertisement calls of these species.

Differences in the sharpness of tuning between tympanal membranes and single auditory receptors suggest that other factors are likely to contribute significantly to receptor tuning. Intrinsic tuning properties of receptors, local mechanical properties of the membrane to which receptors are attached, or both of these factors might be involved (review: Ball et al. 1989). In crickets and katydids, damage to, or even removal of, the tympanal membrane or other mechanical structures does not alter the frequency selectivity of auditory receptors (Oldfield 1985; Rössler et al. 1994). Moreover, the tuning of auditory nerve fibers in larval instar katydids and of auditory interneurons in larval instar crickets is similar to that of adults, even though the tympanal membranes are lacking (Ball and Hill 1978; Rössler 1992b). Similar comparisons of peripheral tuning to that of receptors suggest intrinsic tuning of auditory organs in cicadas (Daws and Hennig 1995/96). *Otoacoustic emissions*, which are thought to arise from active, intrinsic amplification mechanisms within receptors, are sounds that are generated within the ear either spontaneously or in response to acoustic stimulation; these weak sounds propagate from the inner to the external environment through the tympanic membrane. Otoacoustic emissions have commonly been described in vertebrates (review: Köppl 1995), including anurans, and have recently been found in the locust ear (Kössl and Boyan 1998).

In anurans, Lewis (1992) argues that the combination of electrical and mechanical tuning (e.g., tonotopy, traveling waves) provides frogs and other vertebrates with peripheral filters that operate over a wide range of intensity and still preserve good temporal resolution. Primary auditory neurons in anurans have Q_{10dB}-values as high as 8, but the average is between about 1 and 4 (= Q_{3dB} of about 5 to 20), as in most nonspecialized higher vertebrates (Wilczynski and Capranica 1984). Temperature affects the tuning of hair cells in both auditory papillae, primary neurons innervating the amphibian papilla, and spontaneous auditory emissions (Stiebler and Narins 1990; van Dijk and Wit 1995; Lewis and Narins 1999) (see chapter 4 for behavioral evidence for temperature-dependent tuning). These observations support the hypothesis that electromechanical tuning—a metabolically active process—of hair cells contributes to peripheral tuning.

Quantitative differences between hair-cell tuning and that in primary fibers indicate, however, that other mechanisms are also involved, especially with regard to the tuning of receptors in the basilar papilla (Lewis and Narins 1999; Purgue and Narins 2000a,b; Smotherman and Narins 2000). First, the pattern of acoustic energy flow in the inner ear results in different frequency responses of the contact membranes overlying the amphibian and basilar papillae; the peak sensitivity occurs at lower frequencies for the amphibian papilla than for the basilar papilla (Purgue and Narins 2000a). Second, the characteristic frequencies of a sample of primary neurons innervating the basilar papilla in the same animal are very similar, and there is no indication of tonotopy (Capranica and Moffat 1983; Zakon and Wilczynski 1988). Third, the tuning of basilar papilla fibers is little, if at all, affected by anoxia or temperature changes; frequency preferences based on differences within the range of the basilar papilla are also little affected by temperature (see chapter 4). Finally, auditory neurons innervating the basilar papilla do not show tone-on-tone suppression or responses to difference tones. The absence of the latter phenomena and the other properties listed are consistent with the function of a simple resonant structure.

Another source of discrepancies between laser measurements and estimates of frequency sensitivity in the auditory system is the over- or underrepresentation of receptors tuned to particular frequency bands (review: Ball et al. 1989). Recall, for example, that the closely related katydids *Tettigonia cantans* and *T. viridissima* have songs with similar spectra, with the subtle difference that the low-frequency band in the first species is broader and includes lower frequencies than those found in the songs of the second species (see chapter 4, fig. 4.2B). Even though both species have the same number of auditory receptors, females of *T. cantans* but not females of *T. viridissima* preferred signals with the conspecific spectrum. The preference appears to be based on the greater number of receptors tuned to the band of lower frequencies within the low-frequency peak in comparison with *T. viridissima* (Schul 1999).

In grasshoppers, laser holography shows that the different parts of the tympanic membrane have different resonant frequencies (Michelsen 1971b). Because of the interaction of these membrane vibrations and that of Müller's organ itself, Stephens and Bennet-Clark (1982) and Breckow and Sippel (1985) conclude that the different tuning properties of auditory receptor cells seem to be based partly on the presence of tympanal resonances and partly on resonances of the Müller's organ. Resonances of Müller's organ might be influenced by the different positions of the receptor cells. These ideas are supported by a comparative study of grasshoppers by Meyer and Elsner (1996), who measured tuning at the level of the auditory nerve and estimated the vibrational amplitude of the tympanic membrane as a function of stimulus frequency. For nine species a good correlation exists between the tympanal nerve response and frequencies generating the maximum amplitude of vibration near the site of attachment of receptor cells tuned to low frequencies (group III cells) (fig. 5.10A). However, no such correlation is found between tympanal nerve response and the frequency response of the membrane near the site of attachment of the high-frequency tuned receptors (group II cells) (fig. 5.10B). Here the auditory nerve response was maximal to frequencies that were nearly double those that generated the highest vibrational amplitude in the tympanal membrane. Meyer and Elsner (1996) suggest that the different location of the dendrites of the high-frequency receptors compared to that of the low-frequency receptors might explain this discrepancy. Alternatively, nonlinear, intrinsic tuning mechanisms or a differential dependence of receptor response on different aspects of membrane movement (amplitude, velocity, or acceleration) might be part of the explanation.

Matched Filtering in Insects and Anurans?

Many species of insects and anurans have regions of enhanced sensitivity (lower threshold) in their audiograms that correspond to frequencies emphasized in one or more of the signals in their repertoires (figs. 5.10, 5.11, 5.12, 5.13; Nocke 1972; Meyer and Elsner 1996, 1997; Gerhardt and Schwartz 2001). This general match was the origin of Capranica's proposal that the peripheral auditory system of anurans represents a matched filter system (e.g., Capranica and Moffat 1983; chapter 4). The main advantage of such a system is the potential reduction of masking interference from abiotic noise (e.g., wind, wind-blown plants, rushing water) and signaling by other syntopic species that produce different carrier frequencies (chapters 8 and 11). Tuning to conspecific signals can also affect sexual selection (chapter 10) and contribute to species recognition (critical discussion in chapter 4; Meyer and Elsner 1997; and below). Here we show that a great range of variation among species

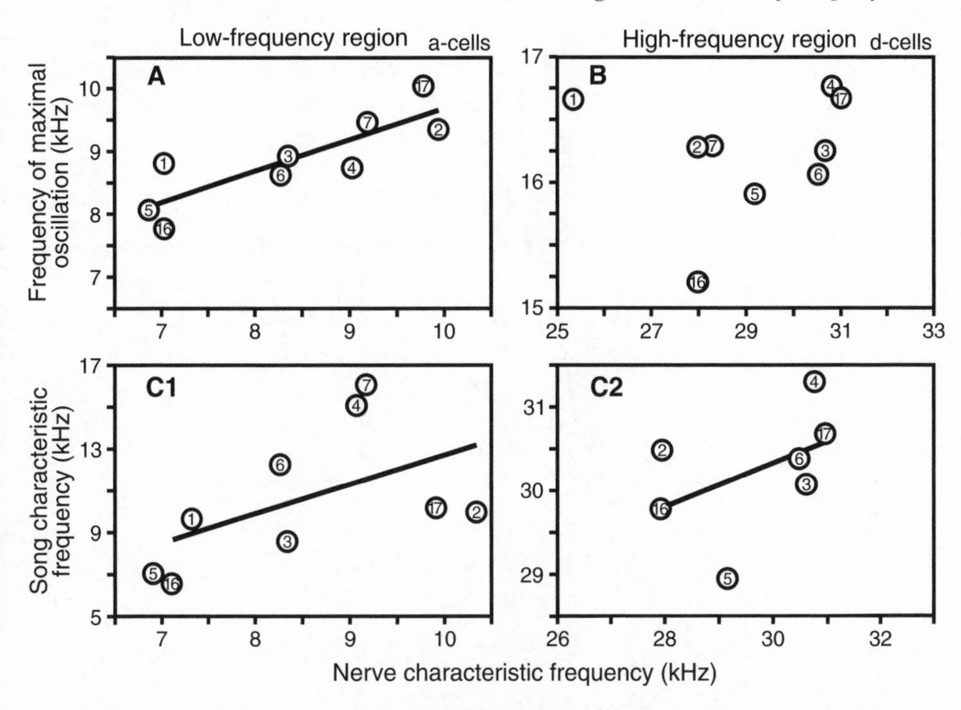

Figure 5.10. Correlations between membrane vibration, tympanal nerve activity, and carrier frequency in acridid grasshoppers. The numbers refer to different species. (A) Scatter diagram and fitted linear regression line for frequencies (range 7–10 kHz) causing maximal tympanal membrane oscillation at the site of attachment of group III cells (Y-axis) and characteristic frequencies of the tympanal nerve (X-axis). (B) Scatter diagram for frequencies (range 25–31 kHz) causing maximal tympanal membrane oscillation at the site of attachment of group II cells. (C1) Scatter diagrams and fitted linear regression line for the low-frequency spectral peaks in the calling songs of nine species of acridid grasshoppers against the frequency showing the "best" response of the tympanal nerve in the same region of the spectrum. (C2) Scatter diagrams and fitted linear regression line for high-frequency spectral peaks in the songs of seven species plotted against the frequency showing "best" response in the same part of the spectrum. Modified from Meyer and Elsner 1996.

exists in the sharpness of peripheral tuning and in the degree to which values of average minimum threshold match the average carrier frequencies in conspecific signals at the population or even species level.

Most estimates of the frequency sensitivity of the peripheral auditory systems of insects and anurans are based on auditory thresholds at the level of the tympanic or auditory nerve or at some higher level in the central auditory system. Such estimates are typically based on some kind of averaged response such as multi-unit recordings from the whole auditory nerve in insects or multi-unit recordings or evoked potentials from the central auditory system in anurans (e.g., Loftus-Hills 1970; Lombard and Straughan 1974; Meyer and Elsner 1996, 1997). Assessment of the distribution of characteristic frequency in samples of primary auditory neurons is another common method (e.g.,

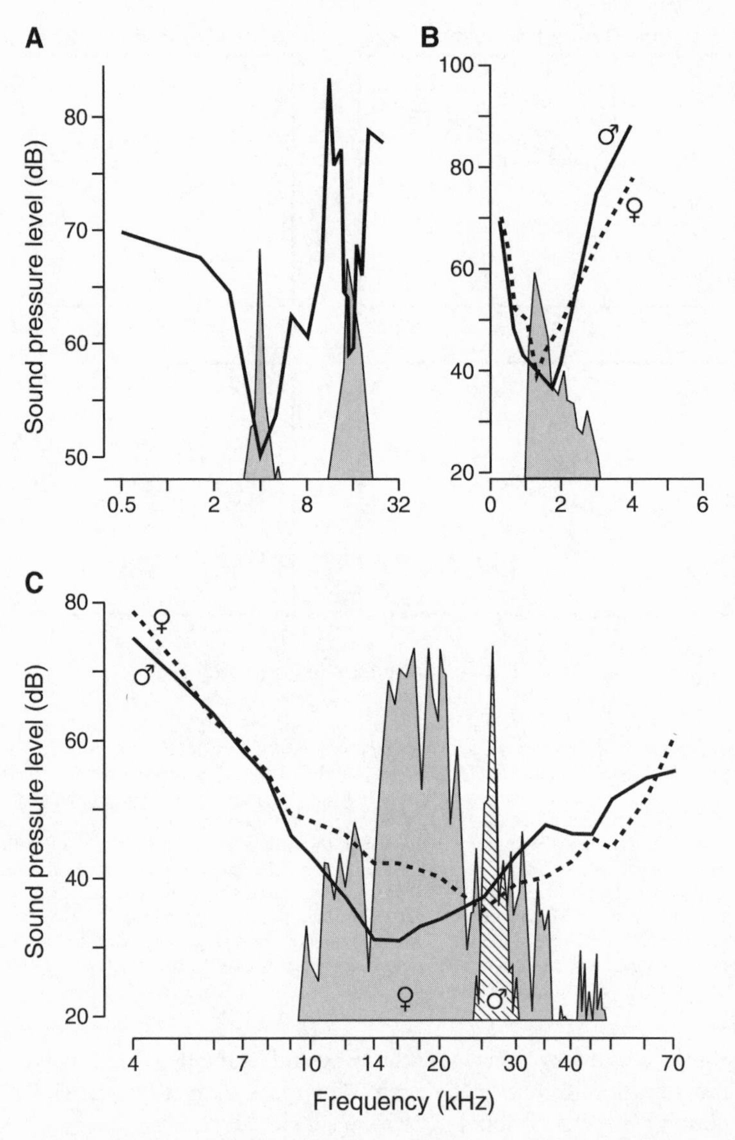

Figure 5.11. Audiograms (*solid, dashed curves*) derived from summed recordings of the tympanal nerve compared with the power spectra (*gray and hatched areas*) of conspecific songs. (A) Field cricket *Gryllus campestris.* Modified from Nocke 1971, 1972. (B) Periodical cicada *Magicicada septendecim.* From Huber et al. 1990. (C) Katydid *Stictophaula ocellata.* Notice differences in the broadband tuning between the sexes. From Heller et al. 1997.

Michelsen 1968; Esch et al. 1980; Capranica and Moffat 1983; Oldfield et al. 1986; Zakon and Wilczynski 1988; Stölting and Stumpner 1998). As discussed above, these physiological estimates do not distinguish between the contributions of peripheral structures (tympanic membranes and pathways to these membranes or directly to inner ear structures) and inherent tuning properties of receptor cells. Moreover, methodological differences can affect these estimates significantly. For example, some estimates of thresholds are obtained by free-field playbacks through a loudspeaker, and others with a closed-stimulation system, such as an earphone sealed on a frog's tympanic membrane. For anurans at least, closed-stimulation tends to exaggerate the low-frequency sensitivity in comparison to free-field stimulation (Pinder and Palmer 1983). If free-field stimulation is used, then the frequency response of the peripheral system will be influenced by the location of the sound source because the relative contributions of direct and indirect sounds on tympanic membranes are both frequency- and location-dependent (chapter 7). Finally, processing that occurs in the ascending pathway (chapter 6) will influence estimates based on recordings from the central auditory system.

At a gross level of analysis, recent surveys of grasshoppers (Meyer and Elsner 1996) and anurans (Gerhardt and Schwartz 2001; fig. 5.12; examples in figs. 5.11 and 5.13) support the idea that tuning in insects and anurans represents some degree of matched filtering. The overall correlation between estimates of maximum frequency sensitivity and the average carrier frequency or frequencies in conspecific calls in a sample of anuran species is especially impressive. The sample includes species with a single carrier band to which either the amphibian or (more often) basilar papilla is most sensitive as well as species with broadband (usually bimodal) spectra, in which case both papillae are involved. In some species, body size affects tuning, and thus the sexes differ because of sexual size dimorphism. Despite differences in neurophysiological methods (single-neuron tuning curves vs. evoked potentials and multi-unit recordings) and the mode of acoustic stimulation (closed systems vs. free-field playbacks), the estimated best excitatory frequency for a species or population was usually within 15% of the average (or midpoint) carrier frequency of the advertisement call. About one-fourth of the data plotted in figure 5.12 do, however, show mismatches of 25% or more. As in insects, species differ in the sharpness of their tuning, which, in general, is not very sharp. Inspection of representative audiograms indicates Q_{3dB} values in the range of 2–3 (fig. 5.13A,B) to as high as about 10 (fig. 5.13C,D); the corresponding Q_{10dB} values are about 2.5–3.

In insects some gross mismatches between the tuning of the whole auditory nerve and the carrier frequency have also been reported (fig. 5.14). In the periodical cicada *M. cassini*, however, the apparent mismatch between the peripheral auditory system and the song spectrum (fig. 5.14A) could not be confirmed by behavioral studies (Simmons et al. 1971). Furthermore, in other

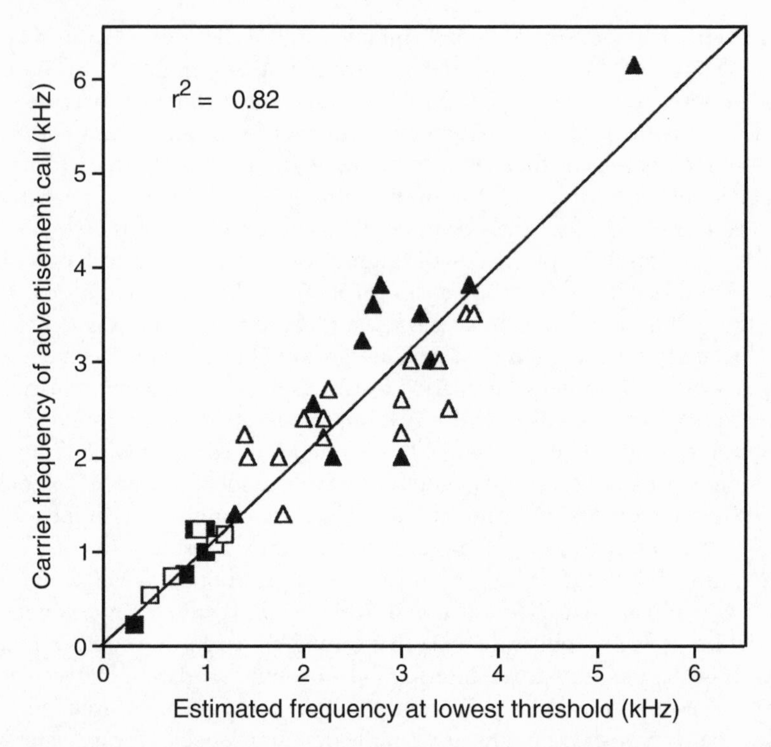

Figure 5.12. Scatter diagram of the mean values of spectral peaks in the advertisement calls of anurans against estimates of minimum threshold, obtained from single-unit data (auditory nerve) (Frishkopf et al. 1968), multi-unit spike data (torus semicircularis) (Mohneke and Schneider 1979), or evoked potentials (torus semicircularis) (Loftus-Hills and Johnstone 1970). Solid symbols show data from studies in which closed-system stimulation was used, and open symbols show data from studies in which free-field stimulation was used. Squares indicate low-frequency sensitivity attributed to the amphibian papilla, and triangles indicate high-frequency sensitivity attributed to the basilar papilla. The line shows where the points would lie if the correlation were perfect. Complete documentation of this figure is provided in Gerhardt and Schwartz (2001). From Gerhardt and Schwartz 2001, fig. 1.

species of cicadas, where a mismatch of as much as 10 kHz was estimated from whole auditory nerve recordings, the tuning of several auditory interneurons with low thresholds (30–40 dB SPL) was found to cover the frequency ranges corresponding to the spectral peaks of conspecific songs (chapter 6; Münch 1999; Fonseca et al. 2000). This result suggests that the tuning of relatively few auditory receptors, which must be camouflaged in whole auditory nerve recordings, are responsible for the frequency selectivity of these interneurons. We propose, therefore, that mismatches based on whole auditory nerve recordings, such as those reported by Mason (1981) and Mason et al. (1999; fig. 5.14B), should be confirmed by additional studies of single auditory receptors and by behavioral studies.

In the katydid *Sciarasaga quadrata*, the hearing system is most sensitive to

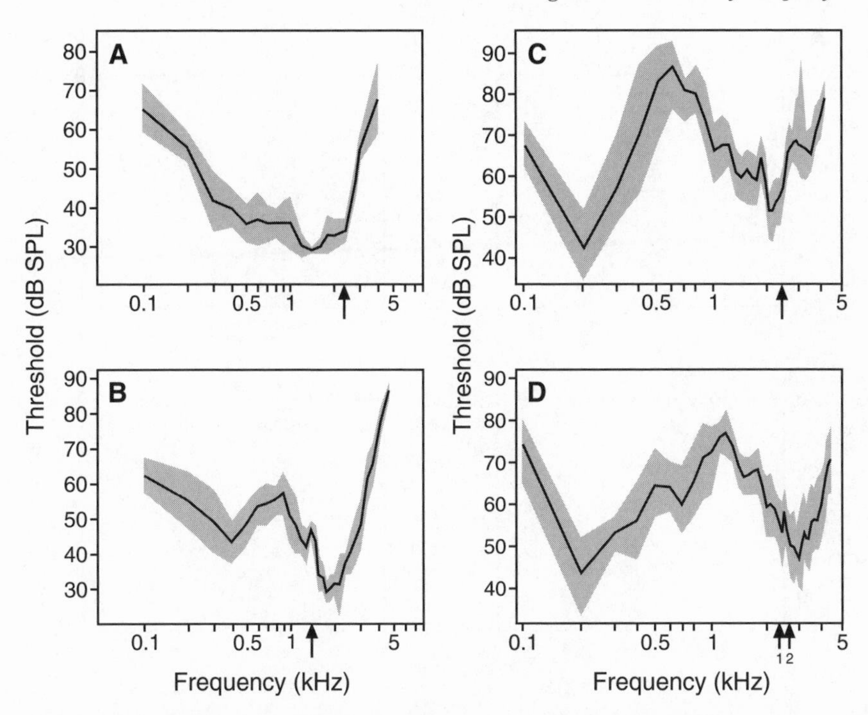

Figure 5.13. Audiograms (solid lines = means; gray areas = standard deviation) based on multi-unit thresholds recorded in the torus semicircularis of four species of anurans. Arrows indicate carrier frequency or frequencies of conspecific advertisement calls. (A) *Rana ridibunda* at 20°C. From Hubl and Schneider 1979. (B) *Alytes obstetricans* at 20°C. From Mohneke and Schneider 1979. (C) *Atelopus chiriquiensis* at 15–18°C. (D) *Hyla regilla* at 15–18°C. Notice that the audiograms of the last two species are very similar, despite the fact that *A. chiriquiensis* lacks a tympanic membrane and middle-ear structures and *H. regilla* has a well-developed tympanic membrane and middle ear. (C) and (D) from Jaslow and Lombard 1996.

frequencies of 15–20 kHz, showing a reduced sensitivity of about 20 dB at the carrier frequency of the call around 5 kHz (Römer and Bailey 1998; fig. 5.14C). However, *S. quadrata* can close its spiracular opening to increase the relative sensitivity of its ears to lower frequencies. Under these conditions, the tuning of the ear better matches the carrier frequency of its call. With partially or totally closed spiracular openings, this species is also able to filter out potentially masking signals that are produced at higher frequencies by sympatric congeners. Römer and Bailey (1998) speculate that the shift in tuning under these conditions parallels a shift in the carrier frequency that might have been an evolutionary response to predation by an ormine fly, *Homotrixa alleni.* By lowering the carrier frequency (over evolutionary time), the song is less likely to be detected by the fly.

As discussed in chapters 4 and 10, frequency preferences (stabilizing or directional) are one of the potential selective forces on the carrier frequency of

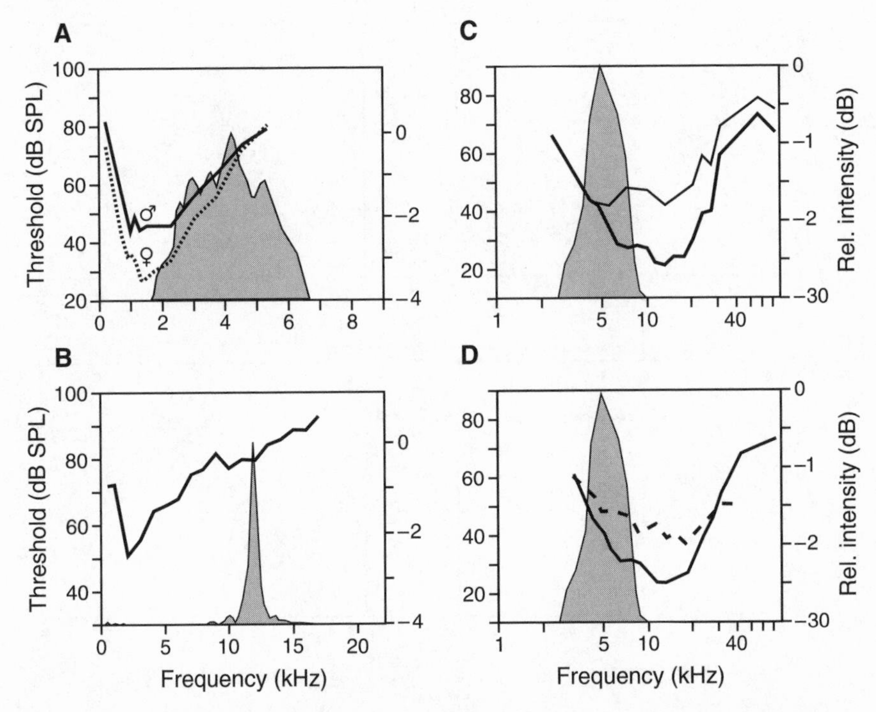

Figure 5.14. Examples of mismatches between whole-nerve (or equivalent) audiograms and conspecific song spectra (*gray areas*) in orthopterans and cicadas. (A) Cicada *Magicicada cassini*. From Huber et al. 1990. (B) Haglid orthopteran, *Cyphoderris monstrosa*. From Mason 1991. (C) Katydid *Sciarasaga quadrata*. Tuning curve of the omega cell, which resembles the tuning of the whole auditory nerve, is shown. Thick, solid line = unmodified, open tracheal system; thin solid line = spiracle blocked completely. (D) Sensitivity of the ear (*dotted line*) of its common predator, the parasitic fly *Homotrixa alleni*. Note that the tuning of both the fly and the katydid in the open-spiracle condition are poorly matched to the spectrum of the katydid's song, whereas by closing its spiracle, the katydid becomes relatively more sensitive to conspecific signals. From Römer and Bailey 1998, fig. 9.

male signals at the within- and between-population levels (Ryan et al. 1990, 1992; Ryan and Rand 1993). Most of the estimates of tuning discussed above, however, are based on threshold estimates of frequency selectivity, and communication in nature takes place at signal and noise levels well above threshold. Nonlinear effects such as saturation are expected to reduce behavioral selectivity, which might explain the lack of frequency preferences in spring peepers (*Pseudacris crucifer*) at relatively high playback levels (chapter 4). Surprisingly, however, elevated background noise levels might enhance the discrimination of signals that have relatively large differences in carrier frequency, presumably by noise-dependent threshold shifts (Schwartz and Gerhardt 1998; see fig. 5.7A). These results again serve to emphasize that audiograms alone can be poor predictors of behavioral selectivity. As shown in box 5.3, which considers a particularly well-studied species (the green treefrog *Hyla*

Box 5.3. Comparisons between Peripheral Frequency Selectivity and Behavioral Preferences: A Case Study

As discussed in chapter 4, frequency preferences in the green treefrog *Hyla cinerea* change as a function of playback level and frequency range (see fig. 4.4). These patterns can be related to some estimates of the tuning and absolute thresholds derived from a large sample of recordings of primary auditory fibers from the eighth nerve (Capranica and Moffat 1983).

As shown in fig. 1, a "low-frequency" population of auditory neurons, whose responses are suppressible by addition of a second, higher-frequency component, is mainly tuned to frequencies below the low-frequency spectral peak in the male's call. A "midfrequency" population of auditory neurons, which are not subject to tone-on-tone suppression, is tuned to the frequency range encompassing the low-frequency peak in the call (0.7–1.2 kHz) (Capranica and Mudry 1993). These two populations of neurons both innervate the amphibian papilla, which in green treefrogs is estimated to contain about 600 hair cells (Wever 1985). However, because of the temperature sensitivity of the tuning of auditory neurons innervating the amphibian papilla, the sensitivity of both populations of neurons might be shifted upward at normal breeding temperatures. (The data presented were recorded from frogs that were as much as 6°C below the average breeding temperature; see Mudry and Capranica 1987b for a discussion.) If so, it is possible that, as in the bullfrog (Capranica 1965), the low-frequency population of neurons might be tuned to the low-frequency peak of the conspecific call, a hypothesis that is supported by behavioral studies of the temperature-dependence of frequency preferences (Gerhardt and Mudry 1980).

A high-frequency population of nonsuppressible neurons, innervating the basilar papilla, is tuned to the high-frequency peak of the conspecific call (centered around 3 kHz). These neurons innervate the basilar papilla, which is estimated to contain fewer than 50 hair cells (Wever 1985). Audiograms based on multi-unit or evoked potentials recorded from the midbrain show three broad areas of enhanced sensitivity that reflect the relative sensitivity of these three populations of primary fibers (fig. 5.9D; Lombard and Straughan 1974).

Furthermore, the thresholds of the most sensitive amphibian papilla neurons in both populations are significantly lower (20–30 dB) than those of the most sensitive basilar papilla fibers (about 50 dB SPL). The latter estimate corresponds well to behavioral estimates of the minimum sound pressure level at which the high-frequency peak (3.0 kHz), in combination with the low-frequency peak (0.9 kHz), begins to influence female preferences for a bimodal stimulus (chapter 4). At high playback levels, significant numbers of

(box 5.3 continued)

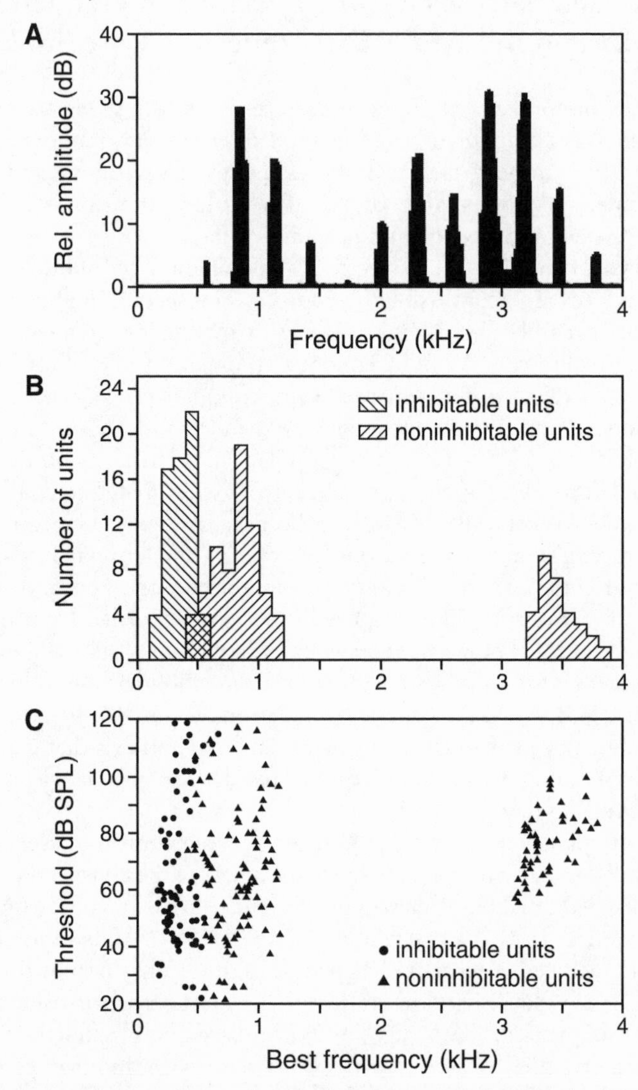

Box 5.3, Figure 1. (A) Power spectrum of the advertisement call of the green treefrog *Hyla cinerea.* Modified from Gerhardt 1974, fig. 2. (B) Histogram showing the distribution of the characteristic frequencies of a sample of primary auditory neurons recorded in the VIIIth nerve. The population with lowest characteristic frequencies could be suppressed by the addition of a second higher frequency tone. (C) Scatter diagram showing the thresholds of primary auditory neurons upon which the histogram of (B) was based. Notice the enormous range of thresholds for neurons with characteristic frequencies below 1200 Hz. From Capranica and Moffat 1983, figs. 7, 8.

(box 5.3 continued)

the most sensitive amphibian papilla neurons that are tuned to the low-frequency spectral peak are likely to be saturated. This could be part of the explanation for the somewhat poorer discrimination of frequency differences in the low-frequency range at high playback levels (85 dB SPL) compared with discrimination at low and moderate levels (65–75 dB SPL) (fig. 4.4A; Gerhardt 1987). On the one hand, this explanation seems unsatisfactory in view of the wide dynamic range that would be provided by the huge range of absolute thresholds (20–120 dB SPL) of neurons tuned to the low-frequency peak. On the other hand, it is hard to believe that such a large proportion of these neurons actually had minimum thresholds in excess of 110 dB SPL.

Recall that the opposite pattern of behavioral selectivity occurred in the high-frequency (basilar papilla) range: discrimination improved with increasing playback level (fig. 4.4B; Gerhardt 1987). One possible explanation for this result is that some minimal number of neurons need to be active and that higher playback levels are therefore needed to stimulate the relatively insensitive basilar papilla neurons, which are also few in number. Finally, whenever a low- or high-frequency peak in a call falls outside the optimum tuning range, the balance of input from the two channels to the central nervous system is unequal. As shown by quantitative behavioral studies, females prefer synthetic calls in which the balance of energy in two spectral peaks is about the same to alternatives in which one or the other peak is significantly attenuated (fig. 4.6; Gerhardt 1981b).

In summary, the audiogram of *H. cinerea* alone does not predict the differences in frequency selectivity that occur at different, above-threshold sound levels. The bases for these different behavioral results lie in how the information from these different frequency channels affects the neurons and neural circuits on which this information converges in the central auditory system, mainly in the torus semicircularis and thalamus (chapter 6). In principle, the behavioral data, in conjunction with quantitative studies of peripheral activity, could predict the optimal range of relative excitation of the two inner ear organs that will best excite these higher neural elements. The convergent excitation will be affected both by the relative intensity and frequencies of spectral peaks in communication signals. Because absolute levels and relative amplitudes of the two peaks will vary with distance, these frogs, as has been proposed for katydids (Römer 1987), could use these changes to estimate the distance to the sound source.

cinerea), differences in frequency preferences at different playback levels might be explained, at least in part, by the significant difference in the absolute sensitivity of the two differently tuned auditory organs.

Sun et al. (2000) used computer simulations to investigate hypothetical matches between audiograms and the spectra of advertisement calls before and after propagation through natural environments. In one species (*Physalaemus pustulosus*), the match improved at the longer distance, and in a pair of subspecies of *Acris crepitans*, the match was poorer at the longer distance. The possibility that tuning could be optimized for a given communication range is intriguing and deserves further research. Given the serious problems of inferring behavioral selectivity from audiograms alone, however, this hypothesis is best tested directly by offering females choices of signals recorded at close and far distances and adjusting playback levels accordingly in each comparison.

Representation of Fine-Temporal Acoustic Properties in the Auditory Periphery

As discussed in chapter 4, insects and anurans show strong preferences in the context of mate choice that are based on subtle differences in fine-time scale properties such as pulse rate, silent gaps within pulses, pulse duration and rise-time, and even waveform periodicity. In chapter 8 we show that insect and anuran signalers often respond very rapidly and precisely to the signals of neighbors or to brief drops in background noise levels. All of these selective responses require these animals to resolve small differences in signal amplitude in the time domain.

Estimates of Temporal Resolution

Temporal resolution is limited ultimately by the response characteristics of tympanic membranes, receptor cells, and auditory neurons. As discussed in appendix 2, part 2, fine frequency resolution is achieved only by a concomitant loss of fine time resolution (see Michelsen et al. 1985). Sharply tuned (high-Q) systems respond to a transient signal with a series of oscillations whose total duration is longer than that of the transient. The response to the onset of the second of two transients thus can be masked by the continuing response to the first transient (fig. 5.15). Estimates of impulse functions of the tympanic membranes of moths and grasshoppers using laser vibrometry show, however, that mechanical factors are unlikely to limit temporal acuity, because transients separated by less than 100 microseconds are still evident in the vibratory pattern of such membranes.

The temporal acuity of single receptors and primary auditory neurons is limited by properties common to all neurons: the time constants of integration, spike duration, and the *refractory period*, which is the time just after an action potential when a second stimulus has no effect or a reduced effect. The last two properties set an upper limit on the firing rate of a neuron and thus

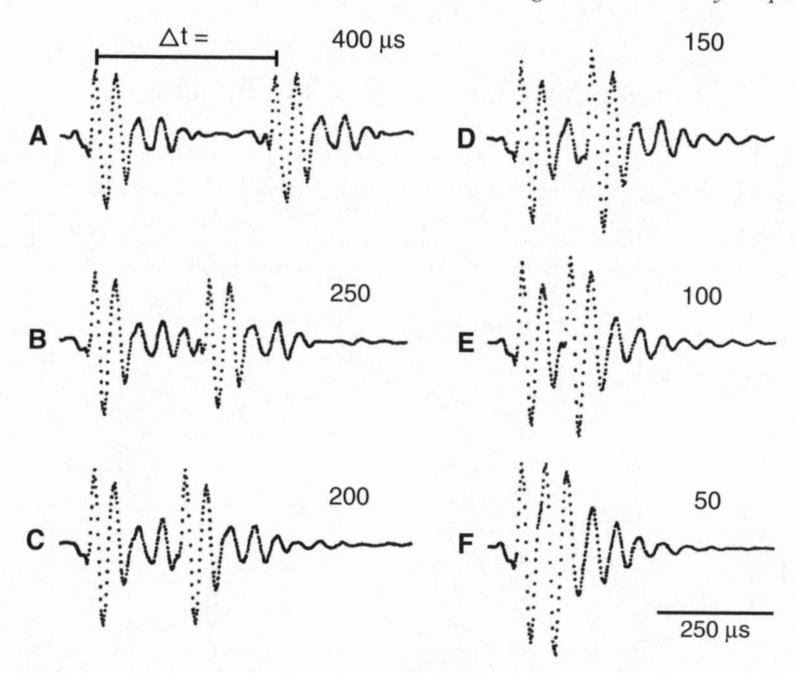

Figure 5.15. Example of the limitation of temporal resolution by tympanic membrane properties. Oscillograms show the vibration response of the tympanic membrane of a noctuiid moth in response to a pair of 16 μs pulses, whose interpulse interval was systematically decreased. Two visually separable responses occur until the interval is about 100 μs or less. From Michelsen et al. 1985.

on the maximum rate of one-to-one synchronization to periodic changes in the acoustic waveform (see below). *Temporal integration* is the summing up of responses to two or more signals over time. At the level of receptors and in behavioral (psychophysical) studies, temporal integration is studied by observing changes in the threshold for detection (or firing rate of neurons) as the duration of a very short signal is increased or as the time (gap) separating two transient signals is varied. Other, indirect methods such as the generation of a temporal *modulation transfer function* (see box 5.4) assume that receptors (or the auditory system as a whole) can be considered as linear systems.

Two kinds of time constants for temporal integration have been reported. Studies using two-pulse stimuli, modulation transfer functions, or both yielded estimates of *minimum integration times* or gap detection thresholds ranging from about 0.25 and 1 ms in two species of anurans, the coqui treefrog *Eleutherodactylus coqui* (Dunia and Narins 1989b) and the leopard frog *Rana pipiens* (Feng et al. 1994). Studies of auditory receptors in insects estimate time constants of the order of 4–8 ms. Gap detection thresholds of < 2 ms have been found in grasshoppers (Ronacher and Römer 1985), but the methodology, threshold criteria, and interpretation of the data are not really comparable to

Box 5.4. Temporal Modulation Transfer Functions

The indirect computation of a time constant for minimum temporal integration makes the assumption that a receptor, and indeed the whole auditory system, approximates a low-pass filter. A low-pass filter transmits low-frequency

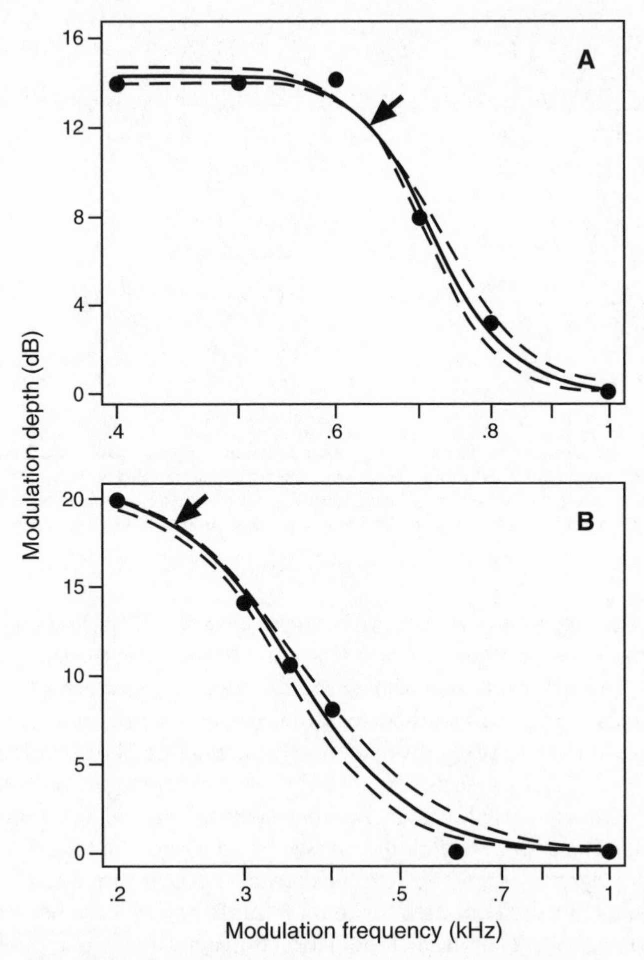

Box 5.4, Figure 1. Sample temporal modulation transfer functions for two primary auditory neurons of a coqui treefrog. The points show the modulation depths and rates of amplitude modulation at which the neurons showed significant synchronization to the stimulus based on an analysis of 200 spikes: (A) the 3 dB down point is about 650 Hz (see arrow), and the minimum integration time using the formula above is 0.244 ms; (B) the 3 dB down point is about 250 Hz (see arrow), and the minimum integration time, 0.64 ms. From Dunia and Narins 1989b, fig. 3.

(box 5.4 continued)

signals better than high-frequency signals of the same intensity. In practice, the slope of the attenuation of a biological filter such as an auditory neuron is estimated by presenting a series of sinusoidally amplitude-modulated signals. The rate and depth of amplitude modulation (peak to valley change in intensity) are systematically varied, and for a series of rates, the minimum depth at which a neuron synchronizes is plotted (appendix 3, fig. A3.1). The criterion for judging whether a neuron effectively synchronizes to a particular stimulus is usually based on a measurement of vector strength (1.0 = perfect synchronization; 0 = no synchronization) and some arbitrary cutoff value (0.4 has been used in studies of primary auditory neurons in anurans by Narins and Hillery 1983). Investigators estimate the 3 dB down point for the function, and from this they compute a time constant using equations from linear systems theory, that is,

$$\text{minimum integration time} = 1/(2\pi * \text{frequency at 3 dB down})$$

the anuran studies (Tougaard 1996). Minimum interspike intervals in insect receptors are of the order of 2–3 ms, so gap detection thresholds < 2 ms must be explained by ensemble and synchronization coding, which is also tacitly assumed by the anuran studies.

The *maximum integration time* (or critical duration) of primary auditory neurons is estimated by increasing the duration of a stimulus until there is no further decrease in the neuron's threshold. Maximum integration times for auditory neurons in the coqui treefrog ranged from about 190–270 ms (Dunia and Narins 1989a). Behavioral studies of this treefrog indicate that the preferred duration of acoustic stimuli in field playback experiments is about 100 ms, thus suggesting that signal detection has not been evolutionarily optimized in this species by changing call duration to match the maximum integration time (Narins 1992a). The interpretation of integration times, whether determined physiologically or behaviorally, is highly controversial, and the reader should consult several chapters in Michelsen (1985) for comprehensive reviews.

Because estimates of temporal resolution based on the auditory threshold and firing rates depend on stimulus frequency, intensity, and duration, these acoustic properties cannot each be unambiguously represented in the firing patterns of single auditory fibers. Still another confounding factor is variation in the adaptation properties of neurons. Insect receptors and anuran primary auditory neurons most commonly discharge throughout the duration of an auditory stimulus. The rate of firing, however, may vary from the beginning of the stimulus to end. In auditory sense cells of insects and primary neurons of vertebrates, the firing rate is usually higher near the beginning of the first stimulus when compared with consecutive stimuli and then declines more or less rapidly to a steady-state rate that is maintained until the end of the stimulus (e.g., Wilczynski and Capranica 1984; Schul 1997). Whereas primary fibers innervating the basilar papilla of anurans show the same basic adaptation pattern as higher vertebrates, neurons innervating the amphibian papilla show a wide variety of patterns, ranging from essentially no adaptation, to a quasiphasic response consisting of a high firing rate at the onset of a stimulus followed almost immediately by a drop to rates that are just above spontaneous rates (Megela and Capranica 1981). As shown in figure 5.16, the whole gamut

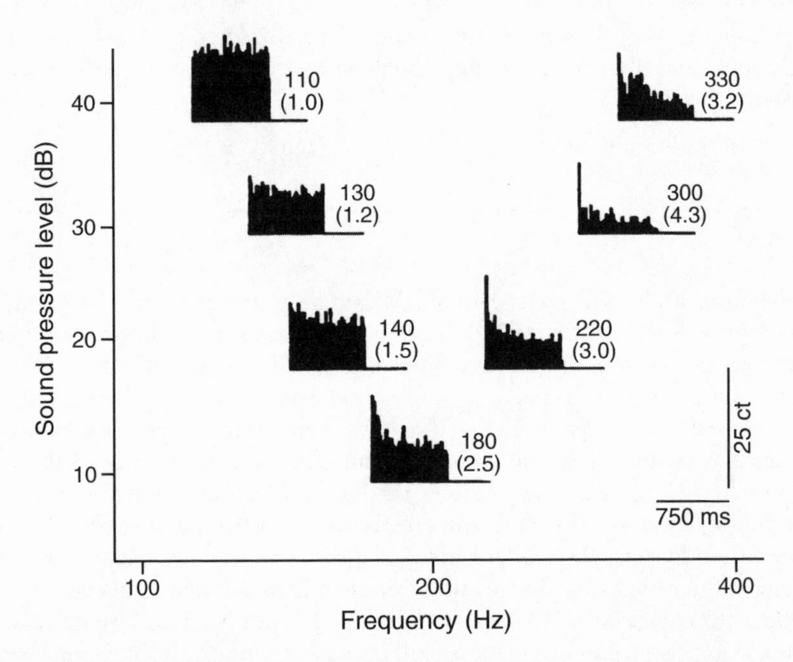

Figure 5.16. Peristimulus histograms (activity of a neuron during presentation of an acoustic stimulus of 1 s duration) showing a range of adaptation patterns of a primary auditory neuron innervating the amphibian papilla of the bullfrog *R. catesbeiana.* Numbers to the right of each histogram show the stimulus frequency (ranging from 110 to 330 Hz), and the numbers in parentheses indicate the adaptation ratio (= maximum response [number of spikes]: steady-state response [number of spikes in final 60 ms of the response]). The vertical axis shows the spike count, and the horizontal axis, time (see inset, *lower right*). When stimulated with a tone burst with a frequency equal to the neuron's characteristic frequency and 10 dB above its threshold (*lowermost histogram*), the adaptation pattern was comparable to that typical of higher vertebrates. At higher intensities and lower frequencies (110–140 Hz; histograms in the left part of the figure), the neuron showed little adaptation but fired at about the same rate throughout the stimulus. At higher frequencies (220–330 Hz) and intensities (histograms in the right part of the figure), the neuron showed a high spike rate at the very beginning of the stimulus, followed almost immediately by a drop in rate by a factor of 3–4. From Megela 1984.

of adaptation functions is sometimes observed in the same neuron, depending on stimulus frequency and absolute intensity (Megela 1984). Lewis and Narins (1999) discuss putative mechanisms, which in part are related to one- and two-tone suppression.

Encoding of Low-Frequency Sounds and Pulse Rate

As mentioned earlier in this chapter, low-frequency sounds and the pulse rates typical of most insect and anuran signals can be encoded in the firing pattern of auditory receptors and primary auditory neurons. In effect, the period of the stimulus is copied because the action potentials occur at a particular phase of a sinusoidal signal or just after the onset of each pulse in a train of pulses (details in Esch et al. 1980; Huber et al. 1980; Morris and Fullard 1983; Capranica et al. 1985; fig. 5.17A,B). Because of a neuron's refractory period, one-to-one coding (one spike per cycle or pulse) will generally be limited to values below about 500 Hz. Even if a neuron does not fire to each cycle or pulse in a stimulus, however, information about the periodicity of a signal can still be encoded if there is a preferred phase of firing. If so, then an interspike histogram will show a series of peaks with intervals equal to the time period of interest (Rose and Capranica 1985). Such synchronization has been documented in primary auditory neurons of anurans in response to sinusoidal signals with frequencies as high as about 0.8–0.9 kHz (fig. 5.17C,D; Narins and Hillery 1983).

No evidence exists that temporal selectivity at the level of the peripheral auditory system correlates with behavioral selectivity for species-typical rates that occur in communication signals. That is, while receptors and primary neurons may show some variation in gap thresholds or in the maximum rate of amplitude modulation to which they show robust synchronization, individual receptors, primary neurons, or whole auditory nerves are not specifically tuned to pulse rates or gap lengths exhibited in signals important for the behavior (see Popov et al. 1974; Rheinlaender et al. 1976; Rose and Capranica 1985; Huber 1990). Within the limits of rate coding and synchronization, heterospecific pulse rates will be encoded just as accurately as conspecific pulse rates, at least at a gross level of analysis.[1] Thus, the decoding of the peripherally encoded temporal information must be accomplished by additional processing at the level of the central auditory system (chapter 6). Moreover, as the data discussed below concerning the resolution of temporal differences suggest, such processing will almost certainly involve averaging over the responses of some subset of the population of peripheral receptors, over multiple stimuli, or both.

---

1. In a recent study conducted from an information-theoretic perspective, Machens et al. (2001) show that coding efficiencies for stimuli with fine-time-scale variations in amplitude were higher when these variations were behaviorally relevant.

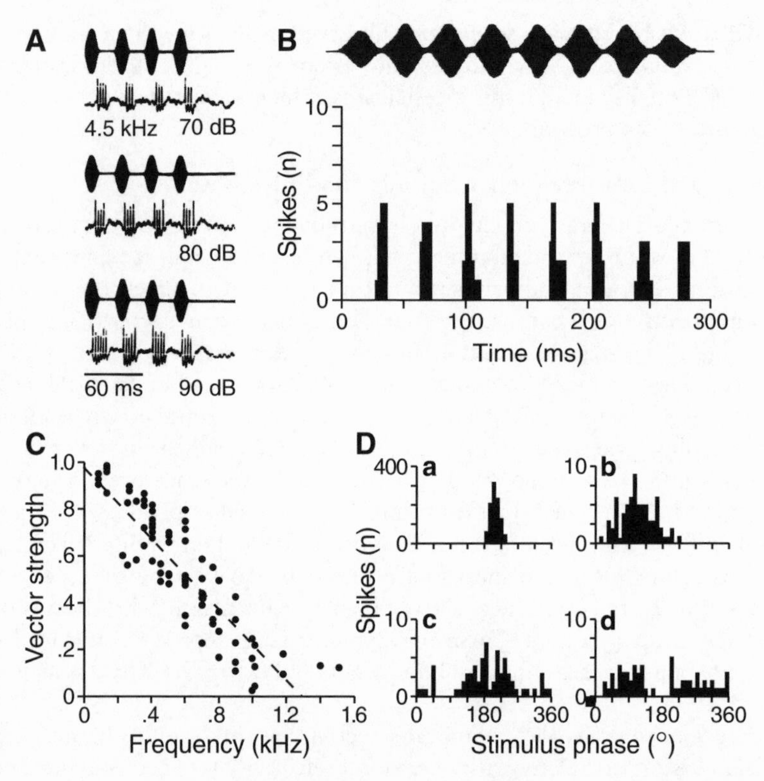

Figure 5.17. Copying of song patterns by single auditory receptor cells or receptor populations. (A) Synchronization to pulses in a synthetic calling song by a single auditory nerve fiber, tuned to 4–5 kHz, at three different sound intensities in *Gryllus campestris.* From Esch et al. 1980. (B) Peristimulus histogram of firing by a primary auditory neuron in a leopard frog (*Rana pipiens*) to repeated presentations of a sinusoidally amplitude modulated (38 Hz) burst of noise (diagrammed above). Notice that the neuron tends to fire at the same time (phase), near the end of each cycle of modulation. From Rose and Capranica 1985. (C) Plot of the vector strength (1.0 = perfect phase locking; 0 = no phase locking) versus stimulus frequency for six primary auditory neurons in the coqui treefrog *Eleutherodactylus coqui.* The neurons were stimulated at different frequencies and intensities. (D) Four samples of peristimulus phase histograms, showing the phase of firing to repeated presentations of modulated tone bursts: (a) 130 Hz at 75 dB; (b) 410 Hz at 85 dB; (c) 700 Hz at 105 dB; (d) 1001 Hz at 105 dB. Based on the arbitrary criterion of a vector strength of 0.4 or higher, the plot in (C) shows that significant phase locking was observed at frequencies of about 800 Hz. From Narins and Hillery 1983.

Encoding the Fine Structure of the Acoustic Waveform in Anurans

The potential for temporal encoding of periodicity and phase relationships in spectrally complex signals by primary auditory neurons has been extensively studied in the bullfrog (Schwartz and Simmons 1990; Simmons et al.1993a,b; Bodnar and Capranica 1994). The temporal periodicity of the waveform as a whole dominates the synchronized firing rate of these neurons rather than the periodicity of the frequencies to which each is most sensitive. Shifts in the relative phase of one or more spectral components, which in turn alter the fine details of the periodic waveform and the degree to which periodicity is reflected in the overall amplitude-time envelope, also affect the accuracy of synchronization and even spike rates in some primary fibers innervating the amphibian papilla (Simmons et al. 1993a,b; Bodnar and Capranica 1994). However, behavioral studies of bullfrogs (Hainfeld et al. 1996), green treefrogs (Gerhardt 1978b; Gerhardt et al. 1990), and barking treefrogs (Bodnar 1996) fail to provide robust evidence for the importance of differences in relative phase alone. Although female barking treefrogs can discriminate between signals in which waveforms differ in both phase and frequency modulation (Bodnar 1996), the biological significance of detecting such subtle differences is unknown. These features of the signals are also likely to be degraded after propagation over relatively short distances.

Encoding of Gaps within Signals and Pulse Rise-Time

Recall that females of the grasshopper *Chorthippus biguttulus* failed to respond to noise bursts that mimicked the pulse trains (syllables) of male songs if gaps of more than 1–2 ms occurred within these trains (Helversen 1972). Ronacher and Römer (1985) examined synchronization in receptor cells in the grasshopper in this context by presenting noise bursts of less than 100 ms, into which they introduced 6–11 gaps varying in duration from 1 to 7 ms. Synchronization within individual receptors reliably encodes gaps as small as 1 ms if their responses were averaged over about 50 stimulus repetitions (fig. 5.18A). By comparison, fewer stimulus repetitions are required to detect gaps of this magnitude when responses are averaged over nine receptors or in the summed action potentials recorded in the auditory nerve. Thus, single receptors cannot reliably encode behaviorally significant gaps if exposed to just a few repetitions of a signal, but summation over many stimulus representations or by multiple receptors can easily overcome this limitation.

As discussed in chapter 4, males of *C. biguttulus* prefer pulses with ramplike onsets rather than pulses with steep rise-times; similarly, females of the gray treefrog *Hyla versicolor* prefer pulses with slow rather than fast rise-times (Gerhardt and Schul 1999). Studying auditory receptors in the locust (receptor properties appear to be nearly identical to those in *C. biguttulus*), Krahe and Ronacher (1993) found that there was considerably greater variation in the response latency of receptors when stimulated by noise bursts with pulses of slow rise-times and rapid decays than when stimulated by time-reversed stimuli (fig. 5.18B). Differences in the degree of synchronization, as quantified in the responses of these receptors to 50 stimulus repetitions, could reliably encode the difference in rise-time. A similar scheme for encoding rise-time in gray treefrogs probably occurs at the level of the dorsolateral nucleus, the first station in the ascending pathway (Schul and Gerhardt, unpubl. data; see chapter 6).

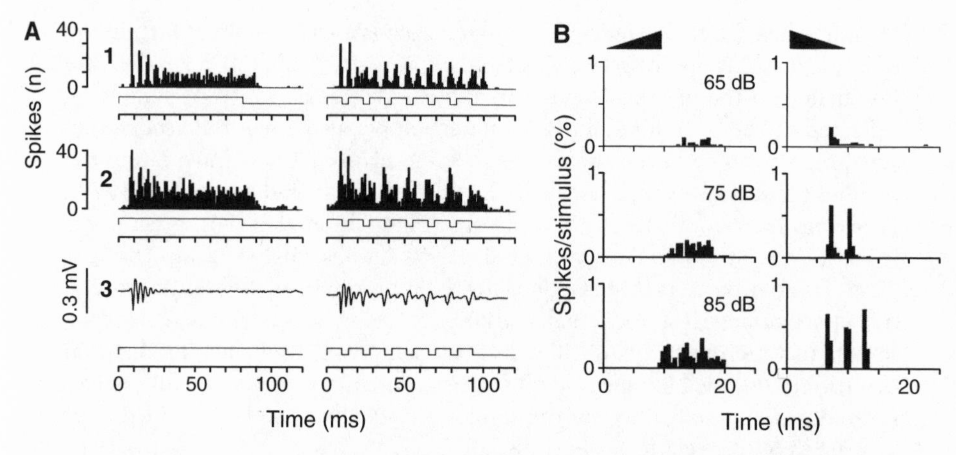

Figure 5.18. Encoding of gaps and rise-time by averaging. (A) (1) Poststimulus time histograms of the responses of a single auditory receptor in the grasshopper *Chorthippus biguttulus* to 50 stimulus repetitions of a noise burst that simulated syllables in male songs. In the histogram to the left, the stimulus (see diagram below) did not have any silent gaps; in the histogram to the right, there were six gaps of 4 ms each per burst. (2) Poststimulus histograms based on the computer-aligned and summed responses of nine different receptors, each stimulated with six stimulus repetitions with no gaps and six gaps of 4 ms each. (3) Summed action potentials from the tympanic nerve based on 20 stimulus repetitions. Taken together these records show that single receptors reliably encode the gap periodicity only when their responses are averaged over many stimulus repetitions. However, summing across receptors (whole nerve recording and the computer-summed responses of 6 receptors) shows reasonably reliable encoding of gap periodicity over fewer stimulus repetitions. From Ronacher and Römer 1985. (B) Poststimulus histograms of a receptor cell in the locust to 50 stimulus repetitions of white-noise pulses at different intensities. The left panels show that the onset of the first spike in response to the slow rise-time pulse (diagram at top) was later, and the scatter in the timing of spikes was greater in comparison with responses to the time-reversed pulse with a very fast rise-time. From Krahe and Ronacher 1993.

Methodological Considerations for Studying Temporal Resolution

As suggested above, the insect could average over signals in single receptors, over the population of receptors to relatively few signals, or some combination of receptors and signal repetitions. What, then, are biologically meaningful numbers of repetitions upon which we should base calculations that provide estimates of temporal resolution? Ronacher and Römer (1985) justified their use of 50 stimulus repetitions per single receptor on the basis of an estimate that approximately 50 receptors in the grasshopper would be active at the sound pressure levels (64 and 76 dB) at which most neurophysiological measurements and behavioral studies have been conducted. Most studies of anurans, however, have used some arbitrary number of repetitions without any justification (e.g., 10: Rose and Capranica 1985; 25: Feng et al., 1994; 20–60: Simmons et al. 1993a,b). One departure from the usual procedure was the method used by Dunia and Narins (1989a,b), who chose a minimum number of spikes for data processing and then noted the duration or number of amplitude-modulated sounds required to elicit this number.

Another, more direct way of approaching this question is to use behavioral studies to estimate the number of repetitions of signals that animals require for reliable discrimination. In the gray treefrog *H. versicolor*, for example, females selectively orient to 4–6 pulse signals with attractive pulse rates or slow rise-time pulses after hearing only one or two repetitions of these attractive stimuli and alternatives with unattractive values (Gerhardt, unpubl. data). Typical advertisement calls each have 14–20 pulses. In the grasshopper *Chorthippus biguttulus*, males recognized shortened versions of the communication signals of conspecific females (Ronacher and Krahe 1998). A three-subunit female signal of 250 ms duration, corresponding to less than a quarter of the natural duration of a phrase, which usually consists of 10–12 subunits, was sufficient to elicit a high proportion of turns toward the speaker. Taken together, these behavioral data suggest that statistics based on the usual number of stimulus repetitions (10–60) will probably overestimate the temporal resolution of individual auditory neurons. Such neural data can be used to estimate the number of receptors or other, lower-order neurons whose activity would presumably have to be averaged by the central auditory system in order to extract fine-temporal information. For example, a recent quantitative analysis showed that a binaural difference of just 1.5 dB was required for a correct turning response by males of *Chorthippus biguttulus* to a 250 ms stimulus (Ronacher and Krahe 2000). Estimates of the differences in spike counts elicited in a pair of receptors (one in each ear) presented with the same stimulus suggest that the insect must integrate the activity of 6–13 receptors to achieve the observed level of behavioral resolution.

Summary and Suggestions for Future Studies

The tympanal organs of insects and the peripheral structures of anurans that are set into vibration by sound waves are situated so that sound waves affect both the external and internal surfaces of the membrane. When sound frequency and source direction are held constant, the tympanic membranes respond linearly over a reasonably large range of stimulus intensity. In insects tympanal membrane vibration is affected during sound production, ventilation, and locomotion in such a way that sensitivity to external sounds is greatly reduced (see box 5.1). In anurans there is evidence that sound-induced vibrations also reach the inner ear organs via nontympanic routes.

Auditory receptors in insects and anurans have basically similar response properties in terms of representing stimulus intensity (dynamic range), frequency, and duration. Because these properties all affect thresholds and the firing rates of primary auditory neurons at higher stimulus levels, the representation of each property at the level of single cells is ambiguous. Labeled-line and ensemble coding are used by the central auditory system of higher vertebrates to disentangle variation in these acoustic properties. Although tonotopic organization of receptors occurs in the peripheral hearing organs of insects and in the amphibian papilla of anurans, there is little evidence that tonotopy is preserved in the brain of insects or in the highest auditory nuclei of anurans (chapter 6). Primary auditory neurons innervating the amphibian papilla of anurans show several nonlinear properties. Some, such as tone-on-tone inhibition and responses to difference tones, also occur in higher vertebrates but not in insects. Others, such as the wide-range adaptation properties of amphibian papilla neurons, are among terrestrial vertebrates apparently unique to anurans.

Reasonably good matches exist between frequency bands of maximum sensitivity and the spectra of conspecific signals, although there are a few notable exceptions that require further study. Peripheral tuning of tymanic membranes is relatively broad, and receptor tuning may involve mechanisms both intrinsic to receptor cells and micromechanical factors. The temperature-dependence of tuning in the amphibian papilla, together with the fact that carrier frequency changes relatively little, if at all, with temperature, means that mismatches will always occur at some temperatures. This phenomenon represents a fundamental constraint on the communication system, and, in general, mismatches in tuning can be sources of sensory biases (see Römer and Bailey 1998 and chapter 11). However, the significance of slight mismatches between carrier frequency and the weak tuning of the auditory system, which have been based almost entirely on data collected at near threshold, have probably been overemphasized. Insects and anurans typically communicate at signal and noise levels that are far above threshold, and the auditory system is notoriously nonlinear. We badly need information on the distances and range of sound pressure levels at which these animals normally make choices.

Primary auditory neurons show reasonably good synchronization to pulsatile stimuli below about 1 kHz, but there is no selectivity for particular periodicities, such as those typical of conspecific signals. One new study suggests, however, that auditory receptors in insects might be more efficient at encoding rapid amplitude variations that occur in natural signals than in other temporally varying sounds (Machens et al. 2001). Gap detection in grasshoppers appears to be based on ensemble coding in the form of synchronized responses of multiple receptors in the auditory nerve. The same is probably true for anurans, but most studies have used arbitrary numbers of stimulus repetitions to study synchronization in single auditory neurons. Quantitative behavioral studies of the minimum temporal differences that are discriminated, the minimum numbers of stimuli that must be heard, and the minimum response latency are needed to guide future studies of temporal encoding.

6

Processing of Biologically Significant Sound Signals in Central Auditory Systems

A MAJOR role of the central auditory system is to extract biologically significant information from the stream of activity elicited in auditory receptors by acoustic signals. As in other animals (e.g., fish: Crawford 1997; songbirds: Dope 1997; bats: Suga 1989), some neurons in higher centers in the brains of insects and anurans show remarkable selectivity to behaviorally relevant (information-bearing) properties of acoustic signals (e.g., Schildberger 1984; Adler and Rose 1998). Such selective neurons are probably nodal points within neural circuits that mediate signal recognition, and their existence raises three questions. First, what intrinsic properties and patterns of connections with other neurons are responsible for their selectivity? Second, to what extent do selective neurons play multiple roles? Do they, for example, participate in the processing of all sounds as opposed to just one kind of communication signal? Third, how does the activity of populations of selective neurons contribute to the choice and execution of appropriate behavioral responses? These questions present a major challenge for neuroethology, and here we can only provide some provisional answers. Perhaps the greatest progress along these lines has been achieved in studies of spatial localization in owls and bats (reviews: Suga 1989; Konishi 1994) and of the jamming avoidance response in weakly electric fish (reviews: Heiligenberg 1991, 1994; Metzner 1993).

The problems of studying acoustic signal recognition are compounded because auditory information does not merely flow sequentially (*serial processing*) from receptors through each successive, anatomically defined element of the ascending pathway to the highest centers in the brain. Rather, as pointed out already by Huber (1962b), the central nervous system can be viewed as a complex organ within which numerous interconnected systems influence one another reciprocally (e.g., insects: Elsner 1994; anurans: Hall and Feng 1987; Luksch and Walkowiak 1998). Moreover, studying recognition is more difficult than studying spatial location because the responsiveness and selectivity of the whole animal are more likely to be strongly influenced by its internal state or reproductive status (e.g., Loher and Dambach 1989; Huber 1990; Loher et al. 1993). The extent to which variation in motivational state influences the pattern selectivity of higher-order neurons is virtually unknown.

Profound differences exist between insects and anurans in the numbers, organization, and anatomy of auditory neurons within the central auditory pathway. In insects, auditory receptors on each side connect to and converge on a smaller number of auditory interneurons (connections also exist between interneurons of each side of the animal). The ratio of auditory receptor axons to interneurons that process and send information to the brain or to the ventral cord is estimated to be about 10:1 in crickets (Schildberger 1994), 4:1 to 8:1 in katydids (Stumpner, pers. comm.), 3:1 to 5:1 in most grasshoppers (Krahe 1997), and about 70:1 in cicadas (Münch 1999). Noctuiid moths, in which a single auditory receptor cell (A1) activates up to seven interneurons, are notable exceptions (Boyan and Fullard 1986). On the one hand, the small number of ascending neurons in insects reflects an economy of information transfer. That is, despite a reduced number of elements, the animal can still perform necessary tasks such as mate recognition, localization, and predator avoidance. Moreover, because of their reduced number and accessibility single, anatomically identifiable neurons can be characterized in terms of how they encode and process behaviorally relevant acoustic properties. On the other hand, the small number of ascending cells represents a bottleneck, which limits the scope for *parallel processing*, in which different ascending pathways simultaneously process different kinds of information about the signal.

In contrast to insects, an enormous proliferation of higher-order auditory neurons occurs in the ascending pathway of anurans and other vertebrates relative to the numbers of receptor cells and primary fibers. An array of anatomically and physiologically diverse cell types exists already in the very first nucleus in the brain stem (Hall and Feng 1988, 1990; Feng and Lin 1996). The ascending pathway consists of more than a half dozen reasonably well-defined auditory nuclei, most of which have reciprocal connections. Auditory neurons within many of these nuclei have been sampled and roughly categorized on the basis of anatomy, response properties, or both (reviews: Hall 1994; Feng and Schellart 1999).

Despite the differences in central nervous system organization, auditory processing in insects and anurans shares several general features. First, as in other animals, inhibition plays a role in sharpening the tuning of some ascending neurons. Second, whereas higher vertebrates typically show a well-developed tonotopic organization throughout the central auditory system, tonotopy in insects is confined mainly to receptor arrays and auditory neuropiles in thoracic ganglia. In frogs, tonotopy has been documented only up to the level of the torus semicircularis (inferior colliculus) and may be absent in higher centers. Third, as already discussed in chapters 4 and 5, auditory processing of sound signals occurs mainly in two or three frequency channels, as exemplified by the enhanced sensitivity of the auditory periphery in different frequency ranges. In this chapter we highlight some of the neural mechanisms that correlate with the behavioral "rules" for dealing with simultaneous excitation of two or more of these channels. Fourth, many higher-order neurons show an increase in selectivity to temporal patterns in comparison with receptors and lower-order cells, a difference that is often associated with a change in the neural code, usually from a synchronization code to some form of rate code.

Anatomical and Functional Overview

Insects

Within the central auditory pathway, neurons have been classified as *local neurons* (LN), whose processes are generally confined to the ganglia where auditory receptor axons terminate or to areas within the brain. *Ascending neurons* (AN) send information to the head ganglia, *descending neurons* (DN) send information to ganglia containing the motor centers, and *T-shaped neurons* (TN) send information in both directions. In figure 6.1 we show diagrams of ascending auditory interneurons that contribute to the selectivity of crickets,

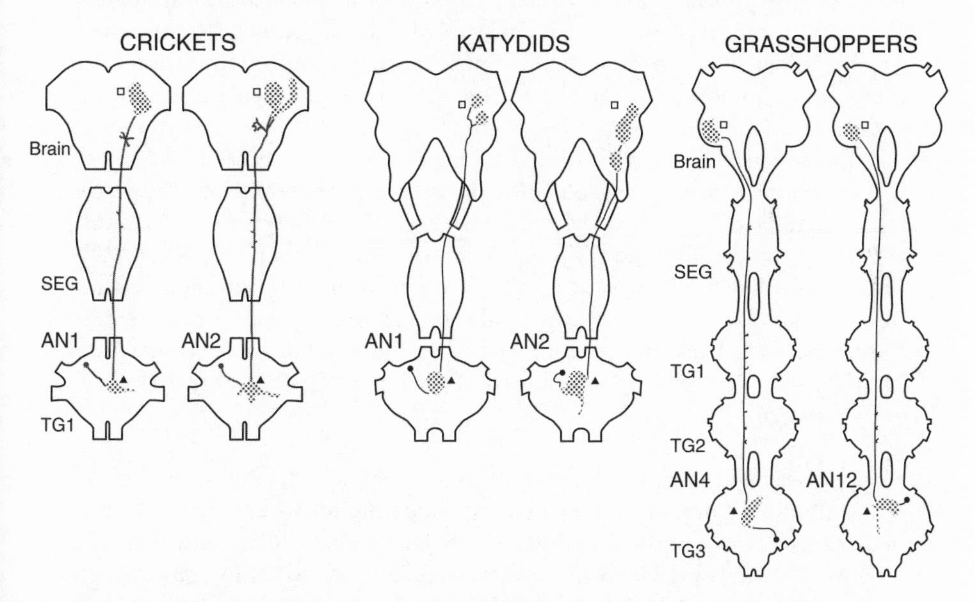

Figure 6.1. Cephalic and thoracic parts of the central nervous system of three kinds of orthopterans showing diagrams of ascending interneurons (AN1 and AN2 in crickets and katydids; AN4 and AN12 in grasshoppers). The input areas are indicated by filled triangles, and the output areas, by open squares. Filled circles show locations of the cell bodies. The interneurons shown are morphologically and physiologically identifiable in every preparation (crickets: Rheinlaender et al. 1976; Schildberger 1984; katydids: Schul 1997; Stumpner 1997; grasshoppers: Krahe 1997; Ronacher and Stumpner 1988). This diagram is greatly simplified, especially with regard to the input and output areas, where the connectivity revealed by electron microscopy can be very complex (e.g., Peters et al. 1986). SEG = subesophygeal ganglion; TG1, TG2, TG3 = thoracic ganglion.

katydids, and grasshoppers for conspecific signals. These neurons are "identified" in the sense that their anatomy and physiology is the same in each individual of a given species. Most such interneurons are present as pairs, reflecting the bilateral organization of the nervous system, and each type has characteristic dendritic fields and specific arborizations in adjacent ganglia and in their termination areas.

In crickets and katydids, only two to four kinds of ascending neurons have been identified, and thus the ascending pathway represents an extremely economical transfer of information from auditory receptors. In grasshoppers and cicadas, more than a dozen cells have been named, and hence the brain can probably select and integrate the input of this modestly larger array of ascending neurons (e.g., Stumpner 1988; Stumpner et al. 1991; Helversen 1997; Krahe 1997). The nomenclature of these neuronal types is confusing because researchers use different criteria (historical, anatomical, or physiological) for choosing names. Some interneurons with the same designation are almost certainly homologs (e.g., most AN1 neurons in crickets and katydids; Stumpner 1997), and others are definitely not. Moreover, there is considerable variability from preparation to preparation in physiological responses. Thus extreme examples of one type of neuron have probably been given different names. Finally, information about developmental origin, an important criterion for establishing homology, is missing for many, if not most cells (review: Boyan 1998).

Crickets and katydids. The axons of auditory receptors terminate within the auditory neuropile of the *prothoracic ganglion*, where they connect to local, ascending, descending, and T-shaped interneurons either mono- or polysynaptically. Receptors provide excitatory input to auditory interneurons, but interneurons may either excite or inhibit one another and can even exert inhibitory effects back at the terminals of auditory receptors (Wohlers and Huber 1982; Römer 1985; Hennig 1988; Schul 1997; Poulet and Hedwig, pers. comm.).

Grasshoppers. In grasshoppers, auditory receptor axons terminate within the first abdominal ganglion and the metathoracic ganglion (TG3, fig. 6.1) in two neuropiles (review: Elsner and Popov 1978; Jacobs et al. 1999; chapter 5). The synaptic connections between receptors and interneurons in grasshoppers seem to be more complex than in crickets and katydids, and many more kinds of local and ascending neurons have been characterized (Elsner 1994 for review; Stumpner and Ronacher 1991a,b, 1994; Römer and Marquart 1984). Auditory receptors are monosynaptically connected to local neurons but not to ascending interneurons (Römer and Marquart 1984). Ascending interneurons receive presynaptic input from different sources, which thereby contribute to their response properties (e.g., Boyan 1999). Despite their similar designations, interneurons in grasshoppers differ significantly from those in crickets and katydids both anatomically and physiologically.

Cicadas. In cicadas the axons from the auditory receptors enter the fused *metathoracic-abdominal ganglionic complex* (MAC). Their terminals form a mirror-image auditory neuropile that reveals the metameric (segmental) organization, and within the MAC the auditory receptor fibers connect to auditory interneurons that branch in different parts of the MAC. More than a dozen auditory interneurons have been described. Most of them are ascending neurons; however, their projection areas within the head ganglia are still unknown (Wohlers et al. 1979; Huber et al. 1990; Fonseca 1994; Münch 1999).

Anurans

The ascending auditory pathway in anurans consists of a series of paired nuclei. Because these nuclei are reciprocally connected, information from each of the two ears is transmitted both ipsilaterally and contralaterally. The inputs from the two ears can also be compared at the same multiple levels (i.e., the same nuclei on the right and left sides of the brain) in the ascending pathway (fig. 6.2; for detailed wiring diagrams, see Hall 1994; Walkowiak and Luksch 1994).

Serial processing in the ascending pathway is based on projections of some neurons from each auditory center (or nucleus) to the next processing center(s). Neurons in a given center do not, however, all project to the same locus in the next higher station, and some neurons send axons directly to higher centers without making synaptic connections with neurons in anatomically intermediate nuclei (Feng and Lin 1991; Walkowiak and Luksch 1994). For example, some neurons in the dorsolateral nucleus project directly to the torus semicircularis. These complex patterns of connectivity are the anatomical bases of serial, hierarchical, and parallel processing of auditory information in the ascending pathway.

Lower and higher centers are also reciprocally connected, thus providing the basis for central feedback loops (Hall and Feng 1987; Walkowiak and Luksch 1994; Luksch and Walkowiak 1998). Although there is good evidence for parallel processing of spectral and temporal information, as exemplified by the different selectivity of the two auditory nuclei in the thalamus, anatomical studies indicate that the outputs of these two nuclei do not converge in higher centers, such as the striatum of the telencephalon, but rather on structures found in the brain stem (Hall and Feng 1987). The thalamus is also connected reciprocally to nuclei in the ventral hypothalamus, which concentrates sex hormones and almost certainly regulates seasonal and other long-term variation in calling behavior and phonotactic readiness (Wilczynski et al. 1993). Although the thalamus of anurans receives inputs from other sensory systems, its anatomical complexity is far less than that of birds and mammals, and many

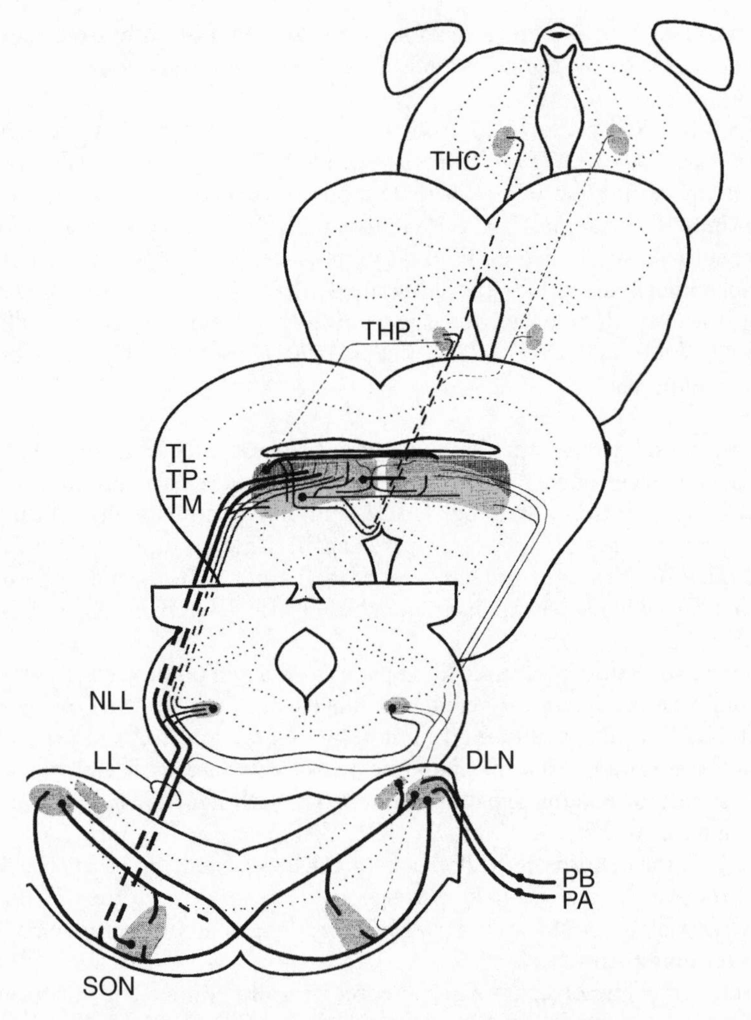

Figure 6.2. Diagram of sagittal sections of the brain showing the major stations in the ascending auditory pathway of an anuran. PB = auditory nerve fibers derived from basilar papilla; PA = auditory nerve fibers derived from amphibian papilla; SON = superior olivary nucleus; DLN = dorsolateral nucleus; LL = lateral lemniscus; NNL = nucleus of the lateral lemniscus; TL = laminar nucleus of the torus semicircularis (inferior colliculus); TP = principal nucleus of the torus semicircularis; TM = magnocellular nucleus of the torus semicircularis; THP = posterior nucleus of the thalamus; THC = central nucleus of the thalamus. Original figure provided by W. Walkowiak.

of the connections to higher centers found in these other vertebrates are lacking (Butler 1995). Finally, as mentioned in chapter 3, neurons in some subnuclei (laminar nucleus and tegmentum) of the torus semicircularis not only appear to integrate the activity of higher centers with auditory input from the periphery, but also project to premotor and motor areas in the brain stem

(Endepols and Walkowiak 1999, 2001). Anatomically at least, these connections suggest the existence of a widely dispersed sensory-motor interface (Walkowiak and Luksch 1994).

Spectral Processing in the Central Nervous System

We address two sets of general questions that arise in considering how the central nervous system of insects and anurans processes biologically important spectral properties. First, what are some mechanisms that might maintain or improve the frequency resolution that is evident in auditory processing at the peripheral level (chapter 5)? Is, for example, tonotopy preserved within the ascending pathway up to and including the highest centers? Can inhibitory and excitatory interactions within the CNS improve the resolution of different frequencies within the usual two or three frequency channels? For insects, how well do tympanal nerve recordings predict the frequency selectivity of auditory neurons within the ascending pathway? Second, what role does the central nervous system play in mediating the different behaviors that are elicited by the stimulation of different frequency channels or by simultaneous stimulation of two or more channels? Recall, for example, that crickets, whose calling songs often have a single, relatively low-frequency carrier, are also sensitive to the ultrasonic signals produced by bats. Many katydids, grasshoppers, and anurans have signals with broadband spectra, whose spectral peaks simultaneously stimulate two or more frequency channels. Simultaneous excitation of multiple channels has a variety of behavioral results, ranging from inhibition and evasive behaviors to the enhancement of signal attractiveness (chapter 4).

Tonotopy

Tonotopy is one solution to the confounding effects of variation in frequency and intensity on spike rate in a single receptor or primary fiber. By means of this form of labeled-line coding, the central nervous system can, in principle, determine the frequency of a stimulus on the basis of *which neurons* are active rather than depending solely on a single cell's firing rate.

Insects. In chapter 5 we showed that receptors in many insects show some form of tonotopic organization, which, in katydids and grasshoppers is preserved at the level of auditory neuropiles within the thoracic ganglia (see fig. 5.4 and fig. 6.3). However, little anatomical or physiological evidence exists for tonotopic organization in the brain of insects (review: Pollack 1998), and it is difficult in principle to see how any fine-grained tonotopy could be maintained simply because of the extensive convergence of receptors and local interneurons on the relatively few ascending or T-shaped interneurons that send information to the brain.

Römer (1987) suggests that tonotopy in katydids might function in the

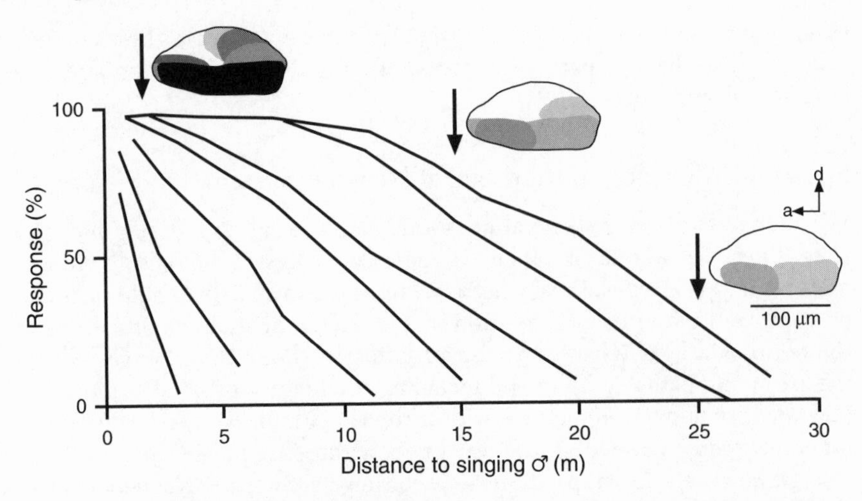

Figure 6.3. Topological representation of distance to the sound source by receptor activity within the auditory neuropile in a katydid (*Mygalopsis marki*). The firing rate of seven receptors is plotted as a function of distance. The arrows next to diagrams of the auditory neuropile indicate three distances (2, 15, and 25 m) at which the locations and relative activity of receptors within the neuropile are shown by the degree of shading (light = low activity; dark = high activity). From Römer 1987, fig. 5.

estimation of the distance to a sound source (see box 5.3 for a comparable explanation for anurans). The absolute thresholds of receptors terminating in the neuropile are correlated with their frequency tuning. Moreover, receptors tuned to low frequencies are represented in the anterior part of the neuropile, and receptors tuned to high frequencies, in the posterior part. Because the high-frequency part of the broadband spectrum of the conspecific song attenuates more than does the low-frequency part in natural habitats, activity of different regions of the neuropile correlates with the distance to a singing rival (fig. 6.3). As discussed in chapter 8, signaling male katydids as well as other insects and anurans tend to space themselves on the basis of the perceived intensity of nearest neighbors. Support for this hypothesis requires more knowledge about how the different interneurons connecting with different parts of the auditory neuropile ultimately convey information about its differential excitation to higher centers. Römer (1987) presents some preliminary data about such interneurons.

Anurans. As discussed in chapter 5, the receptors for low-frequency sounds in advanced anurans are tonotopically organized within the amphibian papilla, and the receptors of the basilar papilla are more sensitive to a substantially higher range of frequencies (Lewis and Narins 1999). This fine- (within the amphibian papilla) and gross-level (between the two papillae) tonotopy is observable within three major stations of the ascending pathway. First, the spatial pattern of tuning by neurons in the dorsolateral nucleus reflects both the

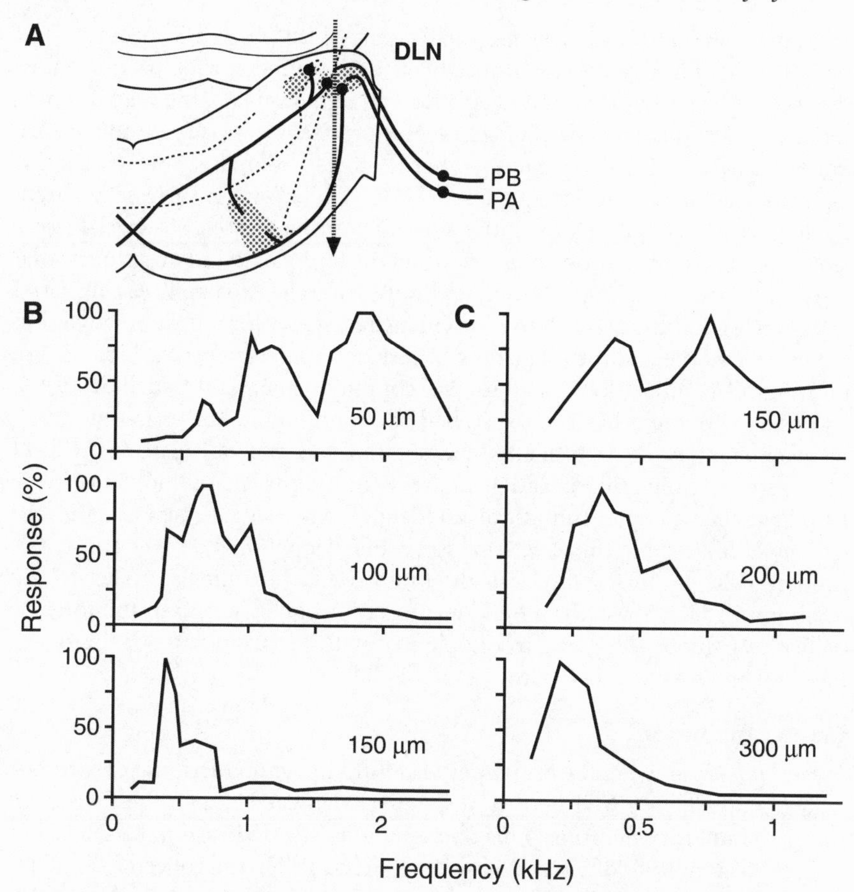

Figure 6.4. Tonotopy in the dorsolateral nucleus (DLN) of the leopard frog. (A) Diagram of the DLN showing approximate locations of inputs of auditory nerve fibers derived from the basilar papilla (PB) and amphibian papilla (PA). The arrow shows the approximate location of a ventrodorsal penetration of a recording electrode through the medial DLN. (B) Diagram showing the systematic decrease in frequency sensitivity with depth, beginning with the high-frequency response provided by the basilar papilla. In (C) the progressive decrease with electrode depth in low-frequency sensitivity provided by the amphibian papilla is detailed. From Fuzessery and Feng 1981, figs. 5, 6.

termination points of the two inner ear organs and the tonotopy within the amphibian papilla (fig. 6.4). The basilar papilla fibers mainly enter the dorsomedial part of the dorsolateral nucleus. The amphibian papilla neurons terminate throughout the rest of this structure, with neurons tuned to low frequencies found ventrally and those tuned to somewhat higher frequencies occurring in more dorsal locations (Fuzessery and Feng 1981; Feng 1986a). Second, at the level of the superior olivary nucleus, a dorsal and lateral concentration of neurons sensitive to low frequencies exists; higher frequencies are represented in the ventral and medial part of this nucleus (Fuzessery and Feng 1983a). Third, although tonotopy occurs within the principal nucleus of the torus semicircularis of advanced anurans (e.g., Ranidae), the spatial representation of frequency is neither simple nor precise. Cells in a central core are most sensitive to low frequencies, whereas the surrounding cortex contains neurons sensitive to high frequencies (Feng 1986b; review: Fuzessery 1988). In more primitive groups (e.g., the Bombinatoridae), extensive spatial overlap occurs between neurons tuned to different frequencies, so even the spatial separation of low- and high-frequency inputs from the two papillae is obscured (Walkowiak and Luksch 1994). No evidence exists for tonotopic organization at the level of the auditory thalamus, where neurons selective for biologically meaningful combinations of tones are common (Feng and Schellart 1999), and so far no study has even searched for tonotopy in the striatum of the telencephalon. A major difficulty in trying to find frequency maps at the level of the torus semicircularis and at higher levels is that many individual neurons have such complex tuning curves that defining a single characteristic frequency is impossible (fig. 6.8; Fuzessery and Feng 1983b). By contrast, the tonotopically organized nuclei in the brain stem of mammals project to the midbrain in such a way as to build a common frequency map in the inferior colliculus (Aitkin 1986), and tonotopic maps also occur in parts of the thalamus and auditory cortex (Clarey et al. 1992).

Lateral Inhibition

Lateral inhibition, whereby excitatory and inhibitory interactions serve to enhance contrasts, is a well-known mechanism that can sharpen the frequency tuning of auditory neurons. This phenomenon is widespread in katydids and crickets (katydids: Oldfield and Hill 1983; Schul 1997; Stumpner 1997; crickets: Schildberger et al.1989; Selverston et al. 1985; Boyan 1981; Boyd et al. 1984; Horseman and Huber 1994a), and some neurons in the ascending pathway of anurans also show this effect (Fuzessery 1988). In figure 6.5A we show an example of a cricket AN1-type neuron whose tuning is sharpened by inhibitory sidebands on the low- and high-frequency side of the excitatory response area (Boyd et al. 1984); a comparable example from the torus semicircularis of the leopard frog is illustrated in figure 6.5B. In insects at least, the resulting tuning of interneurons is seldom dramatically sharper than that of receptors (review: Stumpner and Helversen 2001).

The inhibitory neurotransmitter GABA plays a prominent role in shaping frequency tuning in the midbrain (Hall 1999) as well as the inhibitory areas of such cells in the torus and auditory thalamus (Hall 1994). The sources of the inhibition are not, however, well understood. For the AN1, some inhibitory influence comes from a local interneuron (omega cell) (Horseman and Huber 1994a), but this input does not explain all of the sharpening. No data are available to answer this question for frogs.

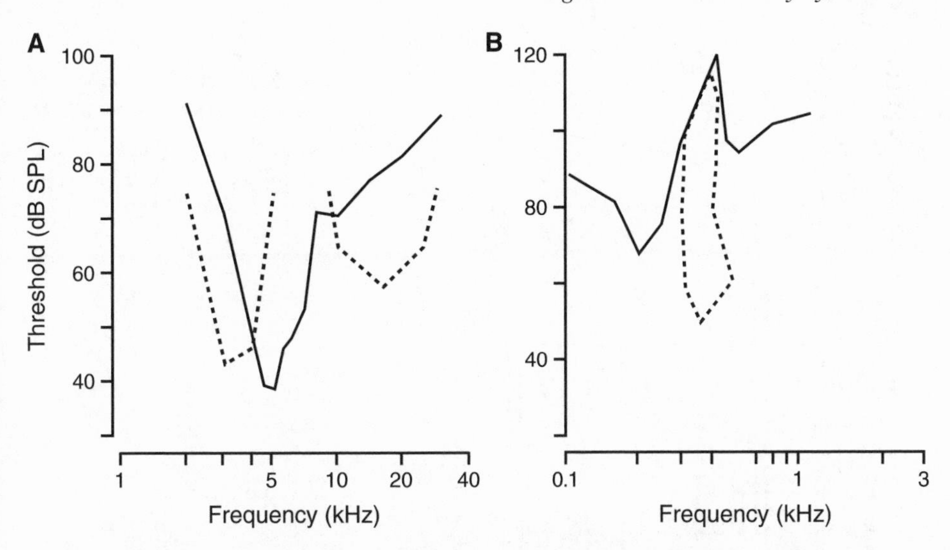

Figure 6.5. Sharpening of tuning by lateral inhibition. (A) Sharpening of the tuning of the AN1 ascending interneuron in the cricket. The excitatory tuning curve is shown as a solid line, and flanking, inhibitory curves, as dotted lines. From Boyd et al. 1984, fig. 1. (B) Excitatory (*solid line*) tuning curve and inhibitory tuning curves (*dotted line*) of an auditory neuron in the torus semicircularis of the leopard frog. From Feng and Schellart 1999, fig. 6.10c.

Frequency Channels in Insects and Anurans: Functional Roles and Interactions

Whereas two or three frequency channels in the ascending pathway are well defined anatomically and physiologically in crickets and katydids and at least partially so in anurans, the situation is not so simple in grasshoppers and other insects. The functional significance of each channel also varies from taxon to taxon, and many insects that do not communicate by sound are nevertheless very sensitive to the ultrasonic signals of bat predators (reviews: Fullard 1998; Greenfield 2001). Some insects and anurans are also highly sensitive to vibrational stimuli, which can be used for intraspecific communication (through plants or the substratum) as well as for predator detection (Kalmring 1983, Kalmring et al. 1983, 1990; Narins and Lewis 1984; Schul, pers. comm.). In both taxa, some neurons are sensitive both to airborne sound and vibration (Kalmring et al. 1978; Christensen-Dalsgaard and Jørgensen 1996b).

Crickets and katydids. The low-frequency channel, which is often dedicated to intraspecific communication, consists of the pair of ascending neurons (AN1) or their homologs (fig. 6.1). In crickets these AN1 neurons are tuned sharply to the carrier frequency (about 4–8 kHz) of the calling song (fig. 6.6A; review: Schildberger et al. 1989), and their sensitivity ultimately depends on connections to a relatively large number of low-frequency tuned receptors (Pollack

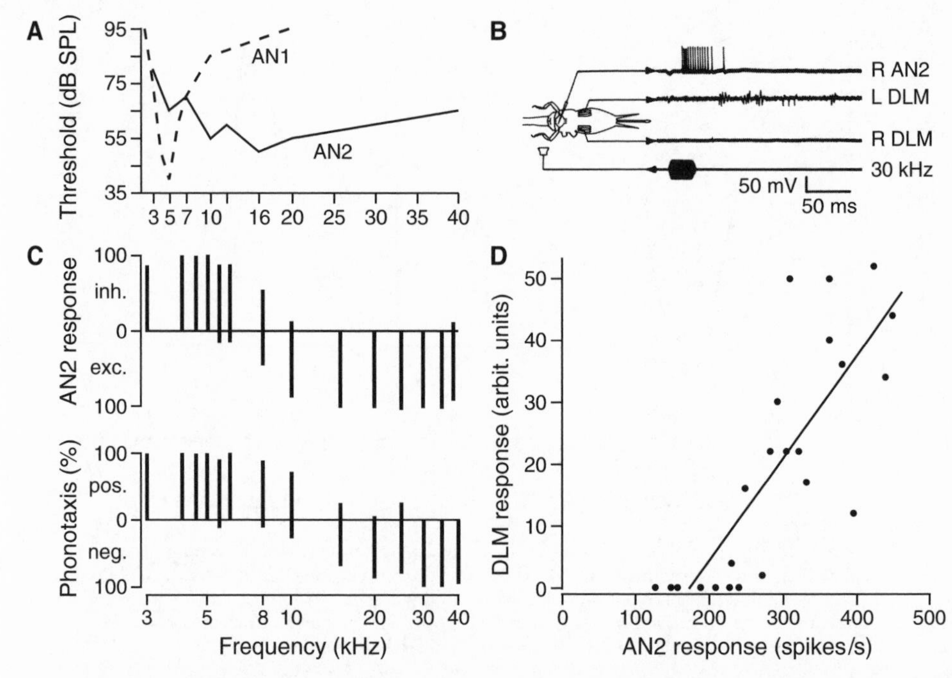

Figure 6.6. Frequency sensitivity of ascending interneurons in crickets. (A) Tuning curves of a typical AN1 and an AN2 interneuron in the cricket. From Rheinlaender et al. 1976. (B) Setup for stimulating and recording from the AN2 interneuron and flight steering muscles (abdominal dorsal longitudinal muscle [DLM]) in a "flying" preparation. The cricket is viewed from below. The lowest trace shows an oscillogram of a 30 kHz tone delivered to the left ear, and the upper two traces show the response of right AN2 and the left DLM muscles. From Nolen and Hoy 1984, fig. 1A. (C) Correlation of the response of the AN2 (= INT-1) auditory interneuron (*top*) with phonotactic steering behavior in tethered flying crickets (*Teleogryllus oceanicus*) as a function of stimulus frequency. From Moiseff and Hoy 1983, fig. 11.14. (D) Quantitative analysis of the response of AN2 in relationship to negative phonotaxis as measured by DLM activity. Notice that AN2 fires at rates greater than about 200 spikes/s when negative phonotaxis occurs. From Nolen and Hoy 1984, fig. 1B.

1994). In katydids such as *Tettigonia viridissima*, *T. cantans*, and *Mygalopsis marki*, AN1 neurons are tuned to the relatively "low-frequency" components (7–16 kHz) that occur in the broadband calling songs of these species (e.g., Römer 1987; Hardt 1988; Schul 1997). Although not homologous with the AN1 of these other katydids, the AN1 of the duetting katydid *Ancistrura nigrovittata* is also well tuned to the relatively low frequencies typical of male song (Stumpner 1997, 1999b).

The roles played by the high-frequency channel differ in emphasis in crickets and katydids. In crickets the high-frequency channel, represented by the pair of AN2 (= INT-1) interneurons, is specialized for predator detection and may secondarily participate in processing conspecific signals such as courtship songs. In katydids, however, the high-frequency channel is usually involved both in the processing and localization of conspecific signals (chapter 7) and in predator detection (see below). This makes sense because the communication signals of katydids usually have broadband spectra that include ultrasonic frequencies. In both European (*Tettigonia*) species and in an Australian species (*Requena verticalis*), simultaneous excitation of the low- and high-frequency channels is required to elicit reliable and accurate phonotaxis (Bailey and Yeoh 1988; Jatho et al. 1994).

Whereas the AN2 interneuron is the dominant component in the high-frequency channel in crickets, other kinds of cells also contribute significantly to high-frequency sensitivity in katydids. For example, TN1 neurons are high-frequency-tuned cells in *Tettigonia* that copy the temporal pattern of conspecific songs and respond to sounds simulating bat echolocation sounds; TN2 and other T-fibers neurons are probably specialized for bat detection (Schul 1997; Faure and Hoy 2000b; see below). A recently described interneuron, designated as AN5, is a major component of the high-frequency channel in the duetting katydid *A. nigrovittata* (Stumpner 1997, 1999a,b). AN5 has an unusual morphology in that its cell body is situated in the seventh abdominal ganglion. This interneuron responds best to the female song frequency and is suppressed by simultaneous stimulation with frequencies below about 25 kHz. Recall from chapter 4 that the overall frequency response of the tympanic nerve in both sexes of *A. nigrovittata* is nearly the same and is tuned to the frequency of the male song. Behavioral experiments, however, show that males are highly responsive to the female song (Dobler et al. 1994b). This result again serves to emphasize that estimates of spectral sensitivity based on the activity of the tympanic nerve alone can be misleading because the existence of AN5 is not evident in whole-nerve recordings.

The role of the AN2 interneuron in predator avoidance in crickets has been extensively studied (Moiseff and Hoy 1983; Nolen and Hoy 1984; reviews: Hoy 1994; Pollack and Hoy 1989). This interneuron derives its sensitivity to high frequencies from relatively few receptors, which have, in comparison to the low-frequency receptors, a disproportionately strong effect on interneurons in the CNS (Pollack 1994). AN2 neurons also respond—albeit at higher sound levels—to the lower-frequency sounds of cricket communication signals. We do not know whether this low-frequency sensitivity arises from connections with less-sensitive low-frequency receptors, or from the secondary sensitivity of the AN2 to low-frequency sound (Esch et al., 1980; Pollack 1998).

Negative phonotaxis and evasive responses to high-frequency sounds are especially reliable in "flying" preparations (fig. 6.6). Nolen and Hoy (1984, 1986) monitored the activity of steering muscles in *Teleogryllus oceanicus* at the same time as they recorded the activity of AN2 interneurons. During stimulation by high-frequency sounds, steering that would result in negative phonotaxis occurred only when the animal was "flying" and when the firing rate of AN2 exceeded about 180 spikes per second. Flight per se does not affect the firing pattern of AN2, but AN2's activity apparently has an indirect influence on descending neurons that are also activated by ultrasonic signals (Brodfuehrer and Hoy 1989). However, negative phonotaxis has also been reported in walking crickets (Pollack et al. 1984), so the activation of flight is not absolutely required for AN2 to mediate evasive actions.

Two-tone experiments show that a sound of 5 kHz, which excites AN1, can also suppress the evasive behavior normally elicited by a sound of 30 kHz, which excites AN2. Under this condition of simultaneous excitation, AN2 may then even enhance mate attraction, perhaps because its activity might be interpreted by the cricket as courtship song (Libersat et al. 1994; review: Pollack 1998). As discussed in chapter 7, studies in which AN2 is artificially hyperpolarized show that this interneuron also contributes to positive phonotaxis to calling-song frequencies, albeit to a much lesser extent than AN1 (Schildberger and Hörner 1988).

Grasshoppers. Grasshoppers have several groups of low- and high-frequency receptors, which, however, vary significantly in bandwidth and tuning, even within the same anatomically defined group (chapter 5; review: Pollack 1998). Low- and high-frequency channels are not so well defined as in crickets and katydids, and ascending interneurons that are connected to these receptors are responsive to a wide range of frequencies below about 15 kHz, corresponding to the broad spectra of their songs in this range. All of these interneurons show complex frequency- and intensity-dependent responses, brought about by excitatory and inhibitory interactions (Römer and Marquart 1984; Römer and Seikowski 1985; Boyan 1999), and the brain seems to evaluate the important features by comparing input from several (or all) of these ascending neurons (Helversen 1997; Krahe 1997).

The preference of females of *Chorthippus biguttulus* for stimuli with bimodal spectra, for example, suggest that the output of ascending neurons with different tuning properties (but still falling within the "low-frequency" channel) converge in higher centers (Helversen and Helversen 1997; see fig. 4.5). The inhibition of the male turning response by the same high-frequency energy that enhances female responses suggests that there are sexual differences in connections between ascending neurons and brain neurons or that the same connections form different kinds of synapses.

Cicadas. As discussed in chapter 5, some cicadas provide examples of species in which peripheral auditory sensitivity seems to be mismatched with respect to the carrier frequency of conspecific signals. However, estimates based on tympanal nerve recordings alone can be misleading, and hence such mismatches could be more apparent than real. In *Tettigetta josei* the tympanal nerve and several ascending interneurons are tuned to about 1 kHz and to a range of about 3–6 kHz, whereas the calling song of this species has a carrier

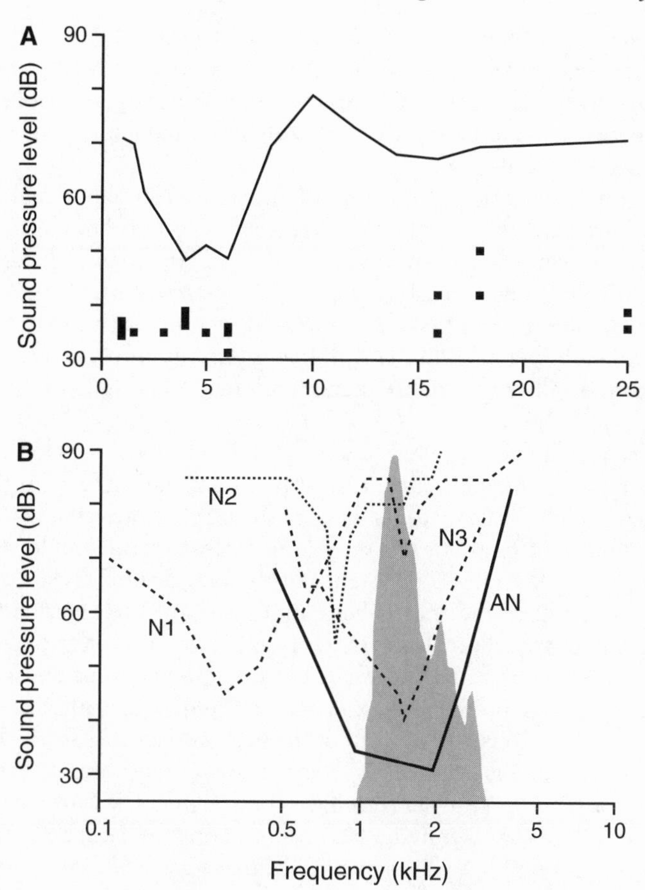

Figure 6.7. Tuning of ascending neurons in cicadas. (A) Mean threshold curve of the auditory nerve (*solid line*) of *Tettigetta josei;* filled squares below show lowest thresholds of thirteen auditory interneurons within and outside the range of the highest sensitivity of the auditory nerve. The neurons with lowest threshold between 16–18 kHz are tuned to the spectrum of the calling song. Modified from Münch 1999, fig. 33.4. (B) Examples of tuning curves of three ascending auditory interneurons in *M. septendecim.* Solid line indicates the threshold curve of whole auditory nerve (AN); hatched area indicates calling-song frequency spectrum. Neuron 1 (N1) is tuned to the timbal rhythm (about 200 Hz), N2 to the rib rhythm (about 700 Hz), and N3 to the carrier frequency of 1–2 kHz. Modified from Huber et al. 1990, fig. 7A.

frequency of about 16 kHz; only a few interneurons are, however, tuned to frequencies of about 16–25 kHz (Münch 1999; fig. 6.7A).

The songs of different species of periodical cicadas that sing at the same time and place differ in spectral and temporal properties (chapter 2). Recall that *Magicicada septendecim* broadcast a continuous buzz within a frequency band of 1–3 kHz, whereas the songs of *M. cassini* are made up of ticks followed by a buzz within a frequency band ranging from 4–9 kHz. Field observations and playback experiments provide evidence that both species respond only to conspecific songs. Moreover, males of *M. cassini* respond preferably to the ticks and try to synchronize their songs with the songs of conspecifics in the chorus, whereas trills and buzzes attract both males and females to the chorus center (Moore, unpubl. data).

Among the ascending neurons recorded in the meta-abdominal complex of *M. septendecim*, a class was found that responds selectively to the conspecific calling and courtship songs (spectral band 1–3 kHz). These cells do not respond to the calling and courtship songs of *M. cassini* (spectral band 4–9 kHz) (fig. 6.7B), despite the fact that the auditory nerve is sensitive to the songs of this species (Huber et al. 1980, 1990). The selectivity of these neurons must thus be based on their specific frequency and temporal tuning.

Anurans. A population-level analysis of central auditory neurons indicates that the low-frequency channel remains to some extent separable from the high-frequency channel at least up to the level of the torus semicircularis. Although some cells found throughout the ascending pathway have simple, V-shaped responses (fig. 6.8A), some convergence of the two channels is already found in the superior olivary nucleus, where about 20% of the neurons have W-shaped tuning curves (fig. 6.8B). At the level of the torus semicircularis, an enormous diversity of tuning curves is found. Some neurons are so broadly tuned that they respond to some extent to any frequency within the hearing range (review: Fuzessery 1988). Still other neurons can be effectively stimulated by a single tone falling within *either* of two widely separated frequency bands; for this reason, these neurons have been termed *OR units* (fig. 6.8C). A few neurons in the torus semicircularis respond only to *simultaneous presentations* of two or more different frequencies (Fuzessery 1988). Such *AND units* (fig. 6.8D) are more commonly found in the posterior nucleus of the thalamus (Hall and Feng 1987; Fuzessery 1988), but most studies of selectivity for spectral patterns at this locus have been based on *evoked potentials* (summed activity of many neurons recorded with a large electrode).

The main impetus for the search for AND properties in the central nervous system was Capranica's (1965) pioneering behavioral studies of evoked calling in the North American bullfrog *Rana catesbeiana.* He found that in a laboratory colony of these frogs, simultaneous presentation of two frequencies representative of the two spectral peaks of the bullfrog's advertisement call (and hence each exciting a different auditory inner ear organ) was necessary to elicit vocal responses by males. Subsequently, Mudry and Capranica (1987a) found that combinations of two or three appropriate frequencies, again corresponding to spectral peaks in the advertisement call, elicited nonlinear, *facilitated responses* from the auditory thalamus. That is, the magnitude of the response to the combination tones was greater than the linear sum of responses to the individual components presented one at a time.

Similar results have been found in the thalamus of the green treefrog *Hyla*

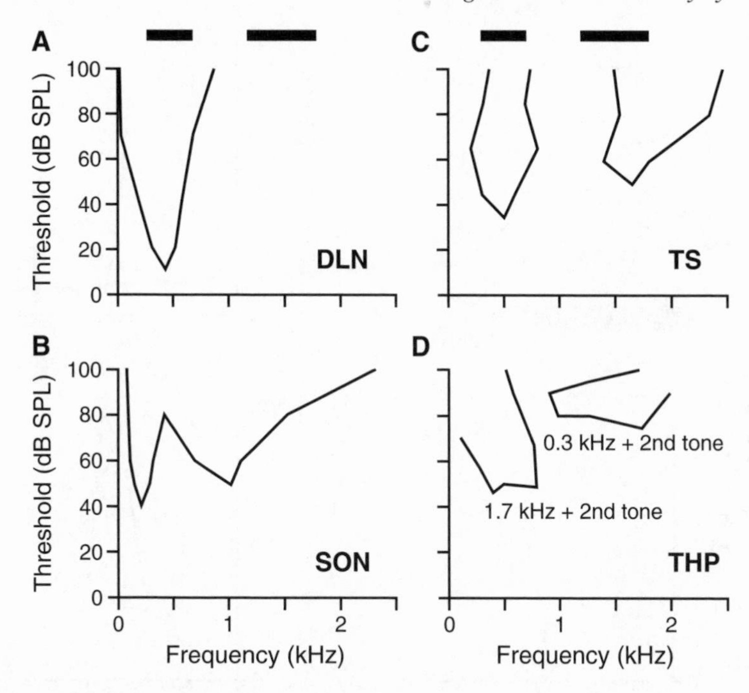

Figure 6.8. Representative tuning curves of auditory neurons in the ascending auditory pathway of the leopard frog *Rana pipiens.* (A) V-shaped tuning curve of a neuron in the DLN (dorsolateral nucleus). (B) W-shaped tuning curve of a neuron from the SON (superior olivary nucleus). (C) Single neuron in the TS (torus semicircularis) with dual-frequency sensitivity; such neurons respond when stimulated by a single frequency that matches either curve (OR property). (D) A neuron in the posterior nucleus of the thalamus (THP) with dual-frequency sensitivity. Many such cells do not fire reliably unless stimulated simultaneously with two tones (AND property), each one matching one of the disjunction response areas. Horizontal bars above show the frequency ranges of the two spectral peaks found in the leopard frog's advertisement call. Modified from Hall 1994, fig. 3.

cinerea, where nonlinear facilitated responses also occur in response to combinations of low- and high-frequency tones (fig. 6.9A; Mudry and Capranica 1987b). As discussed in chapters 4 and 5, females of *H. cinerea* prefer synthetic calls with two spectral peaks to signals with only one peak at playback levels of about 55 dB SPL and higher (up to at least 85 dB) (Gerhardt 1981b; see fig. 4.6). The neural response was greatest over a somewhat higher and narrower range of playback levels (70–80 dB SPL) than in the behavioral studies. More significantly, although the optimal high-frequency peak corresponded to the behaviorally most effective frequency (3.0 kHz), the optimal low-frequency peak of about 0.5 kHz for eliciting the nonlinear response was nearly an octave lower than the most effective low-frequency component in the behavioral studies. This discrepancy is probably attributable to the fact that the behavioral and neurophysiological experiments were conducted at significantly different temperatures. Recall that the frequency preference of

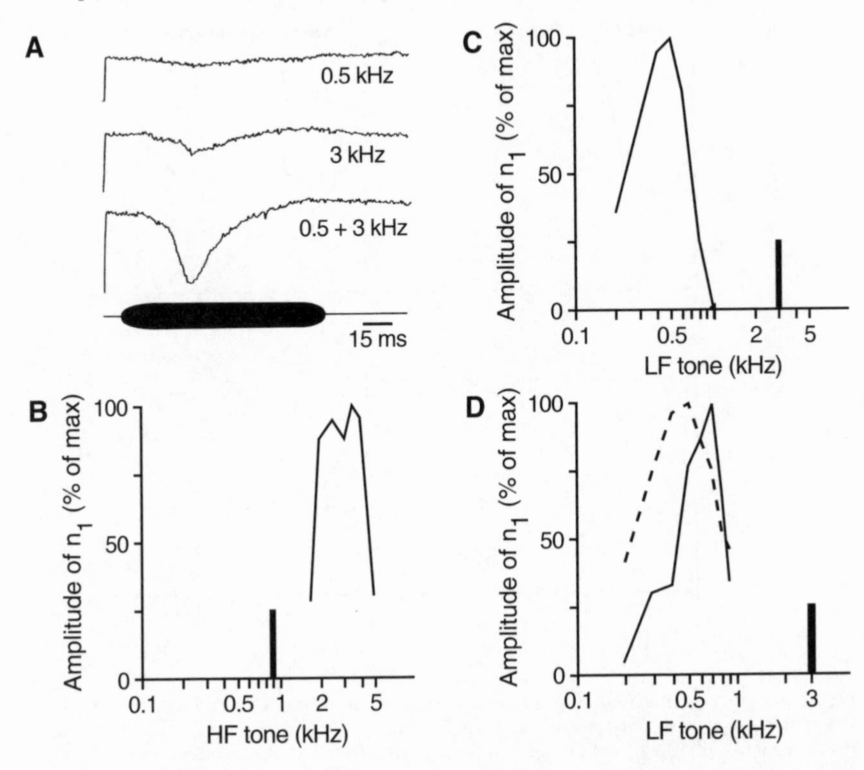

Figure 6.9. Processing of spectral patterns in the auditory thalamus of the green treefrog *Hyla cinerea.* (A) Evoked potentials in response to a low-frequency tone of 0.5 kHz alone (*top trace*); a high-frequency tone of 3.0 kHz alone (*second trace*); a nonlinear facilitated response evoked by the simultaneous presentation of the two tones. (B) Diagram showing the best range of high frequencies for evoking a nonlinear response when the low-frequency peak was held constant at 0.9 kHz (*vertical bar*). (C) Diagram showing the best range of low frequencies for evoking a nonlinear response when the high-frequency peak was held constant at 3.0 kHz (*vertical bar*). (D) Effect of temperature on the optimal frequency of the low-frequency peak of a two-tone combination (the high-frequency peak was always 3.0 kHz; see vertical bar). Note that at 23°C (*solid line*), which is the normal breeding temperature, the optimal low-frequency peak is much closer to the typical low-frequency peak in the conspecific advertisement call (0.8–1.0 kHz) than at 15°C (*dashed line*). The data plotted in (A), (B), and (C) were from preparations recorded at temperatures of 20°C ± 2°C. From Mudry and Capranica 1987b, figs. 1, 4, 8.

females for calls of 0.9 k Hz to calls of 0.5 kHz reverses when the temperature is decreased from the range of normal breeding temperatures (> 23°C) to 16–18°C (chapter 4; Gerhardt and Mudry 1980). The generally lower temperatures at which most of the neurophysiological studies (18–22°C) were conducted might have also raised the threshold for the facilitated response.

Another discrepancy between behavior and physiology at the level of the auditory thalamus is not so easily explained. Female green treefrogs respond phonotactically to each call in a series that can be repeated as often as 0.5 s, and males also rapidly alternate calls with neighbors (chapter 8). By contrast, substantial recovery of evoked responses in the thalamus—even to optimal stimuli—required interstimulus intervals of 50 s or more. Whether these extraordinarily long refractory periods reflect the condition of the preparation (immobilized animals that are not in a state of reproductive or phonotactic readiness) or are a real property of this thalamic nucleus in intact, behaving animals is an open question (see box 6.2).

Temporal Processing in the Central Nervous System

A general concept that applies to temporal processing in the central nervous system of a wide range of animals is that the synchronization code, which dominates in receptors or primary auditory neurons, becomes transformed, usually to some form of rate code, in auditory neurons within the ascending pathway and especially in higher centers in the brain. A second generalization is that the receptors and primary auditory neurons of insects and frogs usually do not show any particular selectivity for the temporal patterns typical of conspecific signals (chapter 5). That is, these cells usually copy the pulses within any signal, provided that pulse rate is below their upper cutoff (but see Machens et al. 2001). Hence, selectivity for temporal patterns must be accomplished by further processing in the central auditory system, usually in the brain (insects) or higher centers in the anuran pathway. These principles are especially well exemplified by studies of some crickets and anurans, where the neuronal data are well correlated with the results of behavioral studies.

Some exceptions to the second generalization include some ascending cells in the katydids and in at least one species of periodical cicada. That is, several ascending neurons originating from the MAC of *Magicicada cassini* show selective responses to the two parts of the song (fig. 6.10; Huber et al. 1990). Some cells copy both the ticks and the buzz; others respond only to the initial tick part but not to the subsequent buzz, which differs from the tick part in the rate of timbal actions. A third group of interneurons responds only to the beginning of the buzz or to the whole buzz. The detailed mechanisms underlying this selectivity are unknown, but these results suggest that at least part of temporal pattern filtering is already accomplished in lower stations of the central auditory system of this cicada.

Neural Correlates of Pulse Rate Selectivity in Crickets and Anurans

As shown in chapter 4, behavioral studies of field crickets have established, at least for two species of *Gryllus*, that preferences based on differences in pulse rate alone can explain selective phonotaxis to conspecific calling songs. Similarly, one species of gray treefrog (*Hyla chrysoscelis*) also appears to use pulse rate alone for advertisement call recognition. Many ranid species also produce pulsatile calls, and interpretations of neurophysiological studies of auditory neurons in higher centers in the ascending pathway usually assume that pulse

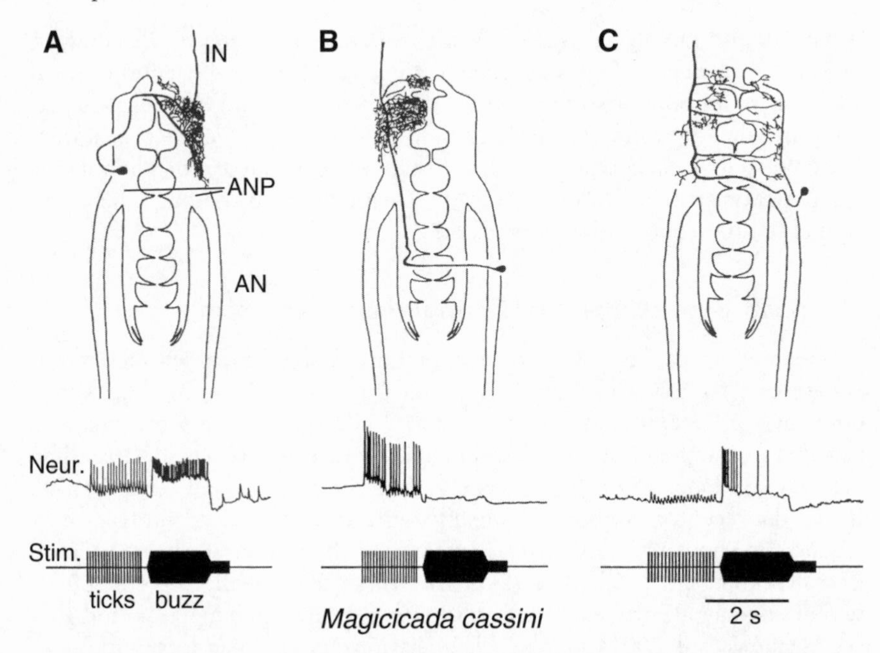

Figure 6.10. Responses of three different MAC neurons in the cicada *M. cassini* to model calling songs containing ticks and buzz. *Upper part,* three different morphologies. AN = auditory nerve; ANP = auditory neuropile; IN = ascending interneuron. *Lower part,* responses obtained from intracellular recordings within dendritic arborizations. (A) neuron copies both tick and buzz. (B) Neuron responds only to ticks. (C) Neuron responds with the beginning of the buzz. From Huber et al. 1990 and unpubl. data.

rate per se is the key property for call recognition. Although this is a reasonable hypothesis, we remind the reader that some species of field crickets, katydids, and treefrogs, whose conspecific signals have distinctly pulsatile structures, definitely do not rely on pulse rate alone for signal identification (chapter 4).

Field crickets. As discussed above, two types of ascending neurons (AN1 and AN2) in *Gryllus* send information to the cricket brain. Although AN1 interneurons, which are tuned to calling-song frequencies, synchronize less well to sound pulses repeated at very high rates (they are "low-pass" filters in this respect) than do receptors, they are still capable of accurate copying of the pulses in synthetic chirps that have even higher pulse rates than those of conspecific chirps (fig. 6.11B). AN2 interneurons, which are sensitive to a broader range of generally higher frequencies, synchronize well only to whole chirps (carrier frequency typical of the calling song) or only to pulses with much longer silent intervals between them. Thus, AN1 must play a key role in the CNS processing of pulse rate because both its spectral and temporal response properties are so well suited for encoding to the key properties of the calling

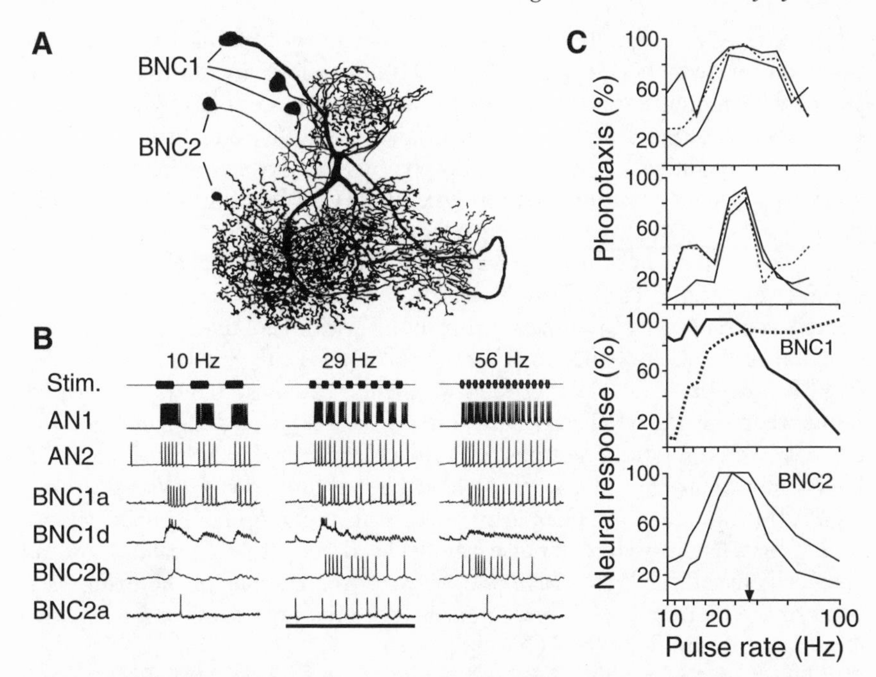

Figure 6.11. Temporal selectivity in the ascending pathway and brain of the cricket *Gryllus bimaculatus*. (A) Overlaid diagrams of five auditory neurons in the left half of the brain. In this sample, low- and high-pass neurons were classified as type BNC1, and band-pass neurons, as type BNC2. From Schildberger 1994, fig. 9. (B) Acoustic stimuli (*first trace*, representative synthetic chirps); other traces show neural responses of ascending (AN1 and AN2) interneurons and brain neurons to these stimuli. Note that some neurons (subcategories within BNC1 and BNC2) with the same morphology have different response functions. BNC2a shows a band-pass characteristic with a maximum response that matches the rate of the conspecific song (*solid line below*). From Schildberger et al. 1989, fig. 14.13. (C) Selective phonotaxis of two females is shown in the upper two panels; each was tested on three consecutive days (shown by the three lines). Lower two panels show neural responses of brain neurons (BNC1 and BNC2) as a function of variation in pulse rate. The two curves in BNC1 show the responses of a neuron with a high-pass characteristic (*dotted line*) and a neuron with a low-pass (*solid line*) characteristic. The two curves in BNC2 show two neurons with a band-pass characteristic. The arrow shows the pulse period (34 ms) corresponding to the stimulus value (29 Hz) in the middle panel of (B). Modified from Schildberger 1994, fig. 9.

song. As shown by Wohlers and Huber (1982), the physiological basis for the difference in temporal encoding between AN1 and AN2 appears to be in the different decay times of their excitatory postsynaptic potentials. Stimulated with sounds having the calling-song frequency, the decay time in AN2 is much longer than that in AN1. Thus AN2 remains above threshold well after the onset of the first pulse in the chirp so that subsequent pulses maintain suprathreshold values of the membrane potential. When AN2 is stimulated with synthetic calling songs with a high-frequency carrier (> 8 kHz), however, the neuron copies the pulses faithfully (Stout and Huber 1981, see also Pollack 1994).

Because AN1 copies the pulse rates of signals with pulse rates below its upper cutoff frequency (e.g., as in fig. 6.11B), the selectivity of the animals for the conspecific pulse rate must depend on neural circuits in higher centers (Schildberger 1984). Local brain neurons, which have been identified in *G. bimaculatus*, are almost certainly key components in a pattern recognition network that decodes the information ascending via AN1 (fig. 6.11A; Schildberger 1984, 1994). In contrast to AN1, most of these local brain neurons usually show poor synchronization to the pulses within the chirps of conspecific song (fig. 6.11B). Rather, selectivity for conspecific pulse rates is reflected in a rate code: brain neurons respond with many more spikes to certain ranges of pulse rate than to other ranges of pulse rate. Some brain neurons (e.g., BNC2b in fig. 6.11B) respond well when the pulse rate is equal to or higher than the preferred conspecific rate; these have been designated as *high-pass neurons.* Other brain neurons respond well when the pulse rate is equal or lower than the preferred rate; these have been characterized as *low-pass neurons.* A third functional type of local brain neuron (fig. 6.11B: BNC2a) shows tuning for a narrow range of pulse rates (typical of conspecific calling songs); such cells have been called *band-pass neurons.* Correlations of the firing patterns of these three types of brain neurons with phonotaxis in *Gryllus bimaculatus* are shown in figure 6.11C.

The anatomical relationships and response latencies of AN1, BNC1, and BNC2 neurons suggest that information is carried from AN1 to BNC1-type neurons, which then converge on BNC2-type cells (Schildberger 1984). The anatomical details are not completely clear, but conceptually the band-pass characteristics of some BNC-2 neurons might arise because their firing rate depends on simultaneous excitation by both low- and high-pass brain neurons (Schildberger 1984; Huber and Thorson 1985). If so, these cells would thus be another example of AND neurons, which in this case operate in the time domain.

A critical evaluation of our current knowledge yields a mixed picture. On the one hand, three features of band-pass neurons in field crickets make them likely candidates for decoding pulse rate information in calling song: (1) They are sensitive to frequencies typical of calling songs: of the seven band-pass neurons described by Schildberger (1984), five were sharply tuned to the carrier frequency of the calling song and two had broadband tuning that encompassed the calling-song frequency; (2) their preferred pulse rate is relatively intensity-independent; and (3) they transform the unselective synchronization coding by AN1 interneurons to a pulse-rate selective rate code. On the other hand, only speculative explanations are available concerning the physiological properties underlying the low- and high-pass characteristics of local brain neurons (review: Pollack 1998). Note, too, that in the behavioral response functions (fig. 6.11C), there is a distinct secondary peak at about one-half of the conspecific pulse rate as the katydid *Tettigonia cantans* (chapter 4; Schul and Bush 2000). This kind of function is not consistent with what might be expected of a band-pass filter constructed by adding the outputs of low- and (especially) high-pass filters. The location of secondary peak is also inconsistent with an autocorrelation model (e.g., Reiss 1964), which predicts peaks at multiples of the preferred rate. Perhaps some form of resonance filter could also play a role in temporal selectivity as suggested by the recurrent inhibition mechanism proposed for the omega neurons by Wiese and Eilts (1985). Moreover, the hypothesized presynaptic convergence on band-pass neurons by low- and high-pass neurons has yet to be definitively proved. Band-pass neurons also do not appear to be directionally sensitive, and thus other parallel systems must be required to evaluate directional information and absolute intensity. Results from other experiments suggest serial processing of sound pattern and direction, at least at the level of AN1 interneurons (Stabel et al. 1989; chapter 7). Presumably, convergence of these other kinds of data with that resulting from processing by band-pass neurons occurs somewhere in the brain (Schildberger 1994) or in the descending pathway. Finally, we might expect neurons that are truly selective for pulse rate either not to fire during the first two or three pulses or to change their firing rate afterward because, by definition, the rate of a stimulus has to be estimated by measuring the interval between at least two successive pulses. This expected property was not evident in the responses of these brain neurons, which fire after the initial pulse and before the occurrence of the second pulse of the preferred stimulus (fig. 6.11B, middle column).

Anurans. The most extensive and systematic studies of temporal processing in the ascending pathways of anurans have been conducted in the Northern leopard frog, *Rana pipiens*, and the grass frog *Rana temporaria* (e.g., Bibikov 1974; Walkowiak 1984; Eggermont 1990; reviews: Rose and Capranica 1985; Feng et al. 1990; Hall 1994; Feng and Schellart 1999). However, temporal tuning in species in three other anuran families has also been investigated. These studies provide an inventory of the filter properties of neurons with respect to the temporal properties of acoustic stimuli as well as estimates of the types and proportions of temporally selective neurons in the ascending pathway (box 6.1).

At least 60% of the population of neurons in the torus semicircularis are temporally selective; of these, about 20–60% of the neurons have band-pass characteristics (e.g., Rose et al. 1985; Walkowiak 1988; Feng et al. 1990; Diekamp and Gerhardt 1995; Penna et al. 1997; Alder and Rose 1998, 2000). Neurons with low- and high-pass filter characteristics are also common, and another type of neuron—*band-suppression*—has also been described. Its response properties mirror those of band-pass neurons in that there is a narrow range of pulse rates that suppress its activity relative to lower and higher pulse rates (fig. 6.12). A great deal of variation exists, and other acoustic properties of sounds can influence the filter category of a given neuron (box 6.1). As in the cricket and other animals, temporally selective neurons are nearly always

Box 6.1. Temporal Selectivity in the Anuran Auditory System: Distribution of Filter Types and Some Caveats

Feng and his colleagues summarize neurophysiological studies of temporal selectivity in the ascending auditory pathway of the leopard frog *Rana pipiens* by documenting neurons with different filter properties (fig. 1). Neurons that show some form of selectivity with respect to the rate of amplitude modulation are found at the level of the dorsolateral nucleus (DLN). Band-pass neurons, some of which respond best to pulse rates (or, more generally, to the rate of amplitude modulation of a signal) typical of advertisement calls, are common in the superior olivary nucleus (SON), torus semicircularis (TS), and central nucleus of the thalamus (THC).

The filter characteristics shown in fig. 1 were all based on the responses of neurons stimulated with pulsed amplitude modulation signals, which are appropriate stimuli because the pulses in the calls of *R. pipiens* are more pulsatile than sinusoidal. However, as emphasized in chapter 4, some species are

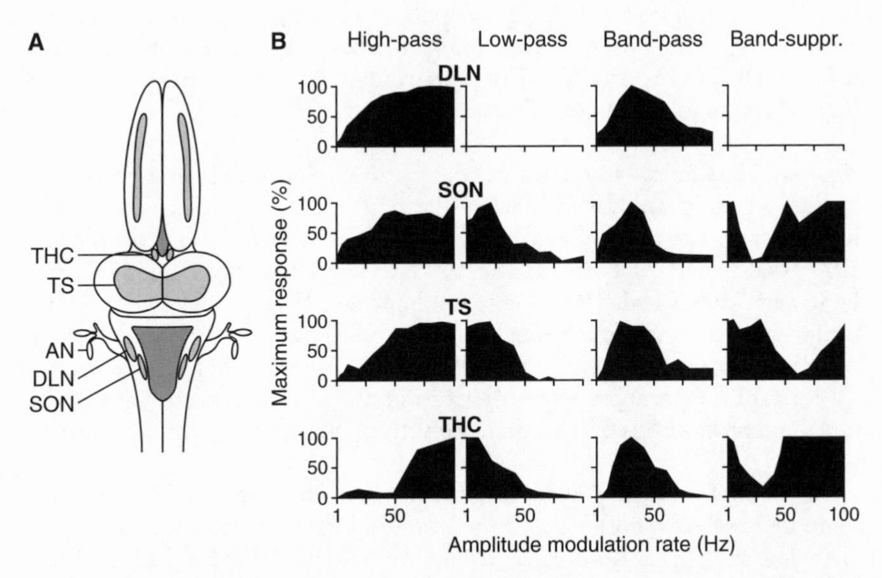

Box 6.1, Figure 1. Summary of temporal selectivity of auditory neurons in the ascending pathway of anurans. (A) Locations of the major auditory nuclei in the ascending pathway. AN = auditory nerve; DLN = dorsolateral nucleus; SON = superior olivary nucleus; TS = torus semicircularis; THC = central nucleus of the thalamus. (B) The distribution of filter types in four stations in the ascending auditory pathway of the leopard frog *Rana pipiens.* For the filter functions, the spike rate is plotted against the rate (Hz) of amplitude modulation of a series of pulsed signals. Whether a neuron is really selective for rate depends to some degree on its adaptation properties and whether it synchronizes well to transient onsets (see next section of this box and fig. 2). From Hall 1994, fig. 6; sample sizes from Feng et al. 1991.

(box 6.1 continued)

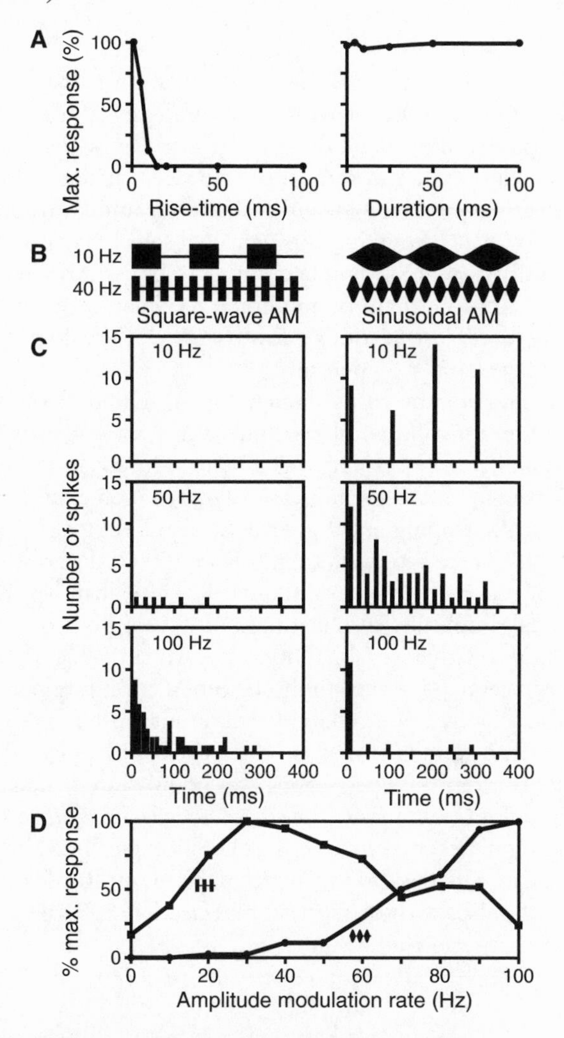

Box 6.1, Figure 2. (A) Rise-time, pulse-duration, and pulse-rate selectivity of an auditory neuron recorded in the dorsolateral nucleus of the leopard frog. Notice that the unit responds only to signals with short (< 10 ms) rise-times, but is not selective at all over a wide range of pulse durations. (B) Diagrams of pulsed (square-wave) AM stimuli (*left traces*) and sinusoidally AM stimuli (*right traces*) with different AM rates used to test the AM selectivity of this neuron. (C) Peristimulus histograms showing the responses of this neuron to stimuli with three different rates of AM: pulsed AM (*left*); sinusoidal AM (*right*). The fact that this neuron is phasic and responded only to pulses with fast onsets and required some minimum recovery time (silent period) explains the different filter functions to sinusoidal and pulsed AM shown in (D). (D) Spike rate (percentage of maximum spike rate) functions of the same phasic neuron in the dorsolateral nucleus of the leopard frog in response to a series of pulsed AM signals (*solid dots*) and to a series of sinusoidally AM signals (*open dots*). This neuron has a band-pass filter characteristic in response to pulsed AM and a high-pass characteristic in response to sinusoidal AM. From Hall 1994, fig. 9.

(box 6.1 continued)

also selective for pulse shape and the joint variation of pulse duration and intervals between pulses. No behavioral tests of *R. pipiens* are available. These neurophysiological studies—and those of *H. versicolor*, which is known to use rise-time, pulse duration, and intervals—are nevertheless valuable because they show selectivity for biologically realistic variations in natural signals. They do not, however, show that this neuronal selectivity is based solely on differences in pulse rate. One simple example of how the adaptation properties of a neuron and the form of amplitude modulation can result in completely different filter classifications is shown in fig. 2. We do not consider this neuron to be a true band-pass neuron.

Even at the level of the torus semicircularis, temporal variables such as the rise-time, duration, and number of pulses in synthetic calls that simulate natural calls can affect the responses of a neuron to variation in pulse rate in unpredictable ways (e.g., Adler and Rose 1998; Condon et al. 1991; Diekamp and Gerhardt 1995; Epping and Eggermont 1986; Feng et al. 1991; Gooler and Feng 1992). For example, Alder and Rose (2000) describe neurons that are selective for relatively low rates; these cells have a band-pass characteristic for sinusoidal amplitude modulation and have a low-pass characteristic for pulsatile signals with a variable duty cycle. To conclude, therefore, that a neuron is truly selective for the amplitude modulation rate per se, we might require that the selectivity be general for different pulse forms and pulse durations (i.e., be independent of pulse-duty cycle and pulse shape). Recent descriptions of the response properties of a population of neurons tuned to relatively high pulse rates in *Rana pipiens* and *Pseudacris regilla* come closest to fulfilling these requirements and have another biologically realistic characteristic: a relatively long latency to fire, which depends on stimulation by a minimum number of pulses with the correct rate (see fig. 6.13 and the text).

more poorly synchronized to the pulses in signals than are lower-order temporally nonselective cells (Rose and Capranica 1985; Hall 1994).

When considering neurons with band-pass characteristics, it is clear that the situation in anurans is even more complicated and controversial than in field crickets. Neurons with true band-pass characteristics are mainly confined to the torus semicircularis and the central nucleus of the thalamus (see also box 6.2). Indeed, only recently have auditory neurons with response properties that come close to meeting all of the expected criteria been described (Alder and Rose 1998, 2000). These neurons, which occur in the medial torus semicircularis, maintain their selectivity for relatively high pulse rates in the

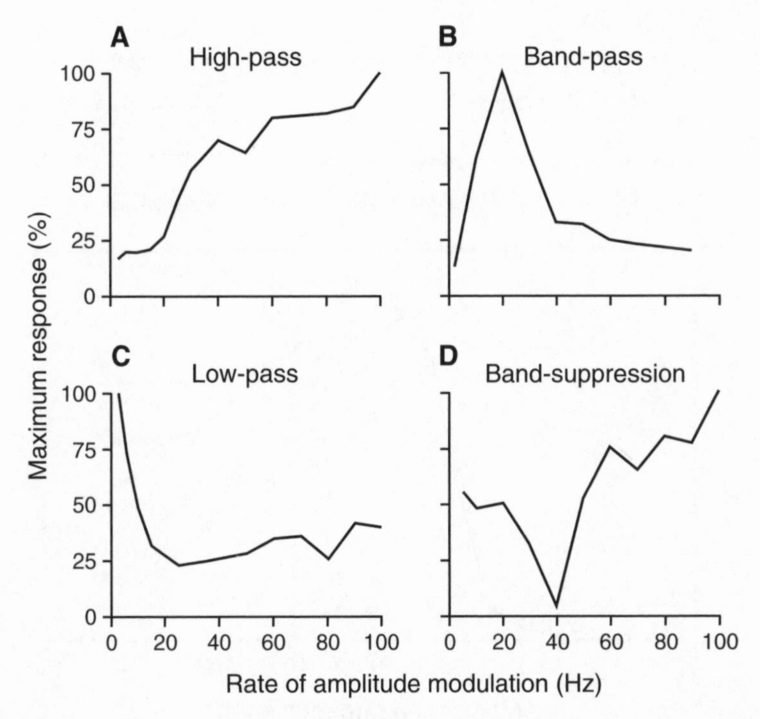

Figure 6.12. Representative amplitude-modulation-rate functions of temporally selective auditory neurons in the torus semicircularis (inferior colliculus) of the leopard frog. From Feng and Schellart 1999, fig. 6.14.

face of changes in pulse shape, pulse-duty cycle, and intensity (fig. 6.13). Moreover, these cells have relatively long latencies and thus do not respond unless a stimulus contains some minimum number of pulses with a particular repetition rate (fig. 6.13A). As mentioned above, this pattern of firing is exactly what is expected if the response of a neuron is truly dependent on evaluating repetition rate. Alder and Rose (2000) also point out that the selectivity of such neurons for long rather than short pulse trains might underlie the selectivity of some species of anurans (e.g., gray treefrogs) for long-duration signals (Klump and Gerhardt 1988; chapter 10).

In general, the tuning bias of band-pass neurons for the pulse rates typical of conspecific signals is much less impressive than in field crickets. Substantial proportions of temporally selective cells respond best to pulse rates that are not typical of any signal in the conspecific repertoire. The distribution of "best" amplitude modulation rates of band-pass neurons in the leopard frog and grass frog, for example, covers a broad range, encompassing the range of all of the signals in the vocal repertoire of these species and beyond (review: Feng et al. 1990). By contrast, the temporally integrating band-pass neurons

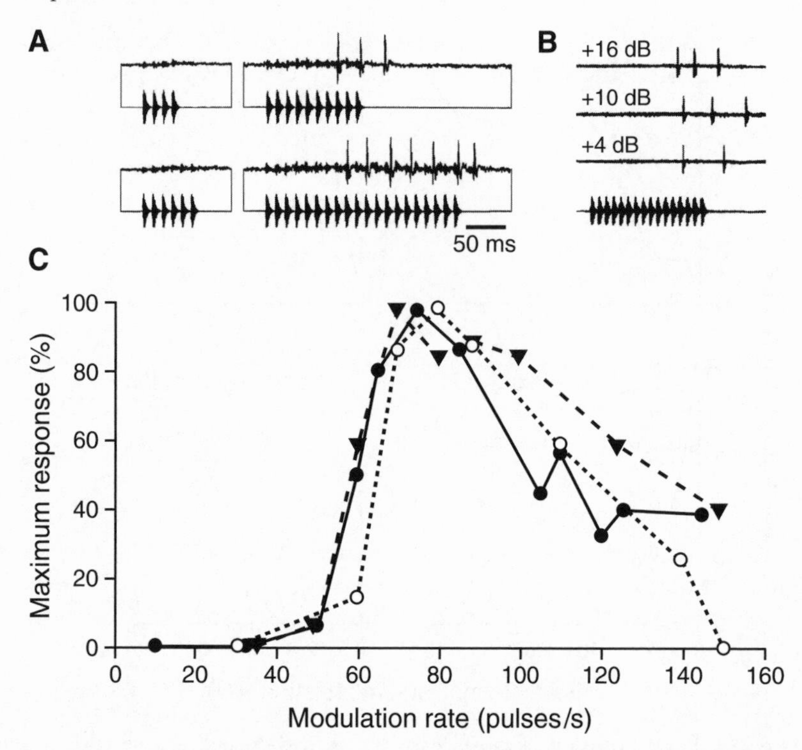

Figure 6.13. (A) Temporal integration by band-pass neurons in the Pacific treefrog *Pseudacris regilla.* Spikes were elicited with a long delay (> 50 ms) and only when pulse trains consisted of more than eight pulses. (B) Intensity-independence of the response. Once the minimum pulse-train duration was presented, the number of spikes elicited did not increase with stimulus intensity. (C) Normalized response level versus the AM rate (carrier frequency = 1.2 kHz). Closed circles = sinusoidal AM; triangles = pulse shape as in advertisement calls, with pulse duration constant at 10 ms; open circles = pulse shape as in advertisement calls, with pulse duration varied with pulse rate to maintain duty cycle of 1.0. From Alder and Rose 1998, figs. 1, 2.

recently found in *R. pipiens* were selective for the pulse rate typical of conspecific aggressive calls but not for that of advertisement calls (Alder and Rose 1998). Neurons showed selectivity for stimuli with sinusoidal amplitude modulation having rates typical of advertisement calls but only low-pass characteristics when more realistic, pulsatile signals were presented (Alder and Rose 2000; see box 6.1). Such low-pass characteristics would of course still be useful for distinguishing between the two call types.

Although significant species differences in best amplitude modulation rate were found between two species of toads, most neurons were not tuned to the range of pulse rates typical of conspecific advertisement calls (Rose and Capranica 1984). In *Bufo americanus,* which produce calls with pulse rates of about 40 pulses/s at 21°C, only 9 of 34 cells had best amplitude modulation rates between 31 and 60 Hz. In *Bufo fowleri,* mean pulse rate is about 120 pulses/s, but only 5 of 19 band-pass neurons had best amplitude modulation rates greater than 100 Hz. About half of the neurons responded best to amplitude modulation rates of 31–60 Hz. Thus, the tuning bias in *B. fowleri* was greater for pulse rates typical of the advertisement calls of *B. americanus* than for the pulse rates typical of conspecific calls. The release calls of these toads vary widely in pulse rate, and cells not tuned to the pulse rate of advertisement calls might play a role in their recognition. However, these signals are produced only infrequently and are directed at males and not females.

A better match between temporal tuning and pulse rate in advertisement calls is found in gray treefrogs. In *Hyla versicolor*, about 18 of 27 band-pass neurons were tuned to the species-specific range (uncorrected for temperature) of about 16–30 pulses/s (Rose et al. 1985). Similar results were found by Diekamp and Gerhardt (1995), with the proviso that the pulse shapes of the test stimuli were modeled after those in conspecific calls. In *Hyla chrysoscelis* from the southeastern United States, 12 of 21 neurons were tuned to the range of 31–45 pulses/s typical of males from that part of the species' distribution. In *H. versicolor*, changing the temperature of a frog resulted in shifts in best amplitude modulation rate of single neurons in the same direction: a correlate of the behavioral temperature coupling we discussed in chapter 4 (fig. 6.14; Brenowitz et al. 1985; Rose et al. 1985). The best match in the tuning of band-pass neurons to the pulse rate of conspecific calls so far reported in anurans was found in *Pleurodema thaul*, where 8 of 9 neurons were tuned to the conspecific rate of 10 pulses/s (Penna et al. 1997).

Although these results are encouraging, we emphasize again that behavioral studies showing how the animals evaluate pulse duration, shape, and intervals should precede neurophysiological studies. Recall from chapter 4, for

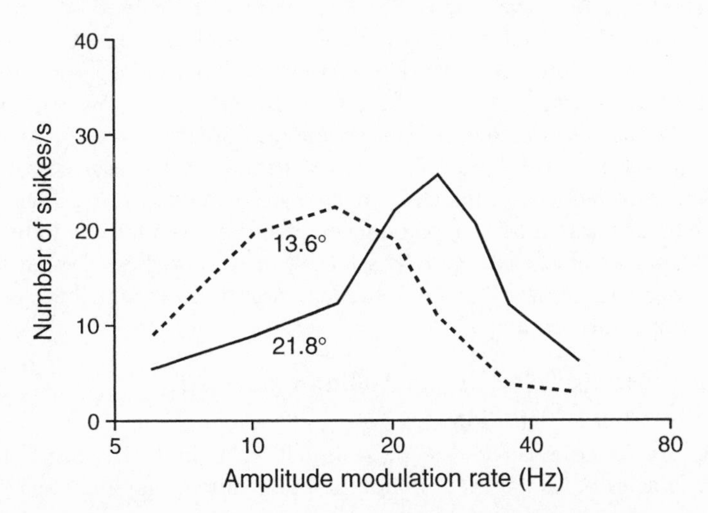

Figure 6.14. Temperature-dependence of the best amplitude modulation rate of a band-pass neuron in *H. versicolor* at two different temperatures. From Brenowitz et al. 1985, fig. 2.

example, that pulse rate alone is not the key property used for signal identification in *H. versicolor*. Even if rise-time is held constant, female preferences are profoundly affected by pulse-duty cycle. Other experiments show that females of *H. chrysoscelis* based their decisions on variation in pulse rate alone, and hence band-pass neurons found in this species might play a more direct role in signal identification.

At the level of the thalamus, band-pass neurons are concentrated in the central nucleus (fig. 6.2; Hall and Feng 1986, 1987). In the leopard frog, thalamic neurons usually showed a narrower range of temporal tuning that better matched the range of pulse rates in advertisement calls than did band-pass neurons in the torus. In the only other study of temporal selectivity at the level of the thalamus, Mudry and Capranica (1987a) found that the largest magnitude evoked potentials in the bullfrog *R. catesbeiana* occurred in response to stimuli with a waveform periodicity of 200 Hz, a typical value for its advertisement call. These results are consistent with but do not prove the hypothesis that the fine details of the waveform encoded by the auditory nerve might be a key feature of signal identification in this species (see Bodnar and Capranica 1994; Simmons et al. 1993a,b; critical discussion in chapter 5). As discussed in box 6.2, however, the thalamus of anurans may not be a locus for the rapid recognition of signals that occurs in behaving animals.

Unlike those in crickets, most band-pass neurons in anurans are not tuned to the carrier frequency of the conspecific call (Walkowiak 1984, 1988). One extreme example is the sample of temporally selective neurons recorded in the torus semicircularis of *Pleurodema thaul* by Penna et al. (1997). Nine of 11 of these neurons were tuned to frequencies well below (around 300 Hz or in the range of 700–1200 Hz) the carrier frequency of the advertisement call of about 2000 Hz. Temporally selective neurons are still likely to be stimulated effectively over reasonable distances by the intense advertisement calls typical of most anurans—provided that their characteristic frequency is not too different from the carrier frequency. Selectivity to conspecific pulse rates would be unlikely to be signaled by maximum firing rate, however, because such neurons would also almost certainly be better stimulated by a variety of other sounds whose frequencies match their characteristic frequency. These facts reinforce the view that pattern recognition in anurans is not likely to be based on the responses of a few temporally selective neurons dedicated solely to this task. Rather, such neurons, along with many others, must contribute to some form of population coding.

Temporal Pattern Selectivity of Auditory Neurons in Katydids and Grasshoppers

Although recognition is based on pulse-rate filtering in the katydid *Tettigonia cantans*, females of *T. caudata* evaluate the pulse-duty cycle rather than pulse rate, and females of *T. viridissima* evaluate both pulse length and intervals

Box 6.2. Is the Auditory Thalamus of Anurans Involved in Selective Phonotaxis?

Casseday and Covey (1996) propose that, despite its different anatomy and connectivity in different vertebrate subclasses, the inferior colliculus (torus semicircularis) is specialized for processing biologically important sounds to which immediate responses are required. Moreover, in several systems it appears that different acoustic properties are encoded and processed in separate pathways before converging on the torus semicircularis. In weakly electric fish, for example, phase (timing) and amplitude information, encoded by different receptor types, converges on the torus semicircularis. Within the torus, neurons integrate this information and generate descending outputs that ultimately determine the sign and magnitude of the jamming avoidance response, a behavior in which a fish raises or lowers its pacemaker discharge rate relative to that of its neighbor (e.g., Metzner 1993; Metzner and Viete 1996). The inferior colliculus is also a major integrating center for the separately encoded interaural time and intensity cues required for sound localization in owls (Konishi 1994) and for the multiple aspects of echoes used by bats to locate and identify prey and to navigate (Casseday and Covey 1996). In each of these systems, rapid modifications of ongoing behavior are required, and the inferior colliculus has strong and direct connections to the appropriate motor systems. Luksch and Walkowiak (1998) provide similar anatomical evidence for anurans.

In weakly electric fish and anurans, there are, moreover, relatively few projections to higher centers beyond the thalamus, and in anurans the long refractory periods of neurons in the thalamus seem to eliminate that neural structure as a site for generating rapid responses. Studies of thalamic auditory activity in freely moving behaving animals are badly needed, however, to learn if these long refractory periods are real or artifacts of immobilization and the fact that most recordings are from males or females not in reproductive condition. Lesion studies so far support the idea that long latencies are not artifacts. Schmidt (1988) reported that extensive lesioning of the thalamus and torus semicircularis did not abolish phonotactic responses in toads (*Bufo americanus*). A recent study of the gray treefrog *H. versicolor* confirmed that females with extensive lesions of the thalamus still responded phonotactically and that they showed the same degree of selectivity for temporal and spectral patterns as do intact animals (Walkowiak, Endepols, Feng, Gerhardt, and Schul, in prep.). In contrast to Schmidt's results, however, even small lesions of torus semicircularis abolished all phonotaxis. The thalamus, with its rich

(box 6.2 continued)

reciprocal connections to the ventral hypothalamus and preoptic nucleus (e.g., Wilczynski et al. 1993), might thus be involved in the integration of information about the animal's reproductive condition with external information from multiple senses in order to regulate the animal's phonotactic readiness over relatively long periods of time.

(chapter 4). Whereas band-pass neurons, such as those described for *Gryllus*, might mediate pulse-rate selectivity in *T. cantans*, a combination of filters, say, modified low- and high-pass neurons, would seem to be required to explain the pulse duration and interval selectivity of *T. viridissima* (Schul 1998).

Ascending neurons AN1 and AN2, which are similar in morphology and physiology to the ascending neurons of crickets, have been described in the katydids *T. cantans*, *T. viridissima*, and *Mygalopsis marki* (Römer 1987; Hardt 1988; Schul 1997). Both neuronal types synchronize to the temporal pattern of the song (fig. 6.15). As in crickets, the fidelity of AN1 in pattern encoding is more pronounced at "low" frequencies in comparison with AN2. Unlike the AN2 in crickets, however, the AN2 in these and other katydids is capable of accurate copying of the pulse patterns of the calling song, which has considerable high-frequency energy, and almost certainly plays a key role providing information needed for sound localization (chapter 7). Moreover, in katydids some T-shaped neurons might be involved in recognizing calling songs. The TN1 neuron of *T. viridissima*, for example, also encodes the time pattern of calling song, at least in its adapted state (fig. 6.15). One possible correlate of the behavioral rejection of the songs of *T. cantans* and *T. caudata*, even in no-choice situations, is that, in its adapted state, TN1 fails to respond to either of these signals. Further understanding of pattern recognition in these insects requires recording from brain neurons.

The large axon diameter of the TN1 interneuron speeds conduction of action potentials and thus probably also contributes to the detection and evasion of predators. Another T-shaped interneuron, TN2, whose morphology and physiology is quite distinct from that of TN1, might be specialized for predator detection (Schul 1997). This kind of specialization of a T-shaped neuron is particularly well exemplified by a study of a North American species, *Neoconocephalus ensiger* (Faure and Hoy 2000b). This cell preferentially responded to the playbacks of short signals that simulated bat echolocation signals than to much longer-duration conspecific songs; the cell also faithfully copied the temporal pattern of the bat signals but not conspecific songs. Auditory interneurons of similar types and with similar function have been

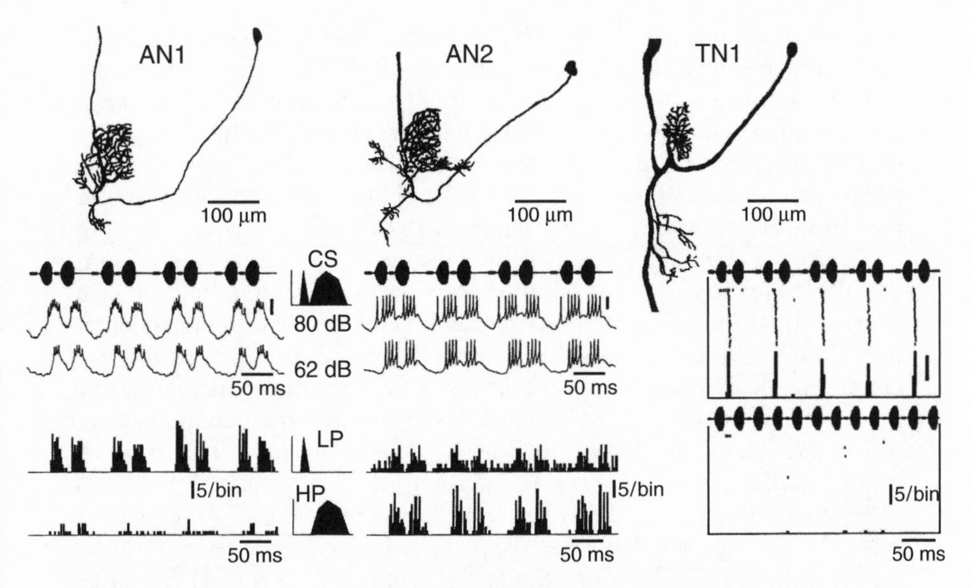

Figure 6.15. Reconstructions of the prothoracic morphology of AN1, AN2, and TN1 interneurons of the katydid *Tettigonia viridissima*. Below each reconstruction are diagrams showing the responses of these neurons in relation to the temporal pattern of conspecific song models having the complete bimodal spectrum. For AN1 and AN2, additional diagrams show that AN1's synchronization to the temporal pattern occurs only at frequencies representative of the low-frequency peak in the song, whereas AN2 can synchronize at both low and high frequencies, albeit more accurately at high frequencies. Although TN1 fires synchronously—usually once—to the double-pulse songs of conspecific males, it fires hardly at all to the higher pulse-rate songs of *T. cantans*. Songs of *T. caudata* have even higher pulse rates. CS = complete bimodal spectrum; LP = low-frequency part; HP = high-frequency part; bin = time window for counting spikes. From Schul 1997, figs. 5, 8, 9.

reported in the haglid *Cyphoderris monstrosa*, a species in a taxonomic group that could be representative of the ancestral state of katydids (Mason and Schildberger 1993).

The network for pattern recognition in grasshoppers is more distributed than that in crickets and katydids, and substantial analysis of temporal patterns occurs already at the level of the metathoracic ganglion (reviews: Helversen 1997; Pollack, 1998). Moreover, the recognition pathway is separated from the localization pathway up to the brain (chapter 7), which is probably the site of integration of pattern and directional information. Brain neurons tuned to conspecific pulse rates have been described but are unidentified morphologically (Römer and Seikowski 1985).

Recall that grasshoppers have more than a dozen different types of ascending auditory interneurons, each with specific filter properties (e.g., Ronacher and Stumpner 1988; Stumpner 1988; Stumpner and Ronacher 1991a,b, 1994; Krahe 1997). For example, some cells in the set of local, ascending, and T-shaped neurons seem to encode impulses within syllables with fast repetition rates, others encode gaps in syllables, and others encode behaviorally important values of pulse (syllable) interval and pulse (syllable) duration. Still other cells are sensitive to higher-order temporal patterns (Römer and Seikowski 1985; Stumpner and Ronacher 1991b; Stumpner et al. 1991). The array of ascending cells thus provides a wide range of information, from which the brain presumably selects the subset required to discriminate among biologically important signals (Stumpner 1988; Helversen 1997; Krahe 1997).

Females of *Chorthippus biguttulus* are selective for particular combinations of syllable duration and intervals found in conspecific calling songs of males (chapter 4). Within the group of ascending neurons found in grasshoppers, AN12 shows phasic responses that are reasonably well correlated with the selectivity of females. That is, AN12 fires reliably to the onset of syllables only if a critical pause duration is exceeded. In this respect, the requirements for firing match the rising part of the behavioral response curve. That is, AN12 invariably fires a burst of spikes at the behaviorally most effective range of pause durations of 20–25 ms when syllable durations are 85–100 ms (Helversen 1972; chapter 4). However, the preferred interval between syllables increases with syllable duration (Stumpner 1988), and recent behavioral experiments also indicate that the response properties of AN12 alone cannot explain the behavioral selectivity of females (Balakrishnan et al. 2001).

Females of *C. biguttulus* are also particularly sensitive to the presence of short silent gaps within syllables of the male's song; the animals cease responding when these gaps exceed about 3–4 ms at 23°C. Although the presence of gaps influences the response patterns of several different interneurons, AN4 responds particularly well to syllables that lack gaps. Moreover, the firing of this interneuron is completely inhibited as gaps in syllables with rectangularly modulated pulses exceed about 5 ms, a pattern that occurs over a reasonably wide range of playback intensity (fig. 6.16). Each onset of a syllable separated by a gap is converted to an inhibitory potential in AN4 (Ronacher and Stumpner 1988; Stumpner and Ronacher 1994). However, AN4 alone is unlikely to be the only ascending neuron conveying information about gaps to the brain, given that the responses of other interneurons are also influenced by gaps and still other interneurons remain unstudied.

Summary and Suggestions for Future Studies

The ascending auditory pathways of insects and anurans differ markedly. In insects the focus has been on understanding central processing of acoustic communication signals by a few individually identifiable interneurons that project to higher centers from the neuropiles where auditory receptors terminate. At the level of the brain, a handful of interneurons in crickets have response properties that correlate reasonably well with both the spectral and temporal acoustic properties of communication signals that are most effective in eliciting phonotaxis. However, we need further study concerning how the

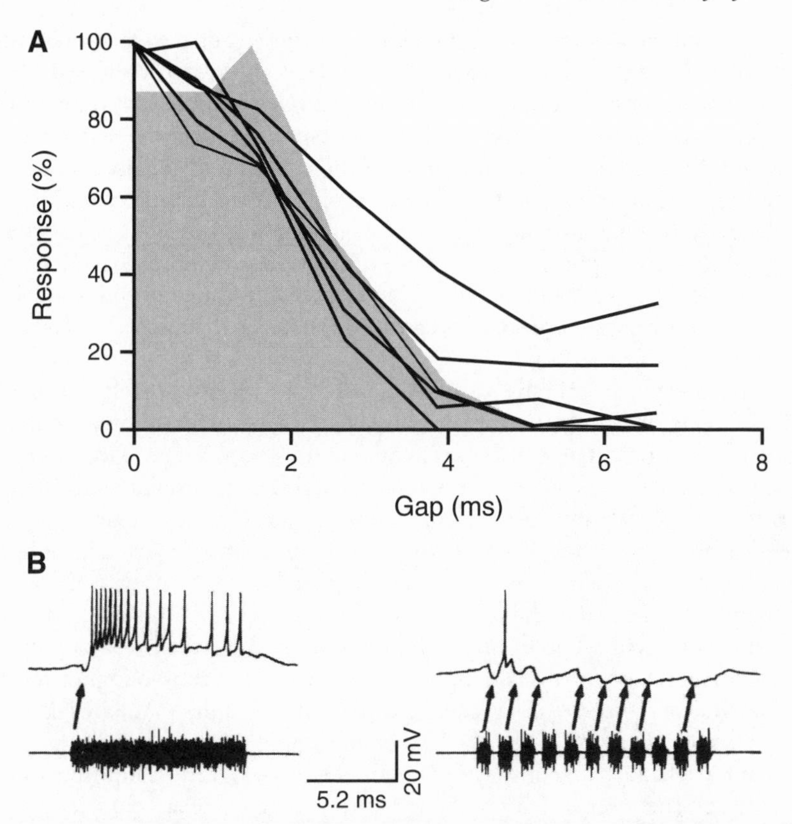

Figure 6.16. Behavioral and neurophysiological evidence for rejection of syllables with gaps in female *Chorthippus biguttulus.* (A) Response of the AN4 neuron as a function of gap width. The behavioral response function (mean of three females) obtained at the same temperature is indicated by the shaded area. (B) *Left,* AN4 responses to syllables without gaps. Sound onset evokes an inhibitory postsynaptic potential (*arrow*), followed by depolarization and the superimposed spike train. *Right,* in syllables with gaps, each onset of sound after a gap evokes an inhibitory postsynaptic potential (*arrows*), and the potentials summate and suppress the neuron's response. Modified from Ronacher and Stumpner 1988 and Stumpner and Ronacher 1994.

selectivity of these cells is derived from various kinds of inputs from other neuronal types, such as the high- and low-pass neurons to which band-pass neurons are presumed to be connected. Furthermore, we are completely ignorant about how the output of band-pass neurons might influence selective phonotaxis, and this important question deserves anatomical and experimental attention.

In the anuran system, our present knowledge suggests where in the ascending pathway transformations in neural codes occur. We also know that the high degree of parallel and serial connectivity revealed by neuroanatomical studies is mirrored somewhat in the specialization of discrete nuclei, at least within the thalamus. At present we have only a catalog of neurons sampled in various auditory centers; the vast majority of these cells have been recorded extracellularly and have not been characterized in terms of their anatomy or their projections to other parts of the brain or to the motor system. The synaptic mechanisms and inputs that are responsible for the selectivity of these neurons are unknown, and none of the cells is individually identifiable from preparation to preparation. However, in vitro studies are beginning to uncover types of auditory neurons that can at least be grouped by anatomy and by physiological responses to electrical stimulation simulating that elicited by acoustic signals (e.g., Luksch and Walkowiak 1998). Other studies show how selectivity for temporal properties can be influenced by a neuron's adaptation properties (box 6.1).

Some neurons in anurans have response properties that correlate well with some aspects of pattern recognition in the whole animal; however, correlations are no more than hypotheses about causal relationships. Moreover, the temporal selectivity of these neurons is seldom robust to simultaneous changes in other temporal properties of sounds, and most often these neurons are not tuned to the carrier frequency of natural communication signals. In both insects and anurans, behavioral studies are required to confirm the AM-rate selectivity of the animal in the face of systematic changes in pulse-duty cycle and rise-time and form. The results could then be used to select test stimuli and better interpret assays of the temporal selectivity of neurons in higher centers in the auditory system. In agreement with other authors (Walkowiak 1988; Feng and Schellart 1999), we doubt that any single type of neuron exists that responds exclusively to conspecific signals. Thus, recognition almost certainly depends on the joint activity of many neurons that are relatively selective for different species-typical features of the signal. Does the temporal selectivity so far described in insects and anurans constitute matched filtering in the time domain? The answer seems to be mainly a matter of personal interpretation. What would be helpful is for proponents of this idea to specify the hypothesis in a quantitative and testable fashion. For example, to our knowledge, no criterion has been proposed about the proportion of temporally selective neurons that would have to be tuned to rates typical of conspecific signals to support this hypothesis.

In studies of brain neurons in insects and anurans, two important assumptions apply: (1) that the sample of neurons encountered at a given level of the auditory system is a representative sample; and (2) that the response properties of each neuron remains constant, so that the array of response properties recorded in the sample represents the response profile of the auditory neurons at the time when an animal is engaged in communication. These assumptions are seldom considered, much less fulfilled. Electrode types and search stimuli influence what neurons will be found (e.g., Alder and Rose 2000), and in the anuran thalamus, recovery times to repetitive stimulation are unreasonably long compared to behavioral response times. As emphasized in chapter 5, behavioral experiments are also needed to delimit the minimum exposure to acoustic signals required for animals to make selective behavioral responses (e.g., Ronacher and Krahe 1998). Such information can help us to evaluate hypotheses that depend on signal averaging over time or over arrays of single auditory receptors or neurons.

Recognition in some systems might be a result not of highly specialized, selective cells in the ascending auditory pathway, but of the integration of sensory processing occurring at the level of populations of neurons in the descending pathway. One example may be the control of the jamming avoidance responses in electric fish, in which the rate of the electric organ discharge could be controlled by the antagonistic (excitation/inhibition) interaction of two descending pathways that have different anatomical and functional connections to the pacemaker circuitry (e.g., Metzner and Viete 1996). In any event, recordings from neurons in behaving animals are critical to verify that the response properties of neurons in immobilized animals are at least similar to those in behaving subjects. In katydids, for example, context-dependent changes in behavioral selectivity suggest concomitant changes in the response properties of single identified neurons (Schul and Schulze 2001). In crickets some descending neurons do not respond to sound unless the animal is behaving (Staudacher and Schildberger 1993).

Further challenges remain at the level of single neurons. For example, we need to discover the neuroanatomical and physiological bases of the selectivity of neurons that are probably involved in signal identification. We also need to study where and how the output of these selective cells affects other neurons and, ultimately, determines or guides the motor programs responsible for appropriate behavioral responses. So far, much more progress along these lines has been made in owls and bats, where there is a tremendous specialization for localizing prey by acoustic cues (reviews: Suga 1989; Konishi 1994). The success of this research, and that of Heiligenberg and his colleagues on the jamming avoidance response in electric fish (review: Heiligenberg 1991, 1994), can be attributed in large part to the intense focus on behavioral studies that defined algorithms likely to be used by the central nervous system. This work then set the stage for finding the neurons and networks that implement these algorithms. For this reason, acoustic communication of insects and anurans offers one of the best systems for understanding the neural mechanisms underlying acoustic pattern recognition. For a handful of species, we have rather precise, quantitative specifications for pattern recognition from behavioral studies (chapter 4), and for other species, we know a lot about how single acoustic properties of behavioral relevance, and combinations of such properties, are encoded and processed in several stations in the ascending pathway. One important goal of this monograph is to encourage new and intensive efforts, comparable to those afforded the owl, bat, and electric fish systems, to acoustic pattern recognition in insects and anurans.

7

Sound Localization

SOUND LOCALIZATION usually depends on binaural cues. That is, unless the sound source is located in the median plane (e.g., immediately in front, behind, or above the animal), sound waves will arrive at one ear before arriving at the other ear (*binaural time cue*) and/or may stimulate one ear to a greater extent than the other ear (*binaural intensity cue*). Turning or moving toward the direction of the sound source (the target axis) usually reduces binaural disparities, and turning farther away increases them. The magnitude of binaural cues depends on such factors as the size of the animal (separation of its ears and the degree of sound diffraction that the body and other structures provide), the frequency of the signal, and whether or not sound waves can reach the internal surfaces of the tympanic membranes. Regardless of how they are generated, binaural disparities can be enhanced by physiological mechanisms.

Behavioral studies of sound localization provide estimates of the binaural cues needed for correct orientation and also show how phonotactic accuracy can vary as a function of the spectral properties of acoustic signals. These data are useful for formulating and testing models concerned with the underlying biophysical and neural mechanisms that consider how peripheral structures and pathways contribute to the directional characteristics of the ears and how binaural differences are processed in the central auditory system. Finally, behavioral studies have demonstrated how information about sound location and biologically meaningful sound patterns can sometimes interact to determine the orientation of the animal's phonotactic response.

Behavioral Studies of Localization Abilities in Insects and Anurans

The following kinds of information about sound localization have been provided by behavioral experiments with insects and anurans: (1) extent to which an animal usually turns away from the direction of the sound source before it makes a corrective turn; (2) direct estimates of the binaural intensity and time differences required for correct turns toward the sound source; (3) evidence that individuals of some species can discriminate different angular deviations of the source location relative to its own position; (4) effects of variation in the spectral properties of signals on localization accuracy.

Estimates of the accuracy of sound localization can be obtained under closed-loop or open-loop conditions. In *closed-loop conditions*, the animal has the opportunity to successively update directional information as it changes position relative to the sound source. These conditions reflect the situation in nature. In arena tests, where the animal moves freely from a starting point toward the sound source, the sound gradient within the sound field surrounding the source provides another cue: the apparent intensity of the signal becomes greater as the animal gets closer to the speaker. Thus, the direction of any particular turn, hop, or head scan might be based not only on directional information obtained at the animal's present position, but also upon information it might remember from previous positions. Although an insect remains at the same point in the sound field on a locomotion compensator (thus eliminating sound gradient cues), the animal can still turn freely (Weber et al. 1981; Schmitz et al. 1982). Thus, if a continuous series of signals are required to elicit walking, the data provide only closed-loop estimates of localization abilities.

In *open-loop conditions*, the animal has no opportunity to obtain additional cues after it begins to move. One clear example is the "flying preparation," in which an insect is tethered and indicates its turning tendency by bending its abdomen while its ears remain fixed in position relative to the target axis chosen by the experimenter (review: Pollack and Hoy 1989). Another method is to allow a tethered insect to manipulate a lightweight sphere in such a way that the sphere's motion indicates how the animal would turn if it were freely walking (e.g., Schildberger and Hörner 1988; Doherty 1991). In Y-maze experiments, the animal can move freely but has to make a decision at a single point during its walk to turn to the left or right (e.g., Rheinlaender and Blätgen 1982). The angular deviation of the Y-maze is varied, and the speaker is sometimes placed in line with the right fork and sometimes in line with the left fork. This method assumes that the animal does not remember any directional cues that it might obtain by orienting movements made prior to its decision point. Even if the animal can move freely, open-loop measurements can still be obtained if the species in question orients to single, brief signals (e.g., Helversen and Rheinlaender 1988; Klump and Gerhardt 1989). It is also possible to minimize potential cues from the sound gradient, even in an arena setup (see below; Klump and Gerhardt 1989; Jørgensen and Gerhardt 1991).

On the one hand, the localization abilities of insects and anurans are likely to be overestimated by laboratory studies because of the simplified acoustic situations in which background noise is minimized and directional cues are not degraded by objects in the environment (Michelsen 1983; chapter 11). On the other hand, insects and anurans in nature localize sounds under closed-loop conditions and have gradient cues as well. Moreover, nearly all studies use measurements of phonotactic movements to estimate accuracy, which will depend not only on the sensory system's ability to determine sound direction, but also on the motor system's ability to control orientation (Wyttenbach and Hoy 1997). Thus, sensory acuity per se is likely to be somewhat underestimated.

Accuracy and Mode of Sound Localization

Observations of the orientation movements of insects and anurans during phonotaxis, even in closed-loop conditions, indicate how the animal obtains directional information. For example, orientation movements typically involve turns or movements to both sides of the target axis. Deviations to the right are often closely followed by deviations to the left because the animals usually overcorrect. This pattern of corrective turns has been called *zigzag walking* or *hopping*. One-eared crickets also scan the sound field and can thereby locate the source of calling songs (box 7.1).

Even if the insect or anuran updates information continuously, the magnitude of an animal's deviations (orientation or movement "errors") from the target axis before making a correcting turn can provide estimates of the minimum binaural cues needed. Insects and anurans seldom turn so that the long axis of their bodies is perpendicular (90°) relative to the target line, a position where we might expect binaural cues to be relatively large, if not maximal. Rather, these animals appear to obtain sufficient directional information in a frontal zone that is usually less than about ±40° with respect to the target axis.

Lateralization and angle-dependent localization. In some species, individuals lateralize the sound source, that is, they determine only whether the sound source lies to their right or left. Other species can additionally determine the magnitude of the deviation of the source location from their body axis. This ability is indicated by a positive correlation between the magnitude of corrective orientation movements and the degree to which the source is located to the right or left side once the minimum deviation needed to make a correct turn has been exceeded. This question is most unambiguously approached with open-loop tests.

The best example of pure lateralization comes from studies of male grasshoppers *Chorthippus biguttulus*, which, when the source is to its right or left, usually makes a rapid turn toward signals simulating the female song. Using shorter-than-normal signals (400 ms rather than 1 s) so that turns occurred after the end of each signal, Helversen and Rheinlaender (1988) achieved open-loop conditions. They also stimulated each ear in a quasi-independent fashion. That is, they located one speaker opposite the right ear and the other, opposite the left ear and adjusted playback intensity so that each speaker stimulated only the adjacent ear. Minimum binaural differences for correct lateralization (turns toward the side of more intense or earlier stimulation) are of

Box 7.1. Sound Localization in One-Eared Insects

We have frequently said that orientation to the sound source is based on binaural cues. If so, then one-eared animals should fail to either orient to a sound source or turn to the side of the remaining ear. Unilaterally deafened male grasshoppers consistently turn toward the intact ear (Ronacher et al. 1986), and one-eared female crickets consistently turn toward the intact side in response to calling songs, and away from that side when stimulated with ultrasound (Moiseff et al. 1978; Pollack et al. 1984). Thus they follow the rule "turn to the side of the more strongly stimulated ear" or away from it in the case of negative phonotaxis. However, a minority of female crickets that had lost one ear by accident or surgery (after amputation of the respective foreleg), either

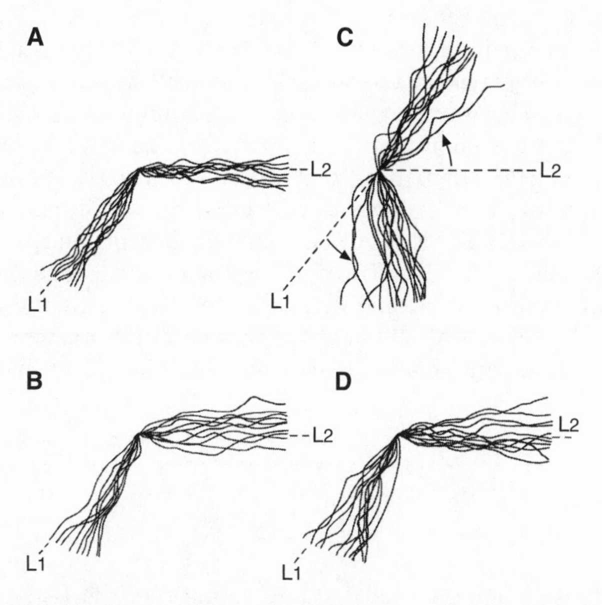

Box 7.1, Figure 1. Phonotactic tracking of a synthetic calling song in one-eared female crickets (*Gryllus bimaculatus*) on the spherical treadmill. Speaker positions (135° apart) are indicated by L1 and L2. (A). Sound-source-dependent walking courses of an intact female. Each trace represents 50 cm of walking. (B) Precise tracking of an adult female that had lost the right foreleg in an earlier larval instar and regenerated the leg to its normal length but without regenerating a functional auditory organ in this leg. (C) Change in walking angle of the same female after amputation of the right foreleg with the ear a few hours before the experiment. Note the deviation from the sound source by approximately 70° toward the remaining left ear. (D) No change in walking courses in the same female after amputation of the previously regenerated right foreleg, demonstrating that one ear is enough for phonotactic tracking. From Huber 1987.

(box 7.1 continued)

early in development or as adults, showed phonotactic orientation (Huber et al. 1984; box 7.1, fig. 1).

As discussed in detail by Huber (1987, 1990), Schildberger (1994), and Schildberger et al. (1989), some phonotactic behavior in one-eared crickets can be explained by changes in the structure and function of central auditory neurons. When deprived from auditory input for some time, central neurons grow new dendrites from their former input area (see first report in Hoy et al. 1978). These dendrites cross the midline, invade the auditory neuropile of the intact side, and change the synaptic efficacy of those target neurons to which they are now newly connected. It seems that with the new organization of the central auditory pathway, those one-eared crickets can follow the rule similar to that used by intact binaural crickets. However, much more research is required to quantify the compensatory changes in the functional properties of these neurons and to learn how the comparison of their outputs at higher levels restores seemingly normal phonotaxis in animals with a single intact ear.

Neural regeneration cannot, however, explain the fact that some female crickets already show phonotactic responses after one or a few hours of unilateral auditory deprivation. In these animals, no structural and functional reorganization takes place. As discussed in Schildberger and Kleindienst (1989) and Schildberger (1994), their orientation requires turning in the sound field, and the current hypothesis is that these animals make successive measurements of intensity before and after turns. In other words, the animals appear to scan the sound field, and this mechanism probably operates along with the normal comparison of intensity between the right and left sides in intact animals.

the order of 1 dB in intensity and 0.5–1 ms in time (fig. 7.1). Once these minimum cues are exceeded, however, the magnitude of an individual's turn is not affected by the magnitude of the binaural differences, which would of course correlate with the degree (angle) to which a real sound source would be located to the side of the animal.

Field crickets such as *Gryllus bimaculatus* are much less accurate in their lateralization performance than the grasshopper *Chorthippus biguttulus*. In Y-maze tests, the minimum angular deviation that elicited correct turns is about ±30° (Rheinlaender and Blätgen 1982); at this separation, the effective intensity difference between the two ears is more than 3 dB. Using flying preparations, Pollack and Plourde (1982) measured the magnitude of steering

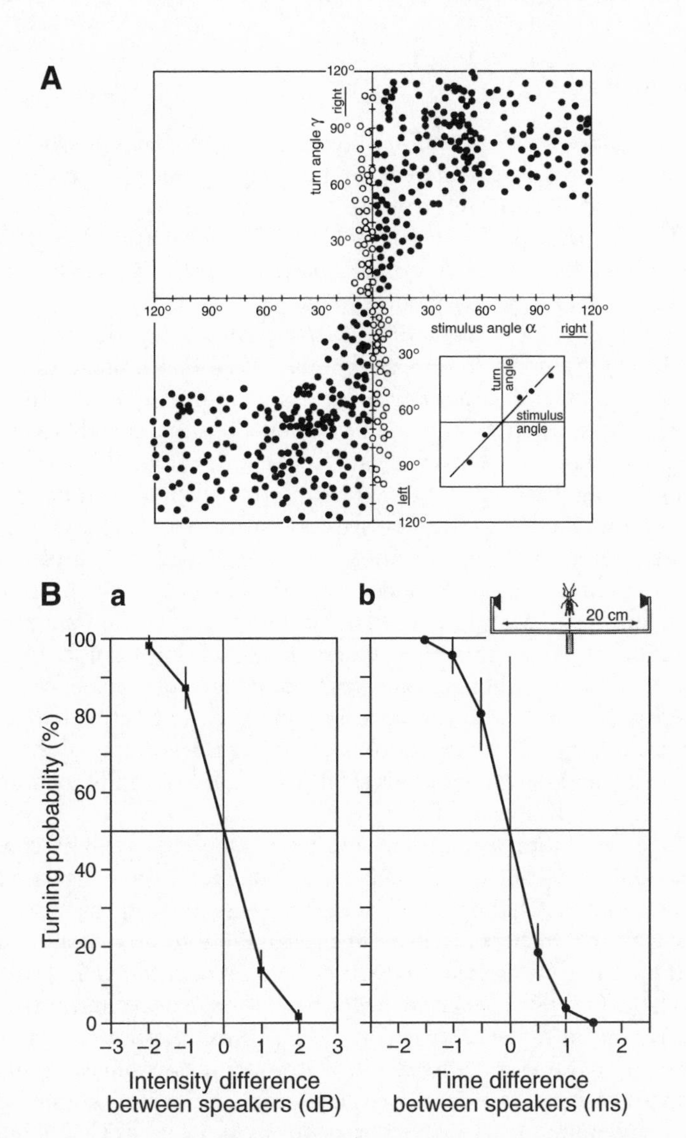

Figure 7.1. Lateralization and minimum binaural cues in the grasshopper *Chorthippus biguttulus*. (A) The relationship between stimulus angle and turn angle is plotted for nine grasshopper males. Each dot represents a single turn. Note the discrepancy between angular-dependent orientation (see inset) and the documented results, indicating pure right-left lateralization. From Römer and Rheinlaender 1989, fig. 56; data from Rheinlaender and D. v. Helversen, unpubl. (B) Minimum intensity and time differences for lateralization: shown is the preference for the louder side in the lateralization performance. *Inset* (*top right of [b]*), experimental setup. (a) Mean probability of turns (% of all turns) of twelve males toward a reference speaker of a constant sound intensity presented on one side, as a function of intensity differences between this speaker and a second one, presented on the other side, the intensity of which was either higher or lower than that of the reference speaker (negative and positive values on the abscissa, respectively). (b) Probability of turns of six males toward a reference speaker as a function of interaural time difference between the female signal from the reference speaker and an equally loud female signal presented from the speaker on the other side. The latter signal was either delayed with respect to the reference signal, or it preceded the latter (negative and positive values on the abscissa, respectively). The distance between each speaker and the animals was 20 cm. From Helversen 1997, fig. 7; data from Helversen and Rheinlaender 1988.

movements in *Teleogryllus oceanicus* in response to playbacks of synthetic calling songs presented from different angles. Steering magnitude increased as the sound source was moved from ±10–50° off the target axis, leveled off at 50–130°, and then declined at 130–155°. Wyttenbach and Hoy (1997) reinterpreted data from this study and suggested that the accuracy of orientation in the frontal range was about ±15°.

Oldfield (1980) reported that for *Teleogryllus* the average size of all turns for angular deviations of 90° or less was about 41° (SD +24°), suggesting little or no ability for angular discrimination (see comparable data from Murphy and Zaretsky 1972). However, other studies found a positive correlation between the speed of turns and angular deviations (Schmitz et al. 1982; Stabel et al. 1989), suggesting that crickets have some ability to distinguish different angles at least under what are effectively closed-loop situations. At present we cannot be sure whether a simple lateralization mechanism adequately explains sound localization (under open-loop conditions) in crickets. This issue is significant primarily for investigators interested in mechanisms of sound localization because lateralization is a very reliable tactic for sound localization. In contrast to the situation in grasshoppers and crickets, unequivocal evidence exists for angle-dependent orientation in at least two species of katydids (*Leptophyes punctatissima*, *Tettigonia cantans*) and one treefrog (*Hyla gratiosa*), all of which were tested in open-loop conditions (Hardt 1988; Klump and Gerhardt 1989; Rheinlaender and Römer 1990; fig. 7.2A).

Nearly all behavioral studies of localization in anurans have been conducted in closed-loop conditions. Although most individuals show zigzag patterns of hopping (thus alternating the "leading" [closest to the speaker] ear), a few individuals crawl for short distances in a remarkably accurate fashion directly toward the speaker (Feng et al. 1976; Rheinlaender et al. 1979). In a small dendrobatid frog (*Colostethus nubicola*), which was tested in its natural environment in Panamá, and in the green treefrog *Hyla cinerea*, tested in the laboratory, measurements of head or body scanning show that these animals obtained sufficient directional cues to make corrective turns in the frontal zone of about ±35°; average jump or orientation errors were about 16–23° (Rheinlaender et al. 1979; Gerhardt and Rheinlaender 1980). In the green treefrog, about one-fourth of the jumps were preceded by head scanning, which improved the accuracy of the jump compared with jumps occurring without previous scanning. The maximum binaural time cue that would be available to a large animal is about 40 μs—when one of the frog's ears is directly facing the speaker (a 90° angle of orientation). However, because green treefrogs rarely turn by more than about 35° from the target axis, they normally experience a maximum time cue of the order of 10 μs. The fact that head orientation (8.4°) is more accurate than subsequent jumps (11.8°) (Rheinlaender et al. 1979) supports the idea that directional (sensory) acuity is likely to be compromised by limitations of the motor system (see also Wyttenbach and Hoy 1997). Jumps obviously require much more motor system involvement than head scans.

Open-loop estimates of localization accuracy are available for barking treefrogs *Hyla gratiosa*. Angular deviations (from the direction of the sound source) of the first head scan or jump were scored after the playback of advertisement calls began (Klump and Gerhardt 1989). No measurement was made if the frog moved prior to or during playback of any call, and frogs were replaced at their original starting point after each movement so that sound gradient cues were also eliminated. The speaker's position relative to the frog's long axis was varied from trial to trial, and, as in katydids, the magnitude of orientation movements was positively correlated with the target angle (fig. 7.2B). These experiments indicate that the frogs can discriminate between lateral source locations that differ by as little as ±15°; lateralization was virtually error-free once the deviation of the source location in the frontal zone was ±15° (Klump and Gerhardt 1989).

Wyttenbach and Hoy (1997) recently used a *habituation-dishabituation procedure* to estimate the minimum audible angle in tethered crickets *Teleogryllus oceanicus* in response to ultrasonic stimulation. Sound was played from the same location until the animal failed to make an evasive movement; changing the location of the speaker sufficiently then restored the response. Minimum audible angles were about 11° in the frontal zone, 45° to the side, and 34° to the rear. These results are not strictly comparable to those discussed above because the method excluded errors introduced by the motor system. Furthermore, in the localization of ultrasound as opposed to calling-song frequencies, the ears are probably acting as a pair of pressure receivers rather than as pressure-difference receivers (see below).

Localization of elevated sound sources. Following up on the observation (May et al. 1988) that flying crickets pitch upward or downward depending on whether a source of ultrasound was below or above them, Wyttenbach and Hoy (1997) demonstrated that *Teleogryllus oceanicus* can discriminate the elevation of the source of ultrasonic sounds. The minimum difference in elevation required for discrimination was about 45° when the source was located in the median plane either in front of or behind the animal. Discrimination of sound elevation is also likely to be found in other insects that normally locate mates or avoid predators in a three-dimensional space.

Arboreal anurans also localize elevated sound sources (Gerhardt and Rheinlaender 1982; Passmore et al. 1984; Jørgensen and Gerhardt 1991). This ability has been quantified for two species (painted reed frog *Hyperolius marmoratus* and gray treefrog *Hyla versicolor*) by measuring three-dimensional jump errors, which simultaneously considers deviations in the horizontal direction (azimuth) and elevation of each jump from perfect orientation toward the sound source (see Passmore et al. 1984 for the derivation of the trigonometric formula for computing three-dimensional errors). In the painted reed frog, the mean three-dimensional jump error was 43°, a value that was considerably higher than the mean error for localization in the horizontal plane (22°;

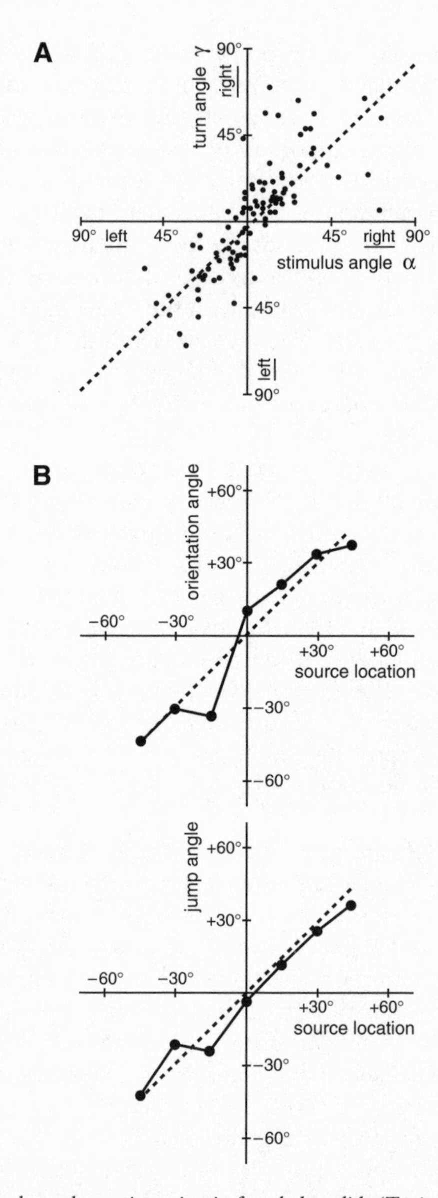

***Figure** 7.2.* (A) Angle-dependent orientation in female katydids (*Tettigonia cantans*). Shown are the relationships between stimulus angle and turn angle of six females. The correlation coefficient of the regression line (*dotted line*) is 0.89. From Rheinlaender and Römer 1990, fig. 8. (B). Open-loop accuracy of orientation and jumps of female barking treefrogs (*Hyla gratiosa*). The dotted lines represent the angles expected if the frogs' orientations were perfectly aligned toward the sound source. Orientation and jump angles (only the first turn or hop was considered) increased with the initial disparity between the frogs' body axes and the speaker position. If the frogs merely lateralized the sound source, then the size of these angles would be independent of the angle of incidence once a critical angle that allows lateralization (here, about ±15°) is exceeded. From Klump and Gerhardt 1989, fig.1.

Passmore et al. 1984). By contrast, in gray treefrogs responding to the normal, full-spectrum call, the three-dimensional error was not significantly higher than the two-dimensional error (23° vs. 19°, respectively), thus demonstrating that these animals locate the elevation of a source about as accurately as they locate its azimuth (Jørgensen and Gerhardt 1991). In the latter experiments, the sound pressure level of the stimulus was adjusted after every hop or crawl to the same value that the frog originally experienced at its release point. Because localization accuracy was no better than in situations where this adjustment was not made, it seems unlikely that the frogs make use of the potential information in the sound gradient. In the green treefrog *Hyla cinerea*, females often show head scanning after lifting their heads when responding to calls played from an elevated speaker (Gerhardt and Rheinlaender 1982), but this behavior is rare in the gray treefrog (Jørgensen and Gerhardt 1991).

Spectral Properties and Sound Localization

In crickets the most accurate sound localization occurs within the range of the dominant frequencies of the conspecific calling songs (e.g., Hill 1974; Oldfield 1980; Hennig and Weber 1997). No phonotaxis occurs in walking crickets (*Gryllus*) on a locomotion compensator at frequencies below 2–3 kHz, and systematic errors in the accuracy of phonotaxis ("anomalous" phonotaxis) occur as the dominant frequency of the model calling song (4–5 kHz) increases beyond 8 kHz (Thorson et al. 1982). These results suggest that the effectiveness of songs with carrier frequencies within the conspecific range might be influenced by the locatability of signals (see fig. 7.8) as well as by their stimulation of the auditory system. Why females of *Gryllus bimaculatus* prefer song models with somewhat lower-than-average carrier frequencies (Simmons and Ritchie 1996) thus deserves further study. As shown in chapter 4, crickets walk or fly away from models with ultrasonic (> 20 kHz) carrier frequencies (Moiseff et al. 1978; Pollack et al. 1984; Nolen and Hoy 1986). The directional cues (as opposed to the sound gradient) that might allow the animal to distinguish between sounds immediately in front of it from those immediately behind are unknown.

In several species, higher harmonics in the calling songs are present, although they are much lower in intensity than the carrier frequency (e.g., Nocke 1972; Popov et al. 1974). Latimer and Lewis (1986) reported that under closed-loop conditions, the addition of higher harmonics to the carrier frequency of the calling song improved the orientation behavior of the cricket *Teleogryllus oceanicus*.

In the grasshopper *Chorthippus biguttulus*, the spectrum of the female's song is rather broad, but males can lateralize accurately using only the low-frequency component (centered at about 7 kHz). Recall from chapter 4 that when high-frequency energy (35 kHz) was added, males made fewer turns, which, however, were as likely to be correct as when the male turned in response to the low-frequency component alone. These experiments thus do not rule out the possibility that males can localize high-frequency sounds. Other species of grasshoppers show negative phonotaxis to very high-frequency sounds simulating bat cries (Helversen 1997).

In katydids, high-frequency components in the conspecific song appear to be much more important for sound localization than are low-frequency components. For example, females of the Australian katydid *Requena verticalis* made more accurate phonotactic approaches to synthetic songs containing either the high-frequency (28 kHz) component or both the low- (16 kHz) and high-frequency components of the male song, compared with their approaches to a synthetic call with the low-frequency peak alone (Bailey and Yeoh 1988; see also Morris et al. 1975). However, the greater attractiveness of the bimodal stimulus could have also influenced these results. Better evidence is available from other katydids producing signals with bimodal spectra, such as *Tettigonia cantans* and *T. viridissima*. Females of these species require both peaks to show reliable phonotaxis, but only the high-frequency peak is used for source localization (Jatho 1995; Schul, pers. comm.).

In two species of anurans, carrier frequency appears to have little, if any, effect on the accuracy of phonotaxis, at least over the range of frequencies emphasized in advertisement calls. In terms of two-dimensional jump errors, the accuracy of jumps by green treefrogs *Hyla cinerea* in response to single-component calls of 0.9 kHz was not improved by adding components (2.7 + 3.0 kHz) that are normally present in the high-frequency band of the advertisement call (Rheinlaender et al. 1979). Females also localized calls having just these two high-frequency components as accurately as they located the single-component call (Klump and Gerhardt, unpubl. data). In terms of three-dimensional jump errors, female gray treefrogs *Hyla versicolor* localized synthetic calls of 1.1 kHz and 2.2 kHz equally well (Jørgensen and Gerhardt 1991). A small decrement in three-dimensional accuracy occurred at 1.4 kHz: 36° versus 24–30° for 1.1 kHz and 2.2 kHz. This result was mildly surprising because even though natural calls contain little energy at 1.4 kHz, laser vibrometry indicates that the tympanic membrane shows the greatest elevational directionality in this region of the spectrum (Jørgensen and Gerhardt 1991).

Peripheral Mechanisms of Sound Localization

Pressure Receivers and Pressure-Difference Receivers

Insect and anurans provide examples of ears that function as *pressure receivers*, *pressure-difference receivers*, or both (box 7.2). A tympanic membrane acts as a pressure receiver when it is driven solely by sound waves impinging on just one side (usually the exterior side). Thus, sound localization depends on the central nervous system's analysis of external binaural differences (time cues or intensity cues arising from diffraction) in the activation of two such receivers.

Box 7.2. Pressure Receivers and Pressure-Difference Receivers

Biologists concerned with the biophysics of sound localization in insects and frogs often consider the physics of microphones as a starting point for explaining the directional properties of ears (e.g., Michelsen 1983; Eggermont 1988). An ideal pressure microphone (receiver) consists of a membrane backed by an air-filled cavity (Beranek 1954). If the receiver (and in an animal, the surface in which the ear is situated) is small relative to the wavelengths of sound, then the membrane can only be driven by sound waves impinging on its external surface. The magnitude of its displacement is independent of the direction of the sound source, giving rise to a so-called omnidirectional pattern (fig. 1A). The ears of many higher vertebrates have both the anatomical structure and directional characteristics that correspond to this physical analog. At higher frequencies (shorter wavelengths), information about sound location from any one point in space can be obtained only by comparing the differences in membrane displacement (intensity) and time of displacement (arrival) between a pair of such receivers.

The membrane of a pressure-difference (gradient) microphone (receiver) can be driven by sound waves from both sides. The extent to which the waves arriving at the internal surface of the membrane influence its displacement (a function of amplitude and phase relative to the external sound waves) greatly affects the directionality of the receiver. Beranek (1954) uses an analogous electrical circuit to derive a so-called directivity index, B, which correlates with different magnitudes and patterns of directional selectivity (fig. 1C). Such electrical-analog approaches have also been commonly used to model directional hearing in anurans (e.g., Aertsen et al. 1986; Fletcher and Twaites 1979). If the sound waves have equal access to the membrane (a "pure" pressure-gradient system), B approaches infinity, and the receiver's directional pattern resembles a figure-eight (fig. 1D). One appeal of the microphone analogy is that the directional patterns of microphones with values of B of 0.5 and 1.0 closely resemble the directional patterns of the ears of insects and anurans, as measured by laser vibrometry and other methods involving recordings of neural activity generated by stimulation of one ear. For simplicity, and following Michelsen (1998a), we use the term *pressure-difference receiver* for all ears in which directionality arises primarily because the membrane can be effectively driven by sound waves arriving both at its external surface and (indirectly) at its internal surface. We emphasize that characterizing the directional patterns of insect and anuran ears is only a first step. A complete

(box 7.2 continued)

understanding of the directionality depends on characterizing the physical structures that guide sounds to the inner surface of the membrane and knowledge about how all of the sonic inputs interact to drive the tympanic membrane as the sound-source location varies (see the text).

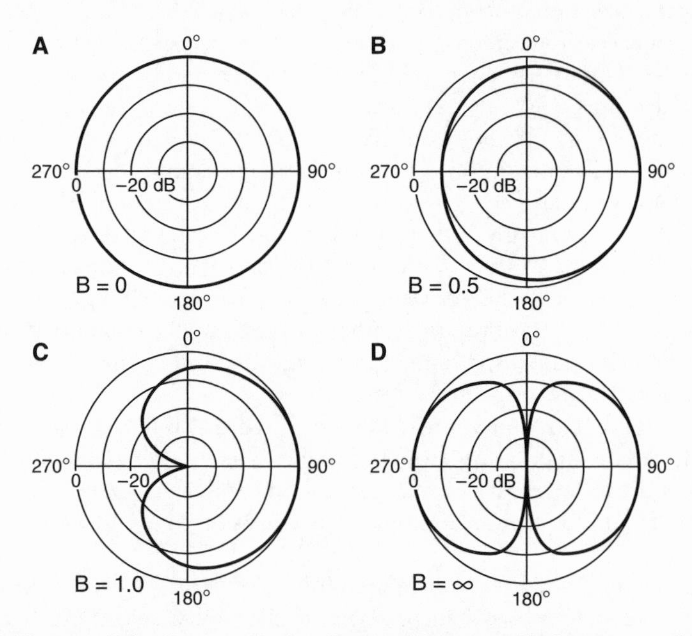

Box* 7.2, *Figure 1. Polar diagrams showing directional responses (displacements, scaled in decibels) for different kinds of microphones, whose diaphragms respond in a similar fashion to tympanic membranes. Here we assume that the wavelength of a stimulating sound of a single frequency is much greater than the dimensions of the receiver. Note that we also label the top part of each diagram as 0°—corresponding to the frontal direction of an insect or frog—whereas acoustical engineers would take the 90° direction as the zero reference. (A) Omnidirectional pattern expected of a pressure receiver, in which the diaphragm can only be driven by sound arriving at its external surface. (B) Ovoid directional pattern expected when the sound waves arriving at the inner surface of the diaphragm are damped about one-half. The tympanic membranes of several species of anurans show a similar pattern. (C) Cardioid directional pattern that occurs when unattenuated sound waves arrive at the inner surface of the diaphragm. The tympanic membranes of orthopterans often show this general kind of directional pattern in response to pure tones. (D) Figure-eight directional pattern characteristic of a system in which externally and internally arriving sound waves have equal access to the diaphragm. At a sound incidence of 0° or 180°, the amplitude and phase is equal on both sides of the diaphragm and its net displacement is close to zero. This kind of pattern has been described for the tympanic membrane of anurans when the mouth is open. Anurans do not localize sounds when the mouth is open, and such a directional pattern would in any case be rather useless for providing reliable directional information. Polar diagrams modified from Beranek 1954.

(box 7.2 continued)

Pressure-difference ears are inherently directional, but insects and anurans still use binaural cues that are available because of the symmetry of the directional pattern (fig. 2). One-eared animals using either pressure-receiver-type ears or pressure-difference-type ears can still localize sounds by moving, which allows them to sample the sound pressure gradient (see box 7.1).

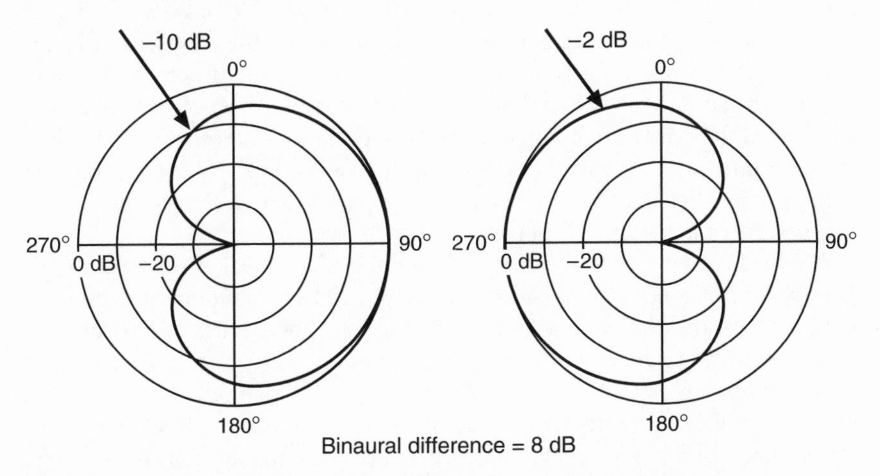

Box 7.2, Figure 2. Diagrams of a pair of receivers with the same cardioid directional pattern. Because of the bilateral symmetry of the directional patterns of a pair of ears, sound arriving from a particular direction will result in binaural differences in intensity.

A tympanic membrane acts as a pressure-difference receiver when its net displacement depends on sound waves impinging on both sides of the membrane. Biologically useful pressure-difference receivers can be technically considered as mixed pressure-gradient systems (a combination of a pressure and a pressure-gradient component; the latter provides all of the directionality) (Beranek 1954; box 7.2). Because the effective transmission of sound waves to the inside of the tympanic membrane and the magnitude of diffraction are highly dependent on frequency, the same system can, in principle, function as a pressure-difference system at relatively low frequencies and as a pressure-receiver system at high frequencies (e.g., Michelsen 1998a). Although a pressure-difference receiver has an inherently directional response, sound localization still depends on the central nervous system's analysis of binaural cues that are available because of the mirror-image symmetry of the directional patterns of the two tympanic membranes (see below and box 7.2, fig. 2).

Nevertheless, the directionality of each ear might aid one-eared crickets that presumably locate sound sources by scanning (box 7.2).

As first proposed by Autrum (1940), the peripheral auditory system and associated structures in some insects are arranged so that each ear might be expected to behave as a pressure-difference receiver (reviews: Michelsen 1983, 1994, 1998a; Ewing 1989; Larsen et al. 1989; Helversen 1997). In chapter 5 we described the anuran peripheral auditory system as a multi-input system, and the anuran ear has been modeled as some form of pressure-difference receiver for more than twenty years (reviews: Eggermont 1988; Feng and Schellart 1999). The situation in anurans is complicated by the existence of extratympanic pathways, which transmit low-frequency sounds directly to the inner ear. As emphasized by Michelsen et al. (1994), the existence of anatomical structures and pathways compatible with a pressure-difference system does not guarantee that the system operates in such a fashion. In contrast to other orthopterans, the use of signals with very high frequencies makes it possible for the ears of katydids to function as pressure receivers.

In the following sections, we consider the directional properties of the peripheral auditory systems of grasshoppers, crickets, katydids, and anurans. Our current knowledge is based mainly on two approaches: (1) recordings from the auditory receptors or primary auditory neurons that receive input from just one ear; and (2) measurements of tympanic membrane displacement using laser vibrometry. In many insect systems, these two methods yield similar estimates of directional patterns, whereas in anurans the considerable discrepancies at low frequencies support the hypothesis that extratympanic pathways are important. Credible biophysical models, which indicate the physical bases of ear directionality, are available for some insects, but our understanding of anuran systems remains primitive in this regard.

Grasshoppers: A Two-Input Pressure-Difference System

The study of *Chorthippus biguttulus* by Helversen and Rheinlaender (1988) included measurements of the directional characteristics of the tympanic membrane based on multi-unit recordings in the auditory nerve in response to synthetic signals simulating female song. In figure 7.3 we show the directional plot of one ear and its mirror image, representing the directional plot of the other ear; the difference between responses of these two plots as a function of the angle of sound incidence is also shown. The binaural difference plot clearly shows: (1) that there is a steep change in dominance of the two ears within about $\pm 15°$ of the midline; and (2) that there is little difference in interaural responses at other, greater angles of incidence. These data are thus consistent with the behavioral data and suggest why these grasshoppers lateralize the sound source rather than resolve different angular deviations.

Although showing a directional pattern that seems biologically useful and is consistent with the behavioral results, the auditory nerve data do not indicate how the directionality is achieved. Recall from chapter 5 that the two

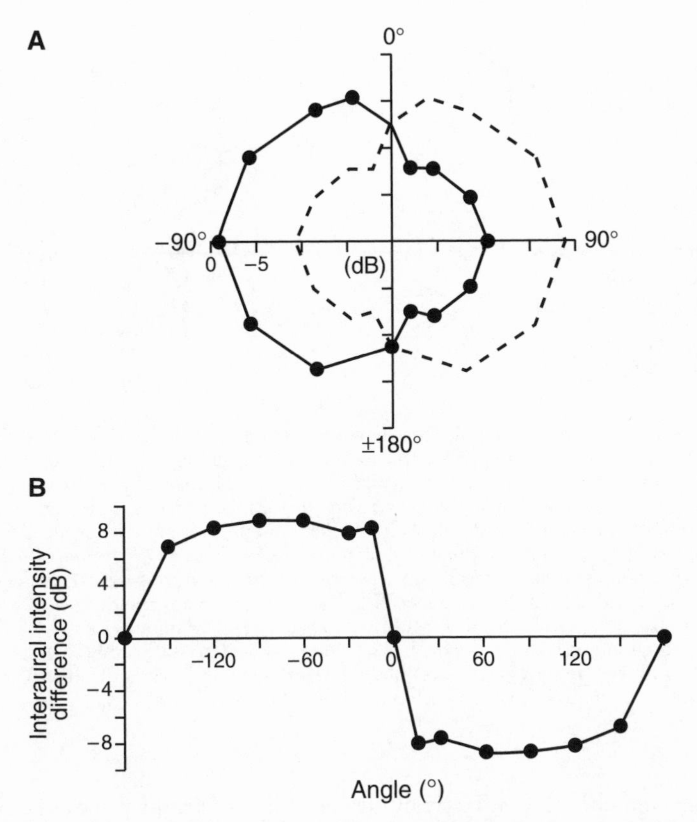

Figure 7.3. Directional hearing in a small grasshopper. (A) Mean directional characteristics of the left and right ear (*solid and dotted lines*, respectively) as measured for four males of *Chorthippus biguttulus* using the spectrum of the female signal. At various angles of sound incidence, the auditory threshold was estimated. (Note: In polar diagrams, points are plotted radially, usually with reference to an omnidirectional pattern in which all points would fall on a circle passing through the four labeled directions. Refer to box 7.2, fig. 1, for diagrams that clarify this method of representing directionality.) (B) Interaural intensity difference as a function of the angle-of-sound incidence, calculated by subtracting the directional characteristic from its mirror image. Refer to box 7.2, fig. 2, for an example. Note the rapid change in intensity differences in the frontal auditory field. The intensity-difference curve was derived from the most sensitive auditory receptors tuned to 7 kHz. From Helversen 1997, fig. 6.

tympanal membranes are coupled internally by air sacs; spiracles located near the ears are also potential inputs (fig. 7.4). Using a combination of laser vibrometry and sound-amplitude measurements obtained by probe microphones, Michelsen and Rohrseitz (1995) estimated the amplitude and phase relationships (time delays) of two sonic inputs to the tympanic membranes in the locust *Schistocerca gregaria* and in *Chorthippus biguttulus*. These estimates were used to model the directionality of the ear at different frequencies and source locations, and the main conclusion is that the grasshopper ear is a two-input system. That is, its directionality can be explained by assuming that each

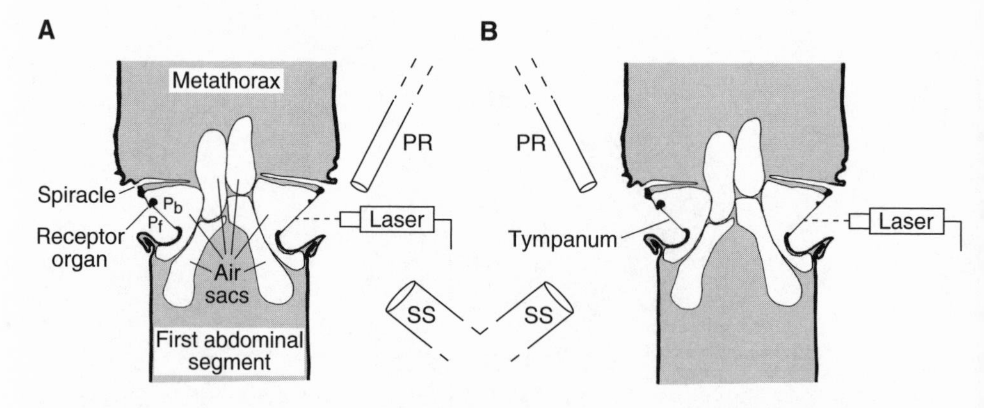

Figure 7.4. Setup for measuring the sonic inputs to the grasshopper ear. Diagrams show a horizontal section of a grasshopper with the locations of the air sacs, spiracles, and tympanic membranes (tympanum in [B]). (A) Each tympanic membrane is driven by sound arriving directly at its external surface, P_f, and sound arriving through the air sacs to its internal surface, P_b. The amplitude and phase of P_b relative to P_f (the transmission gain) is determined by calibrating the tympanic membrane vibrations, set in motion by a local sound source (SS). Laser vibrometry measures tympanic membrane vibration and the sound pressure level at the right ear is measured with a probe microphone (PR). The transmission around the body is attenuated by a wall of beeswax (not shown) that facilitates quasi-independent stimulation of each ear. (B) SS is moved so that it stimulates the left ear, and the sound pressure level is measured at that point while the laser again measures the vibrational input of the right ear, thus providing an estimate of the contralateral input. Modified from Michelsen 1998a, figs. 2.7, 2.8.

tympanic membrane is effectively driven only by sound waves arriving on their external surface and by sound waves that reach their internal surface after entering the contralateral ear and propagating through the air sacs. Other potential inputs, such as the spiracles, can be ignored. The interaction of these two inputs, which depends on the location of the sound source, determines the net displacement of each tympanic membrane.

Michelsen and Rohrseitz (1995) found good agreement between calculated and measured directionality for the locust at all frequencies and for *C. biguttulus* at high frequencies (> 10 kHz) (fig. 7.5). However, the model predicted very poor directionality at 5 kHz in the grasshopper even though behavioral studies show that males can readily lateralize sounds of this frequency (Schul et al. 1999). A second set of measurements based on recordings from the auditory nerve estimated a significantly higher phase delay, which resulted in a better fit with the model at 5 kHz (Schul et al. 1999). The discrepancies between the two studies with regard to localization in *C. biguttulus* might stem from methodological differences, or it may well be that such a model does not apply to this small grasshopper at low frequencies for biophysical reasons. For example, low-frequency sounds might penetrate much of the body surface as well as the ears (Michelsen, pers. comm.).

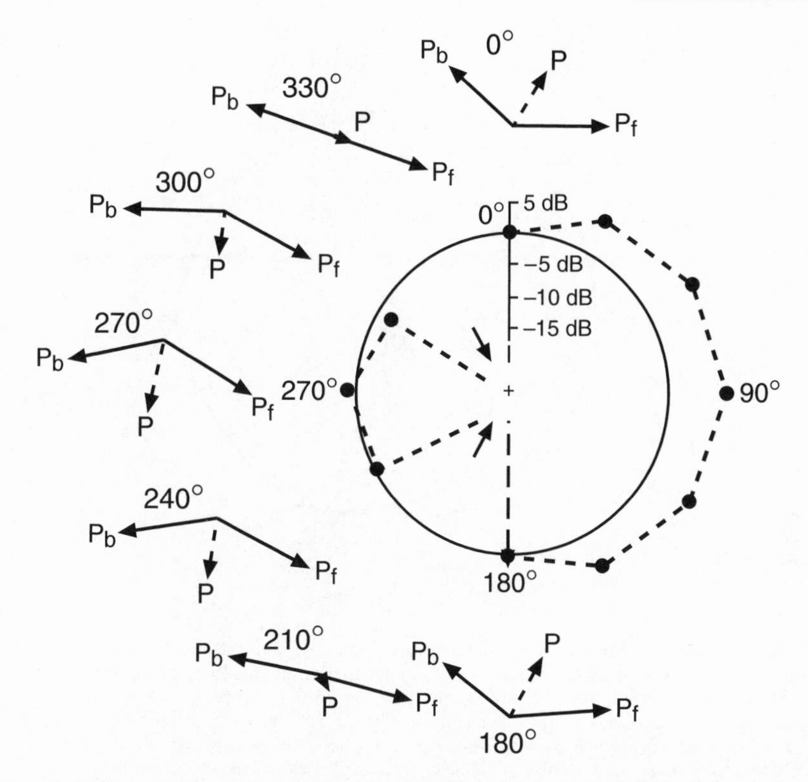

Figure 7.5. Directional pattern calculated for *Chorthippus biguttulus* at 10 kHz. The sounds acting on the external surface (P_f) and internal surface (P_b) are drawn as vectors (relative magnitudes = length of solid lines; phase = angular separation), and the net force acting on the tympanic membrane is shown by the dotted lines (P). Dotted lines in the polar plot show that the overall shape of the polar plot is qualitatively similar to that based on auditory nerve recordings when the ear was stimulated by the natural song (carrier of about 5–7 kHz; see fig. 7.3A). From Michelsen 1998a, fig. 2.10.

Field Crickets: A Three-Input Pressure-Difference System

Crickets typically use communication signals that are lower in frequency and narrower in bandwidth than those of most grasshoppers and katydids. All other things being equal, such signals should allow crickets to communicate over a somewhat greater distance than these other insects. Whereas the harp can be considered the evolutionary invention that made production of these signals possible, another invention—the medial septum (central membrane)—seems to be necessary for receivers to localize such signals.

As shown in chapter 5, there are four possible pathways to each tympanic membrane (fig. 5.1; fig. 7.7B). Recall that sound can potentially drive the tympanic membrane of one ear (1) directly from the outside; (2) from the contralateral ear; (3) from the ipsilateral acoustic spiracle; and (4) from the

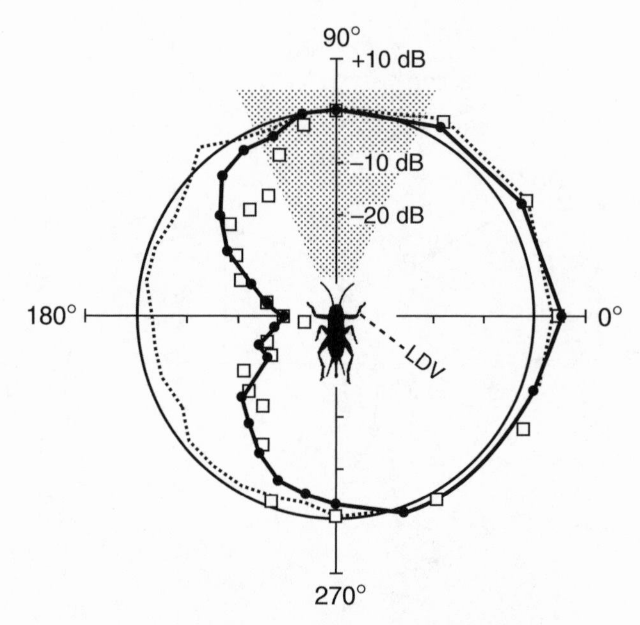

Figure 7.6. Cardioid directionality of a female cricket's ear (*Gryllus campestris*) for a 5 kHz stimulus. Curves were derived from laser vibrometry measurements (data from Larsen et al. 1989) and from recordings of auditory receptors of one ear (data from Boyd and Lewis 1983). Connected closed circles = vibration velocity amplitude (relative to the value for 0 = 90°) when the acoustic spiracles are open; dashed line = with the acoustic spiracles closed, the observed directionality is reduced or abolished; open squares = directional response of one auditory nerve at threshold intensity with the acoustic spiracles open; LDV = direction of the laser Doppler vibrometer located 1 m away. The shaded region indicates the frontal area within which crickets do not consistently turn toward the source; however, they make corrective turns back toward the sound source when the source direction is in the immediately adjacent areas (data from Rheinlaender and Blätgen 1982). From Larsen et al. 1989, fig. 12.11.

contralateral acoustic spiracle. Early studies of *Gryllus campestris* using laser vibrometry and recordings from auditory receptors innervating one ear showed a basically *cardioid directional pattern* (box 7.2; fig. 7.6). These studies also showed that the spiracular openings were likely to be important because blocking them abolished the directional response.

Using methods similar to those described above for the grasshopper studies, Michelsen et al. (1994) made noninvasive measurements of the amplitudes and phase angles of sounds at each of the four sound inputs in *G. bimaculatus*. They then used these values, which were measured at the outside surface of one ear when sound arrived from the frontal region as a reference, to make calculations of the relative contributions of the four inputs to the directionality of the system. One major result was that for the calling-song frequency of about 4.5 kHz, the sound entering from the contralateral ear has a negligible effect on the displacement of the other ear—thus effectively making the cricket ear a three-input system. The calculation of the net forces acting on

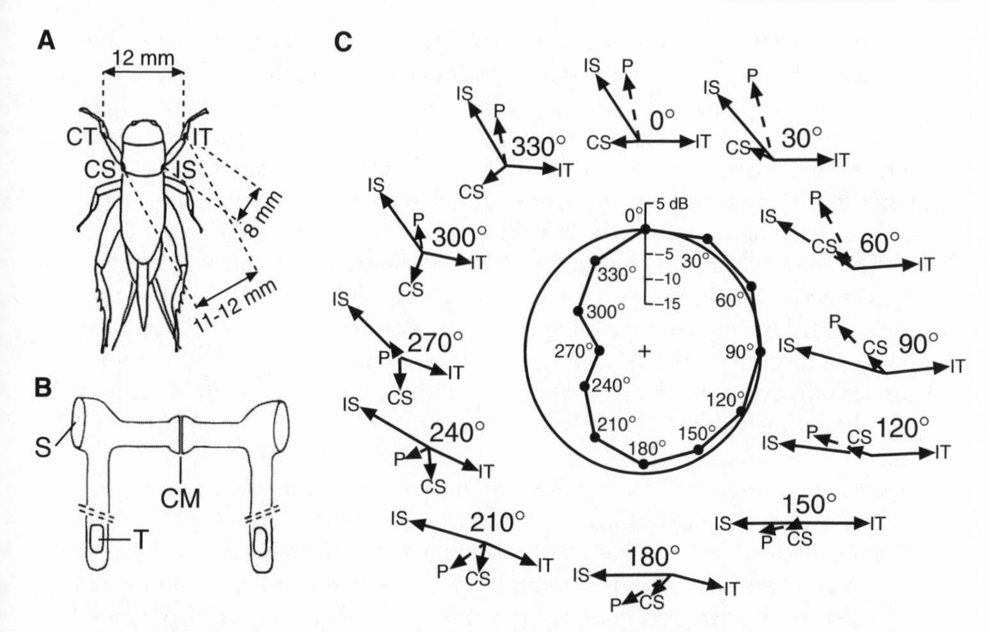

Figure 7.7. Biophysical modeling of directional hearing in crickets. (A) Acoustic inputs to the ears. IT and CT = ipsilateral and contralateral tympanum; IS and CS = ipsilateral and contralateral acoustic spiracle. Values for distances between ipsilateral (IT to IS) and contralateral (IT to CT) inputs. From Michelsen et al. 1994, fig. 1b. (B) Tracheal pathway to show the four-input system and the position of the central membrane (CM) or medial septum; T = tympanal membrane; S = spiracle. From Michelsen and Löhe 1995, fig. 1a. (C) The calculated directional pattern of the force acting on the tympanal membrane of the right ear at calling-song frequency (4.5 kHz). The force is proportional to a sound pressure (P), which is the sum of three vectors: the sound at the outer surface of the tympanal membrane (IT) and the sounds transmitted to the inner surface of that membrane for the ipsilateral (IS) and the contralateral (CS) spiracles. From Michelsen et al. 1994, fig. 10.

one ear is thus reduced to determining the effects of the indirect inputs from the ipsilateral and contralateral spiracles along with the direct effects of sounds arriving at the outer surface of the ipsilateral tympanum. The polar plot, shown for the right ear in figure 7.7C—calculated from the estimates of Michelsen et al. (1994)—is obviously similar to the plots derived by other methods for the very similar cricket *G. campestris* (fig. 7.6). The vector diagrams show how the directionality is achieved. For example, the maximum amplitudes observed at about 30° and 60° arise because the inputs from the contralateral spiracle and ipsilateral spiracle have nearly the same phase. The amplitude null at 270° occurs because the phase angles of the inputs from these two spiracles are about 120–140° out of phase.

All studies indicate that directional cues of the order of 4–5 dB are available in the frontal region, where behavioral studies indicate that these two species of crickets are most directionally sensitive (see fig. 7.6). The magnitude of this difference, which is available as a binaural cue because of the mirror-image symmetry of the two ears (fig. 7.3A; box 7.2, fig. 2), is easily resolved by the central auditory system. The three-input model also predicts that if the carrier frequency is varied, then the position of the *null* (source direction at which the magnitude of the tympanic membrane is minimal) and other features of the directional diagram change. Such a change of the null is qualitatively consistent with the anomalous phonotaxis observed in the tracking behavior of crickets at higher-than-normal carrier frequencies (Thorson et al. 1982; see p. 217).

The critical factor that makes the pressure-difference system of *G. bimaculatus* effective at calling song frequencies is the appropriate delay of the sound arriving from the contralateral spiracle (Michelsen et al. 1994). Only then does the addition of the three vectors provide useful directionality. The anatomical structure essential for this phase shift is the medial septum (central membrane), which is found in the midline of the tracheal system that connects the two auditory spiracles (fig. 5.1; fig. 7.7B; Michelsen and Löhe 1995). A small hole made into the septum changes the phase shift and decreases the directional characteristics of the tympanal membrane vibration (fig. 7.8A). This result provides a mechanistic explanation for earlier studies showing that phonotactic performance deteriorates and that binaural differences in auditory nerve sensitivity decrease after septum perforation (fig. 7.8B; Wendler and Löhe 1993; Löhe and Kleindienst 1994). Whether all crickets communicating with low-frequency sounds have a medial septum is unknown. If this structure is absent in some species, then further study would be needed to demonstrate if and how other components of the system provide an adequate phase delay.

Katydids: Pressure-Receiver Ears

The tympanic membrane of a typical katydid is driven directly from the outside, and sound also reaches its inner surface via a tracheal tube or hearing trumpet (fig. 5.1), an arrangement suggesting a pressure-difference mechanism. However, the input from the tracheal tube, which in many katydids is shaped like a horn with a large opening on the side of the body, dominates (by as much as a factor of ten or more) the external input, at least at the high frequencies (above 10 kHz) typically used by these animals in communication (Lewis 1974; Michelsen 1998a). Thus, the system is equivalent to a pair of pressure-receiver ears, in which binaural intensity and time cues correspond to the external differences in these properties between the two sides of the thorax where the hearing trumpets are located.

Anurans: A Multiple-Input System

As discussed above, the accuracy with which small anurans localize low-frequency sounds and the small extent of scanning movements prior to corrective turns indicate that *external* binaural differences are likely to be minute or even nonexistent. Barking treefrogs *Hyla gratiosa*, for example, accurately locate tone bursts of 0.4 kHz (wavelength of about 1.3 m) (Gerhardt and

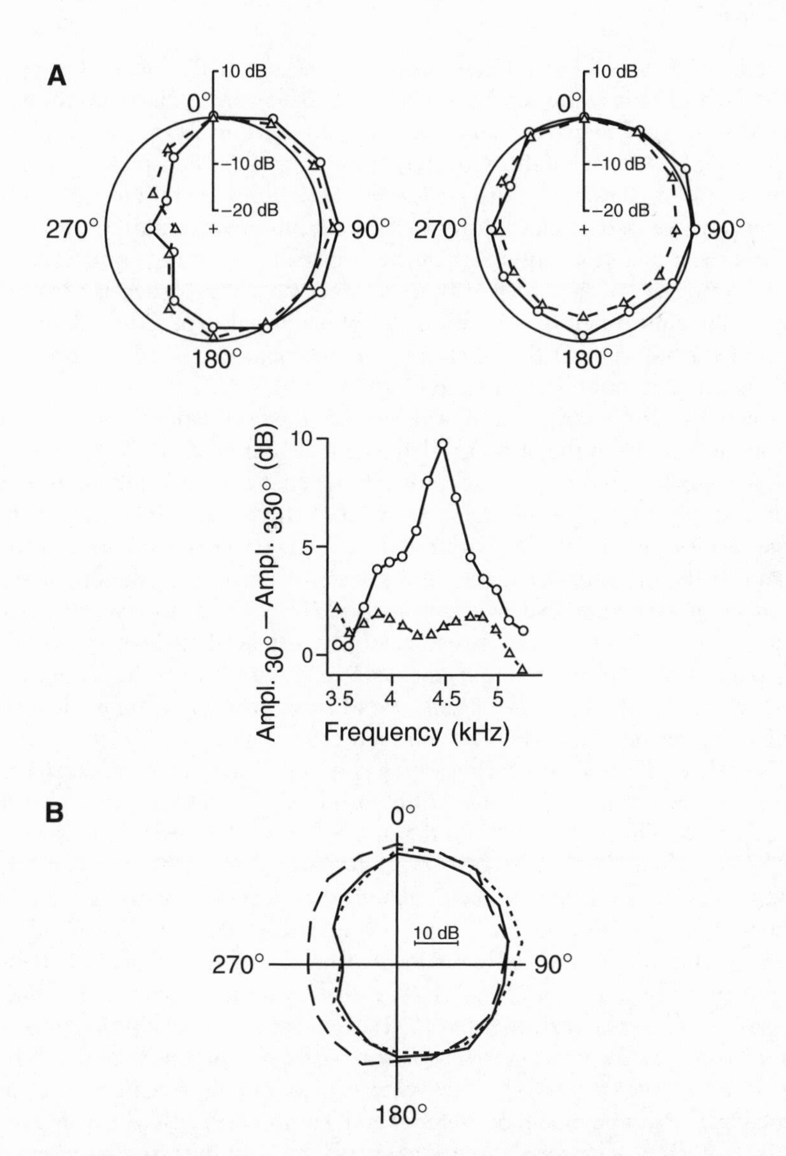

Figure 7.8. The importance of the medial septum for directional hearing in *Gryllus bimaculatus*. (A) *Upper part*, directional dependence of tympanal membrane vibration at 4.5 kHz in a cricket with intact (*left*) and perforated septum (*right*). Solid line = vibrations observed with laser vibrometry; dashed lines = calculated vibrations. From Michelsen 1998b, fig. 4. *Lower part*, frequency tuning of the directional gradient in an intact cricket (*circles*) (i.e., the difference in the eardrum vibration amplitude at the +30° and −30° [= 330°] directions), and with the perforated septum (*triangles*). Notice that after perforation, the difference becomes 0–2 dB at all frequencies. From Michelsen 1998b, fig. 4; data from Michelsen and Löhe 1995, fig. 2b. (B) Influence of septum perforation on the directionality of the ear at 5 kHz and 78 dB SPL. Shown are directional sensitivities of the right ear of an intact animal (*solid line*) after sham operation (*dotted line*) and after septum perforation (*dashed line*), all relative to the value of the intact animal for anterior sound presentation (0°). At 90° the sound source is ipsilateral to the ear. From Löhe and Kleindienst 1994, fig. 2.

Rheinlaender, unpubl. data). The frogs have relatively flat heads (1 cm from top to bottom) and an ear separation of less than 2.5 cm. Similarly, even smaller green treefrogs (*H. cinerea*) readily locate sounds having just the low-frequency peak (0.9 k Hz) typical of their advertisement call. In neither species would usable, if indeed measurable, external differences in interaural intensity be expected at these frequencies. Moreover, the maximum (external) binaural time differences available to the frogs during especially accurate approaches to a sound source are of the order of 10 μs. Higher vertebrates, such as barn owls, can resolve time differences of this order of magnitude (e.g., Moiseff and Konishi 1981), but whether the much simpler auditory system of the anuran does so is an open question (Klump and Gerhardt 1989).

These considerations, and the existence of internal pathways to the inside of the eardrum from the other ear, led Rheinlaender et al. (1979) to hypothesize that small anurans use a pressure-difference mechanism for sound localization. A pressure-receiver mechanism could also, in principle, be used by a large species communicating with relatively high-frequency sounds or broadband signals with high-frequency components. Considering the phonotactic accuracy of a medium-sized treefrog such as *Hyla cinerea*, however, the maximum external intensity difference normally experienced by these animals during phonotaxis—even at frequencies corresponding to the high-frequency peak in their calls (about 3–4 kHz)—would still probably be much less than 1 dB (Rheinlaender et al. 1979; Michelsen et al. 1986).

Models of directional hearing in frogs have, in fact, all assumed that anurans use some form of pressure-difference mechanism (Aertsen et al. 1986; Fletcher and Thwaites 1979b; Pinder and Palmer 1983; reviews: Eggermont 1988; Jørgensen 1991). As data from studies using laser vibrometry and neurophysiological studies have accumulated, the most commonly reported directional pattern is ovoidal in form, with maximal vibrational amplitude occurring when the sound source is located on the ipsilateral side of the axis connecting the ears (figs. 7.9, 7.10A; North American hylids *Hyla cinerea*, *H. gratiosa*, *H. versicolor:* Jørgensen 1991; Michelsen et al. 1986; coqui treefrog *Eleutherodactylus coqui* and grassfrog *Rana temporaria:* Jørgensen et al. 1991a,b; fig. 7.10A; Jørgensen 1991). This pattern most closely resembles that predicted by the simple model of Fletcher and Twaites (1979b), which also accurately predicts that maximal directionality will occur at the resonant frequency of the tympanic membrane. Consider, for example, the directional plot of the eardrum of the barking treefrog *H. gratiosa* (fig. 7.9). At 0.44 kHz (the low-frequency peak in the advertisement call), a difference of approximately 3 dB in vibrational amplitude occurs between about +30° (left = +1 dB) and −30° (right = −2 dB). The mirror-image symmetry of the two ears would yield a 3 dB binaural intensity difference, which should readily be resolved by the central auditory system. As discussed above, treefrogs usually do not turn much more than 30° away from the target axis without making a corrective

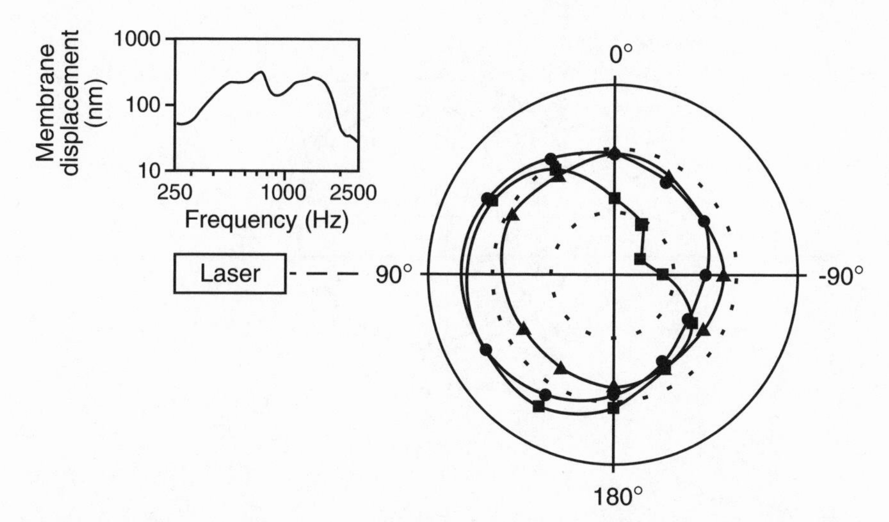

Figure 7.9. Directional response (vibrational amplitude) of the left eardrum of a barking treefrog (*Hyla gratiosa*) as measured by laser vibrometry for three different frequencies: 440 Hz = average low-frequency peak in advertisement call (*circles*); 840 Hz = frequency of maximum directionality (*squares*); 1840 Hz = average high-frequency peak in advertisement call (*triangles*). The dotted circles indicate differences in vibrational amplitude of 10 dB. The inset shows the frequency response of the tympanic membrane displacement when the sound incidence was 90° (ipsilateral to the left eardrum). From Jørgensen 1991, fig. 4.

jump back toward the speaker axis. By contrast, there is virtually no difference in amplitude at these positions in response to a stimulus of 1.84 kHz, which falls within the usual band of high frequencies found in the call.

The model of Fletcher and Twaites (1979b) assumes that the main inputs to the internal surface of the tympanic membrane are from the contralateral ear and nares. Although considerable support for the first input exists (e.g., Rheinlaender et al. 1981), the identity of the other input(s), postulated by all models, is controversial (see discussion in Jørgensen 1991). The following possibilities, which are not mutually exclusive, have been proposed:

1. The entire head and throat of the frog might be acoustically transparent to some extent: blocking both eardrums and the nares did not prevent some sound from entering the mouth cavity (Aertsen et al. 1986). This possibility requires more research.

2. The lateral body wall adjacent to the lungs vibrates well in response to sound and could be a source of input to the tympanic membrane (Narins et al. 1988). However, the directional patterns from anesthetized animals that most closely match those of awake individuals are from preparations in which the glottis, located between the lungs and mouth cavity, is closed and hence likely to greatly attenuate sound input to the inside of the tympanic membrane (Michelsen et al. 1986; Jørgensen 1991; Jørgensen et al. 1991a). Recall, too,

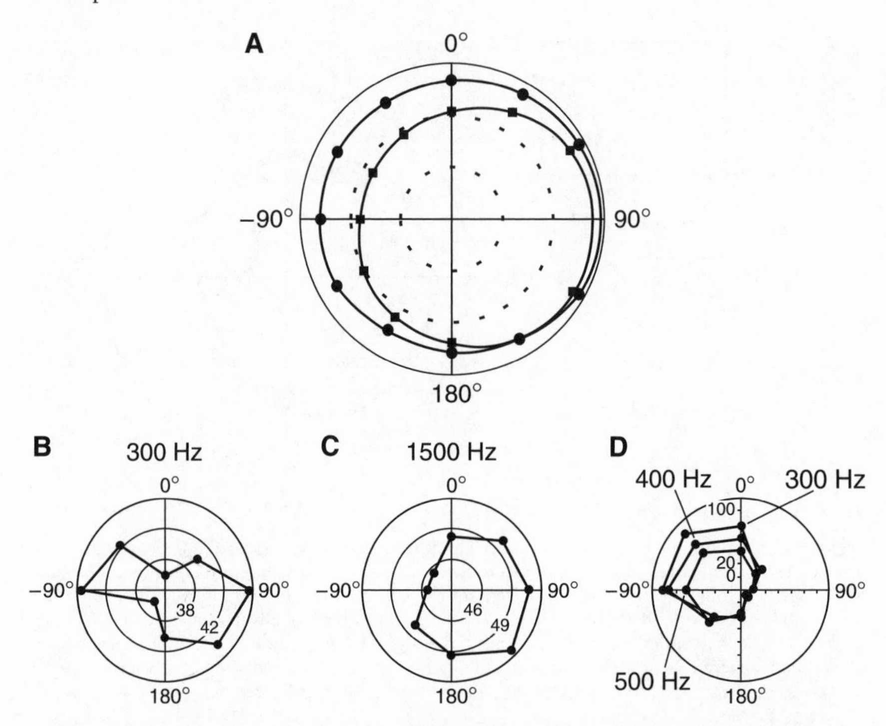

Figure 7.10. Directional plots of the tympanic membrane of the grassfrog *Rana temporaria*. (A) Displacement as measured by laser vibrometry at stimulus frequencies of 400 Hz (*circles*) and 760 Hz (*squares*); the dotted circles indicate 10 dB differences. From Jørgensen 1991, fig. 5. (B) Polar diagram of equivalent dB SPL values (the numbers printed within the diagram) based on variation in the spike rate of a low-frequency auditory nerve fiber. Notice the strong directionality that is very different in form from that of (A). This particular pattern would probably not be very useful for source location in the frontal zone. (C) Polar diagram of equivalent dB SPL values (the numbers printed within the diagram) based on variation in the spike rate of a high-frequency auditory nerve fiber. Although the maximal directional difference was less than that of the pattern shown in (B), the stronger directionality in the front zone would be more useful for source localization and fits with behavioral data on phonotaxis from other anurans. From Jørgensen and Christensen-Dalsgaard 1997a, fig. 10. (D) Phase of eardrum vibrations as a function of direction at 300, 400, and 500 Hz. The numbers on the radial axis are in degrees. Again notice the strong directionality in the frontal zone. From Jørgensen and Christensen-Dalsgaard 1997b, fig. 12.

sound-localization accuracy is reduced in the gray treefrog at the frequency at which the input from the lungs provides the best directionality (Jørgensen and Gerhardt 1991).

3. Direct extratympanic inputs to the inner ear probably exist and have been postulated by many authors (e.g., Feng 1980; Wilczynski et al. 1987; Schmitz et al. 1992; Wang et al. 1996; Jørgensen and Christensen-Dalsgaard 1997a,b; see chapter 5). Whereas directional plots—based on spike rates—of high-frequency neurons have the same basic pattern as that derived from laser

vibrometry in the grass frog (*Rana temporaria*), directional patterns of low-frequency neurons are often very different in form and much stronger than the directional patterns of the tympanic membrane as estimated by laser vibrometry (compare fig. 7.10A with 7.10B and C). The fact that such directionality occurs only at low frequencies also makes it unlikely that input from the "lung-ear," which most efficiently vibrates at much higher frequencies, is a source of extratympanic stimulation (Jørgensen et al. 1991a,b).

Extratympanic effects of low-frequency sound were also shown by Jørgensen and Christensen-Dalsgaard (1997b), who measured the directional response of the grass frog ear in terms of spike timing or relative phase. Low-frequency-tuned neurons show especially strong shifts in phase as a function of sound direction in the frontal region (see also Schmitz et al. 1992). The dynamic range of the phase directionality was greater than that based on spike rates, but it was also more frequency-dependent. The former characteristic might make phase directionality more reliable than spike-rate directionality, provided that the auditory system can process this information in a frequency-specific way. As pointed out by Jørgensen and Christensen-Dalsgaard (1997a,b), much more information about central nervous system processing of directional information is required in order to decide how (and if) these two directional codes are used for source location.

Can Individuals Actively Change the Directional Properties of the Peripheral System?

As just discussed, we are far from defining the acoustic inputs to the two surfaces of the tympanic membrane of anurans, much less identifying the extratympanic pathways used in sound localization. Furthermore, the body wall, buccal cavity, and lungs are dynamic structures that the anuran changes significantly during breathing and might also alter in the context of sound localization. A biophysical solution to directional hearing in these animals must thus take into account how these structures are configured during localization or how the system is still able to deliver reliable directional information in the face of variation in these structures (e.g., Rheinlaender et al. 1981). As pointed out by Jørgensen and Christensen-Dalsgaard (1997a), many of the discrepancies in studies of the directional properties of auditory nerve fibers are likely to reflect differences in the posture and state of inflation of the preparation.

Although the peripheral auditory system of insects seems more stable than that of anurans, stridulation, breathing, and locomotion can alter auditory responses (chapter 5). Since phonotaxis by definition involves movement, then these effects will also have to be considered in order to provide a complete understanding of directional hearing. Furthermore, recall that at least one large species of katydid (*Sciarasaga quadrata*) alters the tuning and sensitivity of its hearing system by opening and closing its auditory spiracle (Römer and Bailey 1998). Closing the spiracle had little effect on the magnitude of binaural intensity differences in this species, which communicates with what is an unusually low-frequency carrier for katydids (5 kHz). In another, small species (*Requena verticalis*), however, a significant difference in binaural intensity was found in comparisons of open- and closed-spiracle preparations (Römer and Bailey 1998).

Central Mechanisms of Sound Localization

The central nervous system has the task of processing the binaural cues provided by the peripheral auditory system and then initiating corrective turns or orientation movements that allow the animal to approach or avoid signals of biological importance. In principle, neurons in the ascending pathway that receive inputs from both ears can compare (1) the spike rates of populations of receptors and auditory neurons located on each side of the animal; and (2) the time of onset of activity (spikes) in these populations of receptors and neurons on each side. The central nervous system might also compare the numbers of receptors or lower-order neurons that are driven by sound arriving from the two ears (Bergeijk 1962; Helversen 1997; Römer et al. 1998). The usefulness of this kind of analysis depends on the symmetry of the input at the peripheral level and in the ascending pathway. Such symmetry is typically assumed rather than tested. There is some evidence that fluctuating asymmetry in an external ear element adversely affects the accuracy of sound localization in the midwife toad (*Alytes obstetricans*) (Bosch and Márquez 2000), but Faure and Hoy (2000a) show that the difference in thresholds of paired T cells in the katydid *Neoconocephalus ensiger* is within the estimated range expected from measurement errors. Faure and Hoy (2000a) also provide a useful review of studies of auditory symmetry and its analysis in insects and vertebrates.

In birds and mammals, information about source location is also represented in spatially organized maps within the central nervous system, as exemplified by neurophysiological and anatomical studies of the barn owl and bat systems (e.g., Suga 1990; Konishi 1994). That is, cells that are driven best by sounds arriving from a particular direction are located in close proximity. The question of topological representation of auditory space in anurans remains open, mainly because of the lack of good studies (review: Feng and Schellart 1999).

Before discussing the evidence that central nervous system comparisons of binaural differences result in correct orientation with respect to the sounds produced by mates or predators, we examine two neural mechanisms by which binaural differences existing at the tympanic membrane are magnified.

Neural Enhancement of Binaural Differences

Intensity cues. A common mechanism for amplifying very small binaural intensity differences is inhibition. In the locust, direction-dependent inhibition

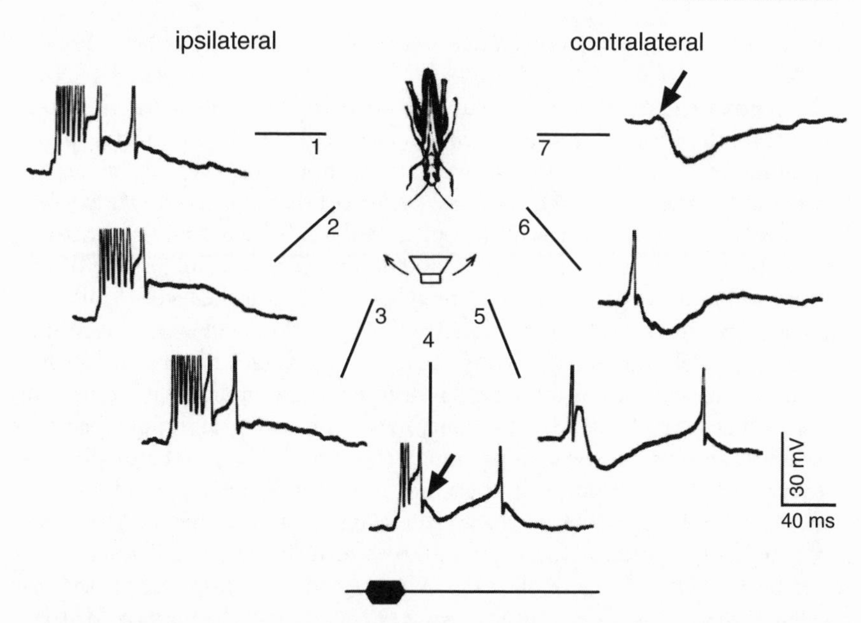

Figure 7.11. Sound-direction-dependent inhibition of an unidentified interneuron in the metathoracic ganglion of *Locusta migratoria.* The neuron is strongly excited by ipsilateral sound pulses at sound frequencies of 20 kHz and a sound intensity that was 20 dB above threshold (speaker positions 1–3) and inhibited by contralateral sound (speaker positions 5–7). Arrow at speaker position 4 indicates the first appearance of an inhibitory postsynaptic potential as the sound source was moved from the ipsilateral to the contralateral side of the animal. Spikes are truncated. The inhibition is assumed to arise from the activity of interneurons on the contralateral side of the animal. From Römer et al. 1981, fig. 4.

was discovered by studying synaptic mechanisms of monaural and binaural processing (Römer et al. 1981). Synaptic inhibition, mostly acting from the contralateral side and depending on sound-source location, shapes the response properties of both local and ascending interneurons (fig. 7.11).

In crickets, interaural inhibition is mainly mediated by a pair of local interneurons, the omega cells that collect the input from many receptors (review: Pollack 1998; Schildberger et al. 1989 and literature therein). The two omega interneurons inhibit one another reciprocally in proportion to how strongly each is stimulated. Hence, if the sound arrives from the right side of the animal and thus more strongly excites the right ear, the omega cell receiving input from this side will fire more strongly and inhibits the more weakly stimulated contralateral omega neuron (Kleindienst et al. 1981). Each of the omega cells also inhibits the two ascending auditory interneurons AN1 and AN2 on the contralateral side (e.g., Selverston et al. 1985; Horseman and Huber 1994a). Similar results were obtained for *Acheta domesticus* where the omega cell inhibits the AN1 homolog, L1 (Stumpner et al. 1995). Recent studies in *Teleogryllus* suggest that the omega neurons interact in a way that preserves interaural response latencies (Faulkes and Pollack 2000; Givois and Pollack 2000).

Katydids are similar to crickets in that the paired omega cells show strong reciprocal inhibition. In *Tettigonia viridissima* there is a steep shift in response magnitude in a narrow (±15°) frontal range; moreover, even at 7.5° to the right or left of the animal, the temporal pattern of the signal is still clearly represented in the firing patterns of the omega cells on the same side as the sound source (fig. 7.12; Römer and Krusch 2000). Thus, the auditory space of this katydid is sharply divided into two lateral hemispheres at the level of this local interneuron. However, the directional pattern is maintained when the contralateral input is eliminated by cutting the leg nerve, suggesting that even without the reciprocal inhibition, external differences in intensity alone provide this katydid with sufficient directionality to discriminate the temporal patterns of spatially separated, overlapping signals (see Schul 1997). What happens at higher neural levels is uncertain because, at least in terms of sound localization, behavioral studies (Hardt 1988; Rheinlaender and Römer 1990) suggest angle-dependent orientation, whereas the patterns of figure 7.12 would seem to be better suited for lateralization. Nevertheless, as discussed in the last section of this chapter, direction-dependent encoding of the temporal pattern has important implications for how mechanisms of sound localization and signal pattern recognition interact.

Neural mechanisms that magnify physical time differences. The small distances separating the two ears of most insects and anurans set an upper limit on the magnitude of binaural differences in time of arrival of short sounds or in the phase of continuous signals. Recall that when intensity is held constant, a difference of about 1 ms is needed to elicit correct lateralization in the much smaller grasshopper *C. biguttulus*. In reality, however, if a sound is located on one side of the animal, both intensity and time differences usually exist. Furthermore, because *neural latency* (delay in firing from the onset of acoustic stimulation) is inversely related to sound intensity, interaural differences in intensity can produce behaviorally useful neural time differences.

Studies of the locust *Locusta migratoria* illustrate this principle particularly well. In this large grasshopper, the onset of the response of single receptor fibers is changed by as much as 6 ms when the sound source is moved around the animal (Mörchen et al. 1978). This neurally generated time cue exceeded the physical time difference as measured at the external surfaces of the ears by a factor of 100 to 1000. These estimates were obtained by recording from pairs of ascending auditory neurons (Rheinlaender and Mörchen 1979; Rheinlaender 1984). For anurans, Feng (1982) estimates that a 1 dB change in sound pressure level results in a shift of 0.1–0.6 ms in the firing latency of primary auditory neurons. Thus, the 3–8 dB binaural differences in intensity arising from the directional symmetry of the two ears would translate to 0.3–4.8 ms differences in the initiation of spikes between right and left auditory nerve

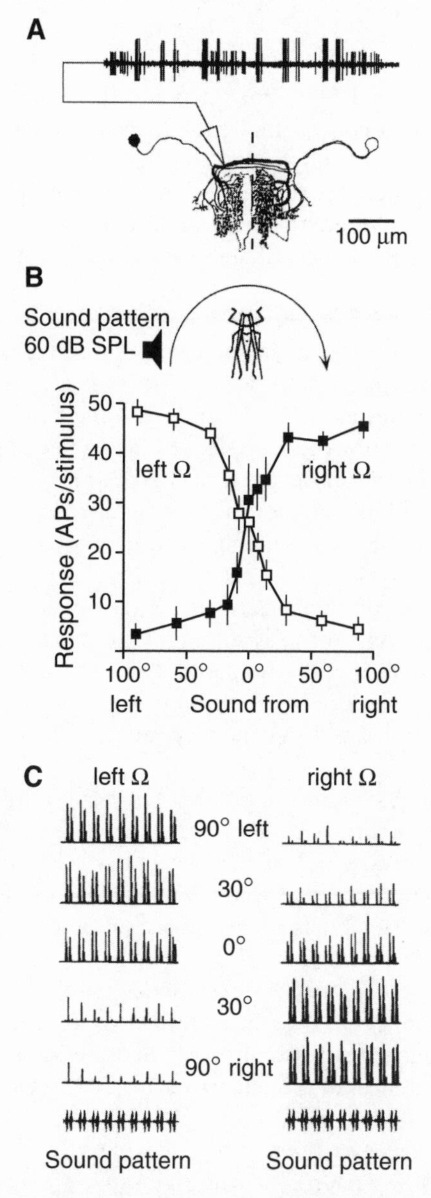

Figure 7.12. Directional responses and representation of the temporal pattern of acoustic signals in the spiking patterns of omega neurons in the katydid *Tettigonia viridissima.* (A) Schematic diagram showing the location at which spike activity of a pair of omega neurons was extracellularly recorded; the origin of the spikes was inferred from the amplitude of the spikes (generally greater for the omega cell with its cell body on the side contralateral to the electrode). The dotted line indicates the midline of the ganglion. (B) Directional responses of a pair of omega neurons from the same animal. The sound source was moved around the animal as indicated by the top diagram. Notice the sharp change in response in the frontal zone. Other pairs of omega neurons showed similar directional patterns, but not all pairs were as symmetrical as the pair shown here. (C) Diagrams showing the reliable encoding of the temporal pattern (see diagram below) by omega cells when the sound was on the ipsilateral side. From Römer and Krusch 2000, fig. 3.

fibers (Feng and Schellart 1999). Binaural neurons in the central auditory system, which are excited by contralateral sounds and inhibited by ipsilateral ones, also show time-intensity trading (Feng and Capranica 1978). In both insects and anurans, the suppression of the response to lagging signals by leading ones must certainly contribute to female preferences for leading signalers (see Römer et al. 1997) that has been demonstrated behaviorally (see chapter 8).

Central Auditory System Comparisons of Binaural Cues

In insects, binaural comparisons in the central auditory system are inferred from orientation movements that occur when the relative activity of paired ascending auditory neurons is known and, in some cases, manipulated. Directional sensitivity has not yet been studied in the few identified brain neurons in crickets with temporal-pattern recognizing properties (see chapter 6; Schildberger 1994). Thus, the loci of directional computations and the integration of information about sound patterns and their location are also unknown.

In anurans, binaural comparisons are inferred from directionally dependent firing patterns of neurons in the ascending pathway that are known to have binaural inputs. Closed-system stimulation (via earphones) has been used to estimate directly the effects of systematically varying interaural differences in intensity, time, or both (e.g., Feng and Capranica, 1978).

Insects. Recall that auditory information is conveyed to the brain by relatively few ascending neurons on each side of the animal (see chapter 6) and that the directional responses of these neurons qualitatively reflect the directional characteristics of the ears (e.g., Boyan 1979a; Wiese and Eilts-Grimm 1985; Schildberger et al. 1989). The strength of excitation on each side of the animal can thus be compared by assessing the outputs of these paired neurons. The simplified wiring diagram of crickets and katydids provides the opportunity not only to demonstrate the role of these paired neurons in sound localization, but also to quantify the extent to which orientation movements are directly related to differences in the activity of these cells (Horseman and Huber 1994b; and below).

Several lines of evidence support the idea that higher centers compare the activity in ascending neurons in the right and left sides of the animal in order to turn in the proper direction. First, phonotactic behavior in *G. bimaculatus* is well correlated with differences in the activity of bilateral and mirror-image ascending neurons recorded in tethered animals walking on a treadmill (fig. 7.13; Stabel et al. 1989). Second, killing one of the two L1 neurons of *Acheta domesticus* using the photo-inactivation technique of Miller and Selverston (1979) introduces tracking errors of as much as 90° (Atkins et al. 1984). Third, manipulations of the membrane potential of ascending neurons, which can increase or decrease their firing rate in response to sound, altered the phonotactic orientation of tethered animals in predicted ways (Schildberger

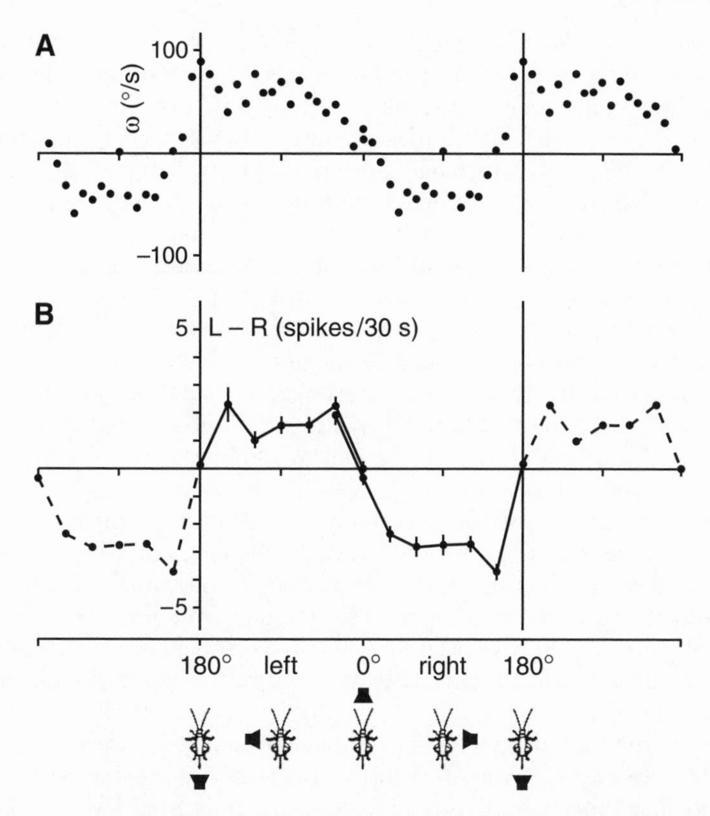

Figure 7.13. Behavioral and neuronal directionality in tethered field crickets *Gryllus bimaculatus* running on a treadmill. The animals were stimulated with model calling songs broadcast at an intensity of 70 dB SPL. (A) Angular velocity of the cricket as a function of stimulus angle (shown in diagrams below). (B) Differences in spike activity of two ascending neurons (AN1, AN2) recorded in the left and right neck connectives with sound presented at the same stimulus angles as in (A). Mean values were calculated over 30 s of running activity, and both plots derived from the same animal. From Stabel et al. 1989, fig. 3; modified by Schildberger et al. 1989, fig. 14.6.

and Hörner 1988). In flying animals, such manipulations of AN2 neurons elicit either positive or negative phonotaxis, depending on the type of manipulation and the carrier frequency of the stimulus (review: Pollack 1998).

Experiments in which membrane potentials can be reversibly manipulated are particularly convincing because the orientation behavior of the animal can be observed before, during, and after application of the current. Recall from chapter 6 that in *Teleogryllus oceanicus*, negative phonotaxis, indicated by abdominal bending away from the sound source in tethered flying preparations, was correlated with activity of INT-1 (AN2 in *Gryllus*). Nolen and Hoy (1984) found that intracellular hyperpolarization (deactivation) of INT-1 abolishes the activity of the contralateral abdominal flexor muscles that are used by the insect to turn away from the sound source. Contrariwise, depolarization (hyperactivation) of INT-1 elicits negative phonotaxis. In studies of positive phonotaxis in *Gryllus*, ascending neurons (AN1, AN2) were recorded from females that were mounted on an air-suspended ball and stimulated with synthetic calling songs from a loudspeaker positioned 50° to the left or to the right of the body axis of the female. Although the animal was fixed, it could turn the ball with its legs, and its turning tendency was measured (fig. 7.14A; Schildberger and Hörner 1988). If the AN1 interneuron on the left side was deactivated by a hyperpolarizing current injection, the animal reversed its turning tendency and turned to the right side, where the inactive speaker was located (fig. 7.14B). This turn was presumably caused by the difference in the activity between the left AN1 and right AN1 interneurons. For AN2, deactivation also changed the turning tendency, although a complete sign reversal did not occur. Data from the selective deactivation experiments support the predictions of a model of the directional characteristic of the summed activity of the AN1 and AN2 pairs (fig. 7.14C; Horseman and Huber 1993).

Anurans. Because the ascending pathway of anurans has many more ascending auditory neurons than that of insects, manipulation or killing of any one cell is unlikely to affect phonotaxis or to change substantially the firing activity of any higher-order cells that receive inputs from both ears. However, binaurally driven cells have been studied in the dorsolateral nucleus, superior olivary nucleus, and torus semicircularis (Feng and Capranica 1976, 1978; review: Feng and Schellart 1999). The majority of these neurons are excited by input from the contralateral ear and inhibited by input from the ipsilateral ear, but a few are excited by inputs from both ears. Many of these cells are sensitive to small binaural intensity differences, ranging from less than 1 dB to 5 dB, and some are sensitive to binaural time disparities of the order of 0.1 to 5.0 ms. Clearly, intensity differences of this order of magnitude are available at the level of the peripheral auditory system, and the latency-shift mechanism almost certainly assures that binaural time disparities of 0.1 ms or greater occur in the central auditory system. Two commonly observed directional patterns are shown in figure 7.15. Both show good frontal directionality, which fits well with the behavioral studies discussed earlier in this chapter.

Interactions between Sound Recognition and Localization

We have previously considered sound localization as a separate problem from sound pattern recognition (chapters 4–6). Although convenient, this separation is unrealistic because both tasks have to be accomplished in order to find a conspecific mate or avoid a sound-producing predator. Indeed, the need to orient toward one of two or more different signals that arrive in close temporal proximity from spatially different sources is a common problem that animals face in nature (see chapter 8). Therefore, the neural analysis of these two features must somehow be integrated or coordinated.

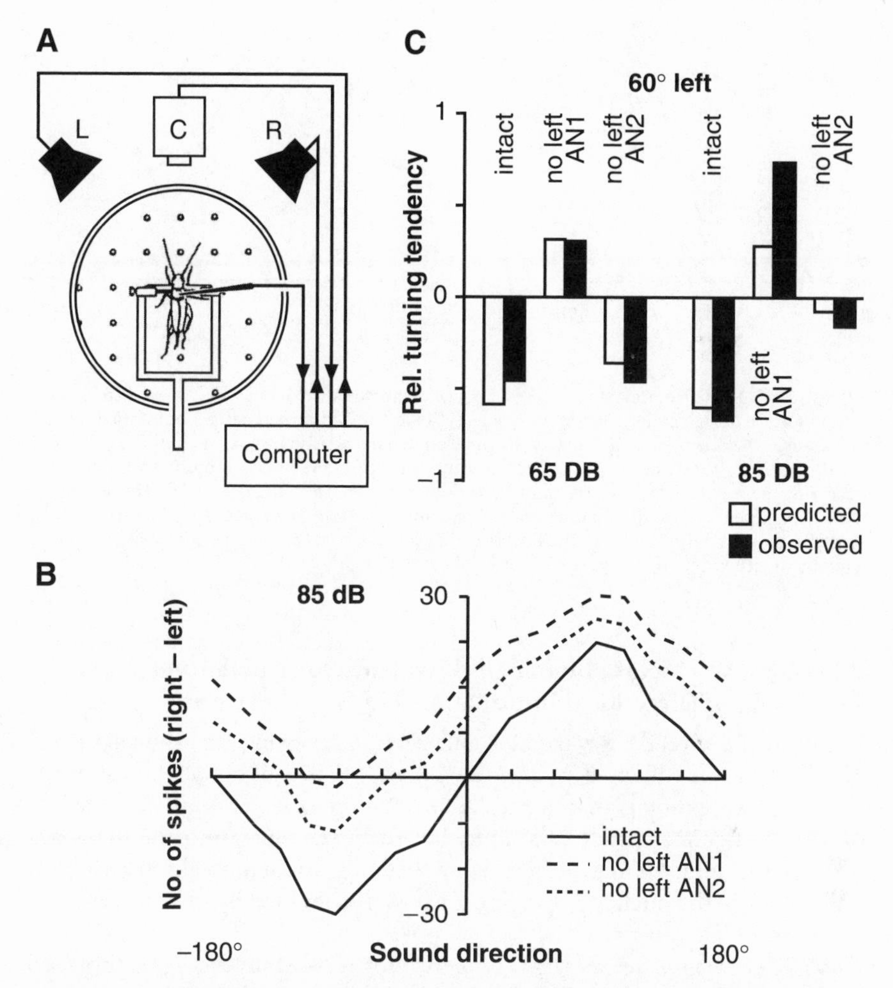

Figure 7.14. Effects of inactivation of single ascending neurons (AN1, AN2) on right-left differences in firing rates and in the turning tendency of crickets. (A) Setup for recording and manipulating identified neurons in the walking cricket. Tethered walking and turning (rotation in degrees to the left or right) is recorded by an infrared camera, C. L and R: positions of loudspeakers relative to the preparation. (B) Differences in the number of spikes between the right and left ascending neurons as a function of sound direction in when the left AN1 or left AN2 were inactivated by hyperpolarizing current. Intact = no current injection. (C) Histograms show the relative turning tendencies of crickets observed when ascending interneurons were inactivated compared with values predicted from the models of Horseman and Huber (1994b). Intact = no current injection. From Schildberger 1994, fig. 7.

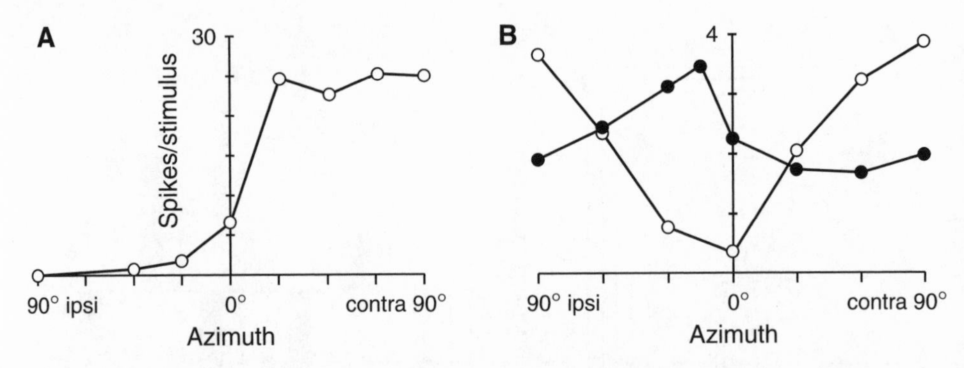

Figure 7.15. Representative directional response characteristics of binaurally driven auditory neurons in the torus semicircularis of a frog. (A) Spike rate is shown as a function of the location of a speaker playing back tone bursts at the neuron's characteristic frequency. In this example, the directional pattern shows strong lateralization, with the shift from inhibition to excitation occurring close to the midline, indicating a narrow section of the frontal field. (B) Two examples of neurons that show maximum (*closed circles*) or minimum (*open circles*) activity when the sound comes from the front; firing rate then decreases or increases, respectively, as the source moves laterally in either direction. From Feng and Schellart 1999, fig. 6.16.

Behavioral and Neurophysiological Evidence for Parallel and Serial Processing of Directional Information and Sound Patterns

Hypothetical circuits for parallel and serial processing are shown in figure 7.16. As we will show below, the essence of the difference in these two modes of processing is that in parallel processing, orientation depends solely on the relative magnitude of stimulation (or its timing) from the two sides (ears) of the animal, whereas in serial processing, orientation also depends on differences in the quality of the signal patterns conveyed by the two ears.

Parallel processing in grasshoppers. Evidence for parallel processing is provided by studies of the grasshopper *C. biguttulus*, in which signal recognition and localization seem to be performed by partly separate channels that include some different sets of neurons. Helversen (1984) studied the female acoustic response to male signals in order to find out if preferences for particular temporal patterns depended on directional hearing. She used pairs of sound patterns, which, when played back from separate speakers, were superimposed in time to form either an effective or an ineffective composite pattern. The reliable occurrence of an acoustic response in the first situation and its absence in the second situation were independent of the locations of the two sound sources (fig. 7.17). Helversen and Helversen (1995) also showed that males turn toward less attractive sounds that provide better directional cues. Thus, the directional decision is independent of pattern quality but is based only on the intensity differences between both sides.

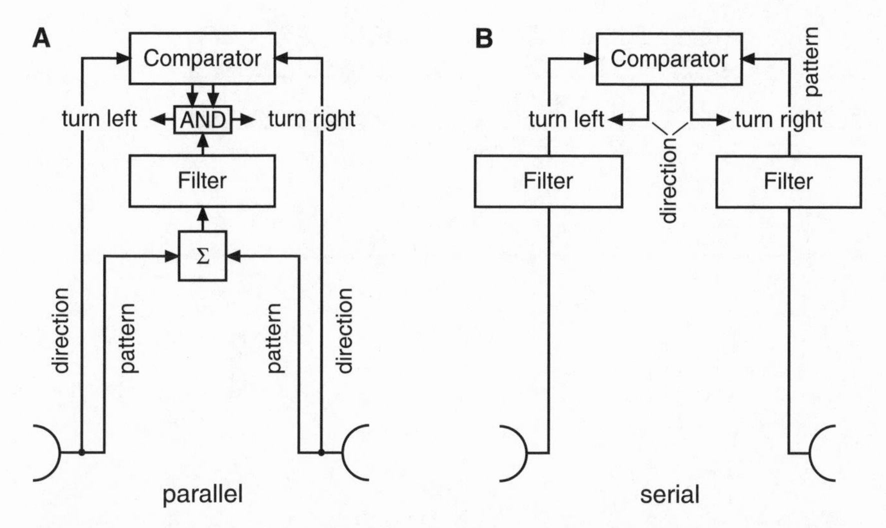

Figure 7.16. Diagrams of circuits, which could explain parallel processing in grasshoppers (A) and serial processing in crickets (B) of sound-pattern and directional information. (A) In parallel processing, the inputs from two ears feed into two channels, one specialized for processing the sound pattern, and the other, for analyzing directional information. Because the channels from each ear analyzing the sound pattern sum, pattern recognition is nondirectional. The two channels carrying directional information and pattern recognition are then compared in higher centers, and the output of the comparator circuit depends solely on binaural intensity and time differences. If the pattern is recognized as adequate, the command for the turn is given (AND). (B) In subsequent or serial processing, localization appears to come after recognition. Each input channel from the corresponding ear passes through filtering networks in the ascending auditory system that evaluates the quality of the sound pattern. The output of the two filters (considered as recognition network) is then compared for the extraction of directional information. From Helversen and Helversen 1995.

The behavioral results are supported by recordings from central auditory neurons. Those neurons that are best suited to encode information about sound direction are poor at encoding the behaviorally relevant temporal features of the song (chapter 6; Stumpner and Ronacher 1991b). Surgical experiments suggest that the division into separate recognition and localization channels probably occurs quite early along the central neuronal pathway within the metathoracic ganglion (Ronacher et al. 1986, 1993; Stumpner and Ronacher 1994). Helversen (1997) provides a model that considers how the different channels involved in pattern recognition and sound localization ultimately combine and process information about sound patterns and location.

Serial or sequential processing in crickets and katydids. As discussed in chapter 6, AN1 and AN2 neurons on each side of the cricket ascending pathway encode information about the sound pattern, and, as shown above, if the representation of the temporal pattern is equivalent on both sides, the animal turns toward the side that is most strongly stimulated. Based on the observation that

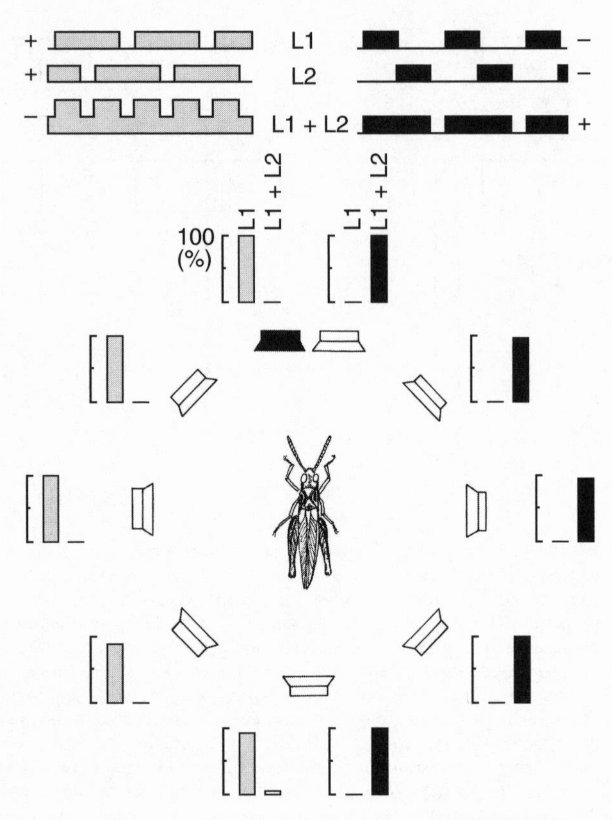

Figure 7.17. Directional-independent summation of temporal patterns by female grasshoppers (*Chorthippus biguttulus*). Diagrams at the top (*gray and black diagrams*) show how two effective temporal patterns (left L1+ and L2+) superimposed from two loudspeakers can be added to create an unattractive pattern ([L1 + L2] −) or how two ineffective patterns (right L1− and L2−) sum to create an effective pattern ([L1 + L2] +). Histograms below show female responses to these temporal patterns when these single components or combinations were played back from speakers 1 m in front of the grasshopper. The diagram below shows that the same results were found when one loudspeaker was fixed in front of the grasshopper (*black symbol*) and the other speaker (*white symbol*) was moved in 30° steps around the animal. In the left half of the circle (*gray histograms*), for example, L1 + L2 was highly ineffective relative to the attractive pattern (L1+) played back from a single speaker, regardless of the location of the second sound source that played L2. In the right half of the circle (*black histograms*), the combination L1 + L2 was always more attractive than the single unattractive pattern (L1−) regardless of the location of the second speaker. From Helversen 1984, fig. 1.

crickets could discriminate reliably between two signals (both acceptable but differing in attractiveness) that were played simultaneously from the two sides, Pollack (1986) suggested that the direction of phonotactic turns is determined by comparing the qualitative (spike patterns) as well as quantitative outputs (spike rates) of two recognition circuits, each located in the corresponding half of the brain.

Direct evidence for serial processing was provided by Stabel et al. (1989), who studied female crickets running on a paired treadmill. When an attractive pattern was played back from a direction horizontal to the walking animals, the crickets show positive phonotaxis to the sound source (fig. 7.18A). But when this same attractive model calling song was played from above the cricket and a continuous tone was simultaneously played from a direction horizontal to the walking animal, the crickets showed a change in phonotaxis away from the source of the tone (fig. 7.18B). This change occurred even though the total amount of stimulation on the side on which the tone was played back was greater.

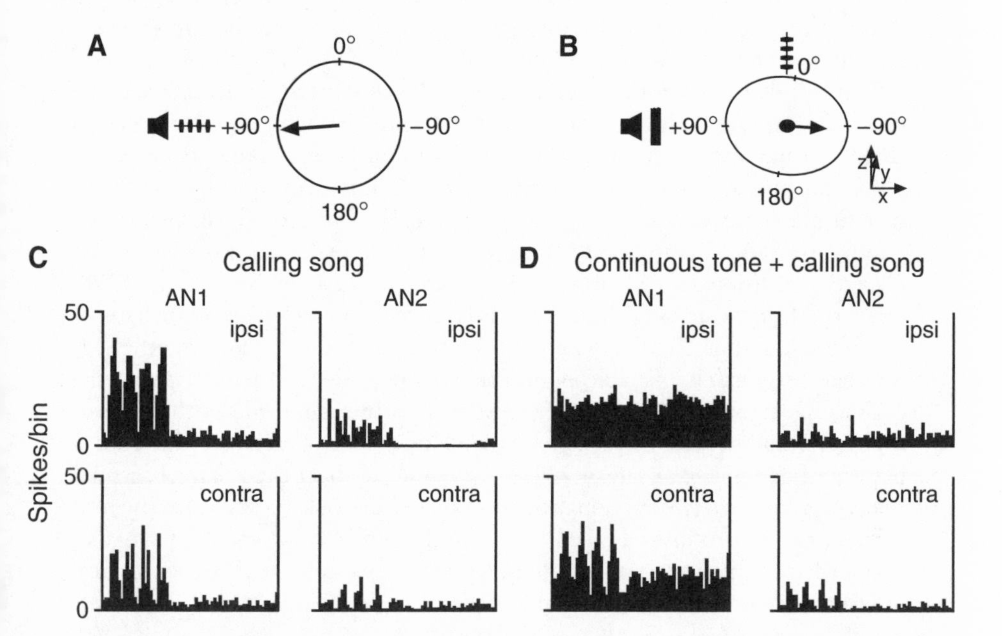

Figure 7.18. Processing of pattern and directional information in female crickets *Gryllus campestris* and *G. bimaculatus*. Results of 2–15 min phonotactic runs of female *G. campestris* under two different sound conditions. (A) Model calling song is played back from the side (speaker and song symbol +90°), the mean vector (*arrow*) indicates phonotaxis toward the sound source. (B) Model calling song is played back from above (song pattern indicated above the circle), and a continuous tone (*black bar*) simultaneously played back from the side (*speaker symbol*); the mean vector (*arrow*) indicates phonotaxis away from the side of the sound source. Modified from Wendler 1989, fig. 2. (C) Peristimulus histograms show the fidelity of pulse and chirp copying in the responses of a pair of ascending neurons (AN1, AN2) in the cricket *G. bimaculatus* when calling song was played back from the sides as in (A). *Upper parts*, responses of both neurons recorded ipsilaterally; *lower parts*, responses recorded contralateral to the sound source. Faithful copying of pulses is evident in all histograms. (D) Responses of both neurons to lateral presentation of a continuous tone of 4.7 kHz and a model calling song from above as in (B). Note that copying of pulses and chirps is masked in the responses of the neurons recorded ipsilateral to the tone (*upper recording*) whereas copying with higher fidelity is seen in the responses recorded from the contralateral side (*lower recording*). Modified from Wendler 1990, fig. 4; data from Stabel et al. 1989.

Extracellular recordings with hook electrodes from the ascending auditory neurons (AN1, AN2) help to explain this counterintuitive result by showing that the more faithful copying of the temporal pattern of the song occurs in the ascending neurons on the opposite side from the tone source (fig. 7.18D). The tone, in effect, interferes with the encoding of the temporal pattern of the stimulus in the neurons on that side. Thus, the rule "turn to the side most strongly stimulated" is only valid as long as an attractive pattern is presented alone. But in the presence of noise (tone), the cricket turns to that side on which the pattern is better represented, for example, with a higher signal-to-noise ratio. This process requires a pattern recognizer or filter system, the output of which is then used to control turning behavior. Therefore, pattern recognition and turning behavior are not parallel processes but must be arranged in sequence (Wendler 1989, 1990).

Comparable behavioral results are available for katydids. Recall that the mean direction of phonotaxis on a locomotion compensator is displaced slightly from the direction of the source of conspecific songs in *Tettigonia viridissima* and *T. cantans* when responding to simultaneous playbacks of conspecific and heterospecific songs (chapter 4). In *T. viridissima* the displacement is away from the location of the heterospecific song, and in *T. cantans* the displacement is toward the location of the heterospecific song. Schul et al. (1998) interpret these results in terms of the difference in effectiveness of the heterospecific signal. For *T. viridissima* the song of *T. cantans* is an unattractive signal that never elicits reliable phonotaxis. Hence, the representation of the conspecific signal in the ascending neurons on the same side as the heterospecific signal is likely to contain elements of this unattractive or ineffective pattern. The opposite pattern in *T. cantans* might reflect the relatively attractive nature of *T. viridissima* songs in no-choice situations. As shown in the next section, the neurophysiological basis for the ability to associate particular temporal patterns with the location of the sound source at the level of local interneurons is well established in *T. viridissima.*

Biological Significance of Parallel and Serial Processing

As pointed out by Helversen and Helversen (1995), the fact that parallel processing occurs in grasshoppers and serial processing in crickets and katydids might reflect a difference in their evolutionary history in the context of communication. In grasshoppers, stridulation appears to have evolved several times independently, whereas the auditory organs are present in many species, even in those that do not stridulate (e.g., Fullard and Yack 1993). This supports the idea that in this group of insects, hearing originally evolved to detect and localize a diversity of signals produced by predators. Undirected evasive responses, such as stopping all movements, could have been the dominant reactions to detecting a predator. Later and in the context of mate finding, parallel processing would make discrimination of the songs of several audible individuals difficult, but a variety of different signals occurring in isolation could be discriminated, recognized, and localized. This idea would be supported by finding parallel processing in insects that do not rely on hearing for intraspecific communication.

In crickets and most katydids, hearing and sound production most probably coevolved in the context of mate attraction (Alexander 1962) and are thus likely to be highly specialized for this task. As shown below, serial processing is advantageous for a receiver in noisy environments because it can first analyze the quality of different signals in its vicinity before choosing a mate and moving toward it. With the later evolutionary appearance of echolocating bats, the ability to detect and avoid predators was probably incorporated into a system already highly specialized for detection and localization of conspecific signals (Hoy 1992; Pollack 1998).

Although the evolutionary scenario outlined above is unlikely ever to be proved satisfactorily, there is considerable evidence that directional hearing in crickets, katydids, and anurans can contribute to signal detection and recognition in noisy environments. This ability is at least analogous to the so-called "cocktail party" effect, whereby human listeners can selectively attend to spatially separated speakers (Cherry 1953).

Two studies provide hints about mechanisms that contribute to communication in noisy environments. First, the contribution of directional hearing is shown by the fact that the right and left omega neurons of the cricket *Teleogryllus oceanicus* selectively encode patterns played from their respective sides (Pollack 1988). Even more impressive evidence of this kind comes from neurophysiological studies of *Tettigonia viridissima* (Römer and Krusch 2000). Simultaneous recordings from pairs of omega neurons showed that the different temporal patterns of overlapping signals were reliably represented in the spiking patterns of the two neurons even when the two sources of these signals were separated by only 15° in the frontal zone (fig. 7.19). Second, in both of these species additional, nondirectional mechanisms involving inhibitory responses of the omega neurons could contribute to selective attention (Pollack 1988; Römer and Krusch 2000; see chapter 8).

Behavioral experiments with anurans suggest that directional hearing can modestly improve the detection and recognition of conspecific signals in the presence of background noise. For example, females of the green treefrog *Hyla cinerea* detect the calls of conspecific males at a lower signal-to-noise ratio when sources of noise are located at 90° relative to sources of signals than when the noise sources are situated right next to the signal sources (Schwartz and Gerhardt 1989). This phenomenon has been termed *release from masking* in the psychophysical literature (Blauert 1983). The magnitude of the improvement is, however, only about 3 dB, and although females are better at detecting either advertisement or aggressive calls when the noise sources are spatially separated, they still do not discriminate between these two signals as they

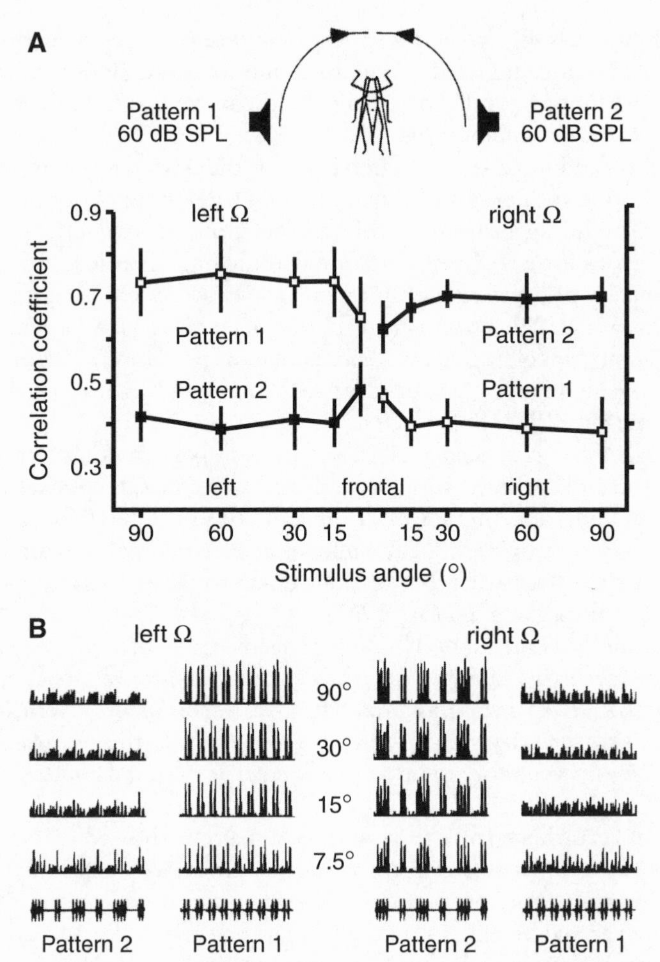

Figure 7.19. Direction-specific encoding of different temporal patterns in katydids. (A) Top diagram showing the setup for simultaneous stimulation with signals having different temporal patterns; the responses of the pair of omega neurons were recorded at the same time. The correlation coefficient refers to the similarity of the temporal pattern of the signal and the spiking pattern and is plotted as a function of the angular deviation of the speaker from the frontal midline. (B) Poststimulus histograms of the responses of both cells, triggered with either pattern 1 or pattern 2. The ipsilateral signal was well encoded even at 7.5°; the patterned response to the respective contralateral signals occurs because of their inhibitory action on the spike discharge. From Römer and Krusch 2000, fig. 4.

do when background noise is very low in amplitude (Schwartz and Gerhardt 1989). In another study, whose results contrast with those discussed above for grasshoppers, the relative attractiveness of a composite pattern to female gray treefrogs (*Hyla versicolor*) depended on the angular separation of two speakers that emitted the two components of the pattern (Schwartz and Gerhardt 1995; details in chapter 8).

Studies of binaurally sensitive neurons at the level of the torus semicircularis of the leopard frog reveal directional effects, mediated by binaural inhibition, on frequency tuning, and on the selectivity for different rates of amplitude modulation (Gooler et al. 1993, 1996; Xu et al. 1994, 1996). However, large effects occurred only when shifts in sound-source direction are far greater than the normal angular differences that the frog experiences as it makes a phonotactic approach, and sometimes not even under these circumstances. Nevertheless, these studies show that sound direction is likely to affect the processing of species-specific spectral and temporal properties.

Summary and Suggestions for Future Studies

Despite their small size and separation of the two ears, insects and anurans accurately locate the sources of conspecific signals or, in some insects, turn away from signals representative of those produced by predators. On the one hand, estimates of the accuracy of phonotaxis and of minimum binaural cues required for lateralization, as studied in controlled laboratory conditions, might overestimate phonotactic performance in nature because the quality of directional cues in the habitat is likely to be much poorer (Michelsen and Larsen 1983; Michelsen 1998a; see chapter 11). On the other hand, animals in nature localize the signals of their mates in closed-loop conditions, where they have the chance to successively update directional information, and nonauditory cues such as visual stimuli that can improve localization are also frequently available (e.g., Honegger and Campan 1989; Huber 1990; Helversen and Wendler 2000). More field experiments are needed to learn how insects and anurans actually perform under complex, noisy conditions.

Insects and anurans have adopted several solutions to the problem of sound localization using the small external binaural cues that are available to them (Michelsen 1998a). The biophysics of the grasshopper, cricket, and katydid systems is fairly well understood. The pressure-difference system of grasshoppers seems to be adequately modeled as a two-input system, whereas three inputs are necessary to explain ear directionality in field crickets. The phase shifting associated with the medial septum of these crickets appears to be an evolutionary invention that allows these insects to communicate with relatively low frequencies. But we do not know whether all cricket species have this invention and whether accurate localization is always restricted to a narrow frequency range that depends on the size of the cricket.

Most katydids appear to use ears with a pressure-receiver mechanism for localization. These insects have evolved an acoustic trachea to amplify the pressure to the inner surface of the membrane. Moreover, exotic structures around the acoustic spiracles of some species probably serve to accentuate external binaural differences even more. The high signal frequencies required to make this work are strongly attenuated in natural habitats, but this disadvantage is countered somewhat by the inaudibility of these sounds to many predators.

Biophysical mechanisms of directional hearing in anurans are still poorly understood, and additional behavioral studies designed to estimate minimum binaural cues under open-loop conditions are badly needed. The ability of some anurans to localize relatively low-frequency sounds suggests that at least small species use some form of pressure-difference mechanism. Multiple potential pathways exist to the internal surface of the eardrum, and one or more of these pathways must contribute to such a mechanism. The functional significance of the sound input via the lungs is still unknown, and the biophysics of extratympanic pathway(s) and the extent to which anurans might be able to modify these and the multiple inputs to the inside of the tympanic membrane are outstanding open questions. We also do not understand how anurans distinguish elevated from nonelevated sounds. We suspect that most of these questions can only be satisfactorily answered by sophisticated experiments with freely moving and behaving animals. After all, the anuran has much control over the configuration of potential internal pathways, and most of the conflicting results from biophysical and neurophysiological studies center on the state and position of immobilized preparations.

In both insects and anurans, interaural inhibition plays a role in preserving or magnifying binaural differences existing at the tympanal or inner ear organs. In crickets and katydids, paired omega neurons mutually inhibit one another and ascending neurons. In grasshoppers, several local interneurons with inhibitory functions are interspaced between auditory input and the contralateral ascending pathways (Helversen 1997). In anurans the majority of binaural interneurons are excited by contralateral stimulation and inhibited by ipsilateral sounds. The trading of time and intensity cues, which can be accomplished by paired sound sources, has been demonstrated in both groups. Of course, in nature, when the animal tries to locate a single source, these two cues augment rather than oppose one another. The inverse relationship between latency and stimulus intensity—a general property of many auditory neurons—is an important mechanism for creating usable neural time delays.

Parallel processing of information about sound pattern and source location has been rigorously documented in one species of grasshopper. Is this a general mechanism in the grasshoppers and other insects in which the detection of predators was supposedly the first evolutionary role of hearing (Helversen and Helversen 1995)? If so, then parallel processing should also be scattered in some other orthopteran groups in which intraspecific communication by sound is absent. Serial processing of sound patterns and source location occurs in crickets and katydids, and probably anurans: the quality of information encoded in ascending neurons on one side of the animal—and not just the activity level—affects the direction of orienting movements. Such direction-specific pattern recognition, can, in principle, facilitate the detection and discrimination of different conspecific and heterospecific signals arriving simultaneously from different directions. In anurans, where this effect has been quantified in two species, the contribution of directional hearing in this context is unimpressive. Neurophysiological studies predict a much stronger effect in crickets and katydids, and behavioral studies are needed to test this prediction.

8

Causes and Consequences of Chorusing

Aggregations of acoustically signaling animals are called choruses. Those species of insects and anurans that form choruses usually do so in areas where the physical resources required by females are concentrated, and chorusing usually occurs during seasonal and daily time periods that are favorable for courtship and reproduction (Wells 1977; Thornhill and Alcock 1983). Because most choruses are located in resource-rich places, we think that other potential benefits of chorusing (increased attraction of females, reduced assessment costs, reduced predation risks), which have seldom been demonstrated in insects and anurans, are better interpreted as consequences rather than as primary causes of aggregated signaling. Indeed, males and sexually unreceptive females may also use chorus sounds as cues for quickly locating patches of resources (e.g., high-quality food plants in desert grasshoppers of the genus *Ligurotettix* [Greenfield and Shelly 1990; Muller 1998] or favorable burrowing sites in mole crickets of the genus *Scapteriscus* [Forrest 1983]).

Chorusing intensifies competition among males to attract mates, to acquire and defend resources needed by females, or both. Choruses are thus focal points for sexual selection, which we discuss formally in chapters 9 and 10. This chapter considers where and when, relative to other individuals, males in choruses should produce acoustic signals in order to increase their chances of attracting a female. Male tactics, in turn, are driven by female requirements. For example, the high background noise levels and temporal overlap of signals that are inevitable features of dense choruses can make it difficult for females to detect, identify, and locate individual signalers. In some species, such masking interference is reduced by fine-scale interactions in signal timing and the spatial separation of signalers within the chorus. In others, masking interference is increased as males compete to time their signals so that they are more audible or locatable than those of neighboring rivals.

Spatial Aggregation of Signalers

Resources

In addition to occurring where oviposition sites and food plants are abundant, aggregations of male insects sometimes also form in areas where virgin females emerge, where females forage, or in microclimatic refuges (reviews: Thornhill and Alcock 1983; Alexander 1975; Otte 1977). In most anurans, choruses form within or along the margins of standing water where females must lay their eggs. Because these aggregations are centered on resource-rich areas, the mating systems of chorusing species could be some form of a *resource-defense polygyny* (Emlen and Oring 1977) or a *resource-based lek* (Alexander 1975). For insects and anurans, the distinction boils down to whether individual males defend material resources that females require (resource defense) or merely defend a space for the clear broadcast of their signals (lek). Well-studied examples of both systems have been described in insects and anurans (reviews: Alexander 1975; Wells 1977; Thornhill and Alcock 1983; Höglund and Alatalo 1995; Sullivan et al. 1995). There is a continuum of resource control by male signalers and opportunity for choice by females, which can base decisions on resources, signals, or other attributes of the male (Bradbury 1985). For example, the final mating decision of a female insect might depend on an assessment of a male's nuptial gift as well as his acoustic signals (Gwynne 1982). In some species, females can make sequential assessments of several males without interference, but in others, females are chased and forced to mate as soon as they approach an aggregation (reviews: Wells 1977; Thornhill and Alcock 1983).

Female Attraction by Choruses

Aggregated signaling might also be explained at least in part by increased stimulation and attraction of females. If, for example, females require aggregated signaling to become sexually responsive, which might be the case even if the location of choruses is not centered on resources, then an individual would have to join a chorus to have a chance of mating. To our knowledge, no study has tested this hypothesis in insects or frogs, but exposure to the calls of individual conspecific males has been shown to have a stimulatory effect on female reproductive physiology in at least one species of anuran (*Alytes muletensis*) (Lea et al. 2001). Exposure to chorus sounds increased steroid hormone levels and the probability of stimulus-evoked calling in male treefrogs (*Hyla cinerea*) (Burmeister and Wilczynski 2001) and is thus likely to have effects on female reproductive physiology as well.

The range of attraction of grouped signalers is also likely to be somewhat greater than that of single individuals. As pointed out by Bradbury (1981), however, the increase in the amplitude of chorus sounds is not a simple multiple of the number of signalers. Nevertheless, the more continuous acoustic output of a large chorus might more effectively attract females at a distance than the less continuous output of solo males or small choruses. Unfortunately, there is little experimental evidence that chorus sounds alone attract female insects or anurans to the breeding site.

The female-attraction hypothesis traditionally predicts that males in large aggregations (choruses or aggregations within choruses) should have greater *per capita* mating success than males in small aggregations, although recent empirical and theoretical considerations indicate that such a positively non-linear increase is not necessarily required for the evolution of lekking in birds (Kokko et al. 1998). In any event, from the female point of view, the costs of mate assessment might be lower if signalers are aggregated rather than dispersed in space and time. A counterpoint to this argument is that female costs could be higher in aggregations if females risk unsolicited matings or if masking interference is so high that females cannot readily detect or discriminate among males (e.g., Wollerman and Wiley 2002).

For anurans, indirect support for the female-attraction model is available from field studies of two species, *Physalaemus pustulosus* (Ryan et al. 1981) and *Bufo valliceps* (Wagner and Sullivan 1992). In both studies, the *operational sex ratio* (= the proportion of mating females to calling males on a particular night) increased with chorus size at a single breeding site that was sampled on different nights. Tejedo (1993) found that the operational sex ratio increased with male numbers but not the number of *calling* males in the natterjack toad *Bufo calamita*.

The results of other studies do not support the hypothesis. In *Hyla cinerea* (Gerhardt et al. 1987), *Bufo gutteralis* (Telford and van Sickel 1989), and *Hyperolius marmoratus* (Henzi et al. 1995), there was no relationship between chorus size and operational sex ratio, and one study of *Bufo woodhousei* (Sullivan 1985) found a negative relationship. In *Bufo calamita*, the number of females arriving nightly at each of three different ponds was unrelated to the numbers of males at these ponds (Tejedo 1993). In *Hyperolius marmoratus*, different suites of environmental variables explained male and female attendance: males avoided conditions favoring high rates of evaporative water loss; and females were most likely to appear when conditions favored high larval growth rates (Henzi et al. 1995). Thus, in species in which the *same conditions* favor calling by males and breeding by females, the operational sex ratio might increase with chorus size for this reason alone and not because of any increased attraction by chorus sounds.

The experimental evidence for the female-attraction model is even less compelling. As discussed by Gerhardt and Klump (1988b), most studies that have attempted to demonstrate the attraction of male and female anurans by chorus sounds alone have not done so in a robust fashion. Moreover, even when playback studies show the potential for chorus sounds to attract females, this does not necessarily happen in nature. Studies of a prolonged breeder, the barking treefrog *Hyla gratiosa*, illustrate this point nicely. The low-frequency peak of the advertisement call propagates well in natural habitats, which is relatively dry terrain that surrounds semi-isolated and scattered ponds (Gerhardt and Klump 1988b). Females that were previously found mating reliably showed phonotaxis to playbacks of chorus sounds recorded at a distance of 160 m from choruses, where sound pressure levels were less than 50 dB (Gerhardt and Klump 1988b). In view of the risks of traveling over a relatively harsh environment to reach a breeding pond, we might expect that females would use chorus sounds to at least determine that reproductively active conspecific males were present at a given pond. However, Murphy (submitted) found that as many gravid females arrived at a breeding pond at the usual times on nights when he removed all of the males before they formed a chorus as when the chorus was allowed to form. Thus, females did not need the chorus sounds to find the pond nor did stimulation by the chorus sounds increase the probability of female arrival. These results suggest that males and females can locate the pond by other sensory cues and that reproductive activity in both sexes might be stimulated by similar conditions. Of course, chorus attraction from a distance might still be important for other species, and more experiments of this kind are needed. The chorus-attraction hypothesis would also be supported by showing that males and females are reliably attracted to playbacks of chorus sounds from places where no breeding pond exists.

Female Attraction to Groups of Signalers within Choruses

Even though the evidence for long-distance attraction to choruses in insects and anurans is meager, the female-attraction hypothesis is circumstantially supported by observations of spatial distributions of signalers *within* choruses. In nonterritorial species, for example, aggregations of calling males within the breeding site are often observed even when other seemingly equivalent calling sites, used on other nights, are unoccupied (Doolan and Mac Nally 1981; Sullivan 1982b; Pfennig et al. 2000; Gerhardt, pers. obs.). The question thus arises: Why do males fail to use these unoccupied sites to increase spacing and thereby to reduce masking interference? Do these small aggregations increase the chances of female attraction, or are there subtle deficiencies in the unoccupied sites?

Evidence for increased female attraction by aggregations within chorusing areas is mixed. In the Australian bladder cicada *Cystosoma saundersii*, in which males form aggregations in the general areas where females tend to emerge, males sing from bushes of varying size for about thirty minutes just after dusk (Doolan and Mac Nally 1981). The bushes do not vary in resources used by females, which are attracted at a distance by the spectral properties of male songs (Doolan and Young 1989; see chapter 4) and choose males freely. Male bladder cicadas sang in groups (pairs, trios, and other multiples) in small bushes more frequently than expected by random settlement (fig. 8.1), presumably, in part, because their songs propagated more effectively from small

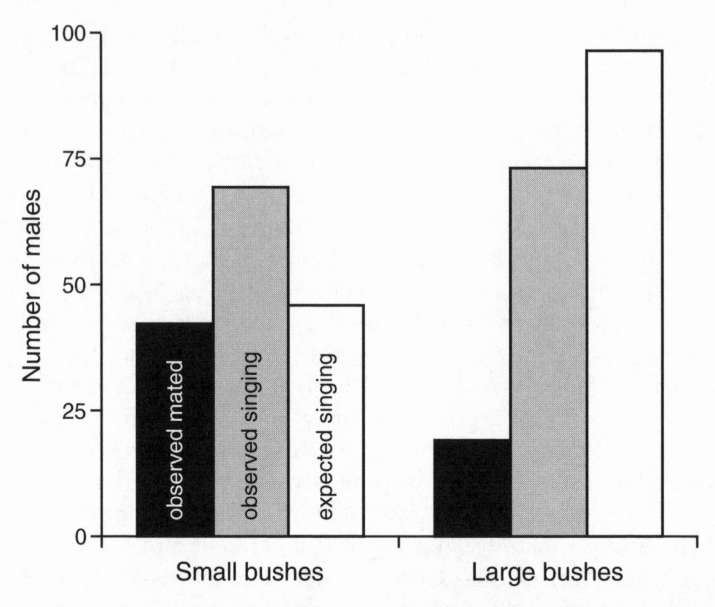

Figure 8.1. Choice of calling sites by males and per capita attraction of females in the bladder cicada *Cystosoma saundersii.* About the same numbers of males were observed singing in small bushes as in large bushes, but because large bushes were more abundant, this distribution of singing males was far different from that expected from random settlement. A greater proportion of the males in small bushes were observed mating (61% of the number observed singing) in comparison to males in large bushes (26% of the number observed singing). From data in Doolan and Mac Nally 1981, table 1.

bushes than from large bushes. Most small bushes, however, did not harbor singing males, and the locations of groups of signalers changed over the course of the season rather than being fixed to a particular set of small bushes. The last observation suggests that males do not aggregate in particularly favorable locations, which have been termed *hotspots*[1] (Bradbury 1985). Groups of singing males in small bushes attracted more females per capita than did isolated singing males. Doolan (1981) also created artificially dense groups of males that attracted more females per capita than groups of normal size. Mating success in artificially enlarged groups decreased, however, because some spacing within aggregations is required by females of this species in order to trigger the final stage of mate choice (see below). Two other field experiments with insects fail to support the female-attraction hypothesis. No difference in per capita attraction of females occurred in katydids (*Tettigonia viridissima*) (Arak et al. 1990) or short-tailed crickets (*Anurogryllus arboreus*) (Walker 1983b)

1. Of course, the general area of signaling can also be considered a hotspot, but referring to this larger scale of analysis detracts from the more interesting question: Why do some males in choruses have much greater success in attracting mates than others?

when females were offered choices between males singing in aggregations versus solo singers. In the green treefrog *Hyla cinerea*, more mated pairs were found in parts of the pond where male density was highest, but the per capita success rate was not higher in these places (Gerhardt et al. 1987).

Playback experiments with insects and anurans also fail to support the female-attraction hypothesis. Cade (1981b) found that the average number of female field crickets *Gryllus texensis* (= *Gryllus "integer"*) attracted per speaker to an array of speakers was less than the number of females attracted to an isolated speaker. One unrealistic feature of this experiment was that the broadcast from the speaker array was simultaneous. Thus, masking interference between the sounds emitted from spatially clustered speakers could have reduced the attractiveness of the array relative to unmasked signals emanating from the isolated speaker (see pp. 280–81). Shelly and Greenfield (1991), studying the desert clicker (*Ligurotettix coquilletti*, a grasshopper), found that about the same numbers of females were attracted to six bushes, each of which had a single speaker playing back conspecific songs, as were attracted to six bushes with three speakers in each. Finally, Schwartz (1994) compared the attractiveness of different numbers of sound sources to previously captured gravid females of the treefrog *Hyla microcephala*. He used a computer-based playback system, which simulated both the spatial and, to some extent, the temporal patterns of calling males within small choruses (fig. 8.2). More females were attracted to larger arrays of speakers (7 vs. 1; 2 vs. 1; 6 vs. 2), but the per-speaker attractiveness of the small (or single) and large arrays of speakers did not differ.

The experiments with desert clickers and treefrogs were, however, unrealistic in one way or another. Shelly and Greenfield (1991) held the sound pressure level of the signals from each speaker constant, whereas male desert clickers in aggregations increased the sound pressure level of their calls by about 4–6 dB over that of solo males. Schwartz (1994) adjusted the average call rate per speaker to the same value as that of males calling in isolation, whereas individual males in choruses usually increased their call rate. We might thus expect that arrays of speakers playing back signals at a higher rate or sound pressure level would attract disproportionately more females than single speakers broadcasting signals at the rates and levels typical of single males. If so, then increased female attraction would be conditional on these behaviors, which, in turn, might simply reflect responses to increased competition. The question then arises: Would the benefits of increased mating success be sufficient to offset the energetic costs of increasing signal intensity or call rate as well as the disadvantage of increased levels of interference?

Predator-Avoidance Hypothesis of Chorus Formation

Another, nonmutually exclusive factor that could promote or maintain chorusing is the presence of predators or parasitoids. Aggregations could reduce individual risks by the effects of geometry (central individuals are shielded by peripheral ones) or by increasing the chances that predators are detected by

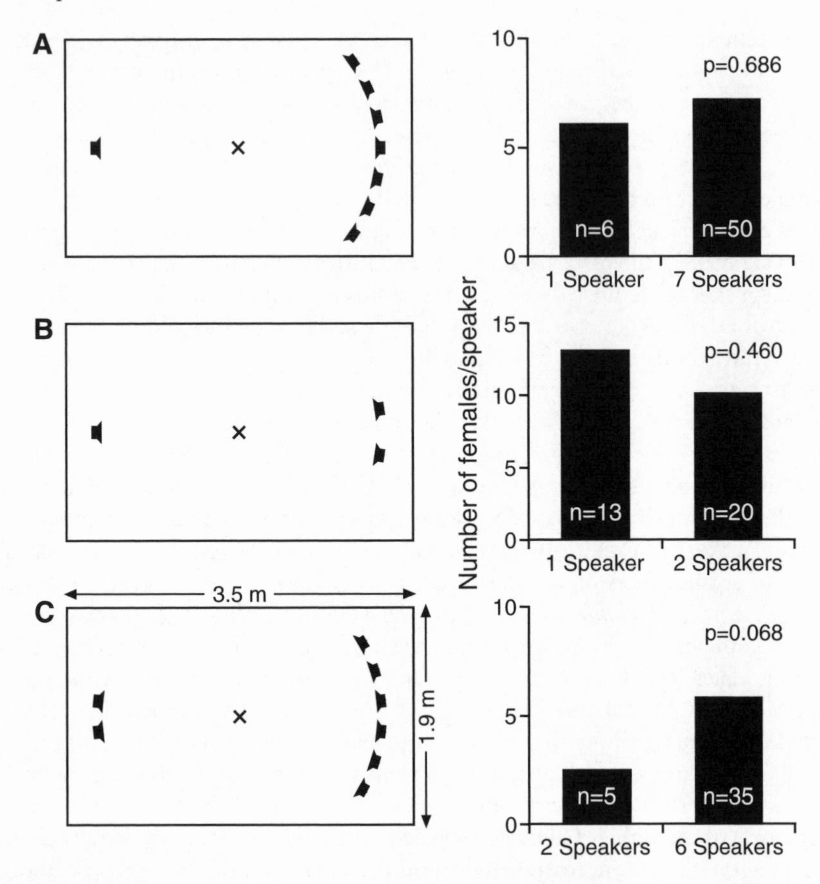

Figure 8.2. Per capita attraction of females of *Hyla microcephala* given choices between single and aggregated speakers, simulating small "choruses" of conspecific males. Diagrams show (A) single speaker versus an array of seven speakers; (B) single speaker versus two speakers; and (C) two speakers versus six speakers. "X" indicates the release point of the females. The histograms indicate the number of females per speaker that chose the multiple speaker array versus a single speaker (A) and (B) or a pair of speakers (C). The samples sizes (absolute numbers of females) are shown on each bar. P-values refer to a binomial test of the null hypothesis that equal numbers of females per speaker would be attracted to the single speaker (A) and (B) or two speakers (C) as would be attracted to the arrays of speakers. From Schwartz 1994, fig. 1.

some individual whose behavior alerts the rest of the group (Hamilton 1971; review: Turner and Pitcher 1986). The noise produced by aggregations might also make it harder for predators to find signaling individuals. Little support exists for any of these specific mechanisms in insects and anurans, although in calling frogs of one species (*Physalaemus pustulosus*), the visual detection of a flying bat model—probably by one or a few individuals—resulted in cessation of calling by the entire chorus (Tuttle et al. 1982). Indeed, a common field observation is that groups of calling insects and anurans often become silent when one or two individuals stop calling after being disturbed by a human intruder. The lower per capita risks in large versus small choruses that have been reported for one species of anuran and for periodical cicadas are probably best interpreted as simple dilution or predator-satiation effects (Ryan et al. 1981; Karban 1982; Turner and Pitcher 1986; review: Williams and Simon 1995).

One direct test of the predator-avoidance hypothesis provides a clear counterexample (Cade 1981b). Whereas clumped speakers did not attract more females of *Gryllus texensis* per speaker than isolated ones, aggregated speakers in some treatments did attract more parasitoid flies than single speakers. These results suggest that crickets in aggregations will actually experience higher risks of predation than solo singers.

Temporal Organization of Signaling: Gross Patterns

Breeding Seasons

Seasonal breeding patterns reflect long-term adaptations, whose origins are central topics in life-history theory (e.g., Stearns 1992). Most species of insects and anurans, even those in the tropics, signal during periods in the annual cycle when conditions are favorable for reproduction and survival of offspring. For insects, breeding periods often coincide with the availability of food plants or female emergence, whereas for most anurans, the availability of water for laying eggs is critical. Photoperiod and temperature also delimit breeding seasons.

Within the general breeding period, weather conditions often determine the timing and intensity of episodes of reproductive activity. This pattern is exaggerated in so-called *explosive breeders* (Wells 1977). For example, in spadefoot toads (*Scaphiopus* and *Spea*) and other desert species of frogs (e.g., *Neobatrachus* and *Cyclorana*), heavy rains are infrequent, and courtship and reproduction, lasting a matter of hours on one night, might only occur every few years at any given locality (Wells 1977; Duellman and Trueb 1986). Although not tightly coupled to particular weather conditions, the occurrence of huge choruses of periodical cicadas is restricted to a couple of weeks separated by intervals of thirteen or seventeen years (Alexander and Moore 1962).

Other species of insects and anurans are *prolonged breeders* whose breeding periods last weeks or months (Wells 1977; Thornhill and Alcock 1983; Duellman and Trueb 1986). In highly territorial ranid frogs, most individual males call on most nights of the breeding season, but in nonterritorial species, a large proportion of the individuals attends only one chorus (Murphy 1994 and references therein).

The number and duration of breeding episodes affect the degree to which the arrival of males and females is correlated, and, in turn, the operational sex ratio. Resources that are available for only a short time, such as temporary pools of water or flowering food plants, can limit the duration of breeding and hence the ability of individuals to monopolize resources or mates. The duration of courtship and mating will also set limits on the opportunity for multiple matings by either sex during any one breeding episode. Taken together, these are the main factors influencing the tactics used in mate acquisition both within and between species (see chapters 9 and 10).[2]

Diel and Circadian Patterns of Signaling

Signaling is typically concentrated in one or more periods during the twenty-four-hour daily cycle (Alexander and Moore 1958; Otte 1977; Walker 1983a; Greenfield 1994a), and there is evidence for circadian rhythms in calling and locomotion by females (review: Loher 1989). In some species of cicadas, crickets, and mole crickets, the calling period lasts for less than an hour (e.g., Doolan and Mac Nally 1981; Walker 1983a), whereas in other species of insects and in many anurans, calling occurs for at least several hours during the night and, under particularly favorable conditions, for entire twenty-four-hour periods (Gelder and Hoedemaekers 1971; Drewry and Rand 1983; Walker 1983a; Loher 1989; Bridges and Dorcas 2000; fig. 8.3A,B). Some insects have two periods of signaling, one during the day and one at night (e.g., Robinson 1990; Heller and Helversen 1993; fig. 8.3C), and some insects even produce a different kind of signal in each period (e.g., Otte 1992).

The change in light level at sunset is the most likely trigger for the natural onset of signaling by the first male(s) (Schneider 1977; Greenfield 1994a), and other males usually join the first male that calls. Playbacks to males of the cicada *Magicicada cassini* initiated sustained chorusing about two hours before the usual starting time (Weber and Huber., unpubl. data). In the katydid *Leptophyes punctatissima*, dual periods of signaling persisted even when playbacks of acoustic signals mimicking the responses of females were presented continuously. The *amount* of calling activity by males was greatly increased by the playbacks during the normal calling periods, which were also somewhat extended during the daylight hours by the acoustic stimulation (fig. 8.3C).

The timing and duration of most nocturnal chorusing is most parsimoniously explained by the presence of diurnal predators or temperature and energetic constraints that affect both signalers and receivers. The onset of signaling often occurs when light levels become too low for diurnal predators to visually locate either calling individuals or phonotactically orienting receivers. In the bladder cicada, which sings for about thirty minutes after dusk, males that sang earlier than the usual time of chorus onset and females that flew during this time period were often captured by birds (Doolan and Mac Nally 1981). Diurnal calling is nevertheless typical of many orthopterans and cicadas (Alexander and Moore 1958; Gogala and Riede 1995; Fischer et al. 1996). Belwood (1990) suggests that diurnal signaling requires special protection

2. Reviews or collections of papers concerned with the evolution of mating systems in insects and anurans, which will not be discussed formally in this monograph, are provided by Thornhill and Alcock (1983), Arnold and Duvall (1994), Sullivan et al. (1995), and Choe and Crespi (1997).

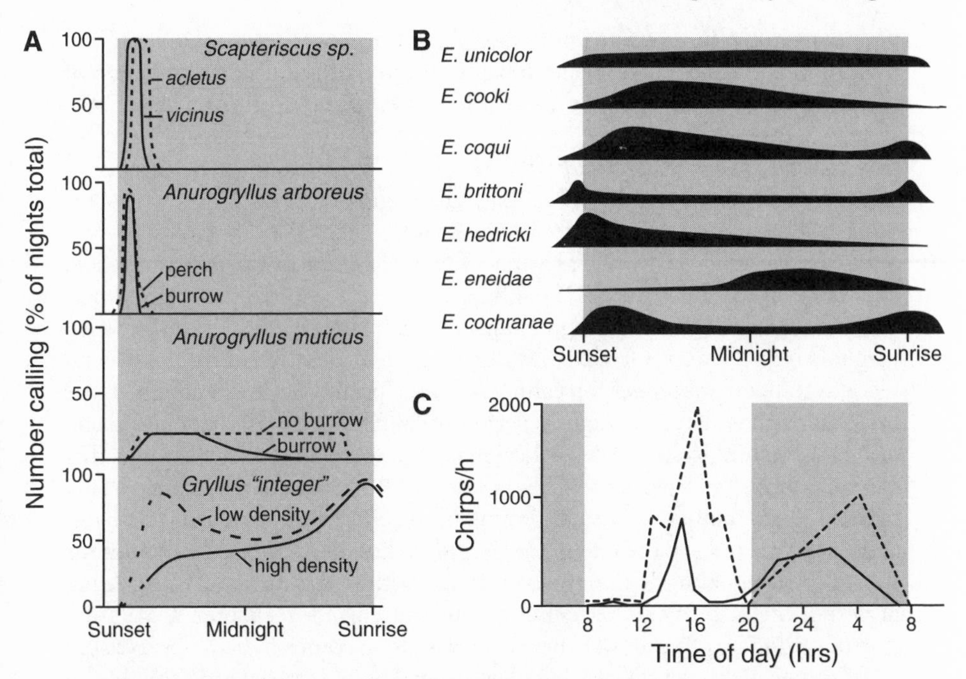

Figure 8.3. Temporal organization of signaling in insects and anurans. Gray areas indicate approximate periods of darkness. (A) Calling periods of mole crickets (*Scapteriscus* sp.), two short-tailed crickets (*Anurogryllus arboreus* and *A. muticus*), and a field cricket (*Gryllus "integer"*). From Walker 1983b, fig. 4. (B) Calling periods of seven species of leptodactylid frogs (genus *Eleutherodactylus*) in Puerto Rico. From Drewry and Rand 1983. (C) Singing activity of males of the katydid *Leptophyes punctatissima* measured over a twenty-four-hour period (*solid line*) and the activity when each call received a synthetic female reply (*dashed line*). From Robinson 1990, fig. 7. Notice that while the activity increased when males heard replies, the periods of time over which males called increased significantly only during the time from 1200 to about 1900 hours. Males also called for somewhat longer during the nocturnal period of calling (about 2200–0400 hrs).

(cryptic colors and morphology or defensive chemicals and aposematic coloration) or behaviors that minimize the risks of predation. For example, species that signal during the day often sing from dense cover, are active flyers, or both. As mentioned above, the periodical cicadas presumably can afford to sing diurnally because their huge numbers saturate predators.

In temperate species, signaling ceases as temperature falls below some critical level or as energetic reserves required for calling become low. Temperature not only affects the duration of calling but also the locomotory performance of male and female insects that fly to calling males (e.g., Walker 1983a). Shifts from nocturnal to diurnal calling have been documented in insects and anurans when nighttime temperatures are cool (e.g., Walker 1983a; Duellman and Trueb 1986; Loher 1989). Temperature can also be too high for continuous signaling during the day, and in tropical environments, the drop in temperature at dusk probably contributes to the initiation of signaling (Römer, pers. comm.). While energy limitations almost certainly determine the end of nightly signaling in some species (chapter 9), in barking treefrogs the end of chorusing is correlated with the cessation of female arrival (Murphy 1999).

Cyclical Unsynchronized Chorusing

Within their daily period(s) of calling, small aggregations of insects and anurans often show a cyclical pattern of signaling that has been termed *unsynchronized chorusing* (Otte 1977; Ewing 1989) or *unison bout singing* (Greenfield and Shaw 1983; Schwartz 1991). These terms apply to relatively gross patterns of chorusing that do not preclude fine-scale timing interactions between neighboring signalers. Groups of males in one part of a chorus call for several seconds to many minutes, stop calling for some period of time, and repeat the cycle many times over the nightly period of calling. The activity of one group of males seemingly stimulates other groups in a chain-reaction fashion that can spread over a larger area. The duration of bouts of calling varies, usually increasing gradually from the beginning of the chorusing period and then decreasing again toward the end of the period (Schneider 1977). The hypothesis that cyclical signaling is a tactic for increasing the total duration of signaling period in the face of energetic limitations is discussed in chapter 9. Stimulation of calling by the calls of other species is also a common field observation.

In some species, certain individuals or groups are consistently the first in the chorus or a small group to begin a bout of signaling (e.g., Alexander and Moore 1958; Whitney and Krebs 1975; Greenfield 1983; Ewing 1989). Such males might also remain leaders or produce especially potent calls that not only attract females but also other males to their vicinity. The existence of such males could be interpreted as support for the *hotshot model* of lek formation (Bradbury 1985). Whitney and Krebs (1975) showed that simulated leaders of the Pacific treefrog *Pseudacris regilla* attracted females more readily than did simulated followers. Pfennig et al. (2000), who studied spadefoot toads (*Scaphiopus* and *Spea*), also found some evidence that males might show selective phonotaxis to conspecific males producing calls with high call rates, which are particularly attractive to females. In the green treefrog *Hyla cinerea*, satellite males, which do not call but sit near callers and attempt to intercept females, also appear to associate preferentially with males producing especially attractive calls (see chapter 9). One way to test the hotshot hypothesis is to move leaders or males with attractive calls to a new position. The hypotheses would be supported if an aggregation of calling males forms around these individuals after they resume calling.

Masking Interference and Interspecific Inhibition

Whatever its causes and associated benefits, the formation of choruses increases mate competition and requires females to detect, recognize, and find individual signalers in the face of masking interference. Males that time signal production to make themselves heard against the generally high background noise levels within choruses should therefore increase their chances of attracting a female.

The potency of masking interference in natural choruses is exemplified by interspecific inhibition, in which signaling by one species suppresses, at a gross time scale, signaling by another species. For example, chorusing by males of the treefrog Hyla microcephala inhibits calling by Hyla ebraccata (Schwartz and Wells 1983a), and singing by the katydid Mygalopsis marki reduces signaling by another katydid, Hemisaga denticulata (Römer et al. 1989). The spectral properties of the songs of the last two species are similar, and in field experiments, removal of individuals of M. marki resulted in three- or fourfold increases in the numbers of singing males of Hemisaga denticulata (fig. 8.4). In the katydid Neoconocephalus spiza, a shift in the diel period of signaling occurs when other congeners with spectrally similar signals are present (Greenfield 1988). Playbacks of the signals of one of these species, Neoconocephalus affinis, not only affected male signaling patterns but also inhibited female phonotaxis (Greenfield 1993). Still another example of interspecific interference involves a cicada (species unspecified) whose chorusing suppressed territorial responses of the poison-dart frog Dendropbates pumilio to playbacks of a conspecific male (Páez et al. 1993).

Fine-Scale Patterns of Signal Timing among Conspecific Signalers

Fine-scale signal interactions occur when signaling by nearby individuals mutually influences the timing of signal production, even if the influence is asymmetrical. That is, after hearing its neighbor's signal, an individual might be inhibited from calling and delay its next signal, or it might produce its next signal at an earlier time than it would have in isolation. In this section, we describe how fine-scale timing of signals relative to nearby signals can sometimes represent a solution to masking interference. An example of fine-scale timing is provided by the Neotropical treefrog Hyla microcephala. Males not only delay the production of relatively long introductory notes, which have a species-typical pulse rate, to avoid overlap, but neighbors also show a remarkable ability to interleave their short secondary notes (fig. 8.5A; Schwartz 1993). Male frogs can also make fine-scale adjustments to changes in background noise produced by conspecific or heterospecific signalers. In the coqui treefrog (Eleutherodactylus coqui), for example, male calling is suppressed by tones. Zelick and Narins (1985) found that a male could produce calls within very short gaps (750 ms), even when they varied the duration of the masking tone in a pseudorandom fashion so the male could not anticipate when the next gap would occur (fig. 8.5B). Another explanation for signal-timing

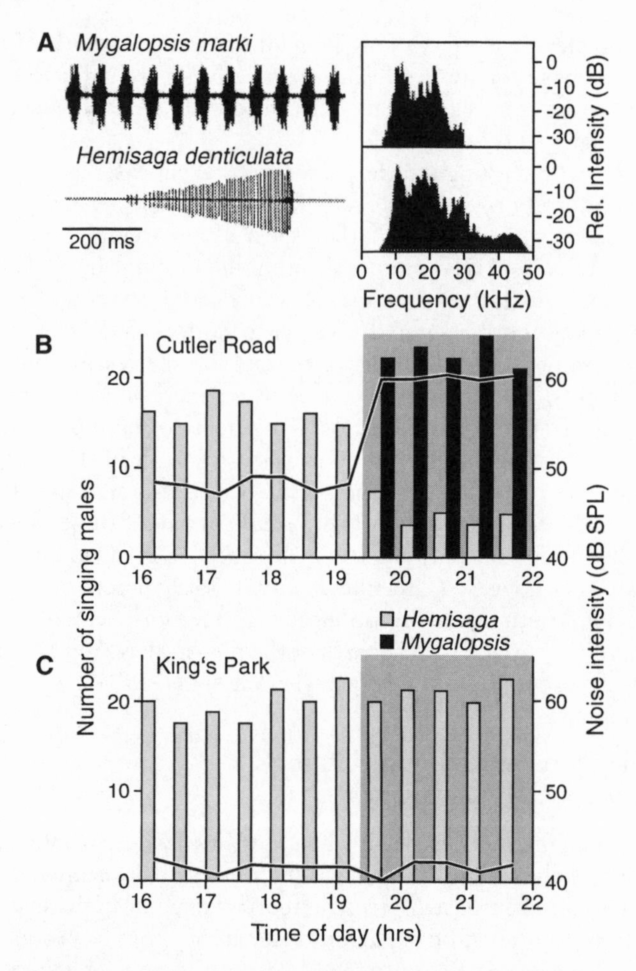

Figure 8.4. Inhibition of calling males of the katydid *Hemisaga denticulata* by the onset of singing by males of *Mygalopsis marki.* (A) Oscillograms and power spectra of the songs of two species. Notice that the spectra but not the temporal patterns of the songs are similar in the two species. (B) Gray bars show the numbers of calling males of *H. denticulata* between 1600 and 2200 h and the reduced number calling (2000 to 2200 h) at a site where large numbers of *M. marki* began calling just after sunset. Black bars show the number of calling males of *M. marki.* (C) Number of males of *H. denticulata* (*gray bars*) calling between 1600 and 2200 h at King's Park, where there were no singing males of *M. marki.* From Römer 1992, fig. 6.10; and Römer et al. 1989, fig. 5, table 1.

interactions is that these patterns are consequences of competition: males time their signals so that they are more likely to be detected or localized by females than are the signals of their neighbors, even if masking interference is increased somewhat (Greenfield 1994a, 2001). Finally, some forms of fine-scale timing might reduce the detection of individuals by predators (see below).

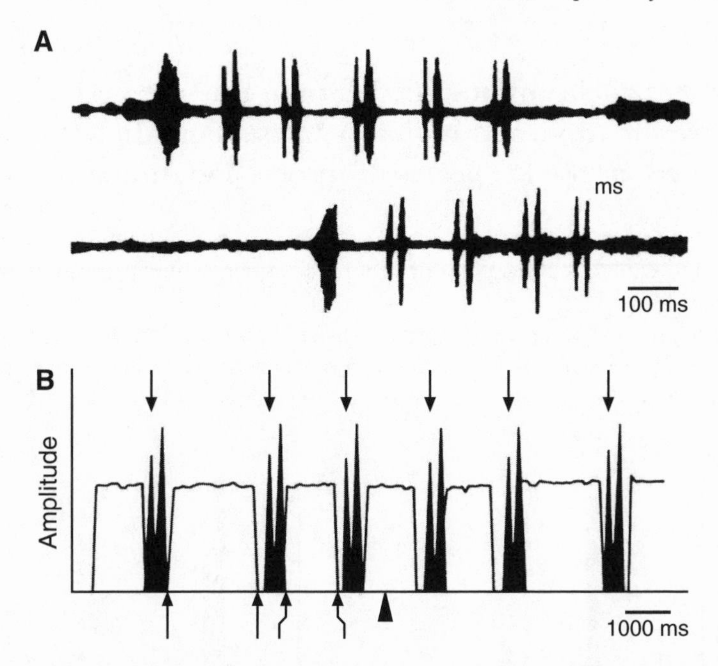

Figure 8.5. Fine-scale timing adjustments in two species of anurans. (A) Oscillograms of the calls of *Hyla microcephala* showing the interleaving of the introductory note and secondary notes of one male (*bottom trace*) with secondary notes of another male (*top trace*). From Schwartz 1993, fig. 7. (B) Diagram showing the production of two-note (co-qui) calls (*arrows*) by *Eleutherodactylus coqui* within 750 ms gaps in masking tones (*triangle*) presented pseudorandomly. From Narins and Zelick 1988, fig. 10.

Patterns of Signal Interaction: Synchrony and Alternation

Fine-scale timing interactions between two neighbors are often characterized by taking the period of repetitive signaling of one animal as a reference and then characterizing the second animal's timing in a relative fashion (e.g., fig. 8.6; review: Greenfield 1994a; see box 8.1). At one theoretical extreme (perfect synchrony), both individuals signal at the same time (relative phase of 0°). At the other extreme (perfect alternation), equal periods of silence are maintained between the signals of one male and those of a neighbor (relative phase of 180°). In fact, few species display long periods of anything approaching either perfect synchrony or alternation. Thus, synchrony can be considered as a pattern in which the signals of the two animals usually overlap in time, whereas alternation occurs when signals only occasionally overlap. Entrainment is a term for a pattern in which acoustic responses of one animal consistently follow those of another animal after a short delay (Greenfield 1994a; fig. 8.6A,B).

We emphasize that considerable variation exists within and between species in the form of signaling interactions. For example, some species show

Box 8.1. Synchrony and Alternation: Data Analysis and Presentation, and Inhibitory Resetting Models Proposed for Some Species of Insects and Anurans

The basic data consist of the times of occurrence of the signals of two neighboring individuals, or the times of occurrence of an artificial stimulus and an animal's signals.

In the anuran literature, researchers usually plot the distribution of signal delays of a focal male's calls relative to the occurrence of a neighbor's calls or an external stimulus. Fig. 1A shows the distribution of call delays of a male

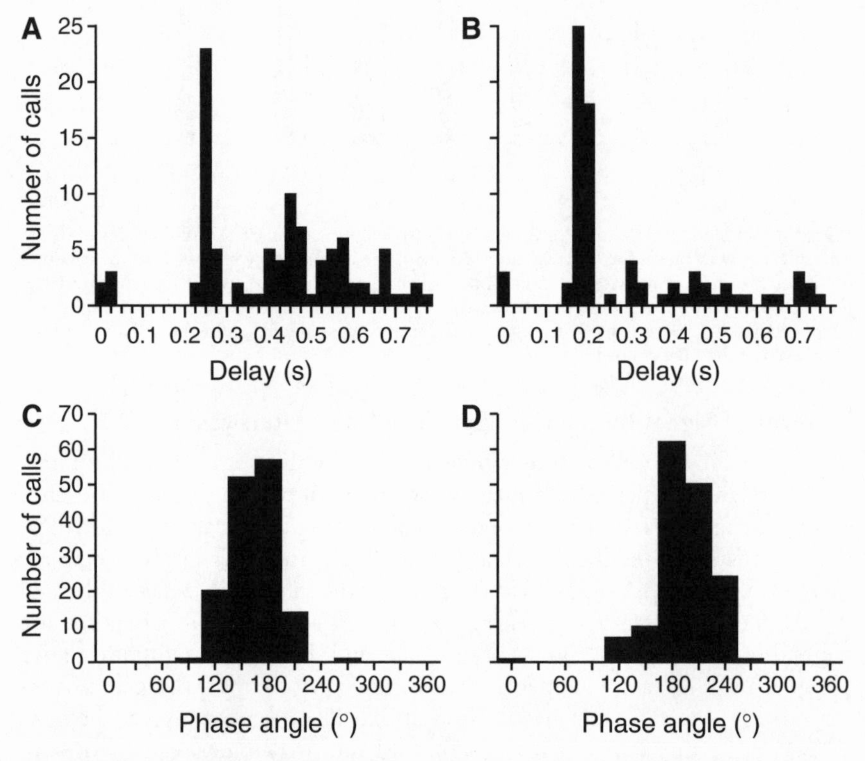

Box 8.1, Figure 1. Histograms show distributions of delays in call production by a male green treefrog relative to the beginning of a stimulus. (A) The delays are plotted relative to the calls of a neighbor. (B) The delays are plotted relative to the time of occurrence of synthetic advertisement calls presented at random intervals (see text). (C) Delays are converted to relative phase angles and are shown for a pair of interacting males; one male is treated as the focal individual and its neighbor as the reference in (C); these roles are reversed in the plot shown in (D).

(box 8.1 continued)

green treefrog relative to the calls of its nearest neighbor. Time zero represents the time when the neighbor's calls began. Notice that the focal male produced very few calls during time period from zero to about 0.22 s, indicating that the neighbor's calls, which had durations ranging from about 0.16–0.20 s, inhibited calling by the target male. Most calls overlapping the external stimulus occurred in the first 0.40 s and had probably been initiated before the neighbor's call was heard. Notice, too, that males gave a disproportionally large number of calls after about 0.20 s compared with other time periods falling within the intercall interval, which is usually about 0.6–0.8 s in this species. This phenomenon is often termed *entrainment.* The distribution of call delays would be much more uniform if the target male were calling independently with about the same period, or if it simply called at any time during the silent intervals between the calls of its neighbor. In figure 1B we show the distribution of delays of a male green treefrog presented with synthetic mating calls. These signals had a mean period equal to that of the local population, but with randomly occurring periods ranging from 0.18 s to 1.6 s, corresponding to the smallest interval permitting nonoverlap to an interval about two times the normal period, respectively (Klump and Gerhardt 1992; see Lemon and Struger 1980 for comparable results for the spring peeper *Pseudacris crucifer*). Notice that the distribution of delays was very similar to that observed when a male was interacting with another frog. Presentation of stimuli at randomized intervals provides a particularly efficient test because of the large variance in stimulus intervals; the calls of the focal animal would much more often overlap with the stimulus by chance if the frog were unaffected by the stimulus. Moreover, these results indicate that the timing of responses does not depend on any assessment or estimate of the average period of the stimulus; rather, the frog adjusts its timing on a period-by-period basis.

In the insect literature, call (stimulus) delays and response delays are used to generate phase-response plots. The response delay is computed by subtracting the period of the target male before stimulation from the period following the external stimulus (T'–T). Both of these times are converted to phase angles by dividing by the unstimulated period (T) and multiplying by 360°. Call delays expressed as phase delays are shown in fig. 1B and 1C for a pair of vocally interacting males of the green treefrog. The strength of the interaction or coupling can be expressed by the length of the mean vector in the phase diagram (r), and the amount of lead or lag in the call cycle by the phase angle (ϕ). In this example, males interacted strongly, and the mean phase angles were close to the 180° angle that is expected if alternation were near perfect. Neither histograms of call delays nor phase-response plots reveal asymmetries

(box 8.1 continued)

in the relative influence of interacting pairs of neighbors as does the method of Wickler and Seibt (1974), which is illustrated in figure 8.6B (see text).

Neural pacemakers are hypothesized to underlie both synchrony and alternation (fig. 2A); each could be characterized by a species- and individually-specific mean period and variance in solo signaling. In this unstimulated situation, when the level of activation of a hypothetical saw-toothed pacemaker exceeds some threshold value, a signal will be produced with an *effector delay* (t) that depends on the time required for the neuromuscular and structural elements of the sound-producing apparatus to produce a signal after neural activation. The level of activation of the pacemaker falls after exceeding threshold and before the signal is actually produced, and then it gradually ascends again to repeat the cycle, generating the solo period (T) of calling.

In the simplest phase-delay model, an external stimulus resets the pacemaker, the level of excitation of which falls back to the basal level and then

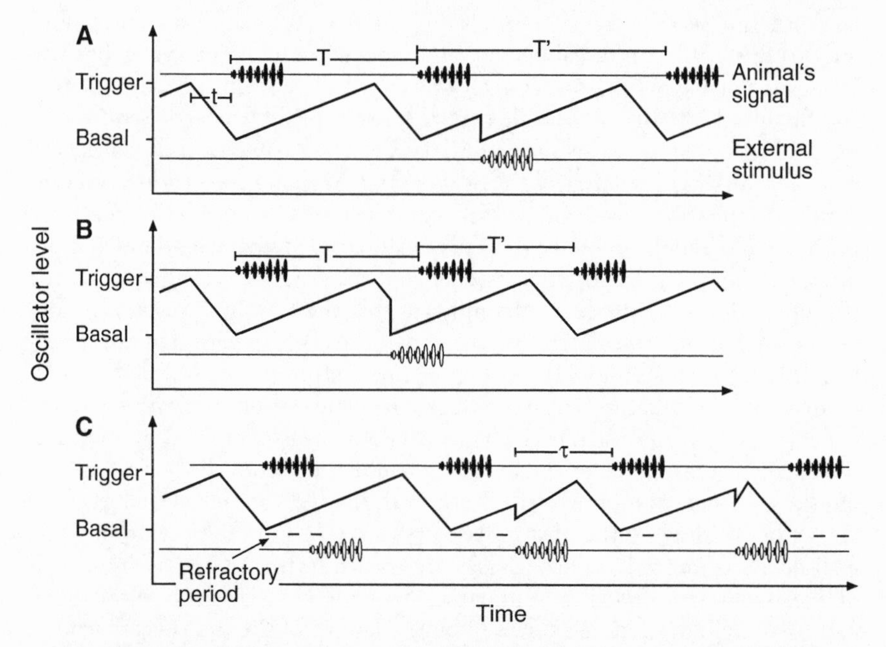

Box 8.1, Figure 2. Diagrams showing phase-delay models of varying complexity. (A) Simple phase-delay model, showing a phase delay (T' = black signals) caused by an external stimulus (white signal); t = effector delay; T = free-running period. (B) Simple phase-delay model, showing a phase advance (T') caused by an external stimulus. (C) Inhibitory resetting model with refractory periods (indicated by dashed lines); τ = rebound interval. See the text for further explanation.

(box 8.1 continued)

begins again to ascend toward threshold. Whether a phase delay or advance of the animal's next signal occurs depends on when the external stimulus occurs relative to the previous signal. If the stimulus occurs soon after the animal's previous signal (fig. 2A), there is a phase delay (T'), which is approximately equal to the unstimulated period (T) plus the time between the animal's previous signal and the external stimulus. If the stimulus occurs late in the natural period and just before the animal would normally produce a signal (fig. 2B), this signal occurs at the expected time, but the next signal occurs earlier than usual (a phase advance: T'), at a time approximately equal to the normal period (T) minus the time between the external stimulus and the animal's previous signal. In this second case, the external signal does not affect production of this next signal because the pacemaker has already triggered its production. However, the second external stimulus resets the pacemaker to the basal level sooner than it would have achieved the basal level in solo signaling (modified from Greenfield 1994a, fig. 1).

Figure 2C shows a modification of the basic model to account for signal interactions in some anurans and insects. First, there is a *refractory period* (Narins 1982), during which the occurrence of an external stimulus does not affect the pacemaker (dashed lines). Second, the external stimulus does not reset the pacemaker to its basal level, and there is a more rapid ascent toward threshold than in the basic phase-delay model. This results in the relatively rapid production of a signal after external stimulation (= *rebound interval*, τ) and the phenomenon of entrainment that is commonly observed in anurans and some insects.

both synchrony and alternation (e.g., Jones 1966a; Moore et al. 1989; Sismondo 1990).[3] In other species, individuals seemingly produce signals at random with respect to their nearest neighbor (e.g., Bailey and Simmons 1991). Indeed, many species of insects and anurans produce long, continuous trills, which inevitably overlap extensively with those of nearest neighbors. In these species there is little or no indication of alternation or synchronization (e.g., toads *Bufo americanus* and *B. terrestris:* Gerhardt, pers. obs.; katydids *Tettigonia viridissima, T. cantans:* Schul 1994; and *Mygalopsis marki:* Bailey, pers. comm.). In still other species, alternation of calling is crude at best, and interacting males instead tend to match the number of notes produced by their neighbor (e.g., Arak 1983b; Gerhardt et al. 2000; chapter 9). Finally, even in species

3. Here it is important to assess the pattern of two interacting neighbors. For example, individual A could be alternating with individual B, and B, with individual C. A comparison of A and C might show that these two individuals produce many synchronously occurring signals.

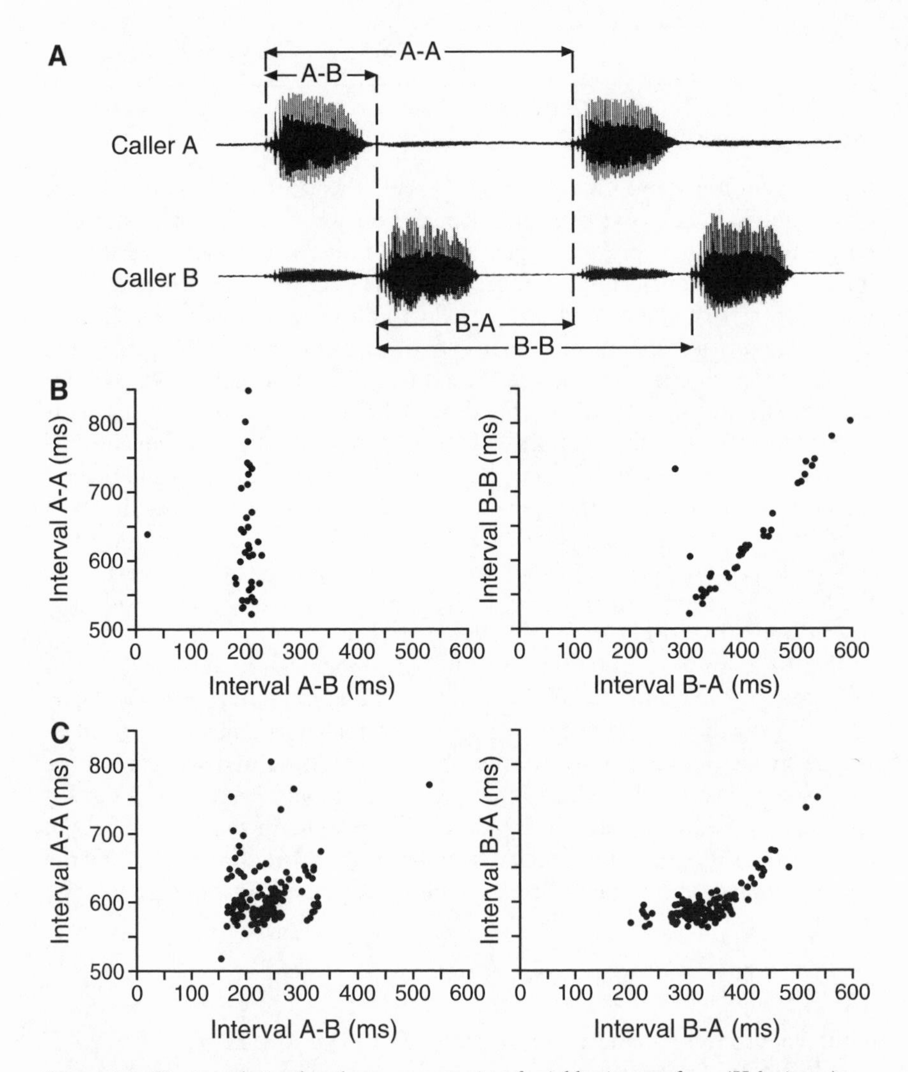

Figure 8.6. Timing relationships between two pairs of neighboring treefrogs (*Hyla cinerea*). (A) Oscillograms in which caller A leads caller B; intervals A-A and B-B are the call periods of the two individuals, intervals A-B and B-A are call delays. (From Klump and Gerhardt 1992, fig. 1). (B) The data plotted are each from one pair of neighbors. Each call period of one caller (A-A or B-B) is plotted against the call delays (A-B or B-A). The timing of one male's calls (the "leader," *left panel*) exerts a strong effect on the timing of the other male's calls (the "lagger," *right panel*). The scatter of points for the leader shows that the delay of its neighbor's calls (interval A-B) was consistently about 200 ms, even though the focal male's call period (interval A-A) varied over a wide range. When the lagging neighbor's call period (interval B-B, *right panel*) was plotted against the delay between its call and that of the leader, however, there was a clear dependency. In other words, the call period of the lagger depended strongly on the timing of the other male's calls. (C) The plots show only a weak, mutual influence of call timing by the two frogs, in which leading and lagging roles switched frequently. In the first male (*left panel*) its call period (A-A) was most often between about 550 and 650 ms regardless of the neighbor's call delay (interval A-B). In the second male (*right panel*), its call period showed a clear dependency of the call delays in the range of 400–550 ms, but not when delays were shorter. Wickler and Seibt (1974) developed this method of assessing asymmetry in the signaling interactions of neighbors. From Klump and Gerhardt 1992, fig. 5.

where alternation is well developed, some individuals seemingly ignore the calls of neighbors, whereas others produce calls that are rigidly time-locked to those of a neighbor, at least over several minutes of calling (e.g., fig. 8.6).

Models of Synchrony and Alternation

Several mechanisms for establishing and maintaining synchrony, alternation, or both have been proposed. In some species of fireflies (Buck 1988), individuals may reset their solo rate during signaling interactions. In the katydids *Mecopoda elongata* and *Pterophylla camelliflora*, for example, males change their calling period in response to external stimuli and maintain the new period (Shaw 1968; Sismondo 1990; Steiner and Römer, pers. comm). In *M. elongata*, males may reduce their solo calling rate and alternate with a slower external stimulus, or maintain a stable "follower" role in response to a faster external stimulus. Males also maintain an elevated rate for some time after stimulation by a faster external stimulus has ceased (Römer, pers. comm.).

In most acoustic insects, however, synchrony or alternation is probably achieved on a period-by-period basis by one of two mechanisms (Walker 1969). The first mechanism involves a rapid acoustic response by an animal to the onset of a signal produced by its neighbor. One example is the irregular and near-synchronous calling of the tropical hylid *Smilisca sila* (fig. 8.7; Ryan 1986; Greenfield 1994a). Neighboring males of the canyon treefrog *Hyla arenicolor* also show extensive overlap of their long trills because one male calls with a short delay after the start of his neighbor's call (Marshall and Gerhardt, unpubl. data). Long bouts of synchrony occurring under certain conditions in the periodical cicada *Magicicada cassini* (Alexander and Moore 1958) might be a third example, but the possibility of mutual adjustments of solo rhythms has not been ruled out.

The second mechanism of period-by-period timing is exemplified by Walker's (1969) work with the snowy tree cricket *Oecanthus fultoni*, which show relatively fine synchrony (fig. 8.8A). Playback experiments indicate that the timing of a synchronizing individual's chirps depends on when an external signal occurs *relative to the individual's previous signal* (fig. 8.8B). Two such timing changes occur in chirp production by the snowy tree cricket. First, after hearing an external (neighbor's) signal during the latter part of a chirp or shortly afterward, the cricket delays its next chirp and lengthens its natural period (*phase delay*). Second, after hearing an external signal in the latter part of its previous chirp interval, the cricket produces its next chirp at the expected time, but the following chirp is produced earlier than usual (*phase advance*), thus shortening its chirp interval. In either case, the result is that the next chirp occurs at about the same time as the next external stimulus, assuming that the period of the external stimulus remains constant.

Male snowy tree crickets achieve and maintain a high degree of synchrony with an artificial stimulus when the repetition rate of the stimulus matches closely that of the individual's solo chirp rate (Walker 1969). Alternation occurs

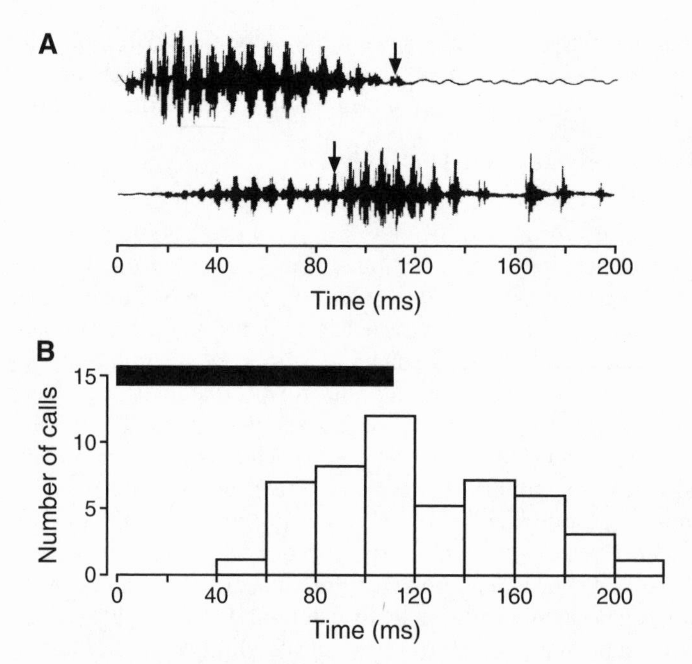

Figure 8.7. Synchronization of calling by the Neotropical treefrog *Smilisca sila.* (A) Oscillograms made from a stereophonic recording showing an example of the response of a frog (*bottom trace*) to the playback of the call of another frog (*top trace*). The end of the stimulus in the top trace and the onset of the male's call in the bottom trace are indicated by arrows. (B) Histogram summarizing measurements of the time of occurrence, relative to the stimulus (*horizontal bar*), of fifty responses by a sample of males. From Ryan 1986, figs. 1, 2.

when the stimulus rate is higher than the solo rate. Above a certain rate, the subject is either inhibited or synchronizes to every other chirp of the artificial stimulus (fig. 8.8D). Similar patterns of response have been observed in an anuran (*Leptodactylus albilabris*), which responds in a phase-locked manner to every fourth or sixth stimulus, when driven by an external stimulus repeated at five to six times its natural solo rate (Moore et al. 1989). In the katydid *Neoconocephalus spiza*, however, males stop calling when the repetition rate of an external stimulus is only modestly higher than that of the target insect (e.g., Greenfield and Rozien 1993).

These kinds of observations in acoustic insects and fireflies led several researchers to consider two general models—*phase delay* and *inhibitory resetting*—that could explain both synchrony and alternation (e.g., Greenfield 1994a,b; Greenfield et al. 1997; Hanson 1978; Jones 1966a,b; Loftus-Hills 1974; Moore et al. 1989; Sismondo 1990). These models assume that there is an underlying oscillator or pacemaker that initiates sound production once it reaches a threshold level of excitation (box 8.1). There is a short (effector) delay after threshold is reached, corresponding to the time required to activate the appropriate muscles. After signal initiation, the excitation level of the oscillator

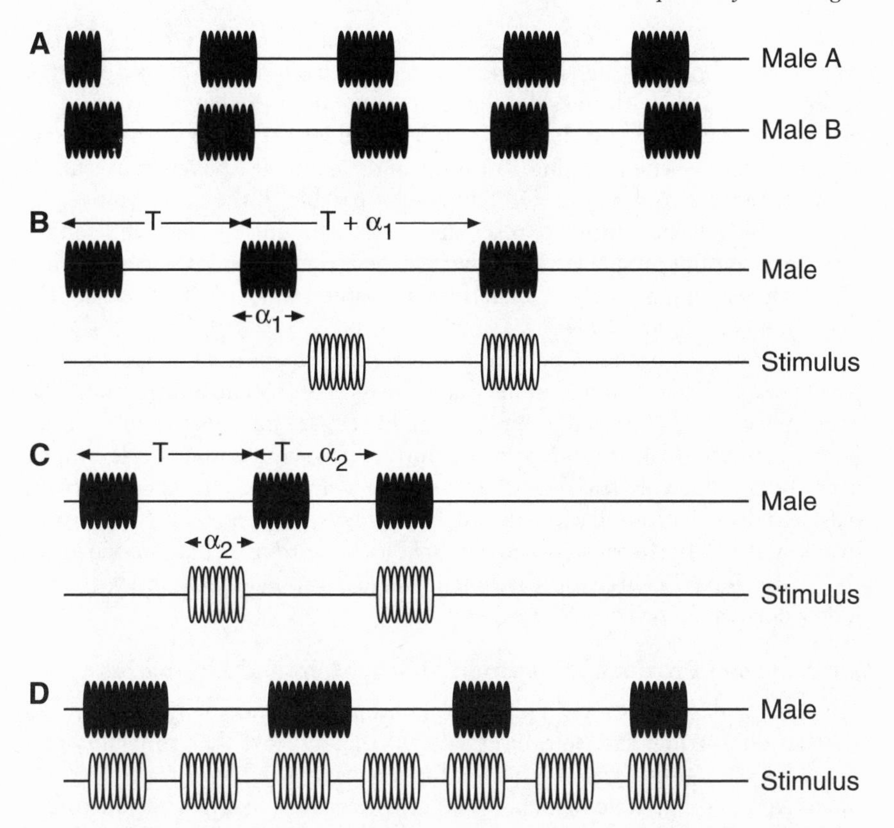

Figure 8.8. Synchronization of signaling in the snowy tree cricket *Oecanthus fultoni.* Diagrams show the timing of cricket chirps (*black*) and artificial chirps (*white*). (A) Songs of two synchronizing neighbors; notice that the same cricket does not always lead. (B) Hypothetical responses of a male to playbacks of artificial chirps. The first two chirps of the male (*top trace*) represent its unstimulated period; after the first external stimulus (*bottom trace*), the cricket shows a phase delay equal to its unstimulated period (T) plus α_1, the interval between the beginning of the cricket's chirp and the stimulus. (C) The first two chirps (*top trace*) again show the unstimulated period of the cricket. When the external stimulus occurs late in this period, the cricket shows a phase advance equal to its unstimulated period (T) minus α_2. (D) Response of a cricket stimulated at a much higher than normal rate; see the text for details. Modified from Walker 1969, fig. 1.

returns to a basal level of excitation. The repetition of this cycle produces an individual's solo pattern of timing.

In the basic phase-delay model, an external acoustic stimulus merely resets the timing of the pacemaker; in the inhibitory resetting model, the external stimulus both inhibits (for at least the duration of the stimulus) and resets the oscillator. In some species, an external stimulus has little or no effect on the pacemaker if the stimulus occurs during the time when the animal is producing a signal and for a short time afterward, thus suggesting a kind of refractory period (Narins 1982). These models suggest that the times required for the pacemaker to trigger the signal and to recover from an inhibitory external stimulus, together with the solo-signaling rate, might predict whether a species shows synchrony or alternation (Greenfield 1994a). For example, synchrony is likely when singing rates are high (e.g., the snowy tree cricket, fig. 8.8) because the song period will be comparable to the time required to process the external stimulus, reset the pacemaker, and produce the sound. Comparisons among species show that synchronization tends to occur in species with fast singing rates; alternation is common in species with relatively slow rates (Greenfield 1994a).

Unfortunately, no neurophysiological studies are available of the putative pacemakers in insects and anurans that show synchrony and alternation. The estimates of 50–200 ms cited by Greenfield (1994a) for effector delay times from pacemaker firing to signal production are perhaps too long. For example, even though duetting katydids do not sustain a rhythmic interchange of signals, females can respond acoustically to a male signal in less than 20 ms (Robinson et al. 1990). Because it will also take time to transmit the information that a signal has occurred to the female's sound-producing circuitry, the effector delay must be even shorter.

Ultimate and Proximate Causation of Synchrony and Alternation

Adaptive explanations for the evolution and maintenance of synchrony and alternation usually focus on timing patterns that increase the probability of a signaler's attracting a mate. Whether the tactics adopted by signalers are cooperative or competitive, female response patterns are likely to be a main determinant of the particular pattern of timing interactions adopted by males of a given species. Female requirements, in turn, are constrained by the sensory mechanisms used to perform these tasks, the kinds of signals (e.g., duration, solo repetition rate, presence or absence of fine-temporal structure) produced by conspecific males, the spatial separation of signalers, and the general acoustic environment. As we have already mentioned, however, the male's sensory and sound-production system, the kinds of signals it produces and environmental factors (spacing, in particular) also influence the timing rules for interacting neighbors in ways that are independent of female requirements. Nevertheless, female preferences might not always be constrained by sensory mechanisms. Females could, for example, prefer leading or lagging signals, not because they are more detectable or locatable, but rather because reliable correlations exist between a male's pattern of interaction (e.g., sustained leadership) and his relative fitness (e.g., Greenfield 1994a).

Cooperative and competitive explanations. Common cooperative explanations for synchrony include (1) preservation and amplification of species-specific temporal patterns; (2) facilitation of the detection of female acoustic replies; and (3) reduction in the detectability or locatability of signalers by predators. We agree with Greenfield (1994a), who concludes that there is little convincing evidence for the first two hypotheses. The sporadic and near-synchronous calling within choruses of the Neotropical hylid *Smilisca sila* might help calling males avoid bat predators that use passive listening to find their prey. Tuttle and Ryan (1982) showed that the bats were more attracted to asynchronous than to synchronous playbacks of calls.

At one level of analysis, alternation can be interpreted as cooperative, especially if females require nonoverlapping signals to assess the temporal fine structure of a signal. Indeed, males failing to alternate would reduce their own chances of mate attraction as well as those of their neighbor, a tactic that would be considered spiteful (Wiley 1983). This functional explanation of alternation is supported by experiments with anurans (Schwartz 1987a). Schwartz used a four-speaker playback experiment to test females of three species of treefrogs. Females were given a choice of two conspecific calls that were presented so that calls did not overlap and another pair of calls that overlapped. Females of *Hyla microcephala* and *H. versicolor*, for which fine-temporal patterns are known to be important for mate choice, preferred nonoverlapping calls to overlapping calls, in which masking interference obscured the temporal fine structure. Females of *Pseudacris crucifer*, whose males produce tonal calls lacking fine-temporal information, showed no preference. Evidence that alternation can facilitate sound localization is available for the grasshopper *Ligurotettix planum* (Minckley and Greenfield 1995).

Synchrony and alternation can also be interpreted as ongoing competitive interactions if females prefer leading or lagging signals. In his pioneering work with synchrony in snowy tree crickets, for example, Walker (1969) emphasized that leading and lagging roles switched frequently between interacting individuals (fig. 8.8A), and Greenfield and Rozien (1993) document the same pattern in the katydid *Neoconocephalus spiza*. Experimental studies of *N. spiza* and several other species of insects that show synchronization or alternation demonstrate that females have strong preferences for the leading signal of a pair (Greenfield et al. 1997). Preferences for leading signals also occur in several species of anurans that show alternation (Grafe 1996b). Two assumptions of the competitive hypothesis are that males respond only to a small set of nearby (or loud) neighbors, and that females choose among the same individuals.

Research with several species of anurans and insects supports these assumptions. In two kinds of anurans (*Eleutherodactylus coqui:* Narins 1992; *Hyla microcephala:* Schwartz 1993; *Physalaemus pustulosus:* Greenfield and Rand 2000) and a grasshopper (*Ligurotettix coquilletti:* Minckley et al. 1995), males better avoided overlap with near neighbors than with more distant males. Thus, in species interacting preferentially with nearest neighbors, the signals of those neighbors might have been the only signals that were reliably detectable above the chorus background. Males might also have more complex decision rules, as shown for *P. pustulosus*, where the number of males affecting a focal individual depends on the relative intensities of the calls of its neighbors (Greenfield and Rand 2000). In green treefrogs (*Hyla cinerea*), females reliably detected the calls of individual males only when the sound pressure level of these signals was at least as great as that of background chorus noise (Klump and Gerhardt 1988b). In a typical chorus, this signal-to-noise ratio exists only at short distances from calling males, which usually space themselves at distances of 0.5–1 m. Thus, a female at any single location could probably detect the calls of only a few nearby males that are likely to be interacting vocally.

The competitive hypothesis predicts that if interacting neighbors have different solo rates, the individual with the faster rate should never decrease its rate to accommodate a slower neighbor. Such reductions could be interpreted as cooperative unless both individuals produced signals at a slower rate, as predicted by inhibitory resetting models, or unless the signals of the individual that slowed consistently led those of the other male. The latter pattern has been observed in the katydid *Mecopoda elongata* (Römer, pers. comm.). Greenfield (1994a) asserts that reductions in solo rate or variance have never been demonstrated adequately in any insect species that shows sustained synchrony or alternation. This hypothesis deserves further testing using experimental playbacks in conjunction with strict operational definitions for "sustained" synchrony and alternation. Indeed, among species that appear to use inhibitory resetting mechanisms, one striking counterexample has been reported. In the katydid *Neoconocephalus nebrascensis*, instead of competing for leadership, some individuals delay signal production and decrease signal duration so that their signals are completely overlapped by those of its neighbor (Meixner and Shaw 1986). Greenfield (1994a) speculates that the delaying insect might be attempting to avoid detection by the leader, thus adopting a kind of satellite strategy (see chapter 9). Males of duetting species of katydids are also likely to have different rules of interaction because females may require some minimum time delay before responding to a given signal (Tauber 2001).

Females of several species of anurans either do not prefer leading to lagging signals (*Centrolenella granulosa:* Ibáñez 1993; *Hyla versicolor:* Klump and Gerhardt 1992) or actually prefer lagging signals to leading ones (*Hyla ebraccata:* Wells and Schwartz 1984a; *Kassina fusca:* Grafe 1999). In the first three species, neighbors tend to alternate calls, whereas in *Kassina*, males deliberately overlap the calls of neighbors as in *Smilisca* and *Hyla arenicolor*. In *Kassina*, females switch their preference to leading signals, however, if the degree of overlap is more than the usual 20%. The mechanisms and selective consequences of these exceptional patterns deserve further study.

Mechanisms underlying preferences for leading signals. Preferences for leading signals have often been considered to be examples of the *precedence effect* (Greenfield 1994a,b; Greenfield et al. 1997). This binaural phenomenon was originally described in humans as a mechanism for suppressing responses to echoes that could interfere with localization of a sound source (Wallach et al.

1949). In our view, the precedence effect probably does not explain many of the preferences for leading signals that have been described in insects and anurans. A defining characteristic of the phenomenon in humans is that the two signals are perceived as a single auditory event (or image), whose spatial location is mainly determined by the directional cues carried by the first sound (Zurek 1987). By contrast, preferences for leading signals in insects and anurans often occur when there is no overlap (e.g., insects: *Ephippiger ephippiger, Ligurotettix coquilletti, L. planum;* anurans: *Hyla cinerea, Hyperolius marmoratus, Physalaemus pustulosus;* Dyson and Passmore 1988a,b; Klump and Gerhardt 1992; Greenfield et al. 1997). Although the precedence effect can occur under some conditions of non-overlap of signals, whether insects and anurans perceive single acoustic images at locations depending on the delay is an open question.

Preferences for leading signals could also be the result of masking, where the neural response to the lagging signal is reduced (e.g., Wyttenbach and Hoy 1993; Römer et al. 1997). Such masking should be stronger when the signals overlap but could still occur if the signals are separated slightly in time (= forward masking). Increasing the intensity of the lagging stimulus by a modest amount (4–6 dB) can abolish the preference for the leading stimulus (Klump and Gerhardt 1992; Snedden and Greenfield 1998). We agree with Grafe (1996b), who concludes that the proper experiments to distinguish between these two mechanisms (masking vs. the precedence effect as described in humans) have yet to be done with insects or anurans. Moreover, both effects might also be operating simultaneously to some extent. Because the term *precedence effect* has been so widely used as a convenient shorthand label for preferences for leading signals, its use in this context is likely to persist regardless of the underlying mechanism.

Spacing within Aggregations

Signaling males of most species of insects and anurans usually maintain some minimum distance between nearest neighbors, whether or not males defend physical resources (Alexander 1975; Otte 1977; Wells 1977, 1988; Greenfield and Shelly 1990). Measurements of both sound pressure level of the signals of nearest neighbors and intermale distances show that sound pressure level varies less than distance, a result interpreted as evidence that call amplitude mediates spacing (e.g., Brenowitz et al. 1984; Römer and Bailey 1986; Wilczynski and Brenowitz 1988). The minimum distance between signaling animals usually decreases with increasing chorus density, but at extremely high densities, the overall spatial pattern may change from uniform to random (e.g., Doolan 1981; Cade and Otte 1982; Robertson 1984; Römer and Bailey 1986; Arak et al. 1990; Dyson and Passmore 1992). Many individuals cease calling at high densities and adopt other mate-acquisition tactics (Wells 1977; see chapter 9).

The "rules" of spacing, just as those of signal interactions, vary among species and even between populations of the same species. Nearly all experimental studies show that at typical distances between neighbors, their signals are well above auditory threshold. Spacing in the Australian katydid *Mygalopsis marki* is an especially good example (reviews: Dadour and Bailey 1985; Römer 1992). Males aggregate in patches of favorable habitat but do not form dense choruses as in many anurans and some other kinds of insects. The regular spacing distance corresponds to an average sound pressure level of about 65 dB, that is, the sound pressure level of a neighbor's song at a given male's position. Differences in the sound pressure level of male songs (correlated with male size) or habitat effects (vegetation characteristics that result in differences in *excess attenuation;* chapters 2 and 11; appendix 3) explained much of the variance in male spacing within and between populations (Römer and Bailey 1986). For example, large males that produced loud songs in one population were spaced at greater distances than were small males that sang less intense songs from dense vegetation in another population (fig. 8.9). A portable neurophysiological setup was used in the field to estimate directly the maximum hearing distance, which was the distance at which playbacks of conspecific song reliably elicited responses in an omega neuron. Maximum hearing distances were, on average, more than twice the typical distance at which males spaced (fig. 8.9; Römer and Bailey 1986; see also Bailey et al. 1993 for comparable data from another insect). Additional neurophysiological and anatomical studies of this species, and the European katydid *Tettigonia viridissima*, support the hypothesis that the profile of auditory activity within the auditory neuropile might encode information about the nearest neighbor's call amplitude and distance (chapter 6; Rheinlaender and Römer 1986; Römer 1987).

Most field studies of anurans also show that spacing occurs at distances at which the intensity of the calls of neighboring males is well above their auditory threshold (e.g., Wilczynski and Brenowitz 1988; Brenowitz 1989; Gerhardt et al. 1989; Dyson and Passmore 1992). Because auditory neurons have elevated thresholds in the presence of continuous noise (chapter 5; Narins 1992a), the signals of neighbors in dense choruses might be only modestly above the background noise and the individual's threshold under these conditions. Aggressive thresholds are positively correlated with the intensity of the nearest neighbor's calls; however, this elevation in threshold is probably not merely a function of sensory adaptation (chapter 9; Rose and Brenowitz 1991; Brenowitz and Rose 1997).

Consequences of Spacing

Spacing of signaling males has consequences for signaling interactions between males and for the effective attraction of females. At one extreme, when the mean absolute distances between signalers are as great as in *Mygalopsis marki*, masking interference should be minimal, and there seems to be little evidence for any kind of fine-scale timing interactions between neighbors

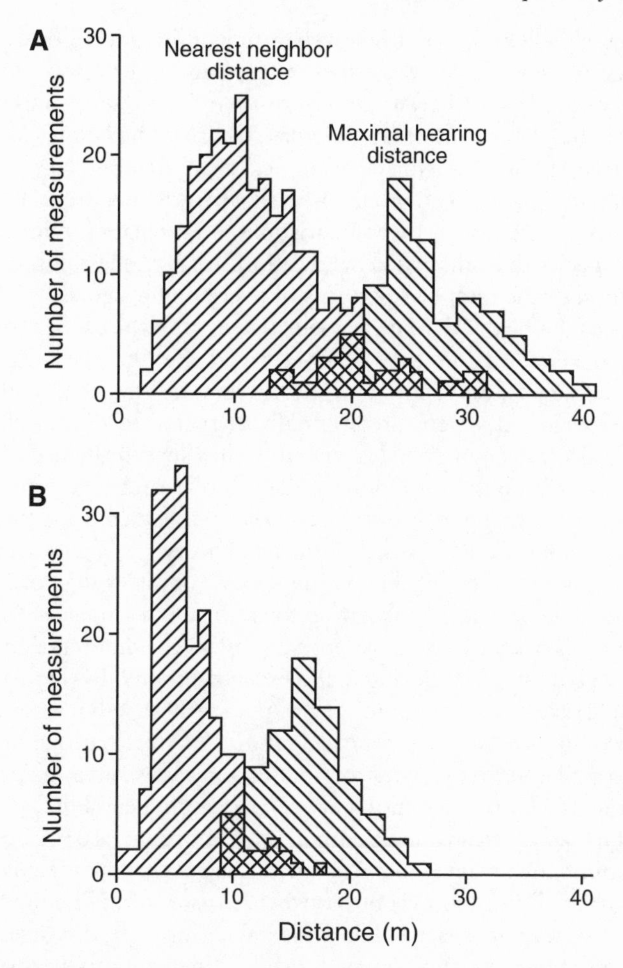

Figure 8.9. Distributions of intermale distances and maximum hearing distances in two populations of singing males of *Mygalopsis marki.* The use of a portable neurophysiological preparation made it possible to estimate directly the maximum hearing distance by measuring the distance from a singing male at which the threshold of the omega interneuron was just exceeded. The average size of males was larger in site I (A), and hence the amplitude of their calls was higher than that of the calls of the smaller males in site II (B). From Römer and Bailey 1986, fig. 1.

(Bailey, pers. comm.) At the other extreme, calling male katydids *Kawanaphila nartee* are so densely packed on food plants, that high levels of masking interference cannot be avoided and timing of signals among neighbors is random (Bailey and Simmons 1991). Whether individuals show consistent patterns of interaction at intermediate distances varies from species to species, but the determining factors are unknown.

Spatial separation improves female discrimination at close range in a number of insects and anurans. As discussed in chapter 7, this so-called cocktail party effect is mediated in part by mechanisms of directional hearing. Essentially, different acoustic patterns that overlap are better resolved when they arrive at the ears from different directions than from a single direction. The highly directional hearing of the katydid *Tettigonia viridissima* discussed in chapter 7 (see fig. 7.12) is a prime example. Recall that two completely overlapping temporal patterns could be resolved when their source locations were on either side of the insect. By comparison, spatial separation in anurans and one insect species to be discussed below affords only modest improvements in the resolution of the temporal patterns of overlapping signals.

Females of the bladder cicada are selective only to the carrier frequency of a stimulus at a distance, but they become selective to the fine-temporal properties of the song at close range (Doolan and Young 1989). Although male and female bladder cicadas were preferentially attracted to bushes where other males were already singing, males within bushes spaced themselves at an average distance of about 1 m (Doolan 1981; Doolan and Mac Nally 1981). As mentioned above, highly dense artificial choruses attracted a greater number of females per male than choruses of normal density, but males in extra-dense choruses had about one-half the mating success of males in choruses of normal density. The probable explanation for reduced mating success is that females do not signal their sexual receptivity unless the difference in the sound pressure level of two neighboring males exceeds about 9 dB (Doolan 1981). If singing males are spaced at about 50 cm, the smallest stable separation, then a female would have to land within about 10–20 cm from one of the males in order to signal her receptivity to him. This response area increases to 20–40 cm at the usual 1 m separation between signaling neighbors.

As in the bladder cicada, male anurans calling from close proximity within a chorus might not attract as many females as males that maintain larger intermale distances. Telford (1985) observed that males of the painted reed frog *Hyperolius marmoratus* spaced themselves at an average distance of 50 cm. When given choices of three identical calls, differing only in the degree to which the sound sources were separated, females chose the most widely spaced sound source, possibly because that source was easier to detect or localize. Dyson and Passmore (1992) offered females of the same species choices among nonoverlapping synthetic calls played back from four speakers. If one call had a lower-than-average frequency, which is normally preferred in two-speaker tests, then females consistently chose the speaker emitting this stimulus only when the four speakers were separated by at least 1 m. Taken together, these experiments show that discrimination of the calls of different individuals might be enhanced by spacing, at least at some chorus densities.

Two experiments show directly that spacing of sound sources provides modest improvements in the detection of the fine-temporal structure of two overlapping signals (Pollack 1986; Schwartz and Gerhardt 1995). Females of the gray treefrog *H. versicolor*, which are very sensitive to differences in fine-scale temporal properties, were placed between two pairs of adjacent speakers.

Each pair of speakers played back calls that overlapped so that the temporal structure of combined signal was unattractive; playback of these signals was then alternated between the two pairs of speakers (fig. 8.10). More females preferentially moved toward one or the other of the pair of separated speakers rather than to one of the pair of adjacent speakers when the angular separation was 120° but not 45° or 90° (Schwartz and Gerhardt 1995 and unpubl. data). However, in a smaller treefrog (*H. microcephala*), females did not selectively respond to calls played back from single speakers separated by 120° (Schwartz 1993). Studies of the cricket *Teleogryllus oceanicus* showed that, as in the gray treefrog, angular separations of interrupting sound sources by 120° improved phonotactic selectivity (Pollack 1986).

In the green treefrog *H. cinerea*, the separation of the sources of individual calls from sources of continuous, broadband masking noise also resulted in improved detection of individual calls compared with tests in which signals and noise came from the same direction relative to the female's release point (Schwartz and Gerhardt 1989). The release from masking was equivalent to an increase in the signal-to-noise ratio of about 3 dB when the source of noise was separated from that of the source of individual calls by 90°.

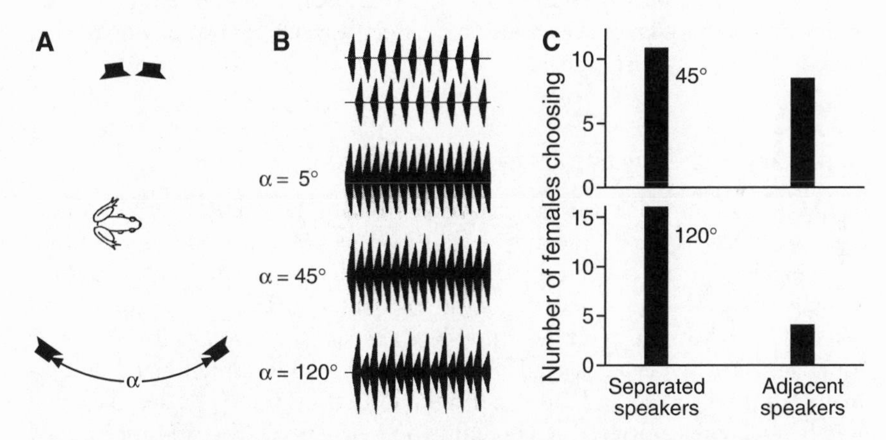

Figure 8.10. Speaker separation and discrimination of overlapping signals in the gray treefrog. (A) Females were placed between two pairs of speakers, close to each other (separation angle about 5°) or separated by angles of 45° or 120°. The diagram of the frog shows the release point of females relative to these two pairs of speakers. (B) Both pairs of speakers were driven by synthetic calls with an attractive pattern but offset in phase so that the combined signal was unattractive (top pair of oscillograms and the oscillogram below labeled $\alpha \approx 5°$). The oscillograms below show how the combined pattern might be encoded by binaural neurons on one side of the animal as the angle of separation between the speakers is increased. (Recall from chapter 5 that binaural cells are excited by arriving from the contralateral side and inhibited by sounds arriving from the ipsilateral direction.) (C) The histograms on the right show the numbers of females that moved either to one of the adjacently located speakers or to one of the spatially separated speakers. A highly significant preference for separated speakers occurred when the separation was 120° but not when the separation was 45° or 90° (not shown). In part from Schwartz and Gerhardt 1995, fig. 2.

A recent study, building on earlier observations by Pollack (1988), shows that a specific neural mechanism evident in omega neurons in crickets and katydids might alleviate some of the problems of masking interference, even if overlapping sounds come from the same direction (box 8.2; Römer and Krusch 2000). These neurons show *selective attention* in the sense that they encode the temporal pattern of the louder of two signals that differ by as little as 2–5 dB. It will be important to follow up these neurophysiological results with behavioral experiments designed to see if this phenomenon translates into better detection in noisy environments.

The above discussion concerns the effects of spacing on the discrimination of signals that differ in spectral or temporal properties. As mentioned in chapter 4, however, dense spacing is expected to increase discrimination of signals that differ only in sound pressure level (Forrest and Raspet 1994; Farris et al. 1997). This occurs because the sound field of an individual with less intense signals will be overlapped by that of an individual with more intense signals when the two animals are close together but not when they are well separated. Thus, the advantage in mate attraction of producing more intense signals would be expected to be even greater in dense choruses. This effect should also contribute to a weakening of female selectivity with respect to signals that differ in other respects (spectral and temporal properties) in dense choruses and would perhaps favor the adoption of alternative tactics by individuals that cannot produce intense signals.

Summary and Suggestions for Future Studies

Insect and anuran choruses typically form during favorable weather conditions within an annual breeding period in places where resources required for reproduction are concentrated. For species that do not directly defend resources needed by females within breeding sites, choruses can be considered leks, and the hotspot hypothesis provides the simplest explanation for their location. Little evidence is available that chorus sounds attract males and females from a distance, and one experimental study even showed that the usual numbers of females arrived at a breeding pond when a chorus was not allowed to form. Additional studies of this kind are needed. Small aggregations occurring within a breeding site might, however, be explained by the female-attraction hypothesis in some species, such as the bladder cicada. The benefits of reduced predation might also favor aggregation in periodical cicadas and some anurans. These effects are probably the result of dilution or predator-satiation effects.

Playback experiments that realistically simulate aggregations of different sizes within choruses could provide strong tests of these hypotheses. Most field observations and experimental studies available to date are either unconvincing in the sense that other factors were uncontrolled or they provide

Box 8.2. A Gain Control Mechanism for Signal Detection in Noisy Environments

A recent study of the katydid *Tettigonia viridissima* (Römer and Krusch 2000) extends an intriguing finding first reported in crickets (Pollack 1988). Namely, in addition to the release from masking afforded by directional hearing, the omega neurons on each side of the animal show *selective attention* to the more intense of two or more signals that arrive from the same direction. This phenomenon is illustrated by fig. 1, which shows that the encoding of the

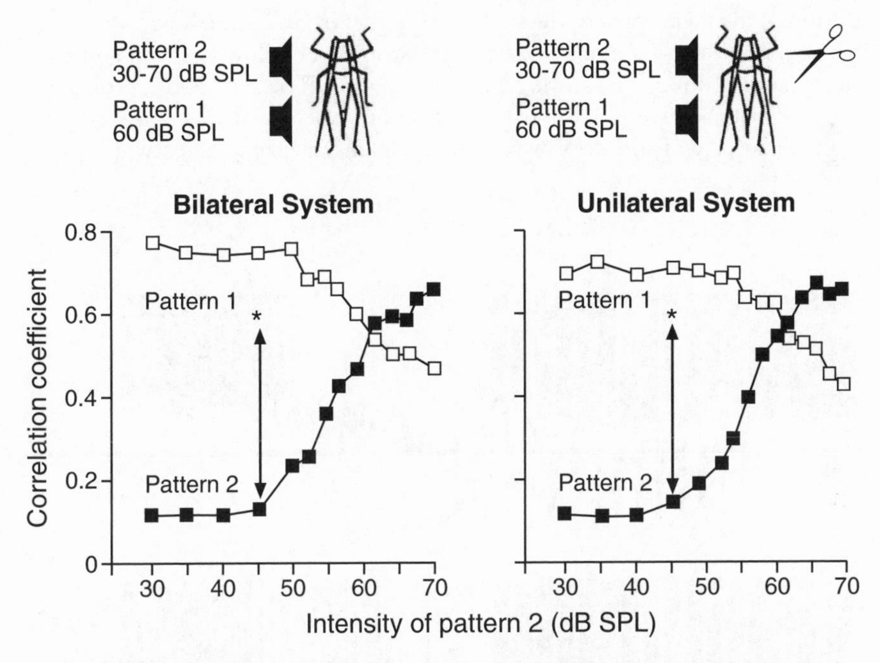

Box 8.2, Figure 1. Representation of two sound signals with different temporal patterns in the spike discharge of one omega neuron when both signals are presented simultaneously from the same side (see diagrams at top). In the left panel the system was intact, and in the right panel the input from the contralateral side was abolished by cutting the leg (auditory) nerve. The magnitude of the correlation coefficient shows the fidelity of copying of the sound pattern. The sound pressure level of the signal with pattern 1 was held constant at 60 dB, and the sound pressure level of the signal with pattern 2 was varied from 30–70 dB. At 45 dB the representation of pattern 2 was quite good if presented alone (correlation coefficients of about 0.6; see the arrows and asterisks). Notice, too, that the fidelity of copying of the more intense signal also declined somewhat as the sound pressure level of the less intense signal reached about 55 dB. The similarity of these patterns in the bilateral and unilateral preparations indicates that the suppression of encoding of the weaker sound was not caused by contralateral inhibition. From Römer and Krusch 2000, fig. 7.

(box 8.2 continued)

temporal pattern of the more intense of two signals suppresses encoding of the pattern of the weaker signal, even if the difference is a matter of 2–5 dB. This effect is not attributable to contralateral inhibition because it is unaffected by cutting the auditory nerve on the contralateral side. Essentially, this mechanism predicts that an individual's signals will be better detected and recognized if (1) its signals are more intense than those of other individuals; (2) it is closer to the female than other signalers; or (3) its signals propagate with less attenuation than those of other signalers, perhaps because of calling from more favorable sites. The mechanism appears to be a slow hyperpolarization of the membrane of the interneuron that begins at the onset of the signal (fig. 2). The time course of the inhibition is very slow (5–10 s) and so is the recovery (up to 15 s) at the end of stimulation. These response properties suggest that there will be intense selection on males to sing as loudly as possible (or to defend sites from which their signals propagate effectively) for as long as possible.

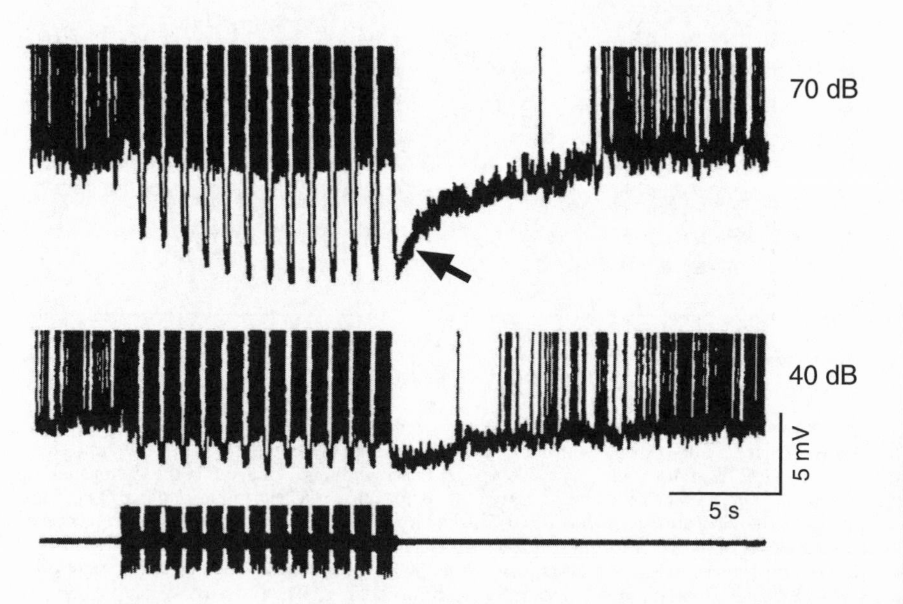

Box 8.2, Figure 2. Intracellular activity in an omega neuron in response to a signal simulating conspecific song (pattern 1) at 40 dB and 70 dB SPL. Spikes are truncated to emphasize the biphasic synaptic events. Note the slow membrane hyperpolarization at the onset of the signal and its slow decay at the signal offset. From Römer and Krusch 2000, fig. 8.

negative evidence. Additional and better-designed playback experiments are still needed, and this approach could also be used to assess the hotshot model of lek formation. The latter hypothesis would be supported if playbacks of especially attractive signals attracted other calling males more effectively than playbacks of "average" males or if males clustered around males with attractive calls after such individuals had been displaced within a natural chorus.

Whatever the causes of aggregated signaling, masking interference is a common consequence for individuals signaling in groups. Fine-scale timing relationships—especially alternation—between neighboring insects and frogs are one of the often-cited solutions. While seemingly cooperative, synchrony and alternation are better explained in terms of competitive interactions, which sometimes increase signal overlap. However, further experimental studies of changes in solo rates during sustained interactions between neighbors are warranted. How is an individual's solo rate affected by its motivational state, energetic reserves, or recent aggressive interactions? If an individual with a fast solo rate were to lower its rate to accommodate a neighbor, what precision and duration of signal interaction between the pair would be necessary to falsify a competitive explanation for synchrony and alternation? It will also be important to learn why there is so much variation among individuals and species in the extent to which the "rules" of inhibitory resetting models lead to consistent patterns of synchrony or alternation. Various phase-delay models have been hypothesized, which can, in theory, explain both synchrony and alternation. Yet there are, to our knowledge, no explicit neurophysiological studies of these putative mechanisms.

Preferences for leading signals are widespread but not universal. The mechanisms underlying preferences for leading signals deserve additional study. In our view, a convincing demonstration of the precedence effect, as originally described for humans, might be accomplished by varying the delay between nearly identical sounds arriving from two speakers separated by less than 180. If phonotactic orientation to locations between the speakers occurred, this could reflect the perception of a single acoustic event, whose apparent location is mainly influenced by that of the leading signal. Alternative (and not mutually exclusive) explanations involving masking also require experimental attention. Finally, because females of some species show no preference for leading signals or prefer lagging, overlapped signals, they can obviously detect and locate both signalers. This might also be true for species in which females prefer leading signals. The question then arises: What benefits, if any, might females gain from mating with a male that is a consistent leader (or follower)?

Spacing within aggregations is a very common phenomenon in insects and anurans, although the minimum distance between signaling neighbors varies enormously within and between species. Spacing in some species is related to the defense of spatially separated resources, but in many species, males attract females directly and defend a calling area that often changes from night to night. There is accumulating evidence that spacing results in at least modest improvements in the detectability of a male's acoustic signals (and structural features of these signals that convey information); both effects should, in turn, increase mating success. Additional studies of the relationship between spacing, fine-scale timing, and masking interference on female preferences are still desirable. For example, there has been no study in which female phonotaxis has been quantified as a function of systematic variation in the spatial separation of sound sources, their timing relationship, and the level of continuous background noise. Yet these factors are universally present in choruses. Neurophysiological studies of one species of katydid suggest that not only directional hearing but also selective inhibition by the louder of two signals arriving from the same direction can help to solve the problem of masking interference in choruses. Behavioral testing of the putative functions of these neurophysiological phenomena is highly desirable.

9

Acoustic Competition and Alternative Tactics

THE PRECEDING chapter showed that chorusing results in competition among neighboring signalers and that two common behavioral responses are fine-scale signaling interactions and spacing. In this chapter we consider additional aspects of competitive signaling. First, we describe special signals and behaviors used to defend territories or to enforce spacing. These signals are diverse in form, and variation in their acoustic properties is often predictive of more overt aggressive behavior. Second, we examine the costs of the signaling competition that takes place over the course of the daily or nightly calling period, the entire breeding season, or both. Because signaling can be energetically expensive, a variety of strategies have evolved to use limited resources most effectively. Finally, we examine alternatives to signaling such as searching and satellite behavior, which are often adopted by individuals that are unlikely to be successful in defending territories or attracting females by calling. As discussed in chapter 8, the presence of acoustically orienting predators and parasites might also favor these other behaviors.

Sexual selection theory provides the framework for understanding competition for mates. Direct competition among males that affects their relative mating success is a form of *intrasexual selection* or *male-male competition.* More generally, successful individuals have traits that allow them to win fights over females, to monopolize resources needed by females, or to outsignal their competitors. Differential attractiveness of males, based on female evaluations of their attributes (including signals), that affects their relative mating success is a form of *intersexual selection* or *female choice.* Whereas both forms of sexual selection target traits that affect mating success, natural selection acts on traits that affect survivorship.

One problem in applying these definitions is that some traits are subject to natural selection and to both forms of sexual selection (reviews: Andersson 1994; Halliday and Tejedo 1995). For example, males might differ in mating success because of differences in their ability to find and subdue females, to participate in more breeding episodes or to signal for longer periods of time than other males (= *endurance rivalry;* Andersson 1994). The traits underlying these abilities are also likely to enhance male survival. Moreover, individuals subject to female choice are often a subset of males that have been successful in male-male competition, and signals that effectively repel rivals may also be especially attractive to females. Finally, variance in mating success can also arise without interactions between males or female choice (Murphy 1998). For example, some males could be better than others at concentrating their signaling effort during times when females are most likely to arrive for mating.

The dominant form of sexual selection in a given species is influenced by its ecology and life history, which often dictate where, when, how long, and how often females breed, and what resources females require that males can economically defend (Emlen and Oring 1977; Wells 1977; Sullivan et al. 1995). Variation in these factors also occurs within species and is likely to account for some of the different results obtained in studies of the same species.

Sexual Competition in Insects and Anurans

The most common form of competition among males that use acoustic signals to attract females is endurance rivalry. Indeed, the most reliable correlates of mating success in anurans are chorus attendance and the duration and rate of signaling within such aggregations (Gerhardt 1994a; Halliday and Tejedo 1995; see chapter 10). The opportunity to participate in choruses usually depends on competition among males to secure areas where resources (food, oviposition sites) required by females occur (Howard 1978a,b; Greenfield and Minckley 1993) or for space that reduces interference by the signals of neighbors (chapter 8). In the katydid *Tettigonia viridissima*, males compete for calling sites at elevations within vegetation that increase the communication range of their signals without increasing predation risks (Arak and Eíriksson 1992). Many species of anurans and some insects use distinctive aggressive signals to enforce territorial boundaries or to maintain spacing. If such signals do not repel rivals, then fights usually ensue (e.g., Howard 1978a). Rivals may also fight directly for possession of a gravid female (e.g., Davies and Halliday 1978).

When competition for territories or the chance to signal is severe, smaller or weaker males are more likely to adopt alternative tactics than larger, dominant males. But if resources or females cannot be easily defended, males of all size classes are likely to adopt alternative tactics periodically, especially at high male densities (reviews: Wells 1977; Thornhill and Alcock 1983; Sullivan et al. 1995). Much debate concerns the extent to which alternative behaviors are promoted by the energetic costs of signaling (e.g., Prestwich 1994) and the risks of predation or parasitism by acoustically orienting species (Cade and Wyatt 1984; Zuk et al. 1995).

Although most sexual competition occurs between males, females compete for males in a few species of insects and anurans. In anurans such competition occurs in species in which males invest in parental care; females actively court males and interfere with the courtship of other females (review: Halliday and Tejedo 1995; Bush 1997). Female reciprocal calling has also been reported in fourteen species of anurans (Schlaefer and Figeroa-Sandí 1998). Although the function of these signals is unknown, their existence raises the question of whether males of some of these species might exercise choice based on female acoustic behavior. In some species of insects, the male's spermatophore or seminal fluids may contain nutrients that can contribute to large egg size and female survivorship (Wagner et al. 2001), and in some low-food conditions, spermatophores become a limited resource over which females compete (e.g., Gwynne and Simmons 1990; Gwynne 1993).

Acoustic Signals Used in Aggressive Interactions

Ritualized displays, including acoustic signaling, commonly allow rival males to avoid direct physical contests that could result in injury (Robertson 1986; Enquist and Leimar 1990). Aggressive signals may contain information about the likelihood that the signaler will attack if his rival does not withdraw. Signals may also indicate a signaler's fighting ability (e.g., Clutton-Brock and Albon 1979; Petrie 1988), but bluffing is also a distinct possibility (Dawkins and Krebs 1978; Grafen 1990). If signals are reliable indicators of fighting ability, then escalation to combat is most likely when rivals are evenly matched (Maynard Smith and Parker 1976). When large asymmetries exist, then males with little chance of winning contests often adopt alternative tactics to acquire mates. Insects and anurans provide examples of these principles and offer opportunities to address many open questions about the honesty of signals, the function of graded signal systems, and the success of tactics in which males do not signal. Another issue that arises in this discussion is the extent to which signalers can continue to attract females while effectively repelling or at least warning their rivals. Some progress has also been made in discovering hormonal factors that affect aggressive behavior and signaling in anurans. Injections of the neuropeptide arginine vasotocin, for example, appear to increase the probability that male gray treefrogs produce attractive signals and take over calling sites (Semsar et al. 1998; Klomberg and Marler 2000). The same substance causes changes in vocal behavior in cricket frogs (*Acris crepitans blanchardii*) that have been interpreted as aggressive responses (Marler et al. 1995; Chu et al. 1998).

Diversity of Signals and Interactions

Signaling males are typically separated by some minimum, species-typical distance, whether or not they defend resources such as oviposition sites or food plants (chapter 8). Aggressive signaling, physical contests, or both often maintain such spacing. In addition to fine-scale timing of advertisement calls with respect to the signals of neighbors (chapter 8), insects and anurans respond in a variety of other ways to signaling by nearby rivals: (1) modification of the repetition rate or duration of advertisement calls; (2) production of distinctive aggressive signals; (3) positive or negative phonotaxis. In many species of anurans, males readily respond to playbacks, which elicit an impressive variety of acoustic responses (table 9.1).

In many insects and some anurans, responses to a rival's calls or to playbacks are confined to phonotaxis (positive or negative) or to modifications of advertisement call structure (e.g., *Gryllus texensis* [= *integer*], a North American cricket: Cade 1981b; Cade and Otte 1992; *Kawanaphila nartee*, an Australian katydid: Simmons and Bailey 1991; *Geocrinia laevis*, an Australian frog: Harrison and Littlejohn 1985; *Acris crepitans*, a North American frog: Wagner 1989a). The most common response to a signaling neighbor or playback is an increase in the duration or rate of advertisement signaling in comparison with solo signaling (reviews: Wells 1988; Alexander 1961). As discussed in chapter 10, females often prefer long-duration signals, signals produced at high rates, or both, and increasing the amount of signaling should also increase a male's chances of detection by females in choruses (Parker 1982; Helversen and Helversen 1994).

Decreases in signaling rates in response to playbacks have also been documented in *Geocrinia laevis* (Harrison and Littlejohn 1985) and two Australian katydids (*Mygalopsis marki* and *Requena verticalis*) (Dadour 1989; Schatral and Bailey 1991b). In the North American katydid *Amblycorypha parvipennis*, neighboring males reduced call rate as the distance between them decreased unless an acoustically responsive female was present, in which case, the males increased call rate (Galliart and Shaw 1991). At high playback levels, males often stop signaling and vacate the calling site (e.g., anurans *Geocrinia laevis*, *Crinia parinsignifera:* Harrison and Littlejohn 1985; Littlejohn et al. 1985; katydid *Mygalopis marki:* Bailey and Thiele 1983). These results are consistent with observations of natural, close-range interactions in which one of two males typically ceases signaling and moves away.

Male insects and anurans may switch from advertisement signals to other signals with a distinctly different structure prior to, during, or after an aggressive encounter. The aggressive ("rivalry") songs of field crickets are usually elicited during antennal contacts between males, after fights, or in response to visual detection of a potential competitor (figs. 9.1, 9.2; Alexander 1961). A recent study of *Gryllus bimaculatus* reports that the frequency of aggressive stridulation, which increased when females were present and was produced only by winners of fights (at least in males that are raised individually instead of in groups), was correlated with male mating success (Tachon et al. 1999). Grasshoppers produce aggressive signals at close range (Young 1971), and, as in anurans, such signals can be elicited by playbacks (D. v. Helversen, M. D. Ritchie, pers. comm.). The aggressive calls of anurans have also been termed "territorial" or "encounter" calls (Littlejohn 1977, 2001; Wells 1988). The latter term has the advantage of not implicitly assigning a function to the signal, but behavioral observations in many species support the idea that signals produced in response to other males or their signals typically precede fighting. In general, male anurans produce fewer advertisement calls and more aggressive calls as rivals move closer or as playback levels are increased by a researcher.

Some anurans produce a few transitional "compound" calls containing acoustic elements of both advertisement and aggressive calls (e.g., *Hyla cinerea:* Gerhardt 1978a; *Hyla rosenbergi:* Kluge 1981; *Philautus leucorhinus:* Arak 1983b; and *Rana nicobariensis:* Jehle and Arak 1998). Some insects (*Ligurotettix planum:* Greenfield and Minckley 1993; *Chorthippus brunneus:* Young 1971) and anurans (*Hyla ebraccata:* Wells 1989; *Hyperolius marmoratus broadleyi:* Grafe 1995) produce a graded series of signals, with advertisement calls at one end of a continuum and aggressive calls at the other end (fig. 9.3). In other anurans (*Geocrinia victoriana:* Littlejohn and Harrison 1985; *Pseudacris crucifer:* Schwartz 1989), aggressive calls are graded in structure even though the advertisement call is distinctly different from any aggressive-call variant. Although the cricket frog *Acris crepitans blanchardii* does not have distinctive aggressive calls, Wagner (1989a) interprets modifications to advertisement-call structure that occur in male-male interactions as a graded signaling system. New evidence suggests, however, that the call changes that occur during vocal interactions or in response to playbacks increase signal attractiveness to females and might not necessarily be signaling aggressive intent (Kime, pers. comm.). Finally, some frogs produce two or more kinds of aggressive calls (e.g., *Rana virgatipes:* Given 1987; *R. clamitans:* Wells 1978; *Crinia* [= *Ranidella*] *parinsignifera:* Littlejohn et al. 1985; and *Litoria phyllocroa:* Gerhardt, unpubl. data). In the last two species, each of the two signals has a different threshold in terms of the behavior of a rival or the SPL of a playback required to elicit it. Graded signals, or different discrete signals with different thresholds, might provide honest information about the probability that the signaler will attack, but this hypothesis has seldom been tested experimentally. Graded signals might also be a partial solution to the dilemma discussed in the next section. Another untested hypothesis is that signals from the aggressive end of the continuum are more energetically costly to produce than signals from the advertisement end of the continuum.

Relative Attractiveness of Advertisement and Aggressive Signals to Females

If males seldom produce aggressive signals unless challenged by a rival, we might expect that such signals would be less attractive than advertisement calls to females or that females might even avoid these signals because of the possibility of being attacked. In general, this expectation is realized for anurans but not insects. In crickets, for example, aggressive songs or other song types

Table 9.1. Responses of Male Anurans to Playbacks of Advertisement Calls

I. Changes in Advertisement Calls

Response to Playbacks	Species	Aggressive Call	Reference
A. Changes in Rate, Duration of Signaling			
Increase call rate; decrease call duration	*Bufo valliceps*	no	Sullivan and Wagner 1988
Increase call duration; decrease call rate	*Hyla versicolor*	yes	Wells and Taigen 1986
Increase note number; decrease note duration, rate	*Rana nicobariensis*	yes	Jehle and Arak 1998
Increase call rate and number of calls per group	*Acris crepitans*	no	Wagner 1989b
Increase call rate, number of calls to match no. in playback	*Crinia georgiana*	no	Ayre et al. 1984; Gerhardt et al. 2000
Decrease call rate, duration, notes per call	*Bufo woodhousei*	no	Sullivan and Leek 1986
	Geocrinia laevis	no	Harrison and Littlejohn 1985
	Rana virgatipes	yes	Given 1987
B. Changes in Call Complexity			
Add secondary notes and tend to match number in playback	*Hyla ebraccata*	yes	Wells 1988
	Hyla microcephala	yes	
	Hyla phlebodes	yes	
Add secondary notes ("chucks")	*Physalaemus pustulosus*	yes	Rand and Ryan 1981
Add multinote clicks, trills at playback levels of 68–81 dB	*Leptopelis viridis*	yes	Grafe et al. 1999
Switch to multinote calls and match no. in playback	*Philautus leucorhinus*	yes	Arak 1983b
Switch to multinote calls and some matching of no. in playback	*Hyperolius tuberilinguis*	?	Pallett and Passmore 1988
Drop first part of two-note advertisement calls	*Litoria citropa*	yes	Gerhardt and Davis, in prep.
	Litoria phyllocroa	yes	
Drop second part of two-note advertisement calls	*Eleutherodactylus coqui*	yes	Narins and Capranica 1978
Decrease number of repeated notes	*Geocrinia victoriana*	yes	Littlejohn and Harrison 1985
C. Changes in Dominant Frequency			
Lower frequency	*Bufo americanus*	no	Howard and Young 1998
	Acris crepitans	no	Wagner 1989a
	Rana clamitans	yes	Bee and Perrill 1996
	Hyperolius marmoratus	yes	Grafe 1995
Lower or increase frequency and increase intensity	*Leptodactylus albilabris*	yes	Lopez et al. 1988

II. Graded Changes in Aggressive Signaling

Response	Species	Reference
A. Continuum Includes Advertisement-like Calls		
Increase length, repetition of introductory notes; drop repeated notes	*Geocrinia victoriana*	Littlejohn and Harrison 1985
Increase call duration, number of pulses and pulse groups; increase pulse duration	*Acris crepitans*	Wagner 1989b
Increase call duration by adding pulses	*Hyperolius marmoratus*	Grafe 1995
Increase duration and decrease pulse rate of introductory note; decrease number of secondary, click notes	*Hyla ebraccata*	Wells 1988
Drop introductory note, lengthen second note, add ticking elements	*Litoria citropa*	Gerhardt and Davis, in prep.
Produce compound calls; increase proportion of aggressive elements with increasing intensity/proximity	*Rana nicobariensis*	Jehle and Arak 1998
B. Continuum of Aggressive Calls Distinct from Advertisement Call		
Increase duration of pulsed aggressive calls	*Pseudacris crucifer*	Schwartz 1989
Increase duration of pulsed aggressive calls (compound calls with advertisement call structure elements produced at low levels of response)	*Philautus leucorhinus*	Arak 1983b
C. Two Discrete Aggressive Calls		
Drop introductory notes; produce long calls with low pulse rate	*Litoria phyllocroa*	Gerhardt and Davis, in prep.
Produce long call with low pulse rate; produce introductory note with uniform pulse rate followed by repeated biphasic notes (two different pulse rates)	*Crinia parinsignifera*	Littlejohn et al. 1985

III. Estimates of Thresholds for Switches to Aggressive Calls

Species	Estimate of Threshold*	Source
Hyla arborea	> 90 dB SPL fast RMS?	Brzoska et al. 1982
Hyla (Pseudacris) regilla	> 87 dB SPL fast RMS? (negatively correlated with nearest neighbor SPL)	Brenowitz 1989
Litoria verreauxii	78–100 dB SPL fast RMS (negatively correlated with nearest neighbor SPL)	Gerhardt, in prep.
Litoria ewingii	mean of 101 dB peak (≅ 91 dB RMS)	Harrison 1987
Philautus leucorhinus	60–69 dB SPL peak?	Arak 1983b
Leptopelis viridis	> 82 dB SPL peak	Grafe et al. 1999b

*fast RMS = "fast" setting on sound level meter; RMS = root-mean-square

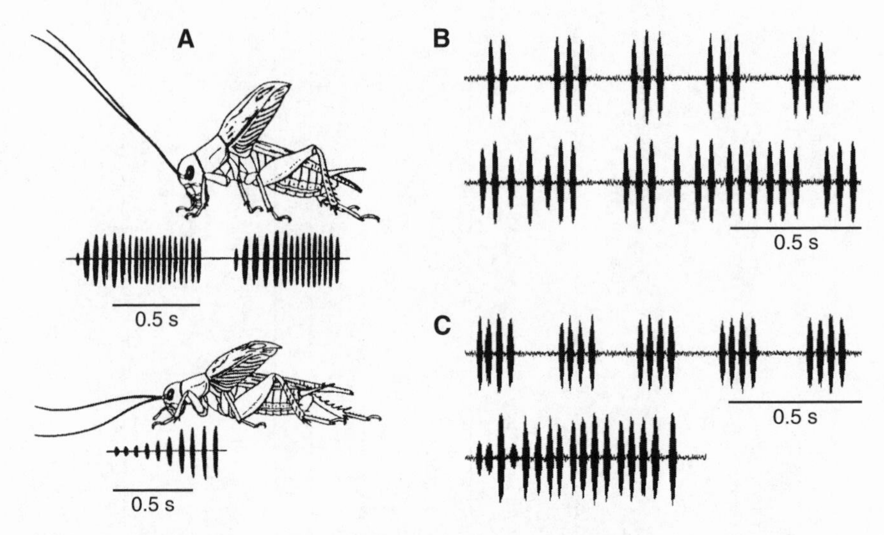

Figure 9.1. Comparisons of advertisement (calling) songs and aggressive (rivalry) songs in field crickets. (A) *Teleogryllus commodus;* the advertisement song is above and the aggressive song, below. Note the different postures and wing positions used in the production of these two song types. (B) *Acheta domesticus;* the upper trace is the advertisement call and the lower trace is the aggressive signal. (C) *Gryllus campestris;* the upper trace is the advertisement call and the lower trace is the aggressive signal. From Loher and Dambach 1989, fig. 2.1; Nelson and Nolen 1997, fig. 1; Ewing 1989, fig. 2.17.

with similar temporal elements (e.g., in *Teleogryllus:* Pollack 1982; Burk 1983) attract female crickets (Weber et al. 1981) or at least do not repel them (Alexander 1962). In a gregarious species, *Amphiacusta maya,* courtship and aggressive songs are so similar in structure that they had to be distinguished by the behavioral context (Boake 1984a,b). Indeed, the so-called courtship song seemingly functions only to increase a male's chances of not being interrupted by other males, and, even so, artificially silenced males mate as often as singing males (Boake 1984a,b).

In studies of seven species of frogs, females have been tested with advertisement and aggressive signals, with signals that combine advertisement and aggressive components, or both (*Afrixalus aureus:* Backwell 1988; *Hyla cinerea:* Oldham and Gerhardt 1975; Gerhardt 1978a; *Hyla ebraccata:* Wells and Bard 1987; *Hyla microcephala:* Schwartz and Wells 1985; *Hyperolius m. broadleyi:* Grafe 1995; *Pseudacris regilla:* Brenowitz and Rose 1999; *Pseudacris crucifer:* Marshall, Humfeld, and Bee, submitted). In all except the last species, females strongly preferred advertisement calls to aggressive calls or to synthetic equivalents of these two call types. In *P. crucifer,* females had, at best, a weak preference for the advertisement call to the aggressive call. Combinations of the advertisement and aggressive call were somewhat more attractive than the aggressive call alone, but not more attractive than the advertisement call alone (Marshall, Humfeld, and Bee, submitted).

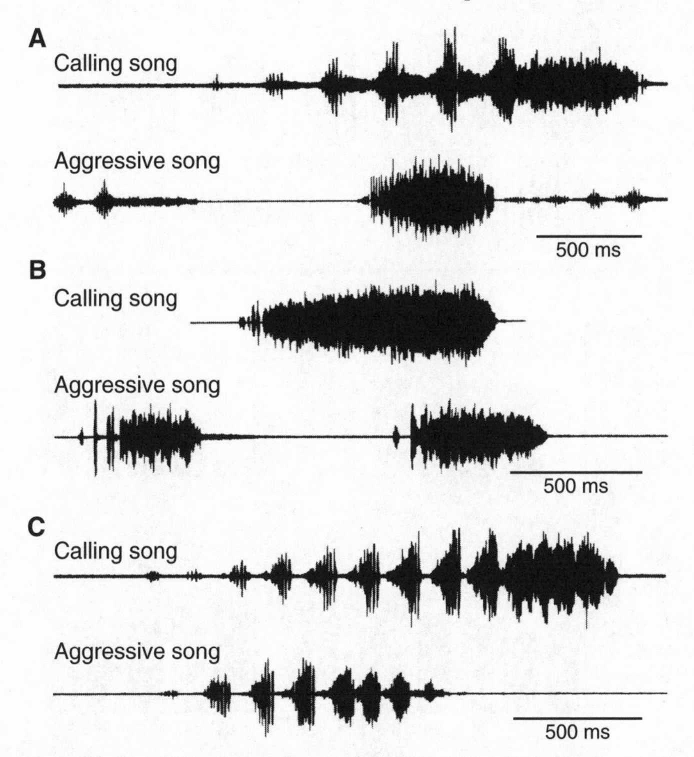

Figure 9.2. Comparisons of advertisement (calling songs) and aggressive (rivalry) songs of three species of acridid grasshoppers. (A) *Chorthippus dorsatus;* (B) *Chorthippus dichrous;* (C) *Chorthippus loratus.* From Stumpner and Helversen 1994, figs. 2, 3, 4.

Males that produce aggressive calls that are relatively unattractive to females face a dilemma: advertisement calls alone may not effectively repel intruders, but the production of aggressive calls could reduce a male's attractiveness to females. Males of some species adopt tactics that not only ameliorate this problem, but can also enhance the attractiveness of their advertisement calls. First, as in the African frog *Afrixalus aureus*, males produce most of their aggressive calls early in the evening when they are establishing calling sites. Because most females arrive later, the chances of repelling females are thereby reduced (Backwell 1988). Second, when first hearing a rival's calls, males might not switch immediately to aggressive calls but rather increase the duration or repetition rate of advertisement calls (examples in table 9.1).

Another kind of structural change that can precede switches to aggressive calls or behavior is an alteration of carrier frequency. As mentioned in chapter 2, males of the white-lipped frog *Leptodactylus albilabris* lower or increase carrier frequency to match the frequency of a stimulus (Lopez et al. 1988). In other anurans, males lower carrier frequency (chapter 2; Wagner 1989a; Grafe 1995; Bee and Perrill 1996; Howard and Young 1998; Given 1999; e.g.,

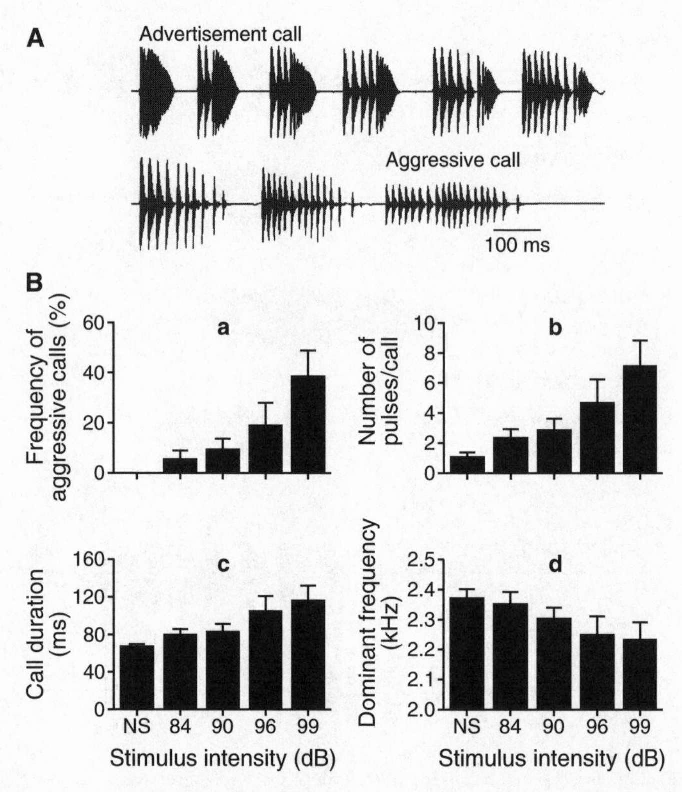

Figure 9.3. (A) Oscillograms of a series of advertisement and aggressive calls produced by the same male of *Hyperolius marmoratus* during a 2-min period in response to playback of advertisement calls. The calls are arranged in order of increasing number of pulses per call and call duration. (B) Histograms with standard deviations showing changes in (a) the proportions of aggressive calls; (b) number of pulses per call; (c) call duration; and (d) carrier (dominant) frequency in response to playbacks of a standard advertisement call at different playback levels. Changes in call attributes occurred in both advertisement calls and aggressive variants. From Grafe 1995, figs. 4, 6.

fig. 9.3B). In *L. albilabris*, call intensity increased regardless of the direction of the change in frequency, whereas in the other species, lowering frequency resulted in a drop in call amplitude. Because aggressive interactions occur at close range, small drops in signal amplitude are unlikely to reduce the effectiveness of the change in frequency, especially if the male successfully mimics a larger male. Females of two species in which males lower carrier frequency (toads, *Bufo americanus*, and some populations of cricket frogs, *Acris crepitans blanchardii*) prefer low-frequency to high-frequency calls if all other properties are held constant (Howard and Palmer 1995; Ryan et al. 1992; Kime, pers. comm.).

If the switch to aggressive calls is influenced by the numbers, behavior, or the identity of potential rivals, then males can adjust their output of aggressive calls according to the perceived level of threat. In whistling treefrogs

Litoria verreauxii, Pacific treefrogs *Pseudacris regilla*, and spring peeper *P. crucifer*, the threshold for switching in response to playbacks of advertisement calls is positively correlated with the amplitude of the nearest neighbor's calls (Rose and Brenowitz 1991; Gerhardt, unpubl. data; Marshall, Humboldt, and Bee, submitted). Having a neighbor with a high-amplitude call is more likely when chorus density is high and the spacing between males decreases. If a male did not adjust his threshold, he would be constantly reacting aggressively to one or more neighbors rather than producing advertisement calls to attract females. Brenowitz and Rose (1997) showed that exposing Pacific treefrogs to playbacks of advertisement calls can shift their aggressive threshold to higher values. Such prior exposure to advertisement calls did not, however, alter the threshold for switching to aggressive calls in response to playbacks of aggressive calls, which the frogs probably perceive as representing an imminent threat. In a similar fashion, chorus density and the order of presentation of aggressive and "attractive" calls can influence the probability that male cricket frogs will alter calls in ways that have been interpreted as aggressive responses (Burmeister et al. 1999a,b; see above).

In both insects and anurans, males of some species produce two-part advertisement calls. One hypothesis is that one part functions to attract females and the other part, to keep rivals at bay. Unfortunately, for most of these species, only one sex or the other has been tested with these complex calls or their constituent parts. In the katydid *Orchelimum nigripes*, for example, males produce two-part calls consisting of ticks followed by a buzz. In aggressive encounters, which are initiated by visual cues, males reduced or eliminated the buzz part of the call (Feaver 1983). The relative attractiveness of ticks and buzzes to females is unknown. Conversely, in the grasshopper *Chorthippus dorsatus*, females preferred the normal, two-part call to either note type in isolation, and signals with the two notes in reverse order were also ineffective (Stumpner and Helversen 1992). How males react to the entire signal or its constituent parts remains untested.

The best evidence for functional partitioning of diphasic calls directed to each sex is available for two anurans. In *Geocrinia victoriana*, males usually produce one introductory note of about 500 ms followed by a series of a dozen or so much shorter (< 70 ms) repeated notes (fig. 9.4A; Littlejohn and Harrison 1985). The introductory notes have a pulse rate that is about one-third of the pulse rate in the repeated notes. Given a choice between playbacks of the introductory note alone and twelve repeated notes from the same call, females responded exclusively to the repeated notes. In playbacks to males, aggressive responses were most effectively elicited by the introductory note alone and less well by the complete call. Males responded by decreasing the rate of production of repeated notes, increasing the duration of introductory notes, and sometimes approaching the speaker. Repeated notes alone were highly ineffective, and only at unusually high playback levels did males alter their calls. Surprisingly, both parts of the diphasic calls of a congener *G. laevis* (fig. 9.4B)

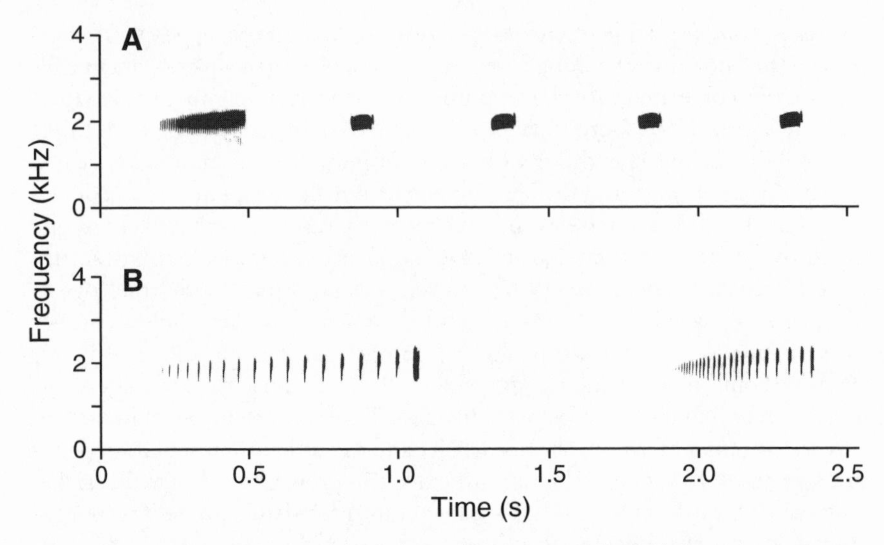

Figure 9.4. Sonograms of parts of the diphasic advertisement calls of (A) *Geocrinia victoriana;* and (B) *G. laevis*. In *G. victoriana,* males produce a long introductory note with a slower pulse rate than those of the shorter repeated notes, four of which are shown in the figure. In *G. laevis,* males also produce long introductory notes that are followed by shorter repeated notes (one is shown) with a higher pulse rate. Modified from Harrison and Littlejohn 1985, fig. 2.

triggered aggressive responses from males as effectively as the complete call, thus indicating that neither part of the call functions specifically as an aggressive signal (Harrison and Littlejohn 1985). Females of *G. laevis* have not been tested with different parts of the advertisement call.

The results for the other species of anuran in which both sexes have been tested are somewhat equivocal. Males of the coqui treefrog *Eleutherodactylus coqui* dropped the "qui" note of the "co-qui" call when confronted by rivals or playbacks (Narins and Capranica 1978), suggesting that the "co" is the aggressive component of the advertisement call. Two-speaker playbacks indicate that the "qui" notes function to attract females. However, soft "co" notes are also produced during close-range courtship, and in aggressive interactions over shelters, males utter multinote aggressive signals consisting of many repeated "qui" notes (Stewart and Rand 1991; Stewart and Bishop 1994). The "co" notes of courtship calls are shorter, and the "qui" notes given in aggressive contexts have less frequency modulation than the corresponding "co" and "qui" notes of the advertisement call (Wells, pers. comm.). In general, species in which males produce advertisement calls with two or more distinct elements deserve much more study, and comparative studies of such systems (and the functions of the different elements) could shed light on the evolution of signal complexity (Helversen and Helversen 1994; see chapter 11).

Males of some species also increase call complexity in competitive situations by adding different acoustic elements, which could also convey some level of aggressive intent as well. These elements enhance signal attractiveness but do not by themselves attract females. This tactic is employed by Neotropical treefrogs (*Hyla ebraccata*, *H. microcephala*), which add additional click notes to their advertisement call in chorus situations or in response to playbacks (see below). In the Neotropical túngara frog *Physalaemus pustulosus*, males add "chucks" to their advertisement calls ("whines") when calling in groups or in response to playbacks (Ryan 1985). Females of *P. pustulosus* prefer whines plus chucks to whines alone, and indeed, the more chucks, the more attractive the stimulus (Ryan 1985). Isolated males presumably do not produce chucks because of the risk of predation by bats (chapter 8; Ryan 1985). It is also possible that adding clicks or chucks to advertisement calls alerts a rival to the possibility that the male may escalate to aggressive behavior if, say, the rival approaches too closely (see below). If, for example, the intensity of playbacks of simple advertisement calls (whines) is very high, males may switch from whines to a distinctive aggressive signal without adding chucks (Schwartz, pers. comm.).

Examples of Male-Male Competition Mediated by Acoustic Signaling

The examples in this section serve to emphasize that male insects and anurans do not simply respond reflexively to signaling by rivals and that playback experiments can be used to discover some of the rules of assessment that determine how a male responds under different conditions and even sometimes to specific individuals.

Aggressive Signaling in Tarbrush Grasshoppers *(Ligurotettix planum)*

The tarbrush grasshopper is the subject of one of the best-documented examples of territoriality and aggressive signaling in acoustic insects (Greenfield and Minckley 1993). Males settle territorial disputes over bushes that serve as both calling sites and food sources for both sexes and advertise their presence with "rasp" calls. These signals attract females and other males, and when another calling male intrudes, resident males produce a more complex call that includes "shucks." Shucks usually precede but sometimes follow rasps.

When males interacted acoustically, four-fifths of all contests were settled without escalation to physical combat, which included biting, grappling, and kicking. Individuals that never produced shuck calls invariably left the contested sites. An index, based on combined measures of rate and duration of shuck calls (= total time signaling), predicted the winner of most acoustic conflicts, but neither body size nor prior residency were strong indicators of victory. Contests in which the two individuals produced signals for a comparable amount of time were prolonged compared with contests in which the signals of males differed, especially with regard to the number of shucks. These results are consistent with the predictions of theoretical analyses of animal conflicts by Parker (1974) and Maynard Smith and Parker (1976), who suggest that when disputed items are relatively plentiful and defeated individuals can anticipate finding another site in the near future, the animal should signal its true fighting ability. Mutual assessments then result in rapid departures by the weaker of two unevenly matched individuals, and prolonged contests occur only when rivals are similarly matched.

Another interesting discovery made by Greenfield and Minckley (1993) was that assessments of acoustic signals of rivals depended to some extent on their alternation. Playback experiments showed that males were much less likely to move toward a speaker or respond with shuck calls when the playback overlapped their own calls than when such signals occurred about 400 ms after the insect's own song. Perhaps the detection of rival songs is masked by the male's own signaling as in other species of insects (chapter 5).

Call Matching in Anurans

In some species of anurans, each frog in an interacting pair may match the numbers of notes produced by its rival. Besides the Panamanian treefrogs discussed below, call matching has been described in four other species of anurans. In a Sri Lankan treefrog (*Philautus leucorhinus*), males produce a single-note advertisement call in isolation (Arak 1983a) but switch to multinote calls when interacting at a distance with a neighbor. During such exchanges, rivals showed a strong tendency to alternate and to match the number of multinote calls produced by their opponent (fig. 9.5A). In the Asian frog *Rana nicobariensis*, matching of multinote advertisement calls was less precise than that in *Philautus*. Males usually did not produce more notes and often produced fewer notes than those in the playback stimulus (Jehle and Arak 1998). In *Crinia georgiana* from southwestern Australia, males did not produce aggressive calls but tended to match the number of notes in advertisement calls produced by neighbors or playbacks (Gerhardt et al. 2000). Matching was more precise than in *Philautus* and *R. nicobariensis*, but males usually produce more than one note in response to one-note calls and fewer than eight notes in response to eight-note calls. Males might also match the total number of notes produced by two or more nearby rivals. In playbacks of four-note calls from two widely separated speakers, for example, males usually produced calls with six to eight notes. This summation occurred even when the SPL of one of the stimuli was as much as 10 dB lower than that of the other stimulus.

The fact that males of these three species do not usually match acoustic stimuli with many notes suggests that these frogs might be energetically limited or that they might also experience diminishing returns of producing extra notes. In the South African frog *Hyperolius tuberilinguis*, for example, advertisement calls consist of one to six click notes, the number of which correlates with chorus density. Males typically produce fewer than four clicks but matched up to four-click calls in playback experiments (Pallett and Passmore 1988). Although females preferred two- and three-click calls to calls with one click, four-click calls were actually less attractive than two- and three-click calls.

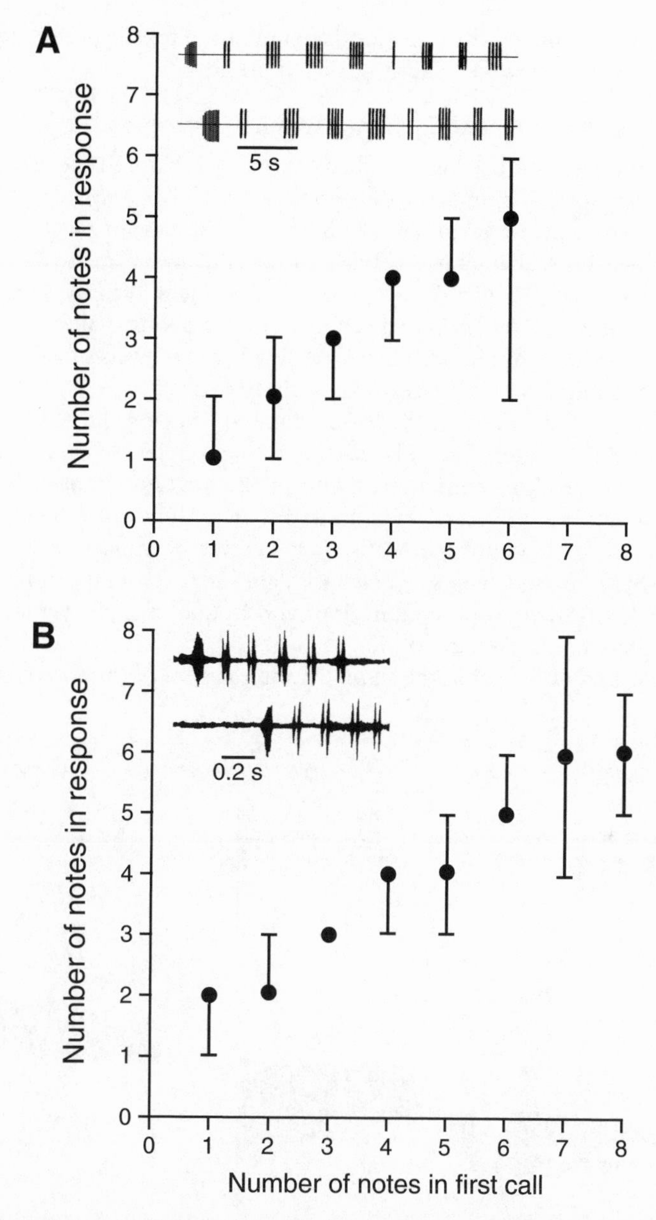

Figure 9.5. Call-note matching in two species of anurans in natural interactions. Plotted are medians (*circles*) and approximate 95% confidence intervals (*vertical bars*) of the numbers of notes produced by males in response to the signals of neighbors with different numbers of notes (*X-axis*). (A) *Philautus leucorhinus* (sample size unspecified, redrawn from Arak 1983a). *Inset,* oscillograms show an exchange of alternating, multinote aggressive calls. (B) *Hyla microcephala* (pooled data for 4 males, 325 calls, redrawn from Schwartz 1986). From Wells 1988, fig. 3. *Inset,* oscillograms show interdigitation of notes in multinote calls given by two males.

Similarly, note number is not a major determinant of female preference in the quacking frog (Roberts and Smith, pers. comm.).

Graded Aggressive Signaling in Panamanian Treefrogs

Vocal communication has been studied extensively in three species of Panamanian treefrogs (*Hyla ebraccata*, *H. microcephala*, and *H. phlebodes*). No other system has been investigated with such an array of elegant playback experiments, which have characterized both patterns of male vocal response and female preferences (review: Wells 1988). Males defend calling sites but not oviposition sites, and females freely choose among calling males. Acoustic interactions also occur interspecifically and involve two levels of aggressive intensity that depend on the perceived proximity of rivals.

When calling in isolation, males of these species produce introductory notes, with a species-specific pulse rate, and one to a few secondary clicklike notes (fig. 9.6A). Males initially respond to the advertisement calls of both conspecific and heterospecific neighbors by increasing call rate and adding click notes, and they alternate calls with nearest neighbors. Schwartz and Wells (1984a) provide extensive playback data for *H. ebraccata* and also show that tone bursts of appropriate durations and frequencies elicited similar vocal reactions (see also Narins 1982).

Subtle differences exist in the signaling interactions of the three species. In

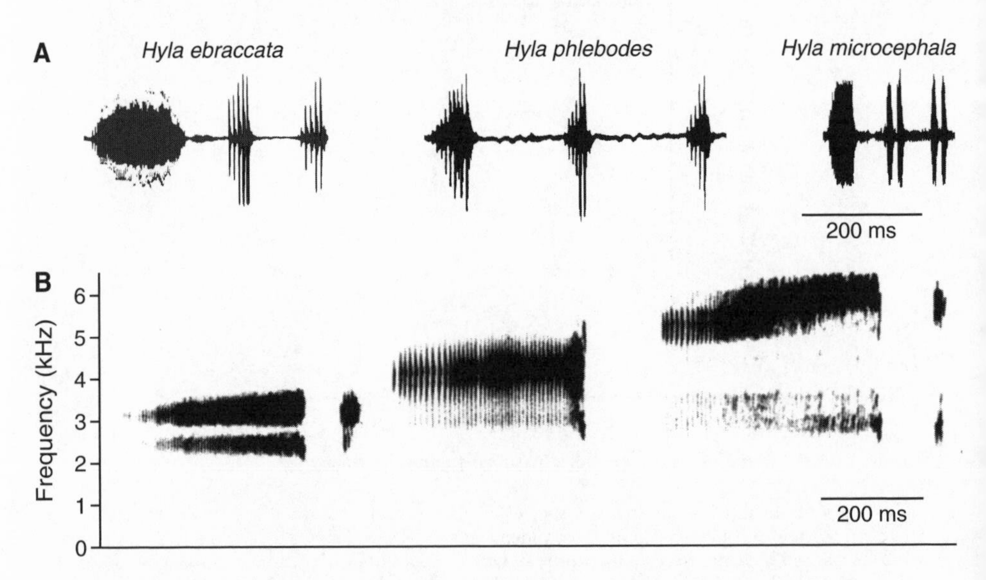

Figure 9.6. Vocalizations of three species of Panamanian treefrogs. (A) Oscillograms of advertisement calls showing introductory notes and secondary "click" notes. (B) Sonograms showing examples of long aggressive calls (plus one secondary click note) of *Hyla ebraccata* (*left*), *H. phlebodes* (*middle*), and *H. microcephala* (with a secondary click note) (*right*). From Wells and Schwartz 1984b.

interactions between neighboring males of *H. ebraccata*, the secondary notes of the first male to call overlap the introductory notes of the second male, which begins his call immediately after the end of the first male's introductory note. Interacting males or males stimulated with playbacks of introductory and secondary notes produced more secondary notes than when calling in isolation, but they did not match the number of secondary notes of the rival and usually produced no more than four secondary notes (Wells and Schwartz 1984a). Pairs of interacting males of *H. microcephala* and *H. phlebodes*, by contrast, usually delayed their response until after the first two or three notes of the stimulus. Males sometimes produced long trains of twenty or more clicks, which were so precisely alternated that overlaps rarely occurred (Schwartz and Wells 1984b; Schwartz 1986). In *H. microcephala* the lack of overlap was facilitated because males increased the intervals between their click notes. As shown in figure 9.5B, males also tended to match the number of click notes in their responses to the number of click notes in a playback stimulus (Schwartz 1986). Even when males of *H. microcephala* and *H. phlebodes* produced many secondary notes, matching was still quite impressive; their ability to do so is probably facilitated by the reaction of the two frogs to each other's signals on a note-by-note basis.

In response to the calls of nearby males or playback of advertisement calls at high sound pressure levels, males of all three species switched to aggressive calls (fig. 9.6B; Schwartz and Wells 1984a,b; Wells and Schwartz 1984b; Schwartz 1986). In contrast to the species-specificity of pulse rate in the introductory notes of advertisement calls, the pulse rates of the aggressive signals are similar in the three species. Not surprisingly, males of all three species respond to the aggressive signals of heterospecific males.

Males of all three species also show graded acoustic responses to playbacks of conspecific advertisement calls of increasing sound pressure level, simulating the approach of a calling rival male. In figure 9.7 we show an example of these changes in the calls of a male of *H. ebraccata* in a natural encounter with another male. The male's first reaction was to increase the duration and rise-time of introductory notes while simultaneously decreasing the pulse rate, number, and duration of secondary notes. As the rival moved closer, the male switched to an aggressive call with a secondary click note, and at still closer distances, the male produced a single, long aggressive call. Preliminary results of experiments in which the duration of computer-generated aggressive calls matched or were shorter or longer than those of a target male show that escalated responses are most likely when the stimulus matches the male's call (Schwartz 1994), a result that parallels that of the studies of aggressive interactions in the tarbrush grasshoppers discussed above.

Neighbor-Stranger Discrimination in Bullfrogs *(Rana catesbeiana)*

Aggressive behavior in the North American bullfrog *Rana catesbeiana* has been studied extensively (Emlen 1968; Howard 1978b; Ryan 1980; Davis 1987,

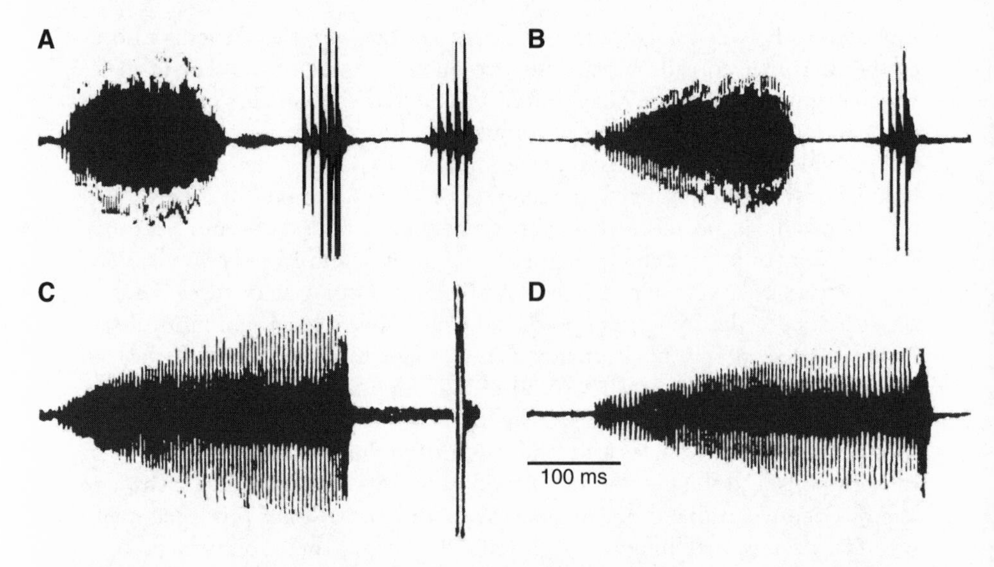

Figure 9.7. Signals produced during a natural encounter between two males of *Hyla ebraccata*. (A) Oscillogram of the advertisement call of *H. ebraccata* with two secondary notes. (B) Two-note call with an intermediate introductory note. (C) Two-note call with a long introductory note and a short click note. (D) One-note long aggressive call. All calls were given on the same night by the same male, and oscillograms are presented in order of decreasing distance from the male's rival. From Wells 1988, fig. 4.

1988; Bee and Gerhardt 2001a,b,c). Unlike the hylid frogs discussed above, male bullfrogs fight vigorously to obtain and defend territories, which may remain stable for days, weeks, or months, depending on environmental conditions (e.g., Howard 1978a,b; Bee, pers. obs.). Females, which usually lay eggs within a male's territory, might assess its quality as well as the male's signals and other attributes. A male's success in winning and maintaining a territory is highly correlated with his body size, and only large territorial males produce advertisement calls consistently (Howard 1978b).

Davis (1987, 1988) showed that, just as in songbirds (reviews: Falls 1982; Stoddard 1996), males discriminate between the calls of familiar and unfamiliar territorial males and also associate familiar calls with particular locations. These conclusions are based on the results of experiments showing that when the territorial neighbor was removed and his calls were played back from their usual position, the resident male usually responded only by countercalling with advertisement calls, produced few aggressive calls, and rarely approached the speaker. However, when the calls of an unfamiliar male were played back from the same location, the resident produced many more advertisement and aggressive calls and often swam toward the speaker (fig. 9.8A). Comparable differences in aggressive responses of resident males to playbacks of the calls

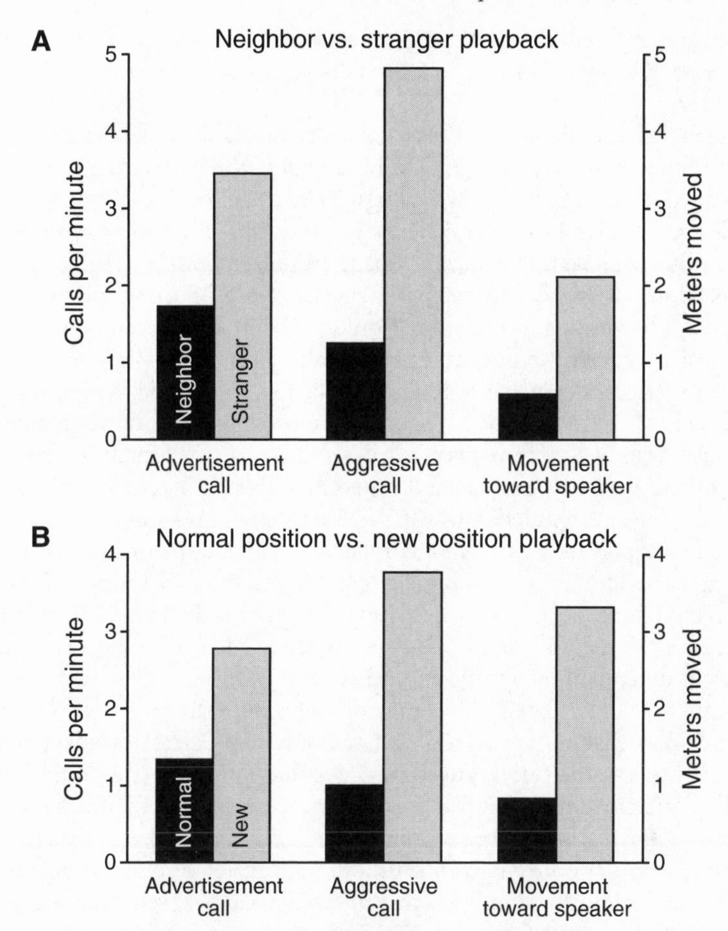

Figure 9.8. Vocal and phonotactic reactions of territorial male bullfrogs (*Rana catesbeiana*) to playbacks of (A) the calls of familiar neighbors and unfamiliar males; and (B) to calls of familiar neighbors played back from their usual position relative to the male's territory and from a new position on the other side of the territory. Males produced more advertisement and aggressive calls and moved farther toward the speaker in responses to the calls of the unfamiliar male than in response to those of the familiar neighbor. Familiar calls played from the new position also elicited greater responses than from the usual position. New original figure based on data in Davis 1987.

of territorial neighbors and those of non-neighbors were recently reported in a terrestrial, Neotropical frog *Colosthethus beebei* in which males have relatively long-term interactions with the same individual neighbors (Bourne et al. 2001). In bullfrogs the resident also showed the same escalation of aggressive behavior when the familiar neighbor's call was played back from the opposite side of the resident's territory (fig. 9.8B). The reduced aggression in response to signals of a familiar neighbor in its expected location has been called the *dear enemy effect* (see Wilson 1975 for a discussion of the historical derivation of this term).

The fact that bullfrogs also respond aggressively to playbacks of synthetic advertisement calls has made it possible to study the underlying mechanisms (Davis 1987). After responding strongly at first, aggressive responses usually waned over a period of several hours of playbacks. If another synthetic call with a different spectral structure was then presented, the resident then renewed its initial aggressive behavior. As suggested by these initial observations, neighbor-stranger discrimination is probably based on habituation, and males also carry over some memory of a habituating stimulus from one night to the next (Bee and Gerhardt 2001a). Males also habituated to the location of a playback of synthetic calls, responding more aggressively when the same stimulus was moved to the opposite side of the territory. Finally, bullfrogs can discriminate between familiar and unfamiliar sounds independently of their location. Males responded with the same level of aggression to an unfamiliar synthetic call played back from a familiar location and from a novel location. Although the familiar call elicited a greater response in the novel location than in the familiar location, it was always less potent than unfamiliar call (Bee and Gerhardt, unpubl. data). The review by Stoddard (1996) discusses examples of individual recognition in other taxa.

An analysis of repeated recordings of males over the course of a breeding season shows that both spectral and temporal properties of advertisement calls differ to a statistically significant degree among males (Bee and Gerhardt 2001b). Carrier frequency (or its correlate, waveform periodicity) is especially well suited for distinguishing among males, and playback experiments show that after habituation to a given synthetic call, males responded aggressively to playbacks of another synthetic that differed in carrier frequency alone by as little as 10% (Bee and Gerhardt 2001c). Such discrimination did not reliably occur when the frequency difference was 5%, but this magnitude of variation in frequency is close to the maximum within-male variation found during a bout of advertisement calls produced by an individual male. Thus, even if male bullfrogs could discriminate differences of 5% or less, aggressive responses would not necessarily be appropriate because they would sometimes be elicited by the calls of an established neighbor.

The dear enemy effect is thought to be adaptive in that individuals might save time and energy that they could expend in attracting females or repelling strange males that are perhaps more likely to attempt a takeover of the territory. However, there has been no definitive test of this assumption in any animal. Given that the carrier frequency of different males often differs by 5% or less, a male probably has to weigh the energy savings against the possibility that not responding aggressively increases his chances of losing his territory to an unfamiliar male with a frequency close to that of his neighbor.

Another, more likely possibility is that males use carrier frequency in conjunction with some other acoustic property to learn the neighbor's call (Bee and Gerhardt 2001a).

Acoustic Bluffing in Cricket and Green Frogs?

Although cricket frogs (*Acris crepitans*) defend only temporary calling sites and green frogs (*Rana clamitans*) defend more permanent territories, males of both species lower the frequency of their calls in response to playbacks (Wagner 1989a,b,c, 1992; Bee and Perrill 1996; Bee et al. 1999, 2000). Because frequency is negatively correlated with body size in these species, this change potentially indicates to rivals that the signaler is larger than he actually is. In both species, large males usually win fights with small males (Wells 1978; Wagner 1989a), and in playbacks to male cricket frogs, the probability that a male abandoned his calling site or ceased calling was inversely related to the frequency of the stimulus (fig. 9.9B; Wagner 1989a).

The interpretation of such drops in frequency as bluffs is not straightforward in either species. In playback experiments, male cricket frogs that "attacked" (moved toward the speaker while still calling) lowered the frequency of their calls, whereas, on average, males that simply called throughout the playback did not (fig. 9.9C; Wagner 1992). Because male size in his sample was uncorrelated with the propensity to attack, Wagner (1992) interpreted his results as supporting the idea that the magnitude of the frequency decrease advertises a male's size-independent fighting ability, even though, as stated above, large males nearly always win fights (Wagner 1989a). However, regardless of their size, males that abandoned their calling sites lowered the frequencies of their calls almost as much as did males that attacked (fig. 9.9C). Thus, an alternative hypothesis is that small males or large males with low fighting ability might initially lower their call frequencies but retreat as soon as any male with superior ability responds aggressively or fails to retreat.

Intriguing relationships also exist between the propensity of a male to alter frequency, the male's own body size, and the apparent size (indicated by carrier frequency) of the opponent. In both cricket and green frogs, some males responded more strongly in response to stimuli with frequencies representative of large opponents than to stimuli representative of medium-sized or small rivals. The greater response to large opponents might be expected if large males are perceived as a greater threat than other males. In green frogs, however, only small males consistently showed this frequency-dependent response, and, moreover, they lowered the frequency of their calls to the greatest extent in response to the large-male stimulus (Bee et al. 2000). The fact that small males were most likely to respond to large-male sounds is consistent with two models predicting that animals with the lowest probability of winning an escalated contest should be most likely to bluff (Bond 1989; Adams and Mesterton-Gibbons 1995). Small male green frogs might have more to

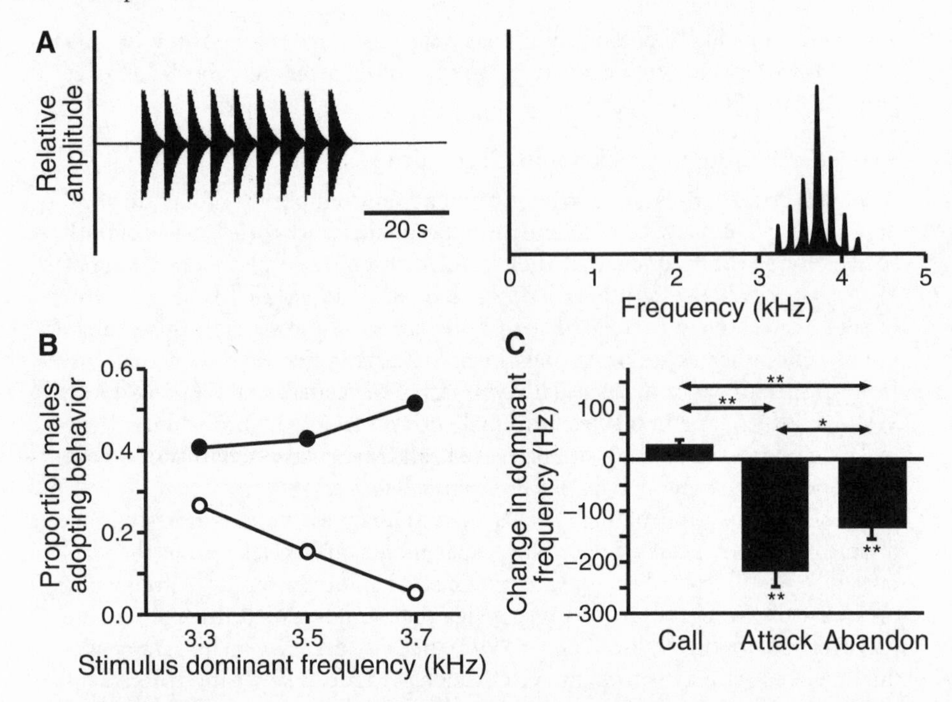

Figure 9.9. Responses of male cricket frogs (*Acris crepitans*) in playback experiments. (A) Synthetic cricket frog call (*left panel*) with a carrier (dominant) frequency of 3.7 kHz as shown in the power spectrum (*right panel*). (B) Probabilities that a male retreated or stopped calling (*open circles*) or attacked (*solid circles*) as a function of the carrier frequency of a playback stimulus. (C) Average change in carrier frequency of males that called throughout a playback, attacked the speaker, or abandoned their calling site; ** and *: statistical significance at the 0.01 and 0.05 levels, respectively. From Wagner 1992, figs. 1, 2, 4.

gain than large ones because acquiring a territory is almost certainly more difficult for them than for large males. The enhanced response of small males to large-male calls is also consistent with the predictions of a third model, which considers the costs of escalation (Payne and Pagel 1996). The model predicts that inferior individuals should escalate immediately when confronted by superior opponents but delay such (escalation) costs in contests with males of the same or lower abilities until the initial (unexaggerated) signals prove to be ineffective.

Endurance Rivalry: Benefits and Costs

After securing a territory or calling site, a male's mating success depends mainly on his ability to attract females. In chapter 10 we discuss how variation in the acoustic properties of advertisement calls affect female preferences. In particular, playback experiments indicate that female preferences are often based on differences in dynamic properties of signals, such as call rate and call duration, which are usually highly correlated with the energetic costs of signaling.

Most studies of anurans (e.g., Gerhardt et al. 1987; Murphy 1994a,b; Halliday and Tejedo 1995) find that chorus attendance is the single best predictor of mating success, probably just because individuals attending multiple choruses increase their cumulative probability of attracting a mate. This relationship is also likely to be true of insects, but field data are meager because of the greater difficulty of observing and keeping track of individuals. One exception is the sagebrush cricket *Cyphoderris strepitans* (Haglidae), in which the mating status of males is indicated by damage to their hind wings, which mating females eat while mating (Snedden 1996). Lifetime mating success of males in an enclosed area of meadow was positively correlated with the number of nights during the season that males were found calling. Whereas abundant evidence is available to show that costs of calling limit chorus tenure in anurans, the extent to which energetic costs affect either short- or long-term signaling in insects is controversial (e.g., Prestwich 1994).

The main goals of our discussion of endurance rivalry are (1) to show that variation in dynamic properties typically indicates variation in the energetic costs of signaling; (2) to evaluate the evidence that energetic costs, which can be considerable, can constrain calling behavior within the daily (nightly) calling period as well as limit chorus attendance during breeding season; and (3) to examine some facts and ideas about how the energetic costs of signaling are met.

Correlations between Signaling Patterns and Energetic Costs

Numerous studies report significant correlations within species between aerobic metabolic costs and signaling rate, duration, or a combined measure of rate and duration (calling effort) (figs. 9.10, 9.11; Taigen and Wells 1989; Pough et al. 1992; Bailey et al. 1993; Prestwich 1994; Grafe 1996a; Hoback and Wagner 1997; Wells 2001). In one species of cricket (*Gryllus lineaticeps*), metabolic costs increase but are not significantly correlated with chirp duration (Hoback and Wagner 1997), and in *Physalaemus pustulosus*, adding chucks to whines appears to incur negligible energetic costs (Bucher et al. 1982). These results are particularly interesting because females of these two species prefer songs with longer chirps and whines plus chucks, respectively (Ryan 1985; Wagner 1996). Anaerobic costs associated with calling are insignificant (reviews: Pough et al. 1992; Prestwich 1994).

Aerobic metabolic rates are also directly correlated with rates of signaling in interspecific comparisons (fig. 9.12). In crickets, differences in metabolic rates during stridulation reflected species differences in song amplitude and signaling patterns (trilling vs. chirping) (Lee and Loher 1993). In *Anurogryllus muticus*, for example, the sound pressure level of its songs are 92–95 dB at 20 cm, and its aerobic metabolic rate is nearly 11 ml O_2/h/g; in a congener, *A. arboreus*, the sound pressure level of the calling song is about 78 dB at 25 cm

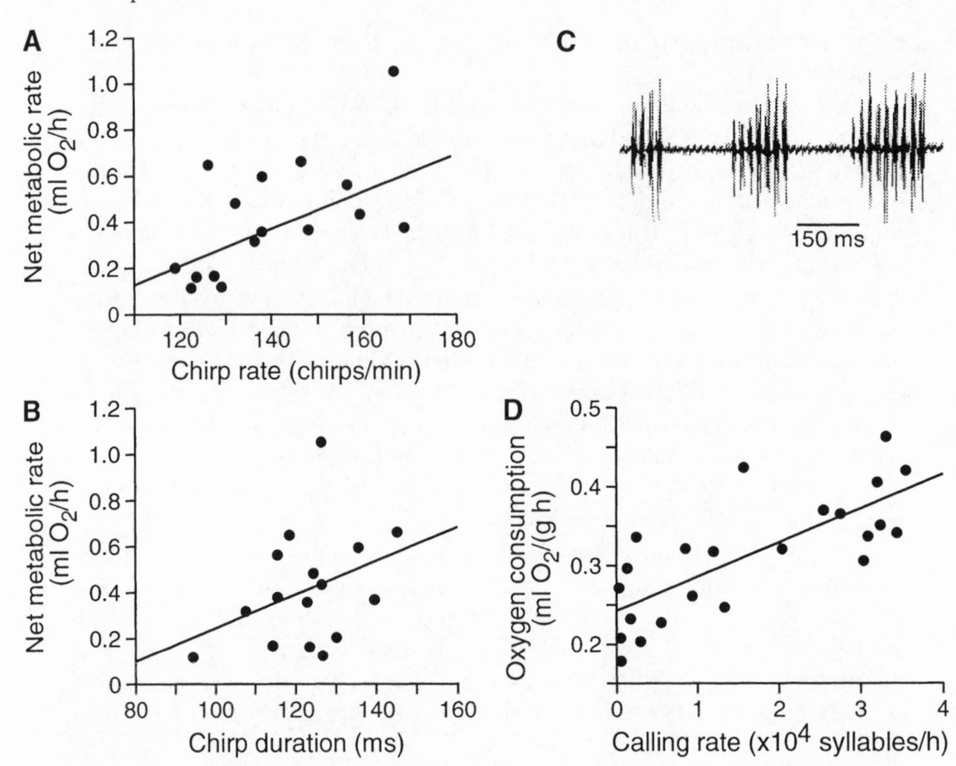

Figure 9.10. Aerobic metabolic costs as a function of variation in dynamic properties of calling songs in a cricket and a katydid. (A) Oxygen consumption and chirp rate in *Gryllus lineaticeps.* (B) Oxygen consumption and chirp duration in *G. lineaticeps;* the data plotted were not corrected for body mass. (C) Three chirps of the calling song of *Requena verticalis* showing variation in the number of syllables per chirp (4, 7, and 9 in the exemplars shown). (D) Mass-specific oxygen consumption in *R. verticalis* and rate of syllable production. From Hoback and Wagner 1997, fig. 1; and Bailey et al. 1993, figs. 1, 2.

and its metabolic rate is about 4 ml/h/g. In *Teleogryllus commodus,* whose songs are produced at about 65–70 dB SPL, the metabolic rate during singing is only about 1.2–1.3 ml/h/g. The latter species produces chirps rather than trills, and hence the reduced workload (production of fewer sound pulses per unit time) probably contributes to its lower costs, at least over the short run. As discussed below, however, the total energetic costs, and hence the extent to which they might constrain signaling over the course of a breeding season, will also depend on the duration of the diel period(s) of signaling (chapter 8) and on the length of the breeding season.

In anurans that signal at high rates, species differences in body size and mechanisms of sound production influence aerobic metabolic costs (review: Wells 2001). For example, medium-sized gray treefrogs *H. versicolor* and *H. chrysoscelis* require one body-wall contraction for each pulse of their pulsed

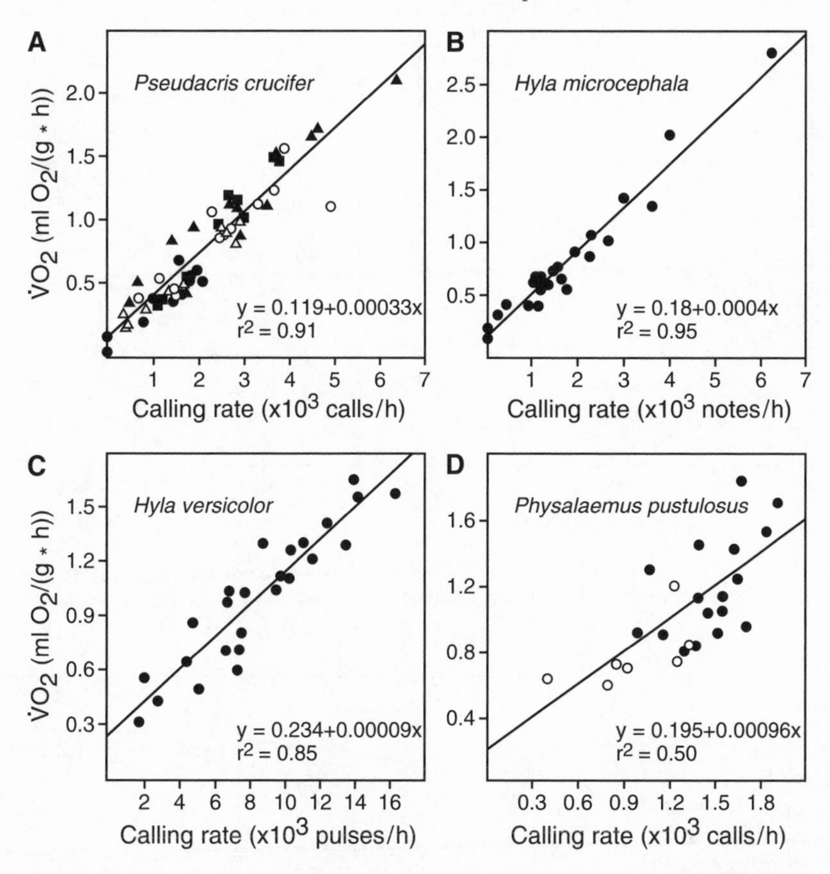

Figure 9.11. Aerobic metabolic rate (mass specific) as a function of variation in dynamic properties of the advertisement calls of four species of anurans. (A) Call rate in *Pseudacris crucifer.* (B) Call rate in *Hyla microcephala.* (C) Pulses per hour (a function of both call duration and rate) in *Hyla versicolor.* (D) Call rate in *Physalaemus pustulosus.* From Wells 2001, fig. 1.

advertisement calls, which contain 8–45 pulses (Girgenrath and Marsh 1997). The calls of these species are more costly than those of small treefrogs that do not produce pulsed advertisement calls (e.g., *Pseudacris crucifer*). Even though males of *Hyla microcephala* produce rapidly pulsed calls, they use passive amplitude modulation, which requires only one sustained muscle contraction per call (see chapter 2); adding secondary click notes, as males of this species do in dense choruses, however, is expensive (Wells and Taigen 1989; Wells 2001). Strong correlations also exist between the calling rates of a species and several aspects of their muscle physiology. For example, mitochondrial volume and capillary length per volume of trunk muscle fiber are much higher in species that sustain high calling rates than in species that call at low rates (fig. 9.13; Wells 2001). For spring peepers *P. crucifer*, significant intraspecific correlations have also been documented between calling rates and

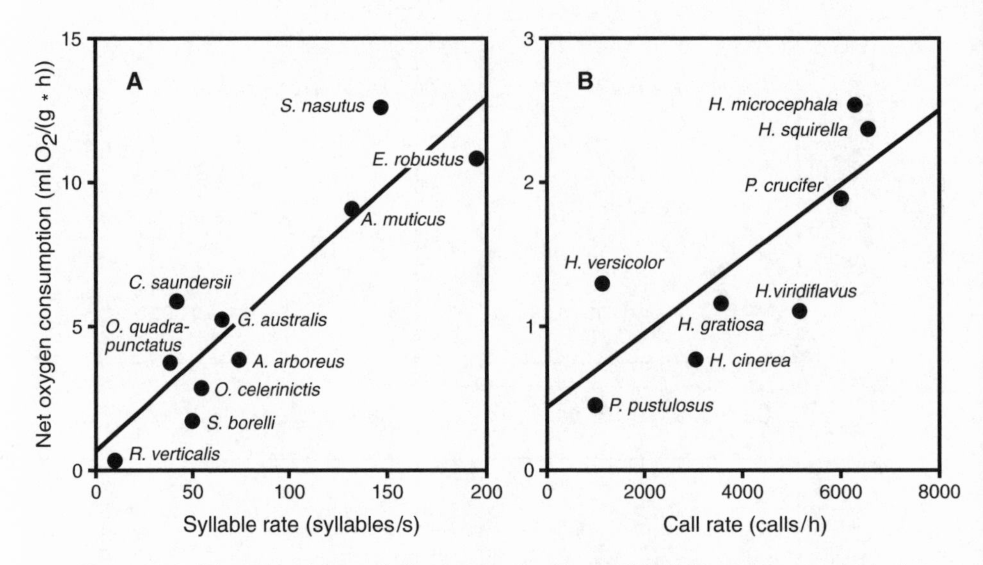

Figure 9.12. Plots of aerobic metabolic costs as a function of (A) syllable rate in ten species of insects; and (B) call rate in eight species of anurans. From Prestwich 1994, figs. 3, 4.

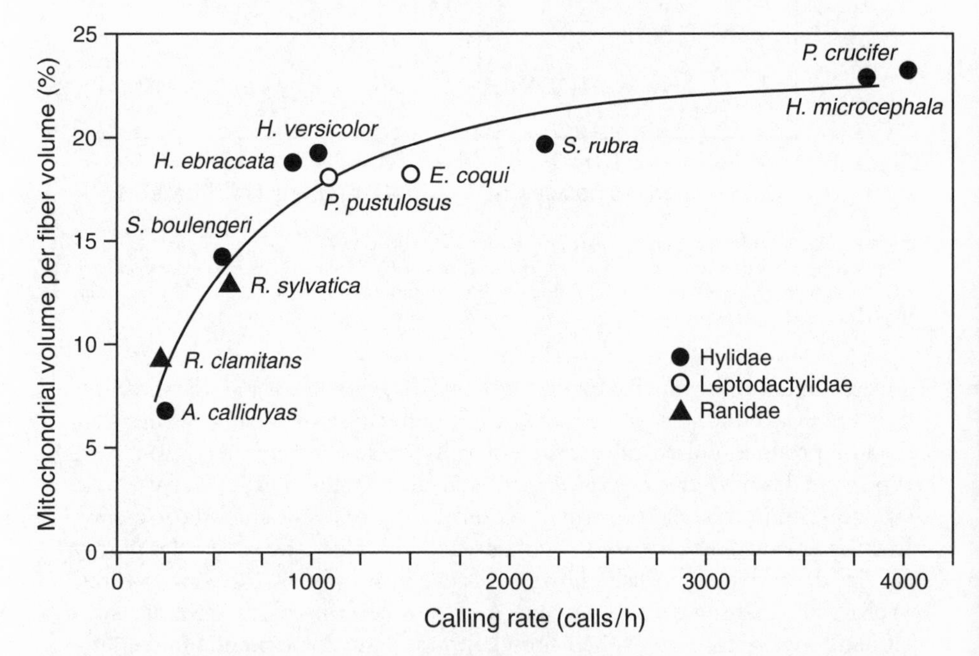

Figure 9.13. Physiological correlates of vocalization and its energetic costs in anurans. Plot of mitochondrial volume within muscle fibers that provide power for calling as a function of call rate. These data suggest that species with high levels of calling have much higher aerobic metabolic capacity than species with low levels of calling. From Wells 2001, fig. 2.

muscle physiology, blood hemoglobin levels and cardiac capacity (Zimmitti 1999).

Energetic Constraints on Signaling

Many studies of anurans document weight loss during the course of prolonged breeding seasons (e.g., Mac Nally 1981b; Arak 1983a, 1988; Cherry 1993; Murphy 1994a,b). In a few species, weight loss was positively correlated with chorus attendance (*Bufo:* Kagarise Sherman 1980, cited in Murphy 1994a; Tejedo 1992; *Rana:* Wells 1978), and in *Uperoleia rugosa*, males that called lost weight, whereas males that adopted satellite tactics gained weight (Robertson 1986). Growth rate was inversely related to calling rate in the carpenter frog *Rana virgatipes* (Given 1988).

Experimental evidence that energetic costs can constrain chorus attendance is available for the barking treefrog *Hyla gratiosa.* Although the breeding season is prolonged, most males attended fewer than one-third of the choruses, which is typical of many other species that call at relatively high rates (Murphy 1994a,b). Aerobic metabolic rates increased about twelve times over resting values during calling (Prestwich et al. 1989), a value equal to the mean for eight anuran species (Prestwich 1994). Males produced, on average, about 8500 calls per night, and Murphy estimated that males usually obtain only a tiny fraction ($\leq 0.10\%$) of the estimated energy needed to fuel a night of calling during a day of foraging. He tested the energy-limitation hypothesis by comparing attendance of males that he fed as they left the chorus with controls that were not fed. Fed males returned to the chorus sooner and for more nights than did controls (fig. 9.14A,B). In one study of the túngara frog *P. pustulosus* (Marler and Ryan 1996), supplemental feeding of males increased the calling activity, whereas in another study, the same effect was not found (Green 1990). These two results cannot be properly compared, however, because neither study quantified rates of calling in males that did or did not receive extra food. Green (1990) also found that fed males did not attend more choruses than controls. Predation risks are, however, almost certainly a more important factor influencing the probability of returning to the chorus in *P. pustulosus* than in *H. gratiosa* (Ryan 1985; Murphy 1994b). Murphy (1994b) also argues convincingly that the proportion of males returning to the chorus is probably not as sensitive a measure of the effects of feeding males as the number of added nights of chorus attendance.

Arguments against energetic limits on signaling in insects center on how easily energy needed for calling can be replenished on a daily basis and on the low costs of signaling relative to other activities. Prestwich (1994) doubts that signaling in insects, even over extended periods of time, is energy-limited because most species are generalist feeders (see also Heller and Helversen 1993). In support of this argument, Prestwich also points out that whereas aerobic metabolic rates during singing in insects increase by about 2–21 (mean = 9)

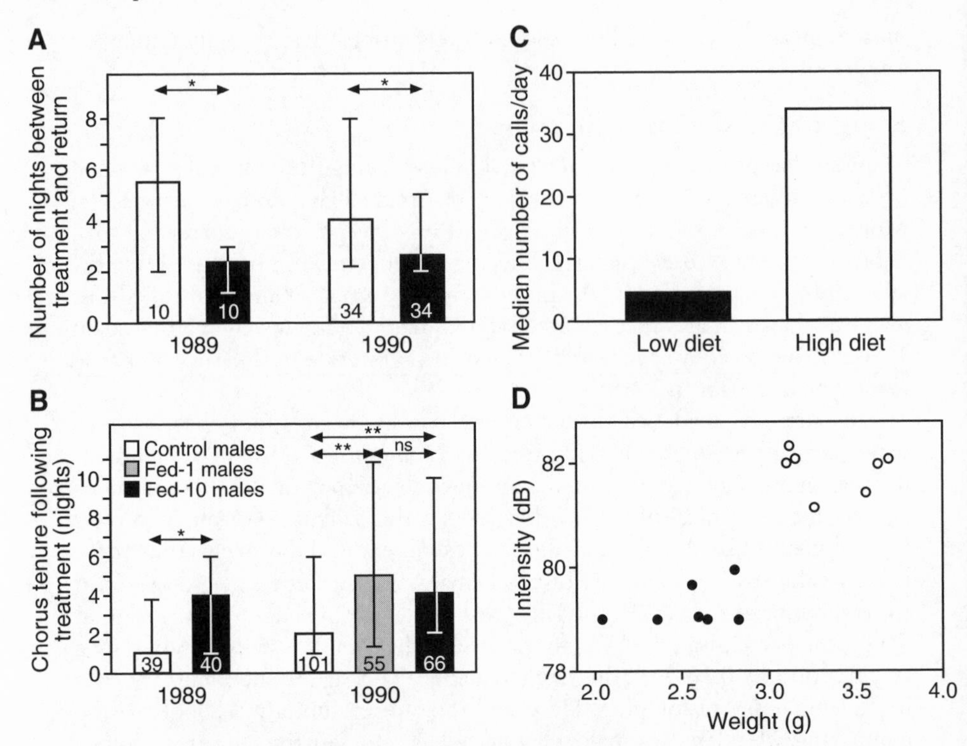

Figure 9.14. Feeding experiments, chorus tenure, and signaling performance. (A) Number of nights (medians ± one quartile) that it took males of the barking treefrog *Hyla gratiosa* that were fed crickets (*solid bars*) and control males (*open bars*) to return to the chorus following their treatment in the feeding experiment. (B) Number of nights that males called in the chorus following their treatment in a feeding experiment (median ± one quartile). Numbers inside bars indicate sample sizes. From Murphy 1994b, figs. 5, 6. (C) Daily call production by males of the katydid *Ephippiger ephippiger* on a restricted diet compared with that of males on a normal diet. (D) Plot showing that the higher weight of males on the normal diet resulted in increased call amplitude. Open column and circles = high (normal) diet; filled column and circles = restricted diet. From Ritchie et al. 1998, fig. 3.

times, metabolic rates in insects increase by 30–200 times during flying. Fighting in crickets is approximately 40 times more costly than stridulation (Hack 1997).

For some anurans, by contrast, the rate of oxygen consumption of vocalizing males often exceeds that observed when males are forced to exercise to peak capacity (Taigen and Wells 1985; Prestwich et al. 1989; Pough et al. 1992). Thus, calling is much more energetically expensive than the usual costs of locomotion, as when the frogs are foraging or moving to or from the breeding site (Walton 1988). As pointed out by Prestwich (1994), this difference between insects and anurans stems from differences between the two taxa in the sound power of their signals and in the number and mass of muscles used in signaling, both of which are generally higher in anurans than in insects.

We are not entirely convinced by Prestwich's arguments, which are contradicted to some extent by some empirical data. First, even if food is immediately available, insects still have to take time to forage to replace the energy expended in prolonged signaling. Second, the durations of flight and fighting are much shorter than bouts of singing, and the net efficiencies of locomotory activities are much higher than the efficiencies that have been estimated for stridulation: 10–20% versus 0.05–6.4% in seven species of insects and six anurans (Prestwich 1994, McLister 2001; but see Bailey et al. 1993 who estimate much higher efficiencies for signaling in a katydid). Evidence is also available that energetic costs can influence both short- and long-term patterns of signaling in crickets and katydids. In two species of crickets (*Anurogryllus arboreus* and *Oecanthus celerinictus*), males devoted about 26% and 56%, respectively, of their daily respiratory budget to stridulation (Prestwich and Walker 1981), and food-deprived males stopped calling after a few days. The daily respiratory contribution to signaling in *O. celerinictus* was higher than that (45%) in two species of anurans (genus *Ranidella*) that deplete their energy reserves for calling during the breeding season (Mac Nally 1981a,b). Finally, species that signal over extensive periods of time have a higher resting metabolic rate than species that infrequently signal (Rheinhold 1999). One hypothesis to explain this correlation is that species with higher resting metabolic rates have evolved to increase the efficiency of metabolism during periods of high activity. This observation suggests that energy limitations might have at least been a factor during the evolutionary history of the species that signal for prolonged periods at high rates.

Feeding experiments also provide direct evidence that signaling effort is affected by available energy in three species of insects. In the katydid *Requena verticalis*, both calling effort and body weight increased with the level of nutrition (Simmons et al. 1992). In *Gryllus lineaticeps*, males on a high-nutrition diet sang at higher rates than did full-sibs on a low-nutrition diet but did not gain weight (Wagner and Hoback 1999). In the katydid *Ephippiger ephippiger*, males on a restricted diet produced far fewer calls per day at a significantly lower amplitude than did males on a normal diet (fig. 9.14C,D; Ritchie et al. 1998). Other species of insects that call at high rates to attract females (e.g., periodical cicadas and wax moths: Reinhold et al. 1998) feed little, if at all, after reaching sexual maturity. Hence acoustic signaling in these species is entirely dependent on stored energy. To the extent that males differ in stored energy or the efficiency of its conversion to signaling, then energetic costs should significantly influence variation among males in chorus participation (e.g., in cicadas) and the duration and rate of signaling.

Fuel for Signaling and Other Physiological Costs

Carbohydrates supply energy for muscle contractions at a faster rate than lipids, but carbohydrate stores are depleted more rapidly than lipid reserves and provide less energy per gram than lipids (Wells 2001). In general, insects and anurans with high calling rates and prolonged breeding seasons meet the energetic costs of signaling with lipids (Wells 2001). By contrast, species with low signaling rates mainly (and sometimes exclusively) use carbohydrates, which are often stored in muscles as glycogen (Lee and Loher 1993; Prestwich 1994; Wells 2001).

These generalizations must be qualified. First, the primary sources of both kinds of fuels are reserves that are stored in the muscles or fat bodies (lipids) and food ingested during the breeding season. The balance depends on calling patterns, the duration of the breeding season, and the availability of food during the breeding season. For example, in frogs such as the wood frog *Rana sylvatica*, which breeds explosively during about a week in early spring when food is unavailable, males rely almost exclusively on stored muscle glycogen, which is rapidly depleted even though calling rates are not particularly high (Wells and Bevier 1997). By contrast, in the green frog *Rana clamitans*, which has a prolonged season in the summer and calls at low rates, males have very small glycogen and lipid reserves that decline only slightly during the season (Wells 2001). These males feed throughout the season and can thereby meet most of their energy costs without tapping stored fuels. In another summer breeder with a prolonged season (the barking treefrog *H. gratiosa*), males also obtain energy for breeding by feeding during the season. However, as discussed above, chorus attendance and, to some extent, their rate of calling on a given night (but not the duration of nightly calling; Murphy 1999) are energy-limited (Murphy 1994a,b). Call rate in barking treefrogs is also much higher than that in green frogs.

A second qualification is that even within individuals of the same species, the ratio of utilization and sources of lipids and glycogen can change with food availability. In spring peepers (*Pseudacris crucifer*) and gray treefrogs (*Hyla versicolor*), for example, males rely on stored lipid and glycogen reserves early in the season when food is scarce, and then later in the season derive most of their energy for calling from feeding. In the reed frog *Hyperolius marmoratus broadleyi*, the proportion of energy derived from lipids varies widely (37–89%), and fed males decreased lipid utilization (and hence increased carbohydrate use) relative to males that had been food-deprived for a week (Grafe 1996a). Lee and Loher (1993) provide comparable data on variation in fuel use within and between species of field crickets.

Third, muscle fatigue, which can be caused by glycogen depletion, might also affect calling performance even if lipids are available and also used as fuel (Schwartz et al. 1995 and references therein). For example, even though lipids are the main fuel for the high call output of the treefrog *H. microcephala*, glycogen depletion from trunk muscles might contribute to an explanation of the cyclical, unsynchronized chorusing pattern of this species. Schwartz and his colleagues estimate that if males called continuously rather than in bouts, the total duration of calling per night would be about three hours rather than four hours (Schwartz et al. 1995).[1] Whitney and Krebs (1975) hypothesized that cyclical chorusing in the Pacific treefrog was caused by fatigue, which they incorrectly assumed was caused by the buildup of lactate (Schwartz et al. 1995). However, glycogen depletion or energy limitations in general are certainly not the only explanations for cessation of calling by this or other species (e.g., Wells et al. 1995; Schwartz 1999). Toward the end of the season, for example, many males that stopped calling or called for shorter than average periods of time still had reasonably high reserves of glycogen.

Finally, as yet unidentified energetic constraints can affect chorus attendance and the duration of the nightly calling period. In the gray treefrog *H. versicolor*, for example, males tend to maintain the same calling effort and aerobic metabolic cost by either producing long calls at a relatively low rate, or short calls at a relatively high rate (Wells and Taigen 1986; fig. 9.15A). Males more often produce long calls, which are preferred by females (Klump and Gerhardt 1987), in dense choruses than when calling from isolated sites, and individuals also increase call duration and reduce call rate when stimulated with playbacks (Wells and Taigen 1986). The nightly calling period of males producing long calls is significantly less than that of males producing shorter calls at faster rates (fig. 9.15B), and long callers also attend fewer choruses (Sullivan and Hinshaw 1992). Wells and Taigen (1986) hypothesized that producing long calls depleted glycogen reserves in the trunk muscles more rapidly than did the production of short calls, but this prediction was not confirmed in a study by Wells et al. (1995). Instead, these authors estimated that glycogen reserves supply only about 25% of the energy needed for a night of calling. More recently, Grafe (1997b) showed that call duration in *H. versicolor* is weakly correlated with the rate of lipid depletion and confirmed that lipids are the main source of energy for calling in this species. More sensitive assays of glycogen levels could still confirm Wells and Taigens's original hypothesis, and the main effect of low glycogen levels might be on muscle performance rather than an energy limitation per se.

Alternative Mating Tactics

Despite the high correlation between mating success and calling effort, chorus attendance, or both, some males adopt tactics other than signaling to secure a mate. Such alternative tactics, the vast majority of which are probably best considered as alternatives within a conditional strategy (Gross 1996; box 9.1), are common in a wide variety of animal taxa (Andersson 1994). The most common nonsignaling tactics in insects and anurans are scramble competition and satellite behavior. In scramble (searching) competition, which

1. There is no evidence that anurans rely on anaerobic metabolism for calling (Wells 2001).

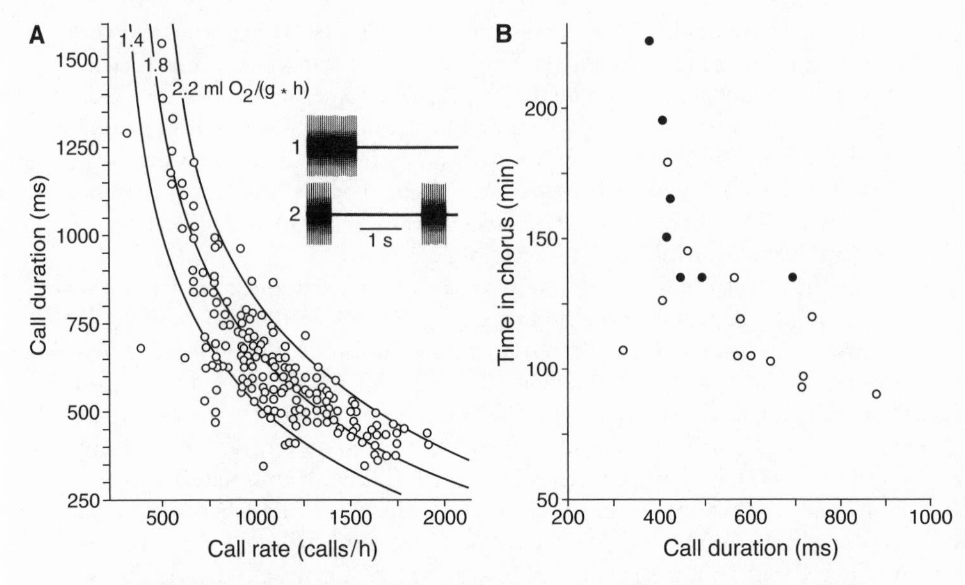

Figure 9.15. Chorus effects and consequences on calling behavior. (A) Plot of call duration against call rate in the gray treefrog *Hyla versicolor*. The rightmost curve is the estimated maximum aerobic metabolic rate sustainable during calling. Higher calling efforts would begin to draw on anaerobic metabolism. From Wells and Taigen 1986, fig. 7. *Inset*, oscillograms showing synthetic alternatives that had the same call-duty cycle but differed in duration by a factor of two; these signals were representative of males calling in choruses of different densities and were used in female-preference experiments discussed in chapter 10. (B) Plot of call duration versus time spent calling during a chorus on a single night. Open and closed circles identify data that were collected in different years. From Pough et al. 1992, fig. 14.19.

most often occurs at high density, males simply move within a breeding aggregation and attempt to force matings or court (sometimes using acoustic signals) any nonsignaling individual they encounter (e.g., Wells 1977; Feaver 1983; Cade and Wyatt 1984; Hissmann 1990; Cade and Cade 1992; Souroukis and Cade 1993; Sullivan et al. 1995).

Satellite behavior is a tactic in which males do not call while situated near a calling male. Satellite males may try to intercept females that are attracted to the calling male, wait for the calling male to vacate its territory or calling site, or both. As with aggressive signaling, much more experimental work has been done with satellite males in anurans than in insects. Table 9.2 summarizes the diversity of behavior within and between well-studied species.

Satellite behavior is not always defined in the same way. For example, although silent male gryllids sit near a calling male and presumably intercept females (Cade 1979; Evans 1983; Sakaluk 1987; Loher and Dambach 1989), noncalling males adopting this tactic and males that actively search (but do not call) have often been lumped in the same category (e.g., Cade and Cade 1992). Moreover, in both insects and anurans, some silent males sitting near

Box 9.1. Alternative Reproductive Strategies and Tactics

As discussed by Gross (1996), strategies and tactics have been variously defined in the literature. Nearly all phenotypic differences in reproductive behavior are probably alternative tactics within *conditional strategies.* That is, the tactic adopted by an individual depends on its status (size, age) relative to rivals or on environmental conditions (e.g., chorus density, nutritional status). Alternative tactics can be maintained by status-dependent selection, and different tactics differ in average fitness. Switchpoints (conditions that cause an individual to change from one tactic to another) are under selection and evolve, and the proximate mechanisms underlying different tactics can be as subtle as differences in the levels of sex hormones during development (e.g., Bass 1992). *Alternative strategies* are based on genetic polymorphisms that are maintained by frequency-dependent selection with the condition that the two or more strategies have the same average fitness. Two possible examples are the set of three strategies (fighting, mimicking of females, and sneaking of copulations) employed by three size classes of male isopods (*Paracerceis sculpta*) (Schuster and Wade 1991) and the winged and flightless morphs of the bladder grasshopper (see the text).

calling ones might simply be resting and not adopting the satellite tactic (Perrill et al. 1982; see below). Because crickets and katydids must first induce the female to climb onto their backs before mating, it would seem that the satellite tactic might be less successful in these acoustic insects than in grasshoppers and anurans, in which males clasp the female.

Still another kind of alternative tactic, which has been considered a form of satellite behavior, has recently been described in a duetting species of the Australian katydid *Elephantodeta nobilis* (Bailey and Field 2000). As in other phaneropterine species (chapter 4), males produce an elaborate calling song and females reply within a specific time window after a trigger pulse. Responding females also move toward calling males. Males that are not producing calling songs are apparently attracted to duets and emit a brief sequence of clicks during the calling male's song and about 200 ms before the trigger pulse. The female then replies earlier than usual, and the calling male may respond to the "satellite's" calls by producing additional trigger pulses. In a laboratory setting, females preferentially moved to the calling song but were attracted to the clicks of the satellite male 20–30% of the time. The conditions favoring this alternative signaling behavior and its potential success in nature still need to be studied.

Several nonmutually exclusive factors may favor satellite behavior. On the

Table 9.2. Satellite Behavior in Anurans

Species	Maximum Number of Silent Males around Caller	Successful Female Intercepts	Density-Dependent	Size Difference	Silent Males Call after Caller Removed	Reference
Bufo calamita	—	Yes	Yes	Yes	Yes	Arak 1988
Bufo cognatus	5	Yes	Yes	Yes	Yes	Krupa 1989
	5	Yes	Yes	—	Yes	Sullivan 1982a
Bufo compactilis	—	—	—	Yes	—	Axtell 1958
Bufo americanus	—	No	—	Yes	—	Gatz 1981a
	—	Yes	Yes	Yes	No	Fairchild 1984
Bufo woodhousei	—	—	Yes	—	—	Brown and Pierce 1967
Acris crepitans	—	Yes	—	No	—	Perrill and Magier 1988
Pseudacris crucifer	—	—	Yes	Yes	—	Forester and Lykens 1986
	—	No	—	No	Yes	Fellers 1979
	—	—	Yes	No	Yes	Lance and Wells 1993
	—	—	—	Yes	—	Gatz 1981b
Pseudacris triseriata	2	—	—	No	Yes	Roble 1985

Hyla cinerea	3	No	—	Yes	No	Garton and Brandon 1975
	—	Yes	Yes	Yes	—	Gerhardt et al. 1987
	—	Yes	—	Yes	—	Perrill et al. 1978
	—	—	—	Yes	Yes	Perrill et al. 1982
	—	—	—	—	Yes	Fellers 1979
Hyla versicolor	3	No	—	No	Yes	Fellers 1979
	—	—	—	Yes	—	Gatz 1981b
Hyla chrysoscelis	2	Yes	No	No	Yes	Roble 1985
	—	—	—	—	Yes	Fellers 1979
Hyla minuta	3	Yes	No	Yes	Yes	Haddad 1991
Hyla arborea	1	No	—	—	—	Marquez and Tejedo 1990
Hyla ebraccata	—	Yes	—	—	—	Miyamoto and Cane 1980
Hyla regilla	1	Yes	—	—	—	Perrill 1984
	—	—	—	No	Yes	Fellers 1979
Hyla picta	1	—	No	—	—	Roble 1985
Hyla squirella	—	—	—	—	Yes	Fellers 1979
Ololygon rubra	—	Yes	—	Yes	—	Bourne 1992, 1993
Uperoleia rugosa	—	—	Yes	Yes	Yes	Robertson 1986a
	—	No	—	Yes	—	Robertson 1986b
Rana catesbeiana	—	Yes	—	Yes	—	Howard 1978a, 1984

one hand, adopting an alternative tactic may be the only possibility to obtain matings for small, young, or weak individuals in species in which larger (older?) established males hold territories and attack other males that signal within or near the territory. Well-studied examples falling into this category include bullfrogs and other ranid frogs (Howard 1978a). In desert grasshoppers (genus *Ligurotettix*) (Shelly and Greenfield 1991; Greenfield and Minckley 1993), late-eclosing males are likely to be satellites, even though calling is a significantly more successful tactic. On the other hand, energetic costs of signaling, masking interference in large aggregations (see chapter 8), or selective pressures from acoustically orienting predators or parasitoids (see chapter 11) may favor the temporary adoption of alternative tactics by any male in the population.

Theory predicts that alternative tactics could exist as a balanced, genetically based polymorphism if the long-term success of alternative tactics is equal (Maynard Smith 1982; box 9.1). There is little hard evidence for genetically based alternative or mixed strategies in any animal (Gross 1996). For example, although artificial selection experiments demonstrate heritable differences in the duration of nightly calling in field crickets, such selection does not create a class of males that are always silent (Cade 1981). In the bladder grasshopper *Bullacris membracioides*, one class of males does not have the option of signaling because these males lack wings and the stridulatory apparatus (fig. 9.16; Alexander and van Staaden 1989; van Staaden and Römer 1997). These males apparently find mates by searching for females on food plants. Male-male competition occurs within the different morphotypes. That is, winged, singing males compete with other winged, singing males, and flightless males fight with other flightless males. We do not know if these different tactics are merely consequences of different developmental conditions that lead to the gross differences in the two phenotypes or whether these are two genetically determined morphs. The first explanation is suggested by the fact that wingless males can (rarely) transform into winged, singing ones (van Staaden, pers. comm.).

The empirical data for anurans suggests that satellite behavior is usually conditional on a male's age, size, physical condition, signaling ability, and the density of calling males (table 9.2). For example, satellite males of the green treefrog *Hyla cinerea* are highly successful in intercepting females (Perrill et al. 1978), but many males called when the calling male with which they had been associated was removed. Males that did not switch or called for only a short time were relatively close to another persistent caller; satellites also seldom called when the calling male was replaced by a speaker playing back advertisement calls (Perrill et al. 1982). Satellite males tended to be smaller than calling males in dense choruses than in choruses of moderate or low density (Gerhardt, pers. obs.). One testable explanation is that small males cannot produce sufficiently intense calls that females can readily detect against the chorus background noise (Gerhardt and Klump 1988a). This idea is supported by

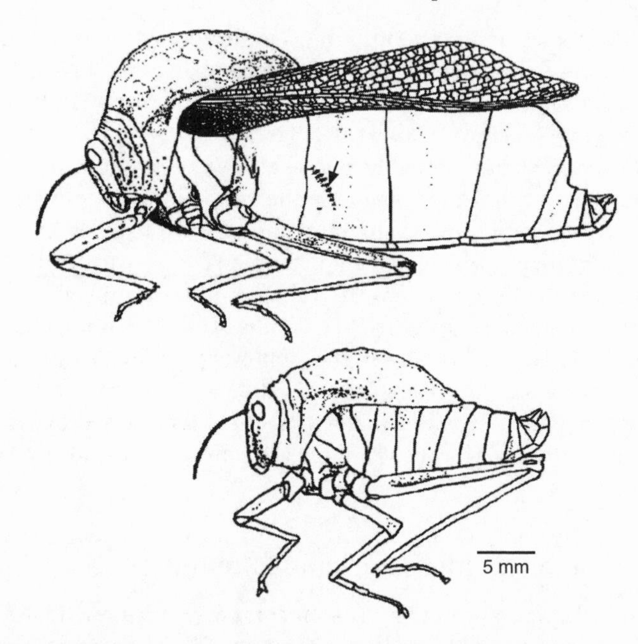

Figure 9.16. Diagrams of two male morphs of the bladder grasshopper *Bullacris membracioides.* (A) Adult male with strongly inflated abdomen and stridulatory apparatus (*arrow*). (B) Alternative male, lacking both wings and stridulatory structures. From van Stadden and Römer 1997, fig. 1.

experiments with natterjack toads (*Bufo calamita*), in which smaller, satellite males preferentially associated with speakers playing back high-intensity calls (Arak 1988). Such calls are also more attractive (and probably more readily detectable) to females than low-intensity calls.

In the green treefrog *H. cinerea*, satellite males are more likely to associate with speakers playing back synthetic calls simulating attractive advertisement calls than they are to associate with speakers playing back less attractive (or aggressive) calls (Gerhardt 1978a, 1988; Humfeld, unpubl. data). Here the playback levels of the alternative signals were equalized so that detectability was not a confounding factor. Comparable data are available for house crickets (*Acheta domesticus*) (Kiflawi and Gray 2001), although the prevalence of this tactic and its success in this species remains undocumented. The relative long-term success of this tactic (and scramble competition) vis-à-vis signaling has not been rigorously estimated, however, and such measurements are essential to make conclusions about the evolutionary forces that maintain these tactics or determine switchpoints (Gross 1996; box 9.1).

Energetic costs might also promote satellite behavior, but no study has so far shown that such limitations alone explain the adoption of satellite behavior in any species. For example, the aerobic capacity of the muscles was similar in calling and satellite males of *Pseudacris crucifer* (Lance and Wells 1993).

Nevertheless, males in poor condition (as indicated by their mass or mass to length ratio) were more likely to act as satellites than males in good condition in *Uperoleia laevigata* (formerly *U. rugosa*) and in *Hyla cinerea* (Robertson 1986; Humfeld, unpubl. data).

Scramble competition probably involves lower energetic costs than calling. Searching, however, must also increase the risk of predation by sit-and-wait predators. Furthermore, males that find and capture females are often challenged by other males, which attempt takeovers. Mate guarding then requires considerable energetic costs, and the release calls and movements of struggling frogs may also attract predators. In studies of the common toad *Bufo bufo*, Davies and Halliday (1978) showed that contests over females often lasted for hours; the persistence of attacks was negatively correlated with the size of the male initially clasping the female. Male size was assessed by rivals on the basis of acoustic (carrier frequency of release calls), visual, and tactile (force of kicks) cues.

Summary and Suggestions for Future Studies

Acoustic signals play a prominent role in competition among males to secure matings. Aggressive signals mediate interactions between males that can exclude other males from resources needed by females for reproduction or from suitable places to signal. Advertisement calls attract females, and chorus tenure and signal duration and rate are often positively correlated with male mating success. These generalizations rest mainly on studies of anurans, and additional data of these kinds from insects are needed.

During vocal interactions, males usually first modify the structure as well as the relative timing of their long-range advertisement signals. These changes, which usually involve increases in signaling rate, duration, or complexity, may also enhance the attractiveness of the signals to females. If such changes are ineffective in repelling a rival, then aggressive signaling may ensue. Although aggressive signals or signals with aggressive components are often attractive to female insects, these signals are usually less attractive to female anurans than are advertisement calls.

Some species of insects and anurans produce a graded continuum of signals, ranging from advertisement-like signals to pure aggressive signals. For a few of these species, experimental data show that these gradations are somewhat predictable of subsequent aggressive behavior. Other species produce two-part calls, and for two species there is evidence that one part of the call is directed at rival males and the other, at females. More studies are needed to assess the behavioral significance of aggressive-call variants and two-part calls. At this time, the results across species are too diverse to make generalizations.

A few studies of possible bluffing using acoustic signals reveal intriguing patterns of size-dependent signaling that need to be followed up by interactive playbacks and additional field experiments. Such methods could address the main question posed by the available data: How does a male who alters his signals respond when his challenge is accepted? Observational data might be necessary because visual and tactile stimuli could also contribute to the assessment of the honesty of a male's acoustic signals. Moreover, simple observations can also test another prediction of the hypothesis of Payne and Pagel (1996), namely, that males challenging an opponent are more likely to win fights than are males that do not challenge. The interpretation of observations and experiments must also consider the prior experiences of males, chorus density, and the time during the breeding season (Burmeister et al. 1999a,b).

In the bullfrog the recognition of the calls of familiar neighbors (and their "expected" location) has been documented and interpreted as an example of the dear enemy effect. Other species of anurans, and perhaps insects, with relatively stable territories in which particular individuals frequently interact are candidates for discovering additional examples. Future studies should concentrate on the development and mechanisms of neighbor-stranger discrimination, estimates of the duration and maintenance of memories for the signals of particular neighbors, and the fitness economics of the dear enemy effect.

Acoustic signaling in insects and anurans is energetically inefficient and costly. The extent to which energetic costs constrain chorus attendance and signaling rates and duration varies among species, and some authorities argue that insects are unlikely to be at all constrained. The few experimental studies that have assessed the effects of feeding on male chorus attendance and calling performance suggest, however, that variation in energetic resources can contribute to variation in signaling behavior in both male insects and anurans. Additional studies of energetic and other factors that affect within-night and across-season variation in signaling would be welcome, and the next step is to learn if these behaviors generally correlate with physical and genetic fitness (see chapter 10).

Satellite male behavior has been described in many species, but for most of these, critical observational and experimental data are lacking. Silent males situated near signaling ones are sometimes assumed to be sexual parasites when, in fact, there is no evidence that they attempt to intercept females or even that they signal after the "dominant" male has vacated the site. Moreover, few studies have estimated the rate of success of satellite males, even in those species in which satellites are known to attempt interceptions. The least controversial explanation for this tactic is that satellites are males that cannot compete successfully with larger or stronger males for territories or calling sites. Satellites are smaller than signaling males in some species, but the relative size and frequency of satellites often varies in time and space within the same species and is sometimes a function of chorus density. Satellite behavior might also be a response to depletion of energetic reserves needed for calling or an evolutionary consequence of selective pressure from acoustically orienting predators and parasites. The evidence supporting these two hypotheses is scant. More studies, in which the physiological and condition measurements can be correlated with switches between calling and satellite behavior, are badly needed. Feeding experiments, such as those discussed above, could also provide direct tests of the influence of energetic resources on calling propensity.

10

Female Choice Based on Acoustic Signals

THIS CHAPTER focuses on the acoustic properties of signals that are likely to serve as criteria for mate choice. After providing an overview of female choice in insects and anurans, we survey patterns of preference (preference functions) with respect to several kinds of acoustic properties. One generalization that emerges is that preferences based on static, fine-temporal properties are usually stabilizing or weakly directional in form, and preferences based on dynamic, gross-temporal properties are typically strongly directional. Patterns of preferences based on other properties are variable. We next review studies that have attempted to provide evidence for mate choice based on signal variation. In general, these studies show that preferences demonstrated in laboratory studies are not always realized in nature, and even strongly directional patterns of preference are unlikely to translate into strong selection on male signals. Some of these studies also identify factors that confound the expression of preferences in natural environments.

The main theoretical issue surrounding female choice centers on the fitness benefits, if any, that a female might gain by choosing mates on the basis of variation in particular male traits or signals. *Direct benefits* increase a female's fecundity (*quantity* of offspring) or her survivorship. For example, by picking a healthy male, a female may have more eggs fertilized and reduce her risks of infection; conspicuous signals would be favored if they allow the female to find a male quickly and hence reduce her risks of predation. *Indirect (genetic) benefits*, which are much more controversial, depend on the acquisition of genes that affect the *quality* of offspring by (1) influencing the attractiveness of male offspring and preferences of female offspring (*Fisher effect*, *sexy-son*, or *runaway models*) or (2) by increasing the viability of offspring of both sexes (*indicator*, *good-genes*, or *handicap models*). These are not mutually exclusive explanations. The Fisher effect might, for example, be initiated by preferences that result in direct benefits or indirect benefits in the form of good genes. As the genetic correlation between male traits and female preferences, which is expected to arise from assortative mating, becomes stronger, then the male trait becomes exaggerated so much that it decreases the viability fitness of its male bearer as it increases his sexual attractiveness. Andersson (1994) provides a thorough treatment of the theory and a review of the empirical evidence for these processes. Preexisting biases, which we discuss in chapter 11, have been interpreted as another way of explaining the evolution of female preferences (Ryan 1990). In our discussion we focus on the benefits that females might receive by using particular acoustic criteria to choose among signaling males; different kinds of benefits might also contribute to the trend for female choice to exert different patterns of selection on fine-temporal and dynamic properties.

Sexual selection will not lead to evolutionary change in signals unless variation in signals has a heritable basis. Indeed, one of the main theoretical objections to models of indirect selection is that genetic variation in male traits used to attract mates might be depleted by strong female choice (Andersson 1994). This problem might be solved by the capture of genetic variation during the evolution of condition-dependent traits or signals (Rowe and Houle 1996; Kotiaho 2000). Genetic variation in female preference is also required to promote the genetic correlations between signals and preferences that are predicted by such models to arise from assortative mating (i.e., offspring of both sexes inherit genetic factors that underlie both the male trait and female preference). At the end of the chapter, we examine the evidence for repeatable and heritable variation in signals and preferences. Because of the meager information available for orthopteran insects, cicadas, and anurans, we expand our taxonomic coverage in this section by considering some studies of the genetic bases of signal and preference variation in other insects.

Overview of Female Choice in Insects and Anurans

Mate Attraction and Reproductive Success

Attraction to the vicinity of a signaling male does not necessarily result in mating or successful fertilization of a female's eggs by the male, especially in insects. For example, field observations of the katydids *Requena verticalis* and *Orchelimum nigripes* indicated that only about 10% of encounters following phonotaxis resulted in copulation (Bailey et al. 1990; Feaver 1983). In the variable field cricket *Gryllus lineaticeps*, females moving toward an attractive calling song moved away if a relatively less attractive courtship song was then presented in its place (Wagner and Reiser 2000). For most species of insects, females can probably more easily avoid unwanted matings than can female anurans. Recall in particular that female crickets and katydids mount the male and can simply interrupt mating by moving away (Alexander et al. 1997). Furthermore, in some insects, females can mate multiple times over relatively short periods of time and store sperm from different males. A variety of male structures and behaviors have apparently evolved to prevent a female's subsequent mating with other males and hence to increase his chances of fertilization success (Choe and Crespi 1997). At the same time, the delay between copulation and fertilization provides additional opportunities for female insects to exercise postcopulatory or *cryptic mate choice* (Eberhard 1996) by influencing sperm transfer (e.g., Sakaluk and Eggett 1996) or by differential provisioning of eggs fertilized by different males (Choe and Crespi 1997). Finally, females of some species apparently benefit directly by mating repeatedly with the same or different males, gaining nutrients (katydids: Heller et al. 1997) or increases in life span and fecundity (crickets: Wagner et al. 2001). Such benefits might reduce the pressure for selective mating.

Field observations suggest that male anurans in which amplexus occurs are usually successful in clasping females that touch or move near them, even if they have not attracted her by acoustic signals (Wells 1977). Visual cues are used for mate choice in a diurnal poison-arrow frog (*Dendrobates pumilio;* Summers et al. 1999), and a variety of visual signals that probably serve both aggressive and mate-attraction functions have been described (review: Hödl and Amézquita 2001). Chemical cues might play a role in mate attraction in an aquatic frog (genus *Hymenochirus;* Pearl et al. 2000). However, females of most anuran species breed nocturnally and appear to rely solely on sounds to identify and assess males. For example, females of the green treefrog *Hyla cinerea* often attempted to induce mating with a male barking treefrog *H. gratiosa* that was tethered to a source of conspecific calls (Gerhardt 1994a). Prolonged amplexus, which occurs in many species of anurans, can be considered a form of mate guarding. However, because fertilization is external in nearly all species, there is effectively no time gap between sperm deposition and fertilization. Simultaneous multiple mating occurs in some species, and fertilization of eggs within the same egg mass by primary and secondary males has been documented in three species (*Agalychnis callidryas:* D'Orgeix and Turner 1995; *Chiromantis xerampelina:* Jennions and Passmore 1993; *Crinia georgiana:* Roberts et al. 1999). There is no evidence that females can control fertilization by different males, and in *Crinia georgiana,* females mating with multiple mates have reduced fertilization success (Byrne and Roberts 1999). Females can still exercise some postmating choice by varying clutch size. In water frogs (*Rana esculenta* complex), for example, females reduce clutch size when amplexed by hybrids, whose gametes displace those of the parental species (Reyer et al. 1999). This tactic would, however, be risky for most species, in which fitness consequences of mating with different conspecific males are likely to be minor. With some exceptions (e.g., *Crinia georgiana:* Byrne and Roberts 1999), the long duration of amplexus in most anurans nearly always precludes remating by the same female on the same night, and the majority of females probably mate only once during a breeding season (see Perrill and Daniel 1983 for data on treefrogs). Thus, the total number of mating opportunities for individual females is limited. Producing large numbers of offspring of slightly lower-than-average quality should outweigh producing small numbers of relatively high-quality young.

In summary, differences in the ability of female insects and anurans to avoid unwanted matings, to mate multiply, and to exercise postmating control suggest that long-range acoustic signals are likely to be more directly and strongly linked to reproductive success in male anurans than in male insects. Nevertheless, because a male insect must first attract a female in order to be further assessed before and perhaps during mating, males producing the most attractive long-range signals will still be favored by female choice.

Studying Female Choice

Two main approaches generate evidence for female choice based on acoustic signals. In one approach, direct or indirect correlations are sought between mating success and some acoustic property. In a second, complementary approach, playback experiments are used to test female preferences, or, ideally, to estimate preference functions, which quantify the strength of preferences for particular values of acoustic properties. Comparing preference functions to the distributions of acoustic properties in natural populations predicts which subsets of males are likely to be favored by female choice.

Patterns of preference are best determined in a way that takes into account female sampling behavior, which also influences mate choice (Wagner 1998). Single-stimulus playbacks, which assess preference strength in terms of the probability of response or the accuracy of orientation and speed of responses, are appropriate for species in which females usually sample male signals in a sequential manner. "Choice" experiments are suited to species in which females can hear and assess multiple males from the same point. Males calling from relatively widely spaced sites are likely to be sampled sequentially, whereas males in chorus-breeding species are subject to simultaneous sampling. If, however, the signals of a chorusing species overlap extensively, females might have to move close to one or a few males at a time in order to assess a series of individuals (Gerhardt and Klump 1988a; chapter 8). All things being equal, increasing the number of males sampled will add to the costs of female choice, especially if signaling males are dispersed. The willingness of females (choosiness) to incur these costs is likely to vary both within and between species (review: Jennions and Petrie 1997).

Playback experiments are valuable even if their predictions about patterns of mating success are not always confirmed in the field. The net effect of female choice on acoustic signals is almost certainly weak, and many field studies simply lack the statistical power to detect even moderate levels of female choice. Because even weak selection can still have profound evolutionary effects over many generations, negative results from field studies do not necessarily mean that preferences demonstrated in the laboratory are spurious or unimportant. Moreover, if robust studies with large sample sizes still fail to find patterns of mating success predicted by playback experiments (e.g., Dyson et al. 1992), researchers can then seek to identify such confounding factors as acoustic complexity, direct male-male competition, satellite interceptions, and variation in the operational sex ratio (reviews: Gerhardt 1994a; Sullivan et al. 1995). Finally, playback experiments not only generate predictions about sexual selection in contemporary populations, but also can reveal "hidden" preferences for extreme values ("supernormal stimuli') or for acoustic properties that are absent from conspecific signals (Arak and Enquist 1993). Such perceptual biases might reflect past evolutionary history and can also predict future directions of signal evolution (Ryan 1990; Ryan and Rand 1993; chapter 11).

Patterns of Preference and Potential Selection on Acoustic Properties

In appendix 4 we summarize patterns of female preferences based on several kinds of acoustic properties. The data are derived from playback experiments with species in which females usually initiate mating with a male after a phonotactic approach or produce acoustic responses that guide him to her. The results are not all strictly comparable because of the variety of experimental designs and response criteria used by different researchers (see chapter 4).

In figure 10.1 we summarize some of the data from appendix 4. For each of four properties, we show the number of species (or populations) showing a particular pattern of preference (or the lack of a preference). Preferences are categorized as stabilizing or weakly directional when intermediate values of a property are at least as attractive as values from one or the other end of the distribution. Within this category a pattern would be considered as stabilizing if the preferred range of values matched the mean value in the calls of conspecific males in the same population and discrimination against higher and lower values were equally strong. Weakly directional patterns arise in two ways: (1) The preferred intermediate value is somewhat higher or lower than the mean but the preference function is still unimodal; (2) the preference function is asymmetrical in that values at one end of the distribution are similar in attractiveness to intermediate values whereas values at the other end of the distribution are less attractive. Preferences are categorized as strongly directional preferences when values from one end of the distribution or beyond are preferred to intermediate values and to values at the other end of the distribution. Unfortunately, most studies have tested too few values in order to distinguish between stabilizing and weakly directional patterns. Indeed, many studies are even less informative because only the two extremes have been tested against one another. If no preference exists, then the function could still be unimodal (intermediates preferred to extremes), and even if values of one extreme are preferred to values of the other extreme, we cannot know whether the function is weakly or strongly directional (Gerhardt 1991; Wagner 1998).

As in an earlier comparative study of treefrogs (Gerhardt 1991), patterns of preference based on static properties (pulse rate and carrier frequency, which

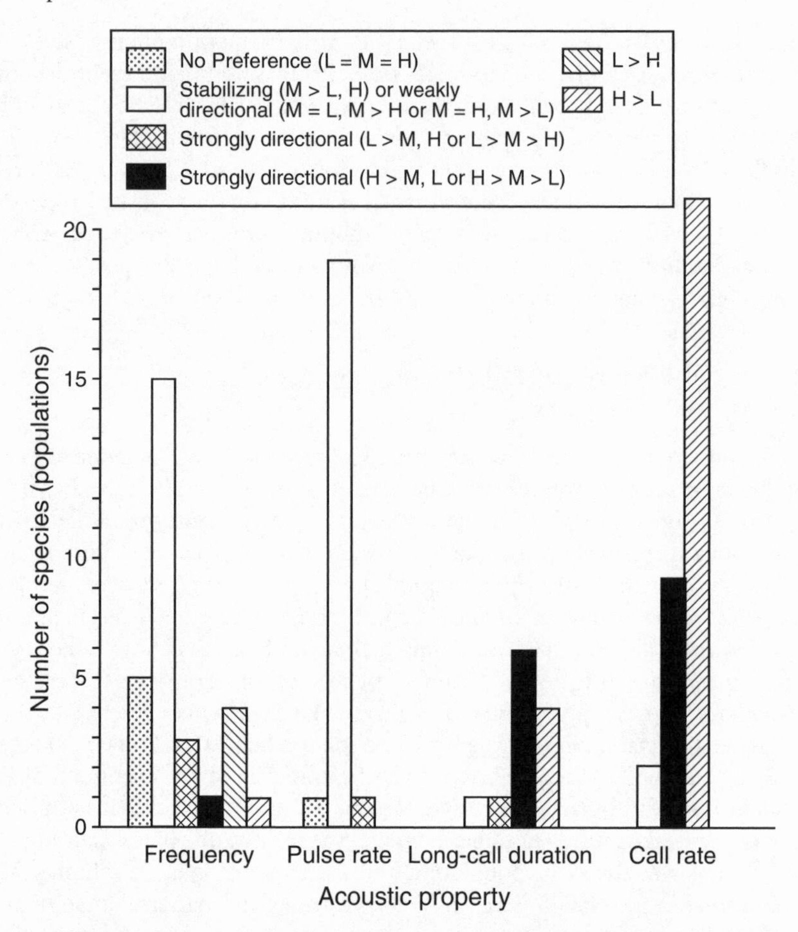

Figure 10.1. Histograms showing patterns of preference based on four kinds of acoustic properties in orthopteran insects, cicadas, and anurans. In all species, females usually initiate matings, and in playback experiments, only one property of the alternative stimuli was varied at a time. L = values of an acoustic signal from the low end of the range of variation; M = intermediate values that were usually close to the mean; H = values from the high end of the range of variation; the equal sign indicates that values were equally attractive; > value to the left of the inequality symbol were more attractive then values to the right. The text provides further discussion of the definitions of stabilizing, weakly directional, and strongly directional patterns and an explanation for why we provide no definition for patterns labeled L > H or H > L.

show low within-male variability: see chapter 2) are most often stabilizing or weakly directional whereas patterns of preference based on dynamic properties (call duration [of "long" calls] and call rate, which usually show high within-male variability) are usually strongly directional (see also Ryan and Keddy-Hector 1992). However, despite the general trend for preferences based on static and dynamic properties to mediate different forms of selection, counter-examples are common, especially in studies of preferences based on carrier frequency. Moreover, patterns of preference based on the duration of short signals, whose within-male variability (when the data are available) differs significantly among species (chapter 2), run the gamut from stabilizing to highly directional.

Female Preferences Based on Static Properties

Carrier Frequency

Patterns of preferences based on differences in carrier frequency are much more variable than those based on pulse rate, even though both are static properties (fig. 10.1). Moreover, in a substantial number of studies, differences in frequency spanning the entire range of variation in conspecific calls did not elicit preferences (e.g., Arak 1988; Cherry 1993; Doolan and Young 1989; Lopez and Narins 1991). Whereas most studies report stabilizing or weakly directional preferences, a few studies provide evidence for strongly directional preferences (Gwynne and Bailey 1984; Ryan et al. 1992; Wollerman 1998; Castellano et al. 2000). Among the latter studies, lower-than-average frequencies were preferred to average frequencies in three anurans (*Acris crepitans*, *Hyla ebraccata*, and some populations of *Bufo viridis*), whereas higher-than-average frequencies were preferred in a katydid (*Kawanaphila nartee*) (appendix 4).

The proximate basis for variability among species in female preferences was anticipated in chapters 4 and 5. Recall that the tuning of the auditory system and female preferences often show a reasonably good match to the general range of frequencies emphasized in conspecific signals. However, auditory tuning also differs appreciably among species in its sharpness, symmetry, and the extent to which maximum sensitivity corresponds to the average carrier frequency or frequencies in communication signals (e.g., Meyer and Elsner 1996; Gerhardt and Schwartz 2001).

Two studies of insects tested a sufficient range of carrier frequency to conclude that the preference function is stabilizing. In the Hawaiian cricket *Laupala cerasina*, females preferred intermediate values to extremes (fig. 10.2; Shaw and Herlihy 2000). The estimated (population-level) mean preference is about 4.76 kHz, which is virtually the same as the mean carrier frequency (4.77 kHz) in the songs of conspecific males; a second estimate, based on the preference functions of fifteen individual females, is 4.94 kHz. Similarly, in the katydid *Phaneroptera nana*, the minimum threshold for the acoustic response of females occurred at the carrier frequency (16 kHz) typical of conspecific calls (Tauber and Pener 2000). Qualitatively similar results are available for field and mole crickets (Stout et al. 1983; Thorson et al. 1982; Ulagaraj and Walker 1975; additional examples in chapter 4).

Preference functions for carrier frequency also tend to be stabilizing or weakly directional in several species of anurans. In the midwife toads *Alytes cisternasii* and *A. obstetricans*, females were offered simultaneous choices of seven

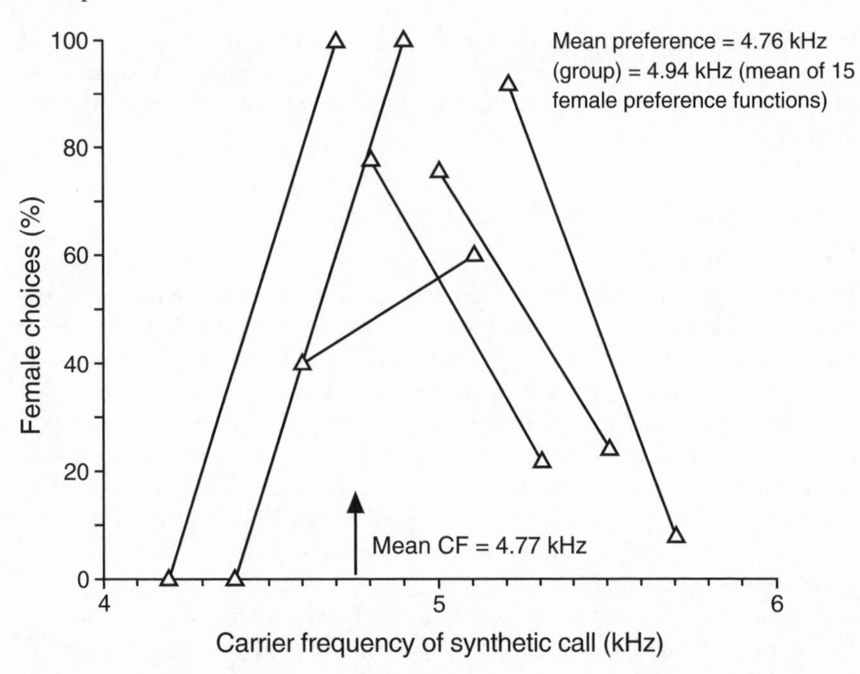

Figure 10.2. Preference function for carrier frequency in the cricket *Laupala cerasina*. The lines connect points showing the percentages of females responding to the alternative synthetic calls with the carrier frequencies indicated. The "sliding window" design used in these experiments estimates the most-preferred carrier frequency for the population as the indifference point (here it is 4.77 kHz) between tests in which females preferred the alternative with the higher frequency and tests in which females preferred the alternative with the lower frequency. Data replotted from Shaw and Herlihy 2000, fig. 3.

synthetic calls that differed in carrier frequency; the arrays of values were either "mean-centered" or represented frequencies at the low end of the range of variation and even lower (Márquez and Bosch 1997a). The latter design was suggested by the results of two-speaker playbacks showing that low-frequency signals are preferred to calls of average and higher frequency, and hence the preference could have been strongly directional or even "open-ended" (Márquez 1995). Considering all of the data, however, the most preferred frequencies in *A. cisternasii* and *A. obstetricans* were only fractions (0.27 and 0.38, respectively) of one standard deviation below the mean carrier frequency in the calls of males (fig. 10.3). Thus, in neither species is the preference strongly directional or open-ended. Females of *Hyperolius marmoratus* showed weak preferences for low-frequency over high-frequency calls, which were abolished by varying the timing relationship of alternative stimuli (Dyson and Passmore 1988a,b). More extensive testing of *Hyperolius m. broadleyi* revealed a distinctly stabilizing pattern (Grafe 1997a).

Frequency preferences in the green treefrog *Hyla cinerea* were originally characterized as stabilizing (Gerhardt 1982, 1987). A more detailed analysis of

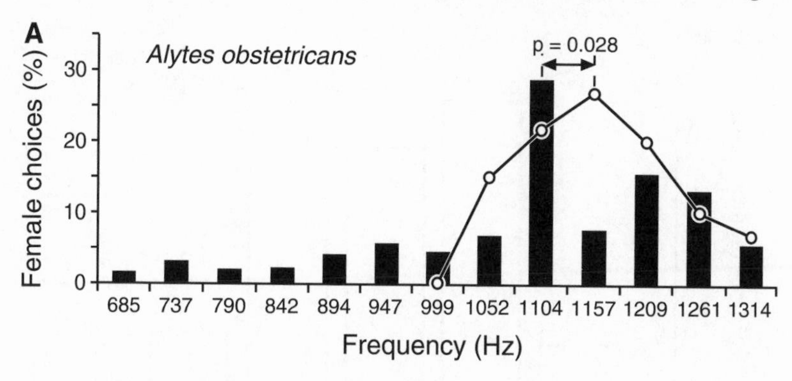

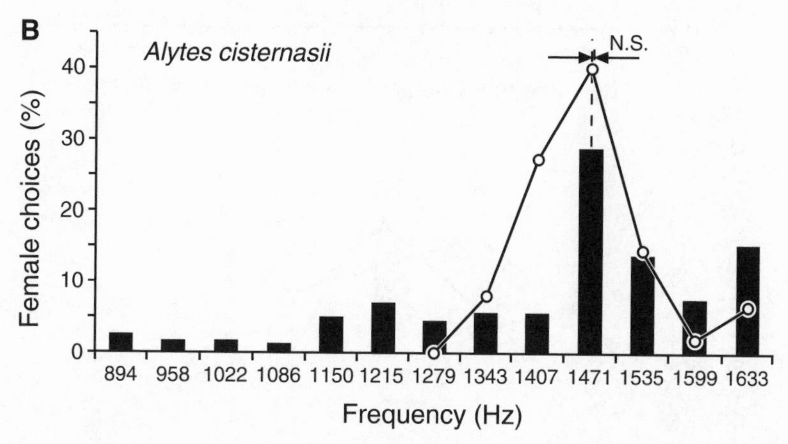

Figure 10.3. Preference functions based on carrier frequency in midwife toads *Alytes obstetricans* and *A. cisternasii.* The bars show the distribution of females attracted to each of thirteen different frequencies. The results are combined from a series in which seven speakers each played back at a different frequency that included values that were mean-centered and another series ("supernormal") that included frequencies well below the mean. Open circles show the distribution of carrier frequency in the population, and vertical dotted lines indicate average female choice and average carrier frequencies. Females of both species preferred low to high frequencies in two-speaker tests, but these multispeaker tests show that the preference is neither open-ended nor strongly directional. The average preferred frequency is less than one standard deviation below the mean frequency in *A. obstetricans* (A) and nearly matches the mean frequency in *A. cisternasii.* (B). The p-value (0.028) refers to the statistical significance of the difference in mean preferred frequency and mean carrier frequency in *A. obstetricans.* NS = not statistically significant. From Márquez and Bosch 1997a, fig. 3.

both two-speaker and four-speaker tests indicates that preference functions are asymmetrical and show a bias toward lower-than-average signals. Discrimination against a low-frequency alternative (0.7 kHz) in favor of a standard call (near the mean of 0.9 kHz) decreased as the playback level was increased; the preference for the standard call over a high-frequency alternative (1.1 kHz) was, however, maintained at all playback levels (see fig. 4.4). Thus, except at low sound pressure levels, females would be expected to discriminate

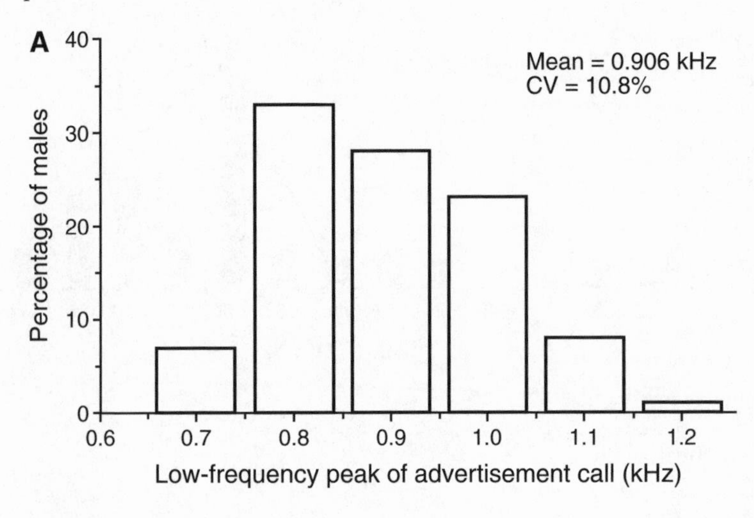

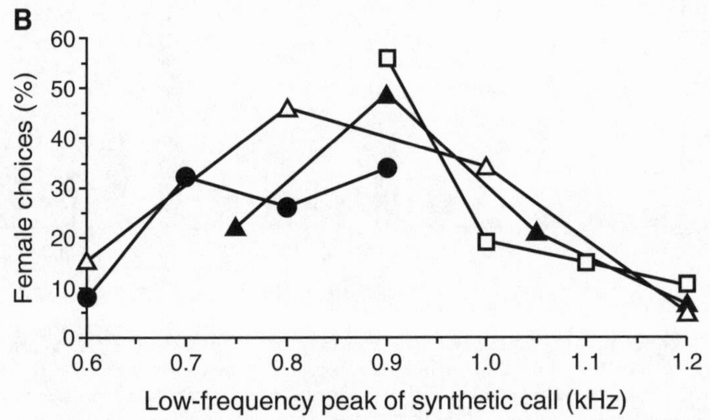

Figure 10.4. (A) Distribution of the frequency of the low-frequency spectral peak of the advertisement calls in a population of green treefrogs (*Hyla cinerea*) from eastern Georgia (U.S.A.). (B) Results of four-speaker playback experiments with females from the same region. The same symbols connected by lines show the percentages of females choosing particular synthetic calls that differed only in the frequency of the low-frequency spectral peak. Data from Gerhardt 1991, unpubl.

more strongly against calls of higher-than-average frequency than against calls of lower-than-average frequency. Moreover, as shown by figure 10.4, females offered simultaneous choices of four alternatives that differed by 0.1–0.2 kHz more strongly discriminated against alternatives of higher-than-average frequency in favor of signals with a frequency near the mean than they discriminated against alternatives of lower-than-average frequency.

The results of multiple-speaker tests with midwife toads and green treefrogs also serve to emphasize that preferences are likely to be somewhat degraded in acoustically complicated, more realistic circumstances. For example, females of *H. cinerea* from Georgia (U.S.A.) never chose an alternative of 0.6 kHz when at normal breeding temperatures, this stimulus was pitted against the standard call of 0.9 kHz in two-speaker tests at a playback level of 75 dB SPL; substantial proportions of females did so in four-speaker tests (fig. 10.4B; Gerhardt 1982, 1991). The idea that acoustic complexity can result in less selectivity is supported indirectly by a study of the painted reed frog *Hyperolius marmoratus* showing that a large-male mating advantage, presumably mediated in part through the correlation between body size and carrier frequency, occurred in low- but not high-density situations (Telford et al. 1989).

Two additional complications in assessing preference functions based on carrier frequency stem from the fact that temperature and body size can affect frequency preferences (see chapters 4 and 5). The implications for intraspecific mate choice of the direct correlation between tuning and temperature have not been formally considered; the prediction arising from behavioral experiments is that males producing lower-than-average frequencies should have increased mating success at lower-than-normal temperatures (Gerhardt and Mudry 1980). Size-dependent frequency preferences have been documented in five species. Large females preferred lower-frequency calls than did small females in cricket frogs (*Acris crepitans:* Ryan et al. 1992), midwife toads (*Alytes cisternasii* and *A. obstetricans:* Márquez and Bosch 1997a), painted reed frogs (*Hyperolius marmoratus:* Jennions et al. 1995), and Australian toadlets (*Uperoleia laevigata:* Robertson 1990). In the last species, playbacks using synthetic calls show that a female usually preferred calls with a carrier frequency that would be produced by males that are about 70% of the female's own weight, provided that a pair of alternative stimuli differed by at least 0.1 kHz. As discussed below, these preferences almost certainly contribute to the strong size-assortative mating pattern in this species.

Fine-Temporal Properties

Patterns of preference based on pulse rate, the most frequently studied fine-temporal property, are predominately stabilizing or weakly directional (fig. 10.1; see also chapter 4). Even though nearly all preference functions based on pulse rate have a basically unimodal shape, considerable variation exists in how well the most preferred value matches the mean value of pulse rate in the population. Again, even if values near the mean are most attractive, the preference function may be asymmetrical in the sense that values from one end of the distribution may be less attractive than values from the other end.

These conclusions are supported and illustrated by several particularly robust studies. In the crickets *Laupala cerasina* and *L. kohalensis,* estimates of the most preferred pulse rate are very close to the mean pulse rates of males in the same populations (Shaw and Herlihy 2000; Shaw 2000). In some tests, however, in which the average pulse rate was not presented, females responded to values well outside the range of variation. This result is consistent with other studies of insects discussed in chapter 4. For example, even though preference functions in field crickets and in the grasshopper *Chorthippus biguttulus* are distinctly unimodal, substantial proportions of females were attracted to signals with pulse rates that deviated significantly from values close to the mean in conspecific songs (Helversen and Helversen 1994; figs. 4.10B, 4.12, 4.14). The gray treefrogs *Hyla chrysoscelis* and *H. versicolor* provide particularly robust anuran examples of strong stabilizing preference functions (fig. 4.19; Gerhardt and Doherty 1988), although the most preferred value is slightly higher than the mean in *H. chrysoscelis* (Bush et al. 2002). In other species, females show relatively weak preferences. For example, although the majority of females of *Bufo viridis* and *Hyla ebraccata* chose signals with values near population means, substantial numbers chose alternatives with pulse rates well outside the conspecific range of variation (Castellano and Giacoma 1998; Wollerman 1998).

Other preferences based on fine-temporal properties involve quantitative differences in fine-temporal structure. In the katydid *Ephippiger ephippiger*, for example, the number of strongly expressed impulses within the main syllables of the song decreases as a function of age after the final molt, reflecting wear of the stridulatory apparatus (fig. 10.5). In two studies of this species, females preferred synthetic songs that simulated those of young males to variants with missing impulses typical of older males (Ritchie et al. 1995; Stiedl et al. 1991). Other, correlative studies found that age was positively related to mating success (review: Zuk and Simmons 1997). We discuss the possible significance of these results below.

Female Preferences Based on Dynamic Properties

Duration of "Long" Signals

Signals that last over about 0.5 s are usually highly variable in duration within bouts of calling (chapter 2), and female insects and anurans usually prefer long signals or signaling bouts to short signals or signaling bouts. Females of the katydid *Scudderia curvicauda*, for example, more reliably responded to playbacks of long phrases than they did to short phrases (fig. 10.6; Tuckerman et al. 1993). In this species, phrase length is strongly correlated with male body size. Females of *Gryllus integer* from California showed stabilizing preferences based on the duration of their very short chirps (Hedrick and Weber 1998; see below), but on a gross time scale, they had strong, repeatable preferences for long, uninterrupted bouts of singing as opposed to shorter, frequently interrupted bouts (Hedrick 1986, 1988).

Females of the grasshopper *Chorthippus biguttulus* responded well to phrases of longer-than-average duration and preferred higher numbers of phrases than males normally produce (fig. 10.7; Helversen and Helversen 1994). Such directional patterns of preference have also been documented for one or the other of the song elements in the biphasic calls of three other members of

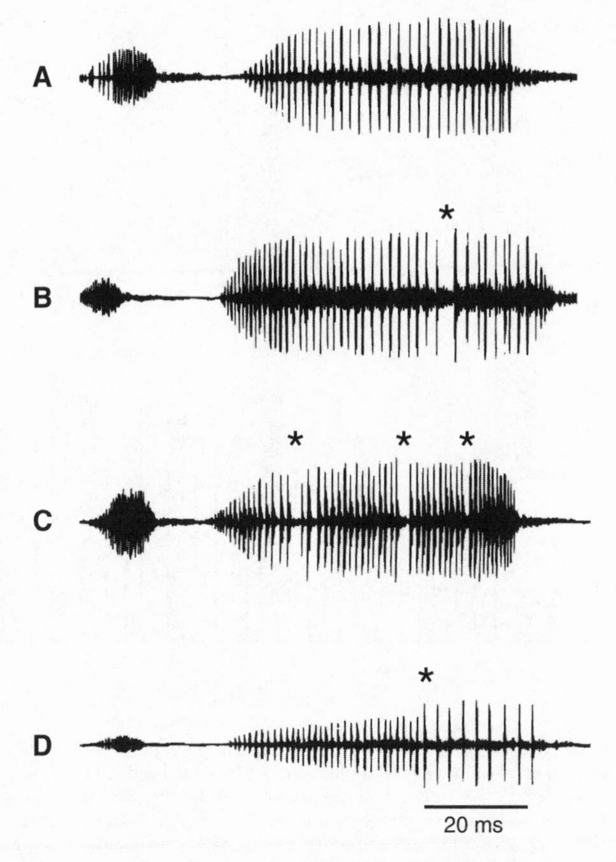

Figure 10.5. Effects of age on fine-scale temporal patterns in the songs of the katydid *Ephippiger ephippiger*. Oscillograms show one syllable (a short pulse made by wing opening and a longer pulse made during wing closing). The oscillogram of (A) is typical of the songs of young males, and the abnormal songs (B)–(D), of older males. Single stridulatory peg impulses in the longer part of the signal are missing in (B), (C) or abnormal (D) and are indicated by asterisks. From Ritchie et al. 1995, fig. 1.

the same genus: the number of A-syllables in *C. loratus* and the duration of B-syllables in *C. dorsatus* and *C. dichrous* (fig. 10.8; Stumpner and Helversen 1994). Females responded best to synthetic signals with values at the high end of the male distribution or beyond. Patterns of preference that can be considered stabilizing or weakly directional, however, are also found in these grasshoppers. In *C. loratus*, for example, increasing the duration of the B-syllable to about three times the normal duration reduced the response probability by about 50%; a similar reduction in response frequency in *C. dichrous* occurred when the number of A-syllables was increased to six or more from the usual one syllable (Stumpner and Helversen 1994).

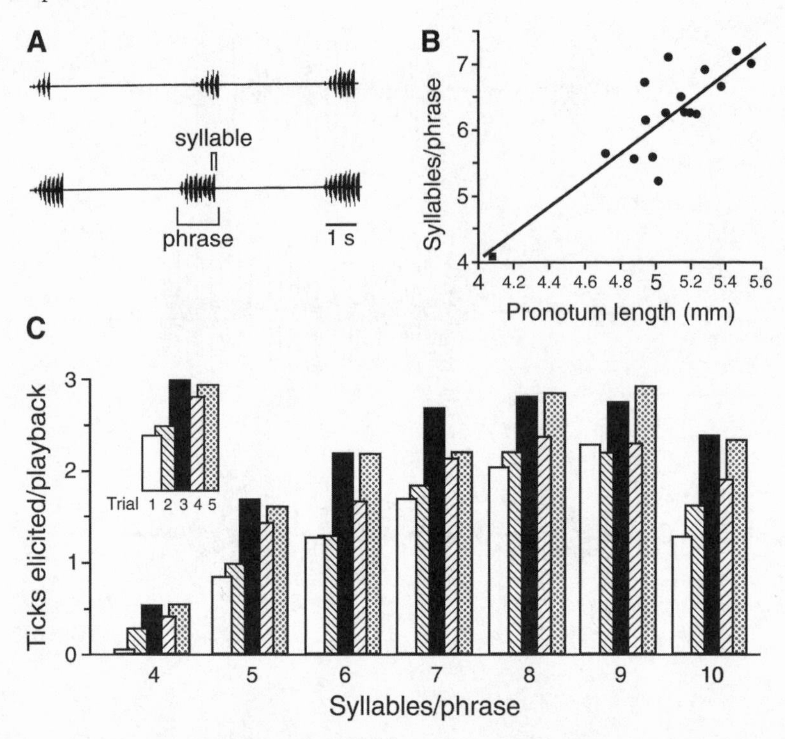

Figure 10.6. Preferences based on phrase duration (syllables per phrase) in the katydid *Scudderia curvicauda.* (A) Oscillograms showing examples of short (*top trace*) and long phrases (*bottom trace*); time base is 1 s. (B) Correlation between phrase duration and male body size (pronotum length). (C) Preference functions based on acoustic responses (ticks) by females to synthetic phrases that varied in the number of syllables per phrase. From Tuckerman et al. 1993, figs. 1, 2, 3.

In all three species of anurans tested with calls of low, average, and long duration (*Bufo valliceps*, *Hyla chrysoscelis*, and *H. versicolor*), preferences are highly directional; in three other species (*Bufo americanus*, *B. viridis*, and *B. woodhousei*), long calls were preferred to short calls (appendix 4). In the gray treefrog *Hyla versicolor*, female preferences for the longer of two calls occurred when synthetic calls differed by as little as one pulse per call (equivalent to a 50 ms [about 5%] difference in duration) (fig. 10.9). In this species and in *H. chrysoscelis*, females continued to prefer the longer of two calls that differed by 50–100% in duration even when the sound pressure level of the longer calls was reduced by 6–10 dB (Gerhardt et al. 1996). The preference function is not open-ended, however. In *H. versicolor*, for example, preferences were significantly stronger when two alternatives had durations well below the mean value in the population than when alternatives differing by the same percentage were above average in duration (fig. 10.10; Gerhardt et al. 2000). Preference functions shown in figures 10.7 and 10.8 are comparable.

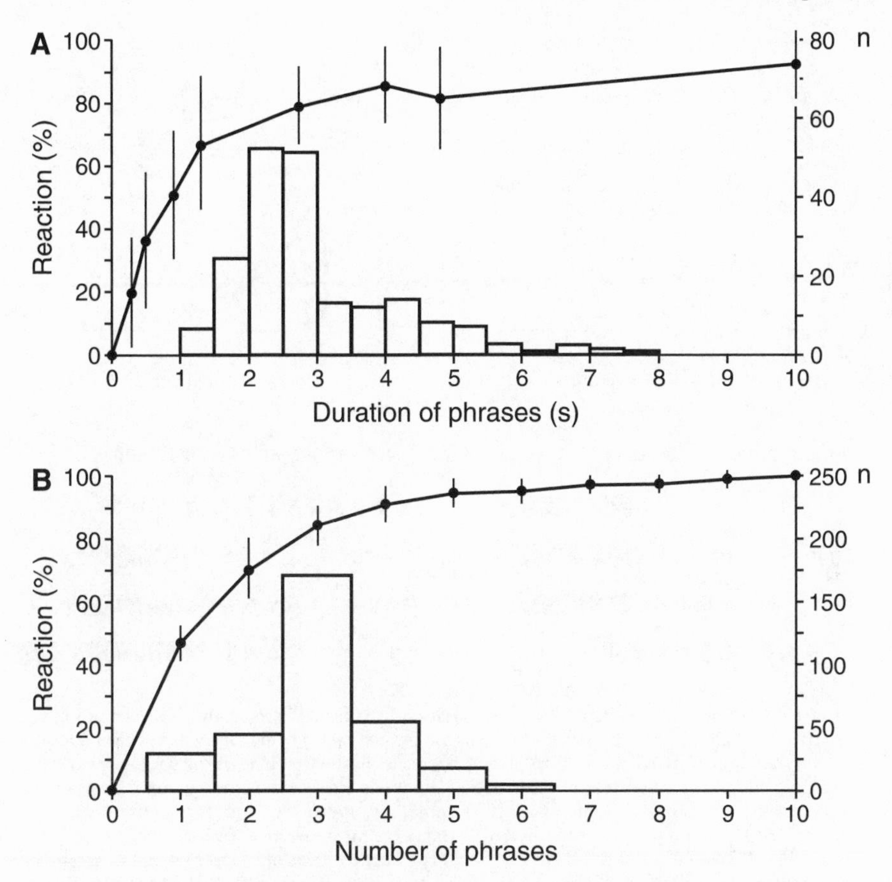

Figure 10.7. Preference functions based on variation in the (A) duration of phrases and (B) number of song phrases in the grasshopper *Chorthippus biguttulus*. Vertical lines indicate standard deviations. Histograms in each panel show the distributions of the two variables in male songs. From Helversen and Helversen 1994, fig. 8.

Rate of Signaling

Experiments with numerous species of insects and anurans show that females prefer signals produced at high call or chirp rates to the same signals produced at low rates. In all seven species that were tested with intermediate values, high call-rate alternatives were also preferred (appendix 4; fig. 10.11; reviews: Gerhardt 1988, 1991; Passmore et al. 1992; Ryan and Keddy-Hector 1992; Bosch et al. 2000). Among the studies summarized in appendix 4, only that of Hedrick and Weber (1998) found a stabilizing preference function (for chirp interval). Female spadefoot toads (*Spea multiplicata*) from one population preferred high-rate calls, but females from another population did not (Pfennig 2000; see chapter 11). In tests of a pair of cryptic sibling species of crickets,

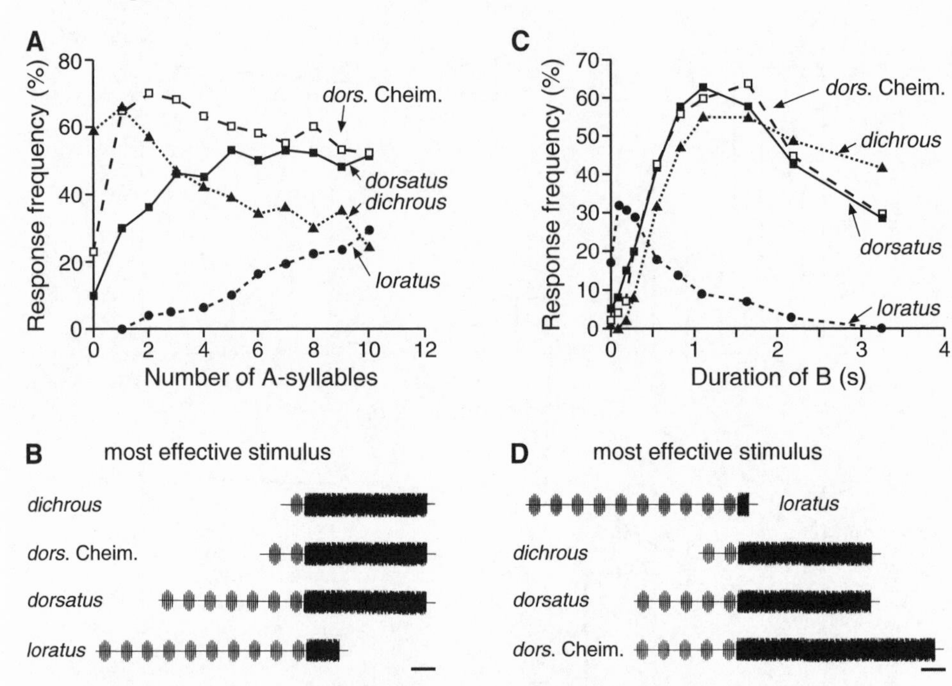

Figure 10.8. Preference functions based on variation in different parts of the diphasic calls of three species of acridid grasshoppers. (A) Response frequencies (normalized) to synthetic songs in which the number of syllables (A-syllables) that make up the first part of the song was varied; geographic variation was evident in comparisons of female responses from two populations of *Chorthippus dorsatus* (*dorsatus* [Leutenbach, Germany]; *dorsatus Cheim.* [a locality in northern Greece]). (B) Diagrams of the synthetic signals that were most effective for each species (or population). The duration of the B-part was held constant for tests of *C. dichrous* and *C. dorsatus* but was reduced to the normal duration for *C. loratus*. The usual number of A-syllables were 4–6 in *C. dorsatus* from Germany, 6–7 in *C. dorsatus* from Greece, 1 in *C. dichrous*, and 8–9 in *C. loratus*. The maximum number of syllables used to test females of *C. loratus* was 10, so they might prefer even more A-syllables. (C) Response frequencies (normalized) to synthetic songs in which the duration of the second part of the song (B-syllable) was varied. (D) Diagrams of the synthetic signals that were most effective for each species (or population). The number of A-syllables was held constant at species-typical values. Females of *C. dorsatus* (both populations) and *C. dichrous* maximally responded to songs with B-part durations much longer than those produced by conspecific males. Females of *C. loratus* often responded to signals without any B-part. Time bars = 200 ms. From Stumpner and Helversen 1994, figs. 10, 11.

females of one species (*Allonemobius fasciatus*) even preferred the higher chirp rates found in the calling song of the other species (*A. socius*) (Doherty and Howard 1996). Finally, in the variable field cricket *Gryllus lineaticeps*, females preferred higher chirp rates in both calling songs and courtship songs (Wagner and Reiser 2000).

Preferences based on differences of as little as about 15% in call rate occur in green and barking treefrogs (Gerhardt 1987; Murphy and Gerhardt 1996), and female green treefrogs *H. cinerea* also preferred calls played back at twice

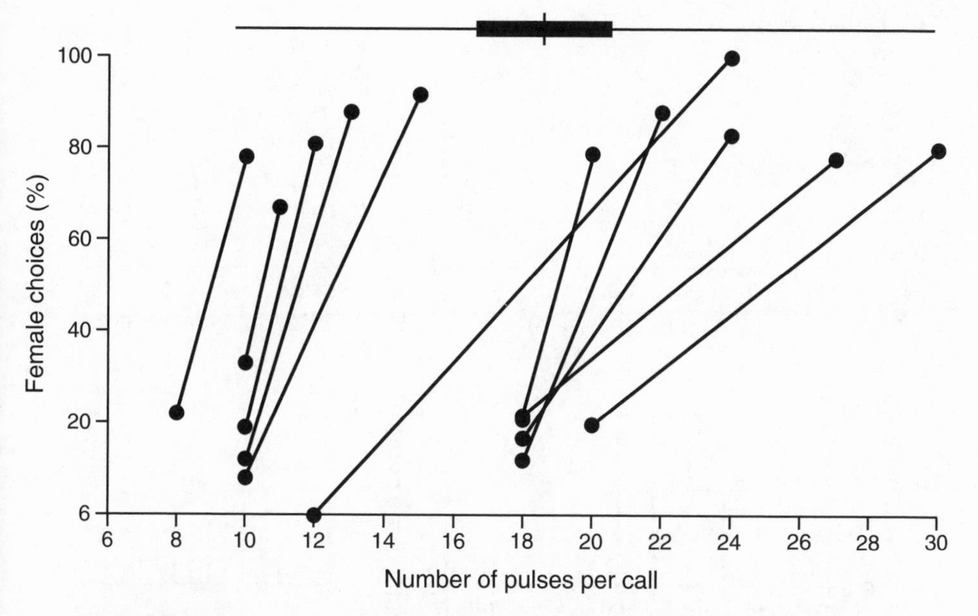

Figure 10.9. Preference function based on differences in the number of pulses per call (= call duration) in the gray treefrog *Hyla versicolor.* Lines connect points that show the percentages of females choosing each of the pair of alternative stimuli. Shown above is a horizontal line representing the range of variation in the mean number of pulses per call in males from the same population; the filled box shows ±1 standard deviation of the mean, which is indicated by the vertical line. From Gerhardt et al. 2000.

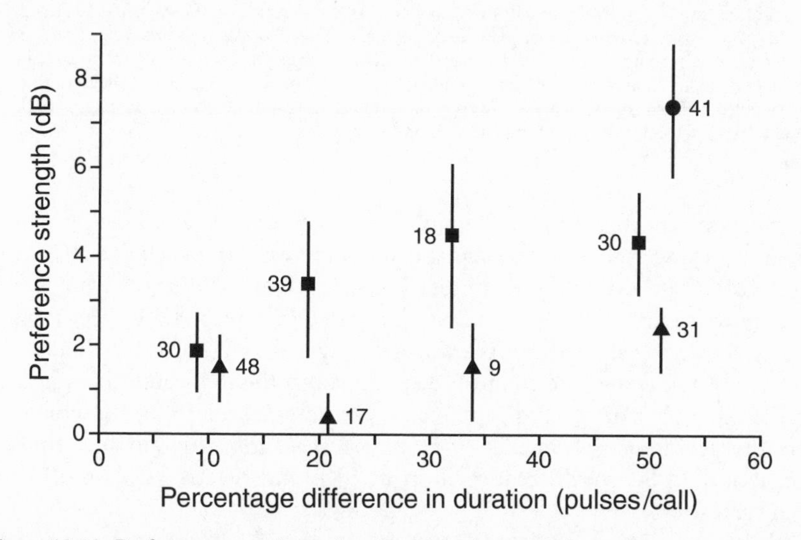

Figure 10.10. Preference strength as a function of absolute and relative duration in the gray treefrog *Hyla versicolor.* Squares = the short-call alternative had a duration of 475 ms (10 pulses/call); triangles = the short-call alternative had a duration of 875 ms (18 pulses/call); circle = results for a test of calls of 375 ms versus 575 ms (or 8 vs. 12 pulses/call, respectively). Preference strength is the largest difference in sound pressure level favoring the shorter call at which females continued to choose the longer alternative. From Gerhardt et al. 2000.

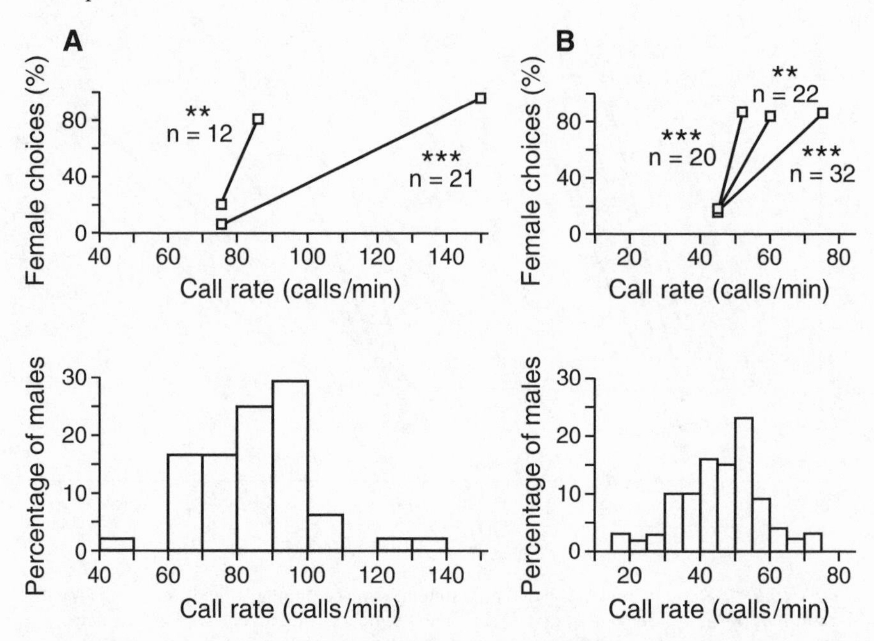

Figure 10.11. Preference functions for call rate in two species of frog. (A) *Top panel,* preferences of females of the green treefrog *Hyla cinerea* for two pairs of alternatives that differed in call rate. Lines connect pairs of alternatives and indicate the percentages of females that chose each stimulus. *Bottom panel,* distribution of the mean call rate of conspecific males from the same populations from which females were collected for testing. From Gerhardt 1991, fig. 5. (B) *Top panel,* preferences of females of the painted reed frog *Hyperolius marmoratus broadleyi* for three pairs of alternatives that differed in call rate. *Bottom panel,* distribution of the mean call rate of conspecific males from the same populations from which females were collected for testing. Asterisks indicate statistical significance of preferences (2-tailed binomial test: $^{*}p < 0.05$, $^{**}p < 0.01$ $^{***}p < 0.001$). From Grafe 1997a, fig. 6.

the normal rate (fig. 10.11A). Doubling the rate once again, however, abolished the attractiveness of the signal completely, even in no-choice tests, thus indicating that beyond a certain rate the signal is no longer recognized. Female gray treefrogs *H. chrysoscelis* and *H. versicolor* continued to prefer high-rate to low-rate calls (within the normal range of variation) that differed by a factor of two when the sound pressure level of the low-rate calls was 6 dB greater that that of the high-rate calls (Gerhardt et al. 1996). By comparison, the preference of females of *H. microcephala* for a higher-rate stimulus was abolished by a 3 dB reduction in its SPL and reversed by a 6 dB drop (Schwartz 1986).

Joint Variation in Duration and Call Rate

Signaling males often vary both the duration and rate of signaling during a bout of calling. As shown in chapter 9, males of the gray treefrog *H. versicolor* tend to decrease call rate when they increase call duration and vice versa (fig. 9.15; Wells and Taigen 1986). Female gray treefrogs *H. chrysoscelis* and *H. versicolor* preferred calls that were twice as long (100%) and repeated at half the rate as short alternatives (see fig. 9.15A, *inset*, for examples of alternative stimuli; Gerhardt et al. 1996; Klump and Gerhardt 1987), and recent tests show that such a preference occurs even when the difference in duration is only 50% (call-duty cycle equalized) (Schwartz et al. 2001). In the katydid *Requena verticalis*, females preferred short-duration chirps presented at a high rate to long-duration chirps presented at a slow rate, even though the call-duty cycle (acoustic stimulation time per cycle) favored the latter signals (Schatral and Bailey 1991a).

In all of the experiments summarized in appendix 4 involving choices between signals differing in call duration, call rate, or both, the values of these properties in alternative stimuli were fixed so that, unlike signal production by real individuals, no call-to-call variation existed. Gerhardt and Watson (1995) showed that females of *H. versicolor* did not prefer variable signals to invariant signals if the alternatives had the same mean values. Comparable data are available for the painted reed frog *Hyperolius marmoratus* (Jennions et al. 1995). More species obviously need to be tested in similar experiments before we can conclude that variation per se within signaling bouts is generally unimportant.

Preferences and Preference Functions Based on Signals of "Short" Duration

As discussed in chapter 2, the within-male variability of short signals (< 0.5 s) differs widely among species. Patterns of preference based on variation in short calls (or the number of pulses [syllables] or impulses in chirps or phrases) also show no general trend. In some species, females show little or no selectivity; in others, preferences are distinctly stabilizing; and in still others, preferences are strongly directional. Indeed, studies from the same experimental setup have sometimes yielded mixed results (Gray and Cade 1999a,b; Wagner et al. 1995).

In insects, stabilizing preferences have been reported in the grasshoppers *Omocestus viridulus* and *Chorthippus brunneus* (Butlin et al. 1985; Eíriksson 1993), in three species of field cricket *Gryllus integer*, *G. texensis*, and *G. campestris* (fig. 10.12; Elsner and Popov 1978; Hedrick and Weber 1998; Gray and Cade 1999a,b), in the bladder cicada *Cystosoma saundersii* (Doolan and Young 1989), and in monosyllabic populations of the katydid *Ephippiger ephippiger* (fig. 10.13B; Ritchie 1996). Weakly directional patterns of preference were found in another study of *Gryllus texensis* (Wagner et al. 1995), and unimodal, but distinctly directional preferences for syllable number occurs in polysyllabic populations of *E. ephippiger* (fig. 10.13C; Ritchie 1996). In the last species, each main syllable lasts for about 50 ms, and males of polysyllabic races produce 2–5 (mode of 4) at intervals of about 50 ms; the preference curve peaks at 5 syllables. In another katydid, *Phaneroptera nana*, females responded

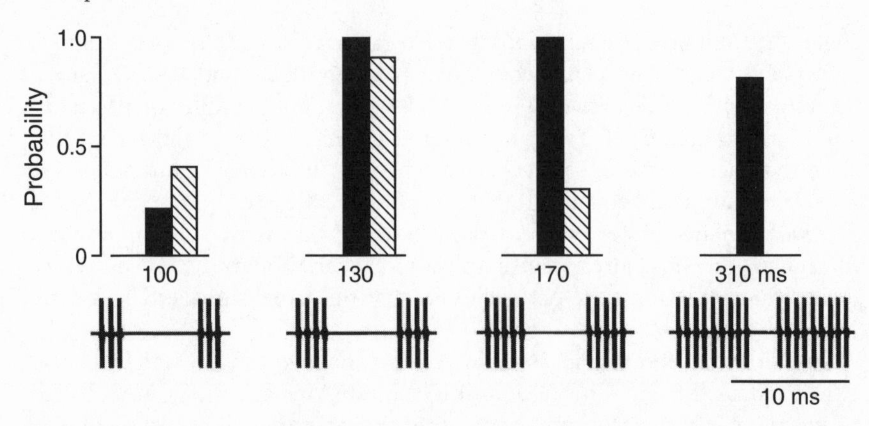

Figure 10.12. Preferences of field crickets for variation in the number of pulses (syllables) per chirp. Histograms show the probability of response to the synthetic models shown in the oscillograms below. Solid bar = *Gryllus bimaculatus;* striped bar = *Gryllus campestris.* From Popov and Shuvalov 1978.

to pulses with durations well beyond the range of variation, and they preferred songs with a greater number of pulses per call in choice tests (Tauber and Pener 2000; Tauber et al. 2001). Strongly directional preferences based on the number of pulses per chirp are reported for the field crickets *Gryllus bimaculatus* (fig. 10.12; Elsner and Popov 1978) and *G. lineaticeps* (Wagner 1996). Male house crickets *Acheta domesticus* that were larger and had higher mating success than small males produced songs with high numbers of pulses per chirp (Gray 1997). In the Hawaiian cricket *Laupala cerasina*, pulse duration was the most variable call property measured, and female preferences were not only strongly directional but might also be open-ended (Shaw and Herlihy 2000).

Among anurans, females generally show little selectivity for short calls that differ in duration. In female spring peepers *Pseudacris crucifer*, the range of values (100–300 ms) that was as attractive as the standard call (duration close to the mean of about 160 ms) is broader than the range of variation in male signals (Doherty and Gerhardt 1984). Other studies of anurans report no preferences based on the duration of any pairs of synthetic calls falling within the range of variation (*Bufo rangeri:* Cherry 1993; *Alytes muletensis:* Dyson et al. 1997). Female green treefrogs *Hyla cinerea* discriminated only against very long calls (420 ms, which is beyond the range of variation) and very short calls (100 ms, which is at the extreme low end of the range of variation) (Gerhardt 1987). Preferences based on duration in the midwife toad *Alytes obstetricans* are weak and differ between populations (Márquez and Bosch 1997b; further discussion in chapter 11). A weakly stabilizing preference pattern occurs in *Hyperolius tuberilinguis.* Recall that females preferred calls consisting of two or three short pulses to alternatives with fewer pulses, but preferred three-pulse calls to four-pulse calls (Pallett and Passmore 1988; chapter 8). A preference

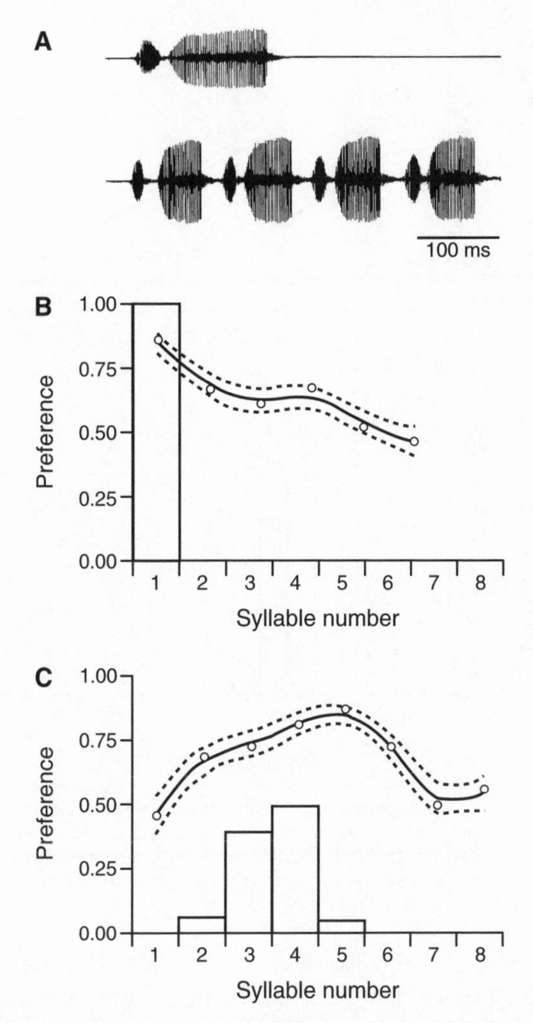

Figure 10.13. Songs and preferences of monosyllabic and polysyllabic katydids (*Ephippiger ephippiger*). (A) Oscillograms showing a syllable (opening and closing movements) that constitutes the song of a monosyllabic race (*upper trace*) and a series of syllables typical of the song of a polysyllabic race. (B) Preference function of females of a monosyllabic race. (C) Preference function of females of a polysyllabic race. Histograms show the distribution of the average number of syllables per song in the two kinds of song races. From Ritchie 1996, figs. 1, 2, 3.

function for call duration in *Hyperolius marmoratus broadleyi* was biased toward short calls (fig. 10.14; Grafe 1997a).

Correlations between Variation in Signals and Male Mating Success

Preference functions predict which males, on the basis of their acoustic signals, should be favored by female choice. Direct correlations between variation

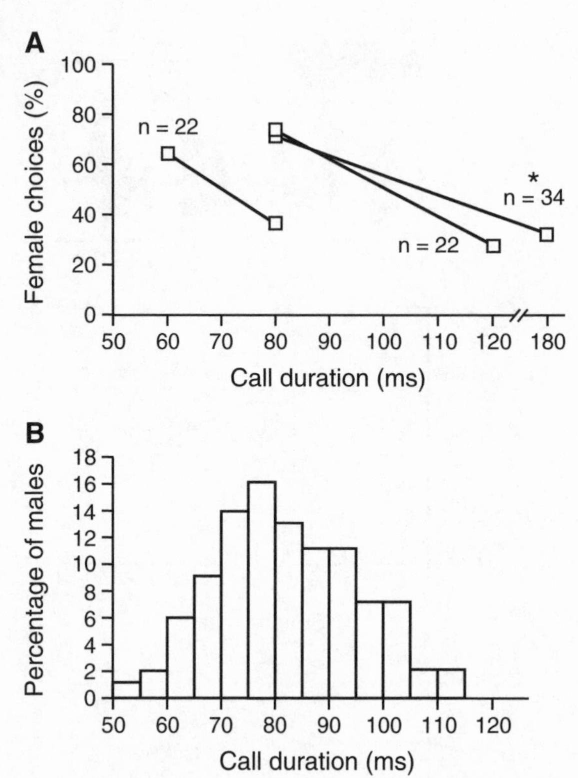

Figure 10.14. Preference function based on call duration in painted reed frogs (*Hyperolius m. broadleyi*). (A) Lines connect values of duration in alternatives and show the percentages of females choosing each stimulus. (B) Distribution of call duration of the advertisement calls of conspecific males in the same population. (* = $p > 0.05$, two-tailed binomial test of the null hypothesis of no preference.) From Grafe 1997a, fig. 6.

in acoustic signals and male mating success provide evidence that these preferences can be realized in nature. Even so, if two or more acoustic properties and other traits of males affect mating success and are correlated, then bivariate analyses can be misleading. Arnold (1994) offers a particularly clear review of multivariate approaches used to generate estimates of *selection gradients*, which essentially quantify the effects of each trait on mating success while statistically controlling for variation in the other traits. Sullivan and Hinshaw (1992) provide an example involving acoustic properties and chorus attendance in gray treefrogs (*Hyla versicolor*). This approach, however, has serious practical drawbacks. Omitting one or more important variables could change the results dramatically, but including too many variables significantly reduces statistical power (Andersson 1994). Low statistical power is probably the main reason that so many field studies have failed to demonstrate significant effects of signal variation on male mating success.

Carrier Frequency

Only four studies to our knowledge have directly considered male mating success in relation to variation in carrier frequency. No significant correlations were found in either green treefrogs (*Hyla cinerea:* Gerhardt et al. 1987) or in two species of glass frogs (*Centrolenella fleischmanni* and *C. prosoblepon:* Jacobson 1985), but sample sizes were relatively low. Although differences in carrier frequency also failed to explain differences in mating success in reed frogs *Hyperolius m. broadleyi,* the methods and sample sizes of this study were adequate to show a significant effect of another call property (Grafe 1997a). Similarly, in the field cricket *Gryllus bimaculatus,* other song properties—but not carrier frequency—were positively correlated with male mating success (see below; Simmons 1988).

Most studies attempting to link variation in carrier frequency to patterns of nonrandom mating have done so indirectly by assessing size-dependent mating success in species whose carrier frequency is correlated with body size. Because mating pairs of insects and anurans often remain together for reasonable periods of time, researchers can easily measure the body size of mated males for comparison with unpaired males found in close proximity. One problem with this kind of study is that unpaired males are not necessarily unsuccessful, and researchers who do not monitor reasonable samples of marked males over substantial parts of the breeding season will not obtain accurate estimates of the relationship between body size and mating success (review: Halliday and Tejedo 1995).

Several other significant problems arise because of such indirect correlations. First, size is often correlated with more than one acoustic property as well as the amplitude of the signal. Teasing apart the relative contributions of multiple acoustic properties and signal intensity has seldom been attempted. In a study of natterjack toads (*Bufo calamita*), large males were more successful than small males. Because females showed no preferences based on carrier frequency, however, the greater amplitude rather than the lower carrier frequencies of the calls of large males was probably the main acoustic determinant of female choice (Arak 1988). Second, correlations between body size and acoustic properties are often relatively weak. In anurans, for example, correlations between body size and carrier frequency are seldom as high as 0.70, and even weaker correlations are typical of insects. Hence, much of the variation in frequency is unexplained by variation in size, and correlations between frequency and mating success are thus likely to be even weaker. Third, the role of carrier frequency in particular and acoustic properties in general in mate choice can be overestimated if large males achieve an advantage by virtue of prior male-male competition for resources or better calling sites. The importance of signal variation can be underestimated if small males successfully adopt alternative mating tactics (review: Sullivan et al. 1995).

Insects. In field or arena studies, large or heavy males have been shown to have greater mating success than small or light males in house crickets (*Acheta domesticus:* Gray and Cade 1999a); field crickets (*Gryllus bimaculatus, G. campestris,* and *G. pennsylvanicus:* Simmons 1988, 1995; Souroukis and Cade 1993) and one species of katydid (*Amblycorypha parvipennis:* Galliart and Shaw 1991, 1996). However, nonrandom mating by size varies from population to population in field crickets, and such variation appears to be explained in part by differences in population density and the operational sex ratio (e.g., the field cricket studies of Cade and Cade 1992; French and Cade 1989; Zuk 1988). Fighting ability, which could reduce the set of males to be assessed by females, is also likely to be positively correlated with male size (e.g., Brown et al. 1996).

In field crickets, carrier frequency is negatively correlated with body size (harp size) in *G. bimaculatus* and *G. campestris,* and females of *G. bimaculatus* preferred songs with carrier frequencies near the population mean to higher-than-average frequencies (Simmons 1995; Simmons and Ritchie 1996). The large, more successful males of *G. bimaculatus* studied by Simmons (1988) did not, however, produce lower-frequency songs than less successful, small males. Even though this negative result is probably attributable to low statistical power, the same study showed that the songs of large males were significantly more intense and had higher pulse rates and lower chirp rates than the songs of small males. Thus, the contribution of frequency preferences to mate choice in this population must be relatively small at best. Similarly, in *A. domesticus* the number of syllables per chirp, and not carrier frequency, appears to be responsible for differential mating success of large males (Gray 1997). Finally, the songs of large, more successful males of the katydid *Amblycorypha parvipennis* were more intense but not lower in carrier frequency than those of less successful, small males (Galliart and Shaw 1991).

In general, field and laboratory studies of insects produce results consistent with the hypothesis that differences in signal amplitude at the female's position, which is usually a good indicator of male proximity, are likely to outweigh any preferences based on differences in carrier frequency (e.g., Bailey et al. 1990). Differences in amplitude can also reflect differences in acoustic output by different males or beaming effects, and, as already mentioned (chapter 8), the distance between signalers can affect intensity-discrimination by receivers (Forrest and Raspet 1994). As Zuk and Simmons (1997) and Brown (1999) conclude in their reviews, the data linking variation in carrier frequency and other acoustic properties to differential mating success in orthopterans is meager and confounded by other attributes of males that also correlate with body size.

Anurans. Field studies show that large males have greater mating success than small males in midwife toads (*Alytes cisternasii* and *A. obstetricans:* Márquez 1993), túngara frogs (*Physalaemus pustulosus:* Ryan 1985), some populations of toads (*Bufo americanus:* Howard and Palmer 1995; *Bufo calamita:* Arak 1988), and some populations of treefrogs (*H. arborea:* Márquez and Tejedo 1990; *H. chrysoscelis:* Morris 1989; *H. ebraccata:* Morris 1991; *H. gratiosa:* Murphy 1999). Other studies of some of these same species did not find size-dependent mating success (*B. americanus:* Kruse 1981; Wilbur et al. 1978; *B. calamita:* Denton and Beebee 1993; *H. arborea:* Friedl 1992; *H. chrysoscelis:* Godwin and Roble 1983; Ritke and Semlitsch 1991). One recent study of two Australian treefrogs (*Litoria chloris* and *L. xanthomera*) found that small males had greater mating success than large males (Morrison et al. 2001). Size-assortative mating success has been found in the Australian toadlet *Uperoleia laevigata* (fig. 10.15B; Robertson 1986, 1990), a South American treefrog (*Ololygon rubra:* Bourne 1993), and one (Licht 1976) of five studies of the toad *Bufo americanus* (review: Howard et al. 1994).

In all of the species listed above, females usually initiate matings, and, with the exceptions of *Hyla arborea* and *Ololygon rubra,* data are available concerning the preferences of females for synthetic calls differing only in carrier frequency. Females in *Physalaemus pustulosus, Hyla ebraccata,* one population of *H. chrysoscelis,* and one population of *Bufo americanus* prefer low- to high-frequency calls. In *P. pustulosus,* maximum call rate was also correlated positively with mating success (Green 1990), and females of *H. chrysoscelis* released near pairs of calling males more reliably chose males producing calls at high versus low rates than they chose males producing low- versus high-frequency calls (Morris and Yoon 1989). Thus, call rate as well as carrier frequency could contribute to the higher mating success of large males in some populations. Females of *H. chrysoscelis* from other populations did not show significant preferences based on differences in carrier frequency that spanned the natural range of variation (Gerhardt and Tanner, unpubl. data).

As mentioned above, size-dependent preferences in *U. laevigata* are consistent with the pattern of size-assortative mating observed in that species. The correlations between the weights of males and females in pairs (fig. 10.15A) was, however, substantially higher than that between body weight and the carrier frequency of the advertisement call (fig. 10.15B), and there was much scatter in the data. Thus, additional, nonauditory cues based on short-range assessments are likely to have contributed to the female's final choice of a male.

In general, consistent patterns of size-dependent mating in species in which females can freely choose mates are less common than random mating by size. In studies of three species, no pattern of size-dependent mating success was found despite very large sample sizes that provided the statistical power to detect quite small effects (e.g., *Hyla chrysoscelis:* Ritke and Semlitsch 1991; *Hyla cinerea:* Gerhardt et al. 1987; *Hyperolius marmoratus:* Dyson et al. 1994).

As discussed in the section on endurance rivalry (chapter 9), mating success in anurans is often best predicted by chorus attendance. Females are unlikely to stay in breeding ponds long enough to monitor the attendance of different males, but if some property of the call were correlated with chorus attendance, then females could use this information to pick a male likely to

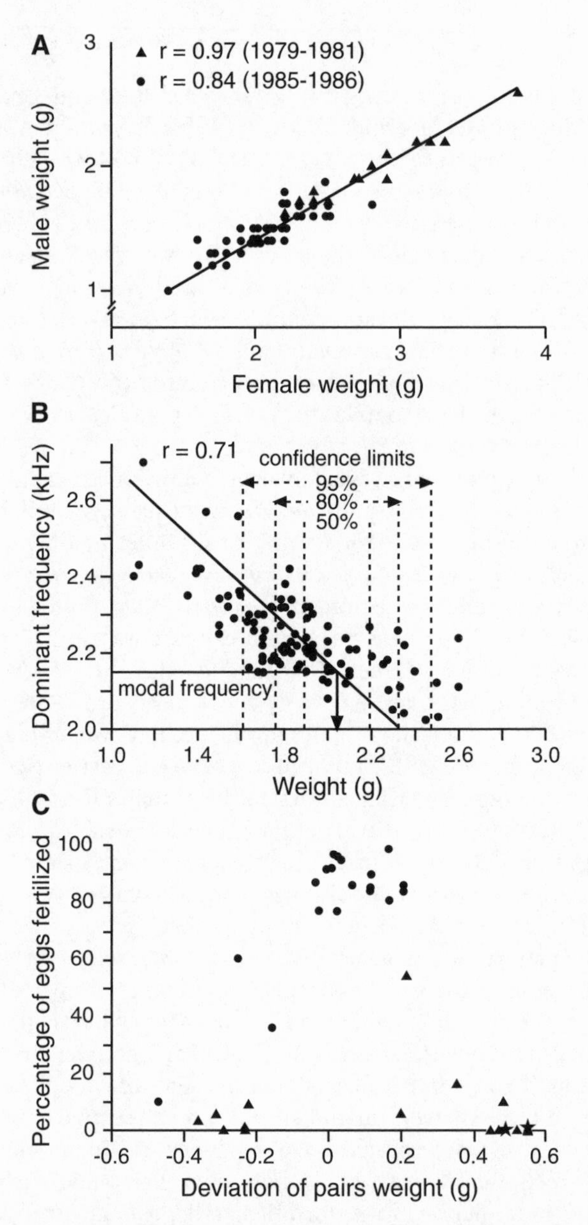

Figure 10.15. Carrier frequency and size-assortative mating in *Uperoleia laevigata.* (A) Assortative mating by body weight; data are from two multiple-season sampling periods (dates and statistics are shown). (B) Negative correlation between carrier frequency and body weight. Note that the scatter is substantial despite the relatively high correlation coefficient. (C) Consequences on fertilization success of size-assortative mating. Females that were paired with males that were smaller or larger than they would have chosen had reduced fertilization success. In (A) and (C), triangles = data from 1979–81; circles = data from 1985–86; stars = females that drowned because they mated with males that were much larger than themselves. From Robertson 1986, fig. 8; 1990, figs. 2, 3.

attend many choruses. Whereas five studies reported positive correlations between attendance and male body size (hence, carrier frequency is a potential cue), seven other studies did not find such a relationship (review: Halliday and Tejedo 1995). In a recent study of the cricket *Gryllus texensis* (= *integer*), large and middle-aged males called for longer periods of time during the diel cycle than did small, young males and very old males (Bertram 2000).

Dynamic Properties

We are unaware of any field studies of orthopterans or cicadas that have assessed male mating success in relation to variation in dynamic acoustic properties alone. Studies of field and sagebrush crickets in arenas found positive correlations between proportion of time calling (gross time scale) and mating success (e.g., Cade and Cade 1992; Cade and Wyatt 1984; Snedden 1996; Souroukis and Cade 1993). Other confounding factors that might have affected mate choice were uncontrolled, however, and correlations with mating success were also influenced by density and the operational sex ratio.

For anurans, increasing numbers of studies have found positive correlations between call rate and mating success: toads (*Bufo calamita:* Arak 1988; *B. rangeri:* Cherry 1993; *B. woodhousei:* Sullivan 1983); the coqui treefrog (*Eleutherodactylus coqui:* Lopez and Narins 1991); painted reed frogs (*Hyperolius marmoratus* and *H. marmoratus broadleyi:* Grafe 1998; Passmore et al. 1992); spring peepers (*Pseudacris crucifer:* Sullivan and Hinshaw 1990). Playback experiments with females of all of these species also revealed preferences for high to low call rates.

The study of *H. marmoratus* by Passmore et al. (1992) was particularly robust. First, they used playback experiments to establish that females of the painted reed frog *H. marmoratus* preferred calls with higher (≥ 20%) call rates. Second, they used event recorders to monitor simultaneously up to eight males at a time on many nights to amass a data set consisting of about a half a million calls from 111 males. The overall rate of calling was high and sustained during the first 1–1.5 h of calling, but then gradually decreased. Nevertheless, individual males varied considerably in both the duration of their calling periods and mean call rates. Third, Passmore et al. monitored the times during the evening when females arrived and initiated matings; female mating times coincided with the period of sustained calling. The relative call rates of mated males were higher than those of other, unmated males immediately preceding mating and during the five-minute period prior to mating (fig. 10.16B). Comparable results were reported for *H. microcephala* by Schwartz (1994) and for *Hyperolius m. broadleyi* by Grafe (1997a), who compared the call rates of mated and unmated males in small choruses.

Mixed results concerning the effect of variation in call duration on mating success have been reported. In a field study, Sullivan and Hinshaw (1992) failed to find a significant relationship between call duration and mating success in

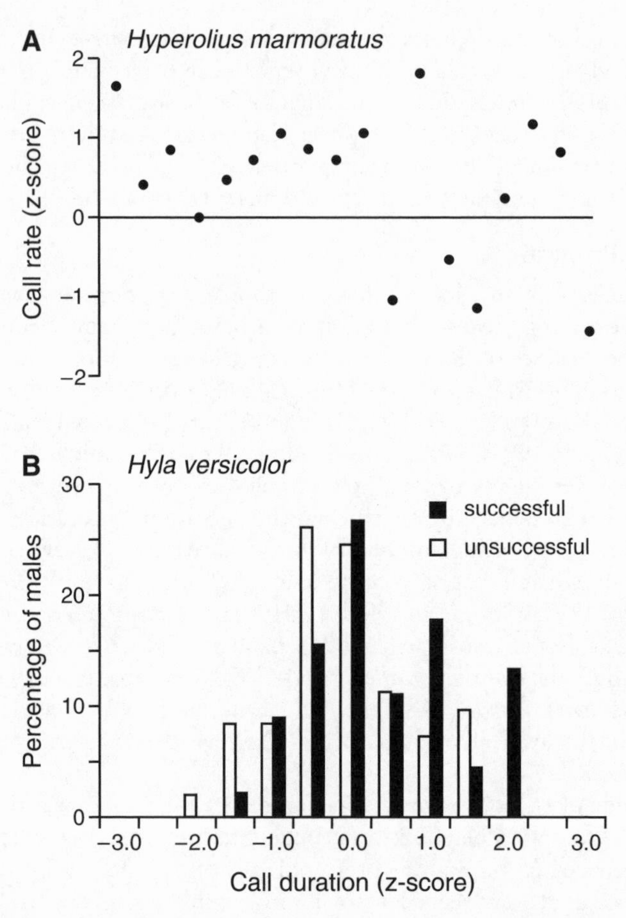

Figure 10.16. Variation in dynamic properties and mating success. (A) Painted reed frog *Hyperolius marmoratus.* Distributions of the mean relative call rates (z-scores) of mated males during the last 5 min preceding mating (0 = population mean of call rate). Notice that the call rates of most males were above the population mean. Modified from Passmore et al. 1992, fig. 8. (B) Gray treefrog *Hyla versicolor.* The relative call duration of successful males that attracted a female (*filled bars*) was higher than that of unsuccessful males (*open bars*) in choruses of up to eight males that were set up in a greenhouse. Z-scores are standard deviation units (0 = the mean). Modified from Schwartz et al. 2001, fig. 8.

the gray treefrog *Hyla versicolor*, but their low sample size and the clumped arrival of females on a few nights probably limited their chances of detecting an effect. More recently, Schwartz et al. (2001) tested females in choruses of males (up to eight at one time) that were set up in a greenhouse pond. The calling performance of each equally spaced male was continuously monitored by a personal computer, and the call duration of males that attracted females was, on average, relatively longer than that of males that did not (fig. 10.16A).

Some evidence is also available that the total amount of calling over the course of a night affects mating success. After two hours of playbacks of prerecorded calls in a large outdoor cistern, twice as many females of the spring peeper *Pseudacris crucifer* were trapped near a speaker that played back recorded calls for 80% of the test period than were trapped near a speaker playing back the same calls for 60% of the time (Forester et al. 1988). The total amount of calling was also the best predictor of mating success in a field study of a dendrobatid frog, *Dendrobates pumilio* (Pröhl and Hödl 1999).

Positive correlations between high values of dynamic properties and male mating success might be expected even if females lack preferences altogether, simply because long signals or signals produced at high rates should be easier to detect in dense choruses than short signals or signals repeated at slow rates (Helversen and Helversen 1994; Parker 1982). Indeed, increased detectability is likely to have contributed to the positive results of some of the field studies just mentioned. If, as we expect, preferences for high values of dynamic properties also contribute significantly to differential mating success, then positive correlations with dynamic properties should also be found in sparse choruses, where background noise levels are low and detection is not a problem.[1] After all, preferences for long and high-rate calls are typically demonstrated by playbacks in laboratory situations, where background noise levels are normally far lower than in nature. Furthermore, even in dense choruses, females can still compare the signals of two to several nearby males, whose calls would be detectable above the chorus background (Gerhardt et al. 1996; Grafe 1997a). Indeed, playback experiments with barking treefrogs (*Hyla gratiosa*) indicate that initially detectable calls are not preferred to calls that are masked by chorus sound once the female moves to a place where the latter signals become detectable (Murphy and Gerhardt 2002).

Potential Benefits of Female Choice Based on Different Acoustic Criteria

Carrier Frequency

By using carrier frequency to select mates, females potentially gain several kinds of benefits. In insects, female tree crickets that mate with larger-than-average males, whose song frequencies they prefer in the laboratory, had increased fecundity (Brown et al. 1996). Simmons (1994, 1995) argues that female preferences in field crickets (*Gryllus bimaculatus* and *G. campestris*) for larger or older males might result in heritable, indirect benefits, but as discussed above, the evidence that females choose large males on the basis of carrier frequency is not especially compelling.

In anurans, mating with males of a particular size may result in direct benefits in the form of greater fertilization success (Bourne 1993; Robertson

---

1. Another perspective is that even in quiet environments, signals must overcome the "internal" noise of female resistance (appendix 1; Wiley 1995).

1990; Ryan 1985). The usual explanation is that the cloacae of the male and female are closer together in pairs that are more similar in size than in mismatched pairs (Licht 1976). In *Physalaemus pustulosus* and *Uperoleia laevigata*, large males fertilized more eggs than small males, but in the latter species, a female forced to mate with a male that was too large had lower fertilization success (fig. 10.15C; Robertson 1990). Field observations also indicated that mating with a very large male sometimes resulted in the female's drowning.

Studies by Woodward and his colleagues (Woodward 1986, 1987; Woodward and Travis 1991) indicate that female choice of larger-than-average males in several species of anurans might gain indirect benefits in the form of superior larval performance, survival, and growth rates. Howard et al. (1994) criticize the design and interpretation of these studies on several grounds: (1) No female choice was demonstrated in the study species; (2) inappropriate criteria were used for selecting sires and judging offspring fitness; and (3) the rearing conditions could have included confounding effects. In their extensive series of experiments with *Bufo americanus*, Howard et al. (1994) found heritable differences in three offspring performance characters (relative weight, relative age, relative numbers), but these traits did not differ between offspring sired by large males and those sired by small males. Comparable results and conclusions were reported in a study of *Bufo bufo* (Semlitsch 1994).

The variability in preference patterns for carrier frequency between species and populations might reflect variability in the reliability of carrier frequency as an indicator of male size, fitness, direct benefits, or some combination of these factors. A proximate-level explanation is that differences in frequency-resolving ability could limit the selectivity of preferences in some but not other species, even if females of all species would benefit by greater selectivity.

Fine-Temporal Properties

The prevalence of stabilizing preference functions associated with pulse rate reinforces the bias of many biologists that the main consequence of intraspecific preferences based on this acoustic property is species identification (Ewing 1987; Bailey 1991; Gerhardt 1991). Indeed, we did not consider mating success in relationship to quantitative variation in fine-temporal properties because no study of which we are aware has ever set out to estimate such a correlation. The only such evidence comes from a retrospective analysis, in which the songs of preferred large males of *Gryllus bimaculatus* were found to have higher pulse rates than those of nonpreferred small males (Simmons 1988). However, we have also emphasized that preference functions based on variation in pulse rate are not always strongly stabilizing and symmetrical and hence potentially result in directional selection in some populations.

Other aspects of fine-temporal structure such as missing impulses might be indicators of age. Female preferences based on age occur in many taxa besides insects and anurans, and a recent model has shown the plausibility of age as an indicator of overall viability fitness (Kokko and Lindström 1996; see also the review by Brooks and Kemp 2001). Many more studies need to be conducted in insects because so far there is no general trend for young or old males to be preferred. Alternatively, songs with missing impulses might not be recognized as conspecific signals (Ritchie et al. 1995).

Dynamic Properties

As discussed in chapter 9, variation in dynamic properties is influenced by the calling activity of rivals, and there are direct and strong correlations with the energetic costs of signaling. Costly signals are featured prominently in indicator (handicap) models (Andersson 1994). Males producing signals with high values of dynamic properties might also be more conspicuous to predators. In one species of field cricket (*Gryllus texensis* = [*integer*]), acoustically orienting parasitoid flies preferred the high chirp rates that also most effectively attracted females (Wagner 1996). In another species (*Gryllus integer*), males producing attractive, long calling bouts displayed behaviors that have been interpreted as reducing predation risks (Hedrick 2000). Thus, variation in dynamic properties is likely, in general, to correlate with energetic and predation costs of mate attraction by males. These signals might thus be reliable indicators of the competitive abilities of males, their success in acquiring the resources that allow them to attend multiple choruses and to maximize their acoustic output, and perhaps their sensitivity to predation risks or their possession of behavioral traits that reduce such risks.

If females gain direct or indirect benefits from mating with a male producing signals with high values of dynamic properties, then the strongly directional patterns of preference based on these properties are not surprising. The best evidence is afforded by studies of two species of anurans in which benefits accrue to females or their offspring when they mate with males that produce advertisement calls with the preferred high values. As reported above, females of the gray treefrog *Hyla versicolor* prefer long to short calls, and in small choruses in a greenhouse, males producing long calls are more often chosen by females than are males producing short calls (Schwartz et al. 2001). Moreover, robust breeding experiments show that both in the laboratory and in the field, offspring of males producing long calls are superior to the offspring of males producing short calls in several larval performance traits that correlate with higher adult survival (Welch et al. 1998; Doty and Welch 2001). Using a half-sib breeding design, these studies eliminate maternal effects by artificially fertilizing two subsets of randomly selected eggs from the same clutches of a series of randomly selected females. Because male gray treefrogs do not defend oviposition sites or provide any material resources to females, the difference in offspring performance must have been derived from genetic differences (indirect benefits) between the male parents. No differences were found in the fertilization success (a direct benefit) of clutches fertilized by long callers and short callers (Krenz et al., unpubl. data).

Evidence for indirect benefits associated with high values of dynamic acoustic properties is also provided by studies of wax moths (*Achroia grisella*), in which there is also evidence for genotype X environment interactions that can, in principle, contribute to the maintenance of genetic variation in these traits (Jia et al. 2000; see below). Additional strong evidence for indirect benefits is available for the fruit fly *Drosophila montana*, although no data are available concerning within-male variability in the preferred trait. Females prefer high-frequency songs with short pulses and preferentially mate with males producing such songs in nature (Aspi and Hoikkala 1995; Ritchie et al. 1998). Offspring of females that choose males producing high-frequency songs (and those with short pulses) have increased survival rates (Hoikkala et al. 1998). In the grasshopper *Myrmeleotettix maculates*, males showed fluctuating asymmetry in the number of pegs per row, and females discriminated against the songs produced by males whose peg numbers had been experimentally manipulated to create asymmetry (Moller 2001). The songs of the manipulated males had shorter chirp durations than did songs of unmanipulated males.

Evidence for direct benefits of preferences for high values of dynamic properties are available from anurans. In spadefoot frogs (*Spea multiplicata*), females that were mated to males producing calls at a higher-than-average rate, which were preferred in that population, had higher fertilization success than females that were mated to males producing calls at a lower-than-average rate (Pfennig 2000). As discussed in chapter 11, this preference was not expressed in a population in which there was the possibility of mating mistakes with another species of spadefoot toad that produces calls at a high rate. Clearly, the costs of mating with conspecific males that are inferior in fertilizing eggs must still be considerably less than the costs of mating with a genetically incompatible mate (Gerhardt 1982). In Australian toadlets (*Pseudophryne bibronii*), males in wet areas called at higher rates than did males from dry areas; the wet conditions are presumably more favorable for egg development in this terrestrial-breeding species (Mitchell 2001).

Finally, as mentioned above, females of the cricket *Acheta domesticus* prefer calling songs with high rather than low numbers of syllables per chirp (Gray 1997). A recent study has shown that such males, which are larger than average in body size, have higher haemocyte loads (Ryder and Siva-Jothy 2000). These cells are important determinants of immune function, and by mating with such males, females could benefit directly because of a reduced risk of acquiring pathogens during copulation and indirectly if offspring inherit increased immune function (Ryder and Siva-Tothy 2001).

In conclusion, the prevalence of strongly directional preference functions for dynamic properties fits with the hypothesis that these properties can be reliable indicators of direct or indirect benefits that females might receive from choosing males producing high values of these properties. Some of the high within-male variation in these properties is almost certainly attributable to purely environmental factors such as the calling behavior of neighbors and variability in the availability of energetic resources needed to fuel signaling (chapters 8 and 9). Nevertheless, significant between-male variation in male signals has often been demonstrated (see below), and, as shown by the breeding experiments just mentioned, some of this variation reliably correlates with differences in benefits to females or their offspring. It is, of course, far too soon to know whether direct benefits are more common than indirect ones, much less conclude anything about the relative importance of Fisher effect and handicap mechanisms.

Phenotypic and Genetic Variation in Signals and Acoustic Preferences

Heritability and Repeatability of Acoustic Signals

Laboratory experiments with insects have demonstrated heritable variation in the amount of signaling in field crickets (Cade 1981a), pulse rate in crickets and planthoppers (Butlin 1992; de Winter 1992; Gray and Cade 1999a), duration of pulses (syllables) in grasshoppers (Charalambous et al. 1994), and duration of calling bouts in crickets (Hedrick 1988). In the planthopper, in which females produce vibrational signals that attract males, the degree of divergence in pulse rate achieved by artificial selection was particularly impressive (fig. 10.17; de Winter 1992). In addition, de Winter (1992) found that two other acoustic properties (signal length and modulation of pulse rate within signals) showed correlated responses to selection based on pulse rate. These studies counter the concern of many theoretical discussions in the last decade that selection in the form of female choice will inevitably deplete such variation. Indeed, because of the many factors that confound the expression of signal preferences in natural situations (chapter 9 and previous sections), the net effect of female choice on particular acoustic properties of long-range signals must usually be very weak selection. Andersson (1994) discusses other mechanisms that have been proposed to maintain variability in the face of selection, including the condition-dependence of male traits under such selection. If the trait is correlated with condition, which may have high genetic variance, then a proportion of that variance will be expressed in the trait itself (e.g., Rowe and Houle 1996).

Heritability estimates for acoustic properties are unavailable for anurans, but analyses of multiple recordings of individuals in natural populations show consistent phenotypic differences in some properties. Such data have been used to estimate repeatability—the proportion of total variance that is due to between-male differences (Gerhardt 1991; Gerhardt et al. 1996; Runkle et al. 1994; Sullivan and Hinshaw 1990, 1992; Wagner and Sullivan 1995). Because repeatability sets the upper bound on heritability, efforts to estimate heritability or additive genetic variance through breeding and artificial selection experiments are probably worthwhile if a signal property is at least moderately repeatable (Boake 1989). If a call property shows little or no repeatability,

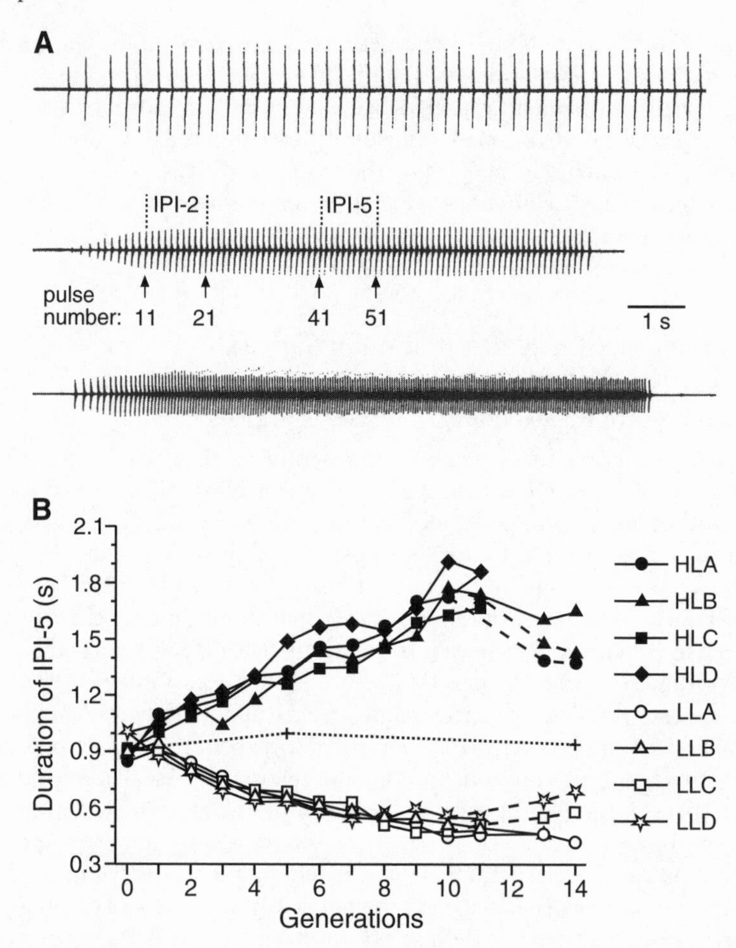

Figure 10.17. Response of an acoustic property to artificial selection. (A) Oscillograms of the calling signals of female planthoppers *Ribautodelphax imitans*, showing variation in interpulse intervals (IPI), a trait that was subjected to divergent artificial selection. Middle trace = unselected lines; top trace and bottom traces = extreme examples from low- and high-selected lines. (B) Plot of changes in IPI-5 during 11–14 generations of selection (*solid lines*) in four low (LLA–LLD) and four high (HLA–HLD) lines; changes after relaxed selection are shown by dashed lines; unselected base population is shown by dotted line. From DeWinter 1992, figs. 1, 2.

then strong selection will probably have little or no effect on its evolution, although females could still benefit from choices based on the property if, for example, variation in the property correlates with environmentally variable, direct benefits. For example, call properties that correlate with energetic costs might be conditional on food availability, which in turn could affect a male's ability to fertilize eggs.

In two studies, repeatability has also been estimated for the relative values of acoustic properties of known pertinence to females (Gerhardt et al. 1996;

Wagner and Sullivan 1995). Unlike biologists who can make multiple recordings of individuals over the course of a season, females of many species of anurans breed only a few times and therefore must base a mating decision on *comparisons* of the signals heard on those occasions. Some males of the gray treefrog consistently produced calls with pertinent properties that were significantly above or below the average values in the chorus on the nights they attended (fig. 10.18).

Inheritance of Acoustic Properties

Most of our knowledge about the inheritance of quantitatively variable acoustic properties is based on analyses of crosses within and between closely related species. The most common result is that acoustic properties are controlled by multiple genes having additive effects and distributed throughout the genome; little evidence exists for directional dominance (see Ritchie and Phillips 1998 for a review of insects and Sanderson et al. 1992 for estimates based on hybridizing anurans *Bombina bombina* and *B. variegata*). Some acoustic properties are, however, like other sexually selected traits, more or less disproportionately controlled by sex-linked genes or maternal effects (Reinhold 1998): pulse patterns in *Teleogryllus* hybrids (Hoy et al. 1977), pulse rate in Hawaiian crickets (Shaw 1996a, 2000), chirp structure in *Acheta* hybrids (Bigelow 1960), echeme length in *Chorthippus parallelus* group hybrids (Butlin and Hewitt 1988), and syllable number in *Ephippiger ephippiger* call races (Ritchie 2000). As discussed by Ewing (1989), sex linkage results in a high potential for rapid evolution of signals because any genetic change will immediately be expressed in the phenotype of the male (heterogametic sex). Such a consequence is unlikely to be the cause of this pattern, however, and sex linkage of any traits influencing sexual behavior might be favored in the hemizygous sex simply because these traits are sex limited in their phenotypic expression (Rice 1984).

Variability in Female Preferences

As mentioned above, estimating variability in female preferences is an important task for researchers because its existence has important theoretical implications. Variability in preference has been documented in many animal species (reviews: Andersson 1994; Bakker and Pomiankowski 1995; Jennions and Petrie 1997). Although several studies have failed to find even phenotypic variation in preference (Butlin 1993, 1995; Gerhardt et al. 1996; Kime et al. 1998), these results might reflect the difficulties of quantifying such variation rather than its absence. As pointed out by Jennions and Petrie (1997), variation in preferences per se might also be confounded by variation in choosiness (= probability that a preference will be expressed if mate assessment or location is costly), sampling tactics, and other factors. The same difficulty is likely to confound efforts to show genetic correlations between variation in signals and preferences, which are predicted by both Fisher effect and handicap models. For example, despite the remarkable divergence in the signals of female

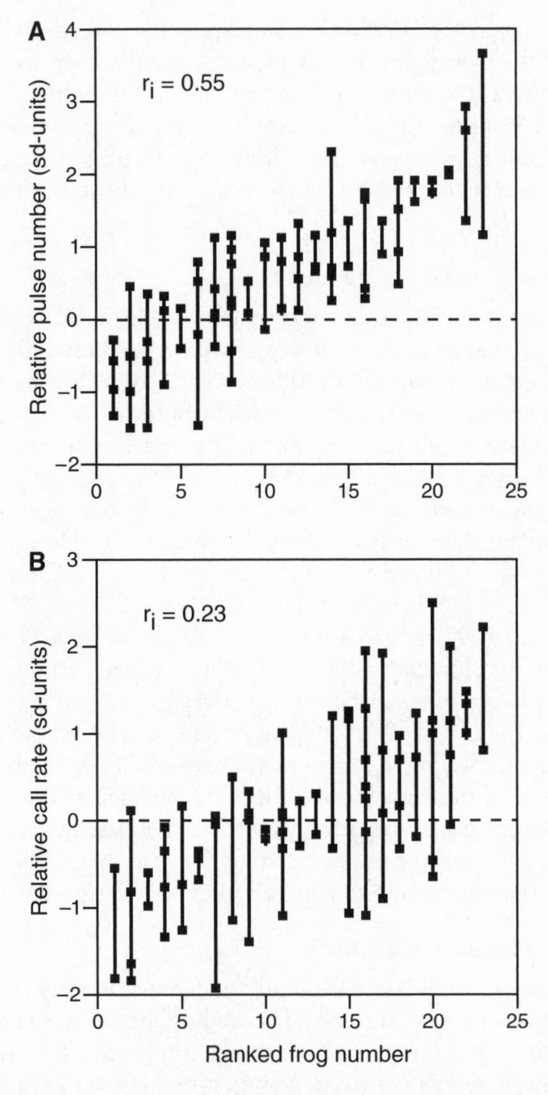

Figure 10.18. Relative repeatability of (A) call duration and (B) call rate in twenty-three males of the gray treefrog *Hyla versicolor.* Individual males were recorded on two or more nights, and the absolute values (squares connected by lines) of the two call properties were converted into standard-deviation units and plotted (on the same vertical line) relative to the mean values in the chorus (0 = dashed line) on the same nights. Some males were consistently above the mean on all nights that they were recorded, and others were consistently below the mean. From Gerhardt et al. 1996, fig. 4.

planthoppers subjected to artificial selection (fig. 10.17), males (the responding sex) in the selected lines showed only weak preferences for the signals of coselected females (de Winter 1992).

In two species of anurans (*H. versicolor* and *Physalaemus pustulosus*), no variation in female preferences was found on the basis of recording two responses per female (Gerhardt et al. 1996; Kime et al. 1998), and in the painted reed frog, repeatable differences in frequency preferences were attributable to size-dependent discriminatory abilities (Jennions et al. 1994). However, evidence for phenotypic variation among females is available for green treefrogs tested multiple times (Gerhardt 1991) and for variation in individual preference functions in barking treefrogs (*Hyla gratiosa:* Murphy and Gerhardt 2000). Another study of gray treefrogs shows that differences in the strength of preferences based on call duration can be highly repeatable (fig. 10.19; Gerhardt et al. 2000). The last study found that merely tabulating the number of binary choices made by a series of females is not a very sensitive measure of variability. Given an easy choice, nearly all females chose the longer alternative in most trials. But when the choice was made more difficult by reducing the sound pressure level of the longer, preferred call, considerable and repeatable variation was detected in the difference in amplitude at which females maintained their preference. Less evidence may be available for variability in female preference than for variability in male signals because preference variability is more difficult to measure.

Studies of orthopterans have also demonstrated genetic differences in preference between species (crickets *Laupala cerasina, L. paraniga, L. kohalensis:* Shaw 2000; katydids *Ephippiger ephippiger:* Ritchie 1996, 2000) or have estimated heritability in female preference (crickets *Gryllus integer* and *G. texensis:* Gray and Cade 1999a; Wagner et al. 1995; grasshoppers *Chorthippus brunneus:* Charalambous et al. 1994). Interspecific crosses in *Laupala* also showed that the shape of preference functions based on pulse rate is inherited and that the peak values (maximum preference) closely matches the mean pulse rates in hybrids and backcross products (Shaw 2000). Crosses and backcrosses between races of *Ephippiger* show variable preference functions with a tendency for moderately strong, but not open-ended, directional selection based on syllable number (fig. 10.20; Ritchie 2000). In contrast to comparable studies of signal properties (see above), neither study found evidence for sex-linked effects on preferences (see also Ritchie and Kyriacou 1996). Taken as a whole, the results of the last two studies—which also show that preferences, like signals, are controlled by multiple genes of small effect—conform to the expectations of models assuming that signals and preferences in these systems coevolve.

Genetic Coupling?

Alexander (1962) proposed that both senders and receivers might inherit a neural oscillator that could be used to temporally pattern the sound pulses of

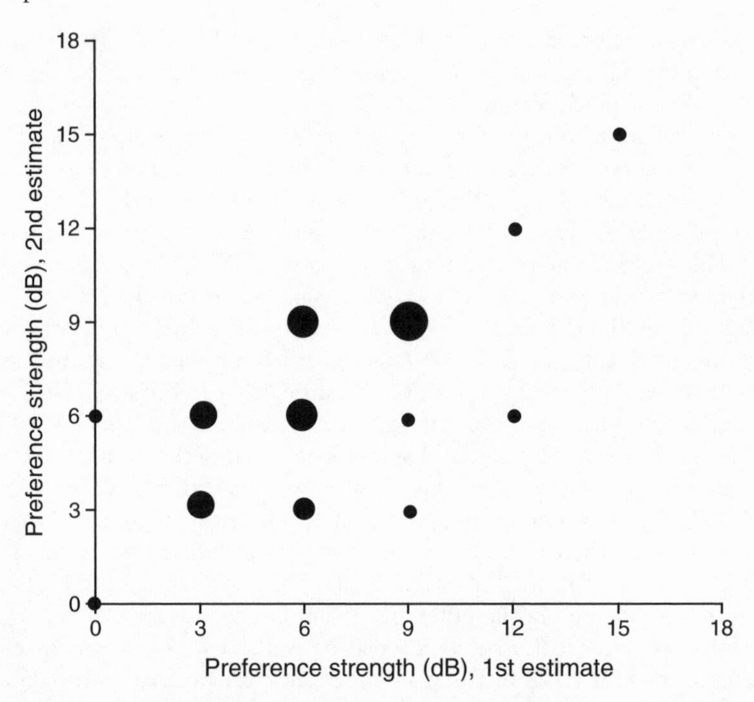

Figure 10.19. Consistency and repeatability of preference strength. Plots of the first and second estimates of preference strength of females of *Hyla versicolor* given a choice between calls differing in duration. The data are based on two estimates obtained from 33 females in the test of alternatives of 8 versus 12 pulses/call. Repeatability was 0.71. Largest circles = 5 females; next largest = 4 females; next largest = 3 females; next largest = 2 females; smallest = 1 female. From Gerhardt et al. 2000, fig. 4.

the sender and as a reference oscillator for signal recognition by the receiver. Senders and receivers would thus be coordinated through evolutionary time since mutations to the common oscillator would change both signal timing and the receiver's internal time reference in the same way. We have already discussed experiments that cast doubt on the existence of a common oscillator in grasshoppers (chapter 4; Bauer and Helversen 1987), but common genes could still influence signals and receiver mechanisms without necessarily specifying a single neural network (Butlin and Ritchie 1989; Boake 1991). Robust analysis of genetic coupling requires genetic crosses over several generations. One such test in *Drosophila* provided strong negative evidence (Greenacre et al. 1993). The principal genetic process that is likely to coordinate senders and receivers is linkage disequilibrium (arising from assortative mating expected by both Fisher effect and handicap models; Bakker and Pomiankowski 1995). By contrast, the essentially pleiotropic effects expected from genetic coupling would, if they occur, play a lesser role in achieving the same result (Butlin and Ritchie 1989).

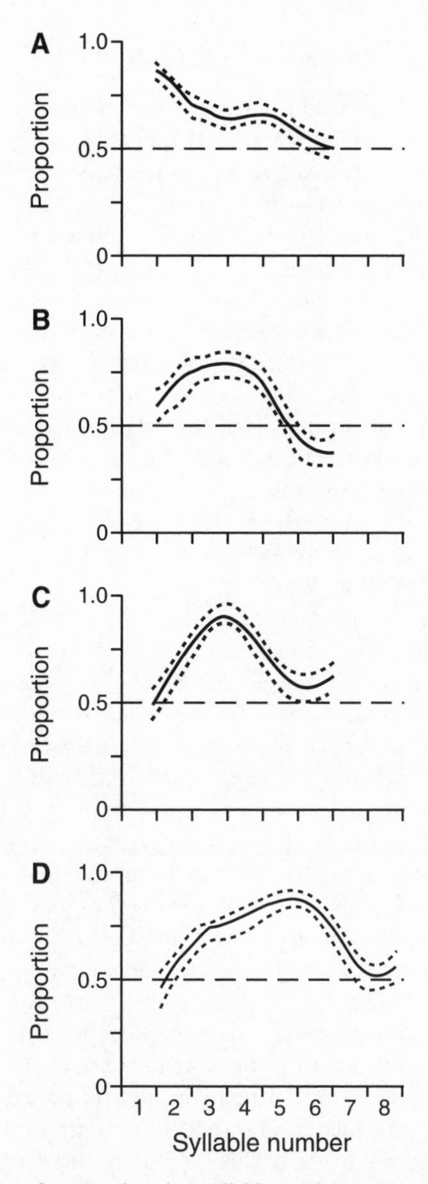

Figure 10.20. Preference functions based on syllable number in song races and hybrids of the katydid *Ephippiger ephippiger*. (A) Monosyllabic race. (B) Hybrid 1 (monosyllabic female X; polysyllabic male). (C) Hybrid 2 (polysyllabic female X; monosyllabic male). (D) Polysyllabic race. The dotted horizontal lines indicate chance phonotaxis. The functions are fitted cubic splines and error intervals (±1 standard deviation) computed from 1000 bootstrap replicates. Females of the polysyllabic race preferred songs with about five syllables, which is one more than the average produced by males of that race. Hybrids preferred intermediate numbers of syllables, but again about one more syllable per pulse than produced by hybrid males. The similarity of the preference functions of the reciprocal crosses suggests little evidence for the influence of maternally derived sex-linked genes. Modified from Ritchie 2000, fig. 2.

Summary and Suggestions for Future Studies

Robust demonstrations of female choice require finding acoustic correlates of male mating success coupled with experiments showing that females selectively respond to signals differing in the same acoustic property or properties. A general trend exists for acoustic properties showing relatively low within-male variability (pulse rate, carrier frequency, some short calls) to be subject to stabilizing or weakly direction selection and for highly dynamic properties (long calls and call rate) to be subject to highly directional selection. The comparative analysis of preference patterns is, however, plagued by the inadequacy of most studies to estimate the exact shape of female preference functions. Future studies should vary the values of acoustic properties in small steps and also assess the affects of varying both the relative and absolute values of alternative stimuli in order to check for nonlinearities in preference functions. Furthermore, by varying the relative and absolute intensities of signals, we can better estimate the strength of preferences and learn if preferences are consistent over the range of distances at which females might assess and compare the signals of different males.

Although having the advantage of eliminating the possible influences of male-male competition and nonacoustic assessments of males, playback experiments are unrealistically simple in terms of the acoustic environment in which females may choose mates in nature. The most important future task is to assess the effects on female preferences of *systematic* increases in acoustic complexity (multiple speakers, background noise, multidimensional call variability). These data can suggest experimental methods and provide estimates of the sample sizes that will be required to detect nonrandom mating patterns in nature. Along these same lines, we need experimental data concerning the time windows over which females assess the signals of different nearby males and how static and dynamic acoustic properties are weighed as the female moves through an ever-changing acoustic landscape.

For insects, studies conducted in enclosures have, at best, identified multiple acoustic and other properties that correlate with mating success. More generally, field observations and experiments with laboratory playbacks conclude that signal properties indicating male proximity are dominant. However, no field study of mating success in insects has yet attempted to evaluate variation in mating success in terms of variation in dynamic properties such as signaling rates and durations.

For anurans a handful of robust studies provide both positive and negative evidence for female choice based on spectral properties of advertisement calls. Most of these studies relied on assessing mating success as a function of body size rather than showing direct correlations with carrier frequency. A few studies showing size-dependent mating patterns, especially size-assortative mating, also demonstrated direct benefits of female choice in the form of increased fertilization success. The great majority of all studies have, however, failed to find consistent patterns of size-dependent mating success in species in which females normally choose freely among calling males. By contrast, most studies examining direct correlations between mating success and call rate have found positive relationships. Monitoring the calling performance (and other signal properties) of multiple males in the field and then assessing their mating success is perhaps the best way of providing robust evidence for female choice based on acoustic signals. Evidence is available for both direct and indirect benefits of choosing males producing high values of dynamic properties, which are reliable indicators of a male's (short-term) energetic investment in mate attraction and which may also reflect overall fitness. Do nonlinear preference functions indicate that females gain more by rejecting males with very low signal values than by choosing males with especially high values?

Whereas the stabilizing or weakly directional preferences based on pulse rate are usually interpreted in terms of species recognition, variation in the strength and symmetry of preference functions suggests, in turn, that the potential for evolutionary change in pulse rate varies among species. Indeed, as discussed in the next chapter, the strength of preference based on pulse rate can vary significantly among populations of the same species. Age-dependent effects on fine-temporal structure in insects might also be indicators of benefits, and experiments designed to demonstrate such effects robustly are badly needed.

Evolutionary responses to the selection generated by female choice are likely because of demonstrable genetic variation in quantitative acoustic properties and preference strength. Both components of the communication system appear to be under polygenic control, with some propensity for a disproportionate influence of X-chromosomal genes on signal structure, but not preferences. Recent studies of signal structure and preferences in hybrid and backcross products are compatible with the hypothesis that insect and anuran communication systems can coevolve in a stepwise fashion. Additional studies using artificial selection on signals or preferences to quantify the magnitude of indirect selection on receivers or senders are desirable.

11

Broad-Scale Patterns of Evolution

Studies of acoustic communication in contemporary populations provide snapshots in time and space. These studies are valuable because they can, in principle, identify the *kinds* of selective forces that are likely to be generally important in generating and maintaining evolutionary change. But these studies alone cannot provide a full explanation for the form of the communication system of any particular species (or even population), whose evolutionary history determines many of its present-day attributes. Geographic variation in communication systems testifies to the importance of recent evolutionary history and highlights the potential for such differentiation to play a role in speciation. The comparative approach, which is usually applied to a group of related taxa, can provide insights about patterns of evolution over longer periods of time.

Differences in communication behavior and underlying mechanisms ultimately arise from locally or regionally occurring mutations and episodes of genetic drift and selection. The prevailing view is that speciation can then occur if gene flow between groups of differentiated populations is restricted or curtailed. However, the outcome of interactions occurring between partially differentiated, incipient taxa that reestablish contact is problematic and a subject of considerable controversy (reviews: Butlin 1987; Littlejohn 1993). Studies of insects and anurans have provided a disproportionately large amount of evidence bearing on this issue, which is usually discussed in the context of reproductive character displacement. Analysis of geographical variation in communication systems within taxa also has the potential to provide data concerning the role of environmental factors. For example, signal variation within the range of a species might reflect selection to improve transmission in different habitats. Signal variation could also result from selection to avoid detection and localization by acoustically orienting predators and parasites, which might be common in some parts of the range and rare in other parts. Endler (1999) provides a thoughtful summary of the environmental factors that can affect signal design and transmission.

The comparative approach contributes to our understanding of broad-scale patterns of evolution in two general ways. First, the comparative approach offers a robust basis for identifying taxa that can shed light on environmental sources of divergence and convergence in communication systems (Brooks and McLennan 1991; Harvey and Pagel 1991). To identify environmental sources of divergent selection on communication systems, for example, comparisons of the communication systems of closely related species living in different kinds of environments are needed. To study convergence, comparisons of the communication systems of distantly related species in the same kinds of environments can reveal common selective factors. The reason for emphasizing selective processes is that the other forces of evolutionary change—mutation and drift—are, by definition, unpredictable.

Second, the comparative approach can sometimes offer hypotheses about the order of evolutionary changes in senders and receivers (reviews: Martins 1996). One common expectation is that these two components will coevolve gradually because senders and receivers interact in ways that have reciprocal effects on their fitness. For example, a very large change in a signal is more likely to make it unrecognizable than more effective, and a very large change in preference could mean that a female does not encounter acceptable signals in the current population. But another possibility is that "new" signals can sometimes tap evolutionarily conserved, preexisting sensory biases, or, contrariwise, "new" preferences can endow a previously irrelevant signal element with communicative significance (e.g., Arak and Enquist 1993; reviews: Endler and Basolo 1998; Ryan 1990). These two scenarios (coevolution and uncoupled evolution of signals and recognition mechanisms) are not mutually exclusive, especially when viewed from the perspective of the entire evolutionary history of a taxon. No matter how or in what order biases or signals originate, reciprocal selection by senders and receivers (and hence the potential for coevolution) commences as soon as communication using a new signal or new bias begins. Moreover, distinguishing between the two scenarios in any particular case depends on assumptions about the relative rates and probabilities of change in senders and receivers in different lineages that are seldom stated and difficult if not impossible to confirm (e.g., Ryan 1996; Endler and Basolo 1998).

In the first part of this chapter, we examine three general sources of environmental selection that might play a role in the diversification or convergence of communication systems: (1) habitat acoustics; (2) acoustically orienting predators and parasites; and (3) masking of signals by those of distantly related species. In the second part, we focus on reproductive interaction between closely related species. In the third part, we consider comparative studies that bear on questions about the origins of signals and receiver selectivity and the extent to which their evolution is correlated. As in chapter 10, our coverage includes some information about insects other than orthopterans and cicadas.

Environmental Selection on the Evolution of Signal Structure

Habitat Acoustics

The hypothesis that habitats impose selection on signals rests on the assumption that males that can attract females from greater distances than other males should have increased mating success. Communication range, in turn, depends on how a signal's intensity and relevant temporal and spectral properties change during propagation through the animal's environment. Operationally, we define communication range as the maximum distance at which a signal reliably elicits appropriate responses from receivers. This definition emphasizes that a signal must be audible and that its relevant properties must not be so degraded that it is no longer effective. If receivers respond phonotactically, then information about the location of the source must also be intact.

Many species of insects and anurans produce intense signals with sound pressure levels exceeding 90 dB at a distance of 50 cm (e.g., Loftus-Hills and Littlejohn 1971; Gerhardt 1975; Prestwich et al. 1989; Bennet-Clark 1997). Other species produce less intense signals (Gerhardt 1975, pers. obs.; Forrest 1991; Lee and Loher 1993; Penna and Solis 1998; Lang 2000), suggesting that selection does not always act to maximize communication range or that individuals of these species cannot produce more intense signals. In anurans, producing intense signals might be more a response to selection for detection in dense, noisy choruses (e.g., Gerhardt and Klump 1988a) than an adaptation for maximizing communication range per se. Individuals of species in which calling males produce relatively soft signals tend to be widely dispersed (Gerhardt, pers. obs.).

The hypothesis that habitat acoustics influence signals and signaling behavior would be supported by showing: (1) that animals within a given habitat signal from places or at times that increase their communication range; (2) that environmental factors influence signal propagation and integrity in ways that affect communication range; and (3) that signal structure, signaling behavior, or both vary within a species (geographical variation) or among closely related species in ways that reflect acoustically relevant differences in the environment. Evidence of the first two kinds is abundant, but few studies have even tried to obtain evidence of the last kind, which would provide the strongest support for long-term evolutionary effects of habitat acoustics on signals and signaling behavior.

Environmental effects on signals. The physical factors that affect signal propagation and integrity in nature are well explained in reviews published more than twenty years ago (Michelsen 1978; Wiley and Richards 1978). Bradbury and Vehrencamp (1998) summarize and extend the information in these reviews. We provide a brief synopsis of the most important principles in appendix 3. Studies of the propagation of insect and anuran signals through natural environments usually show at least one of three expected effects: (1) high-frequency signals propagate less well than low-frequency signals; this effect is usually exaggerated if the sound source and receiver are at ground level (e.g., Keuper et al. 1986; Römer and Lewald 1992; Kime et al. 2000); (2) temporal properties of signals can be degraded at relatively short distances where signal amplitude is still above the auditory (detection) threshold. Such degradation also depends on source-receiver geometry and habitat (e.g., Rheinlaender and Römer 1986; Lang 2000); (3) the usual disadvantage of signaling at ground level can be reversed at night by temperature inversions (van Staaden and Römer 1998; see appendix 3).

Most studies have dealt exclusively with problems of signal detection and identification, but environmental factors can also distort information that is needed for sound localization. Research to quantify these effects is only just beginning. Two studies show that (binaural) amplitude cues are almost certainly useless for sound localization by orthopterans located near the ground in grassland or bushland habitats (Rheinlaender and Römer 1986; Michelsen and Rohrseitz 1997). Michelsen and Rohrseitz (1997) conclude, however, that reliable phase cues might still be available for small grasshoppers using a pressure-difference system under these conditions.

Signaling behaviors that increase communication range. Within a given habitat, the signaling behaviors of some insects and anurans suggest that males attempt to increase communication range. Mole crickets, which construct burrows with acoustic hornlike properties, adjust the dimensions of the horn and the bulb to match each other and their own body size in order to increase their acoustic output (Bennet-Clark 1987; fig. 11.1). Moreover, in one species (*Scapteriscus acletus*), males move inward and outward while producing bursts of sound, suggesting that the insect is searching for a location where the sound pressure is balanced on either side of its forewing or where the burrow is tuned to the resonant frequency of its wings (Bennet-Clark 1987). Detailed measurements and modeling show that the carrier frequency of an Australian species (*Gryllotalpa australis*) is fairly well matched to the second node of the burrow resonance (Daws et al. 1996). Some matching of burrow resonances with carrier frequencies also occurs in some species of anurans that call from burrows (Bailey and Roberts 1991; Penna and Solís 1996).

Aboveground structures can also influence the propagation of insect and anuran signals. Tree crickets achieve two- to threefold increases in the power of radiated signals by chewing holes in leaves, which they use as acoustic baffles to reduce acoustic short-circuiting (fig. 11.2; chapter 2; e.g., Forrest 1991). The advertisement calls of glass frogs (*Centrolenella fleischmanni*) is enhanced by calling from the upper surfaces of large leaves (Wells and Schwartz

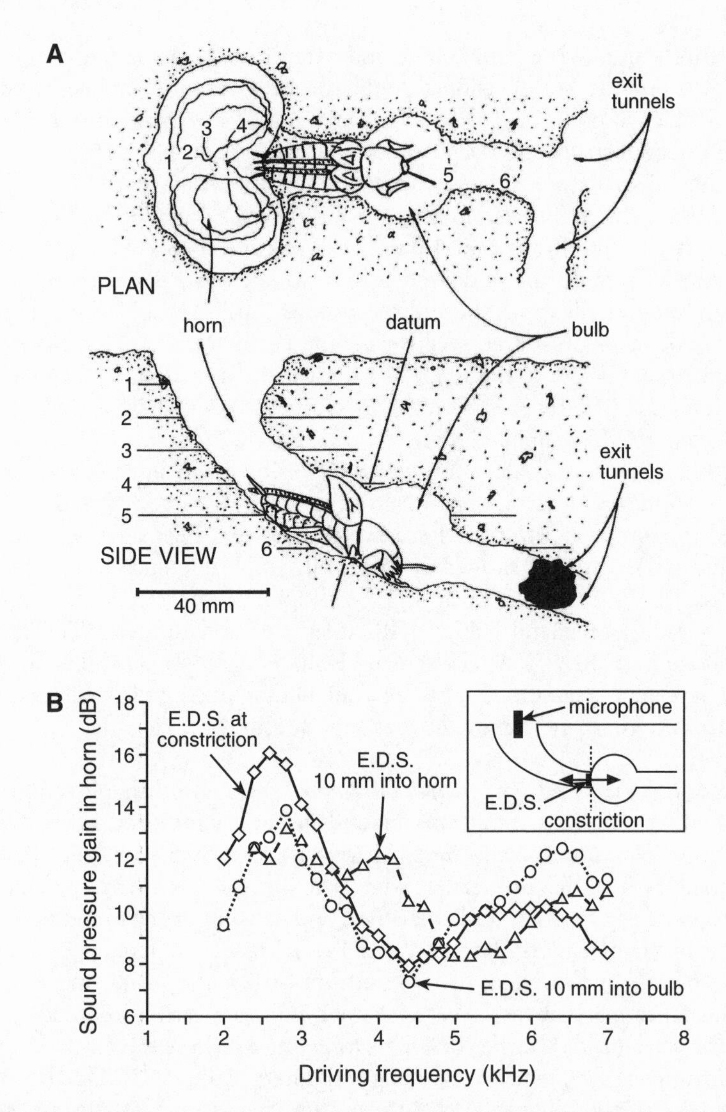

Figure 11.1. Acoustics of mole cricket burrows. (A) Diagrams of the burrow of a mole cricket (*Gryllotalpa vineae*). (B) Sound pressure, expressed as gain in decibels as a function of frequency, measured in a model burrow driven by an electret doublet source (E.D.S.) that was placed in one of three positions: at the constriction; at 10 mm into the horn; and at 10 mm into the bulb. The inset shows the driving and measuring conditions. From Bennet-Clark 1989, fig. 8.12A; and Daws et al. 1996, fig. 4a.

Figure 11.2. Photograph of a singing tree cricket positioned in a cut it has made in a leaf that serves as an acoustic baffle. Original photograph by Michael Boppré.

1982) rather than other kinds of calling sites, but there is no evidence that the frogs pick favorable sites on the basis of any kind of acoustic evaluation.

Differences in estimated broadcast area as a function of the elevation of signalers have been documented in crickets (Paul and Walker 1979) and katydids (Arak and Eíriksson 1992). In the cricket *Anurogryllus arboreus,* the broadcast area of elevated males is, on average, fourteen times greater than that of males singing on the ground (fig. 11.3; Paul and Walker 1979). In the katydid *Tettigonia viridissima,* the estimated broadcast distance of males calling from positions greater than 1 m above the average elevation is at least three times greater (14 m vs. 4 m) than that of males singing at lower elevations in the midst of dense reed beds (Arak and Eíriksson 1992). This last estimate is based on the detection threshold for the high-frequency peak (30 kHz) of the song, which is probably required for accurate localization (chapter 7). As expected, the broadcast range for the low-frequency peak is always greater at a given elevation than the range for the high-frequency peak (see also Keuper and Kühne 1983; Keuper et al. 1986; Römer and Bailey 1986).

Bioacoustic considerations fail, however, to predict either the behavior of males or their mating success in *T. viridissima* or *A. arboreus.* In *T. viridissima,* males did not sing at elevated positions that maximized broadcast distance,

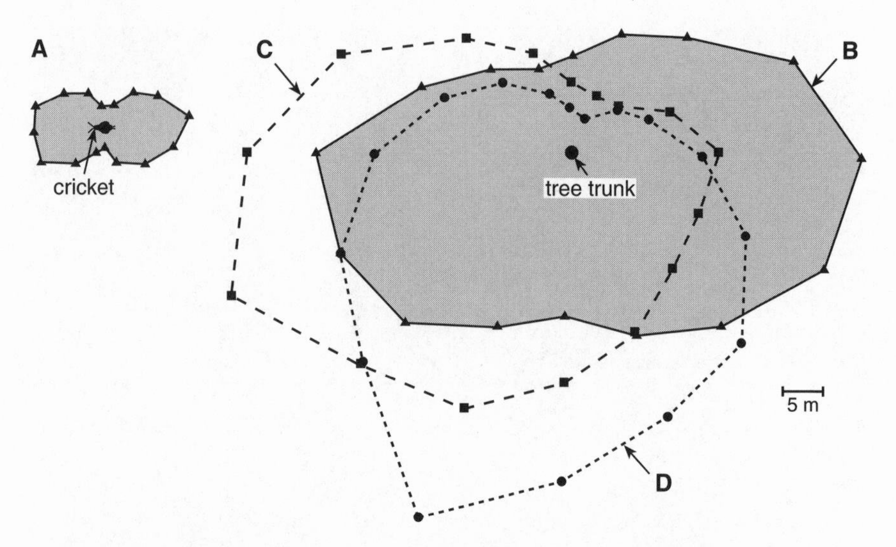

Figure 11.3. Effects of calling sites on broadcast area in a North American cricket. Estimated broadcast areas (= areas at which sound pressure level of the song is above the estimated auditory threshold) of crickets (*Anurogryllus arboreus*). (A) Horizontal broadcast area of a male singing on the ground. (B)–(D) Horizontal broadcast areas of three males singing on trees. The areas are depicted relative to the tree trunk. From Paul and Walker 1979, fig. 1.

perhaps because they might thereby increase predation risks (Arak and Eíriksson 1992). In *A. arboreus*, males calling at higher-than-average elevations on trees had lower mating success than did males calling at lower elevations (Walker 1983b). Why then do some males persist in calling from elevated positions? Although most matings take place in burrows and mated females are then unreceptive to males calling from trees, substantial numbers of older, unmated females in low-density populations are attracted to tree-singing males. But even these males sing for less than an hour before seeking females in burrows (Walker, pers. comm.).

Atmospheric conditions also can affect signal propagation. The male bladder grasshopper *Bullacris membracioides* can increase its communication range by signaling on nights when temperature inversions provide a low-attenuation (cool air) channel near the ground. Under these conditions, sound attenuation is sometimes less than expected by spherical spreading (appendix 3, fig. 3). According to van Staaden and Römer (1997), males appear to restrict most of their signaling to such nights. Propagation of the calls of the barking treefrog *Hyla gratiosa* can also be enhanced near the ground on some nights (Gerhardt and Klump 1988b), but there is no evidence that males of this species restrict chorusing to the conditions that favor this phenomenon. Moreover, as discussed in chapter 8, reproductively active females of this species arrive at the breeding pond even when choruses are prevented from forming. Thus, the attraction of females from relatively long distances might not even be a function of calling in this species.

Signal structure and habitat. In our view, showing that signals and habitat characteristics co-vary in ways that maximize communication range would provide the strongest evidence that habitat acoustics affect the evolution of signal structure (and behavior). Correlative evidence of this kind is available for some birds (e.g., Gish and Morton 1981; Wiley 1991; Van Buskirk 1997). Species living in forested habitats tend to produce songs with relatively low rates of amplitude modulation that are less subject to degradation by reverberation than are fast modulations (appendix 3). By contrast, birds in open habitats tend to produce songs with rapid trills. Because random amplitude fluctuations, which are caused by pockets of rising warm air, occur at low rates (< 30 Hz) (Wiley and Richards 1978), many successive periods of modulation can be transmitted without distortion.

Similar comparisons among species in insect and anuran communities have so far provided little evidence supporting the hypothesis that habitat acoustics influence the evolution of acoustic signals or even call-site selection. Keuper et al. (1988), who studied katydids, and Zimmerman (1983), who studied communities of Amazonian anurans in open and forest habitats, found no significant correlations between signal properties and breeding habitat. More recently, the transmission properties of the calls of groups of anuran species in different natural habitats have been assessed in terms of excess attenuation (Penna and Solís 1998) or degradation, as estimated from cross-correlations of signals recorded near to and at various distances from a speaker (Kime et al. 2000). No significant correlations between habitat and signal transmission or degradation were found, and the signals of several species propagate better in the breeding habitat typical of other species found in the same general region than in their own preferred breeding habitat. The lack of habitat effects among the ground-calling species is particularly telling because degradation was, as expected, significantly greater at ground level than at elevated positions. The general conclusion of all of these studies is that habitat acoustics explain little of the observed signal diversity.

The best data showing habitat effects on the structure of anuran signals comes from studies of geographical variation in the calls of the cricket frog *Acris crepitans.* Recording males in an extensive transect through Texas (U.S.A.), Ryan and Wilczynski (1991) found a general pattern of clinal variation in carrier frequency and in gross temporal properties such as call rate and duration, call groups, pulse groups, and numbers of pulses per group. Males of one subspecies, *Acris c. blanchardii,* whose breeding sites are typically found in grasslands in the central and western parts of Texas, generally produce calls of lower frequency, longer duration, and slower rates compared with males of the other subspecies, *A. c. crepitans,* which lives in pine forests in eastern Texas.

However, males of the first subspecies found in isolated, forested habitats tended to produce calls similar to those of the eastern subspecies (fig. 11.4). Although carrier frequency, which is correlated negatively with body size, also shows east-to-west clinal variation, habitat effects are still evident after controlling statistically for variation in body size (Ryan and Wilczynski 1991). Despite these general trends, a multivariate analysis reveals enormous and seemingly unpredictable variation in call structure among grassland populations of *A. c. blanchardii* (fig. 11.4B).

Although the calls of both subspecies propagate with less degradation in grassland habitats than in forested habitats, the calls of *A. c. crepitans* showed less degradation in forested localities than did the calls of *A. c. blanchardii* recorded from grassland populations (Ryan et al. 1990). These conclusions are based on cross-correlations of power spectra of calls rerecorded at 1 m from a speaker and the same calls rerecorded at greater distances (4, 8, and 18 m). Unfortunately, no data are reported on the attenuation of the signals or on the degradation of specific temporal properties that are also known to be important for discrimination (Nevo and Capranica 1985). Even simple measurements, such as changes in the depth of amplitude modulation (e.g., Ryan and Sullivan 1989), are more straightforward to interpret, and playback experiments show that variation in this property can affect signal attractiveness in other anurans (e.g., *Hyla cinerea;* see chapter 4).

Responses to degraded signals. Showing effects of habitat on acoustic signals is just the first step in testing hypotheses about the role of habitat acoustics in the evolution of signal design. The only way to be sure that various physical measurements reflect biological reality is to conduct behavioral tests. Behavioral effects of signal degradation have been well documented in birds, which can use this information to estimate the distance of their singing rivals (e.g., Naguib 1996). Simmons (1988) speculated that the differential response of male field crickets (*Gryllus bimaculatus*) to degraded and nondegraded songs functions in the same way.

A systematic study of habitat effects on receiver behavior has been conducted with the acridid grasshopper *Chorthippus biguttulus* (Lang 2000), which typically communicates at ground level in an acoustically difficult environment where excess attenuation and temporal degradation are severe. The communication distance, defined as the maximum distance at which a playback elicited the turning response of the male, differed significantly among three environments—the laboratory, a tall grass habitat, and a short grass, rocky habitat (fig. 11.5A). The grasshoppers occur in the last of these habitats. The communication distance was less in all situations than the detection distance ("perception distance" in Lang's terminology), at which the sound pressure level exceeded the threshold of the most sensitive auditory receptors. This result indicates that somewhat more than minimal stimulation of the auditory system is required to elicit the behavioral response. Furthermore, the data show

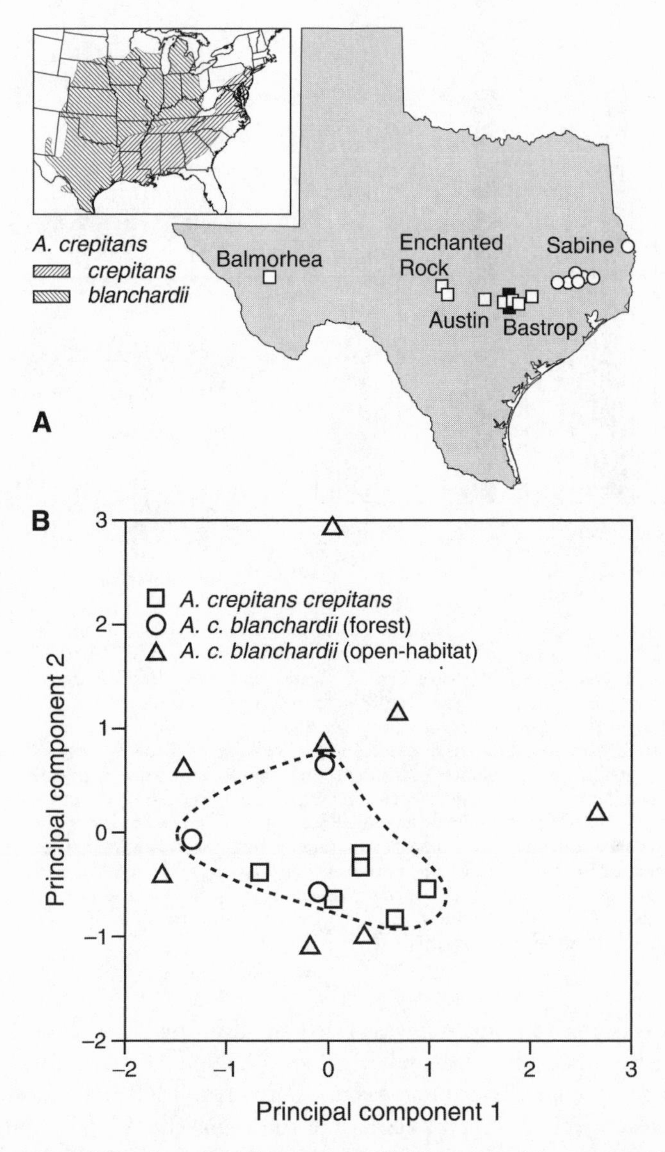

Figure 11.4. Correlations between call structure and habitat in North American cricket frogs. (A) Map showing sites within Texas where males of the cricket frog *Acris crepitans* were recorded. Circles = localities for *A. crepitans crepitans* in forested habitats; open squares = localities for *A. c. blanchardii* in open habitats; filled squares = *A. c. blanchardii* in forested habitats. *Inset*, distributions of the two subspecies within the eastern and central United States. (B) Plot of principal component scores based on spectral and temporal properties of the calls of cricket frogs recorded in Texas. The dotted line encloses the signal space of frogs of both subspecies that were recorded in forested habitats. Notice that call variation in grassland populations of *A. c. blanchardii* is enormous and unpredictable. From Wilczynski and Ryan 1999, figs. 11.2, 11.3.

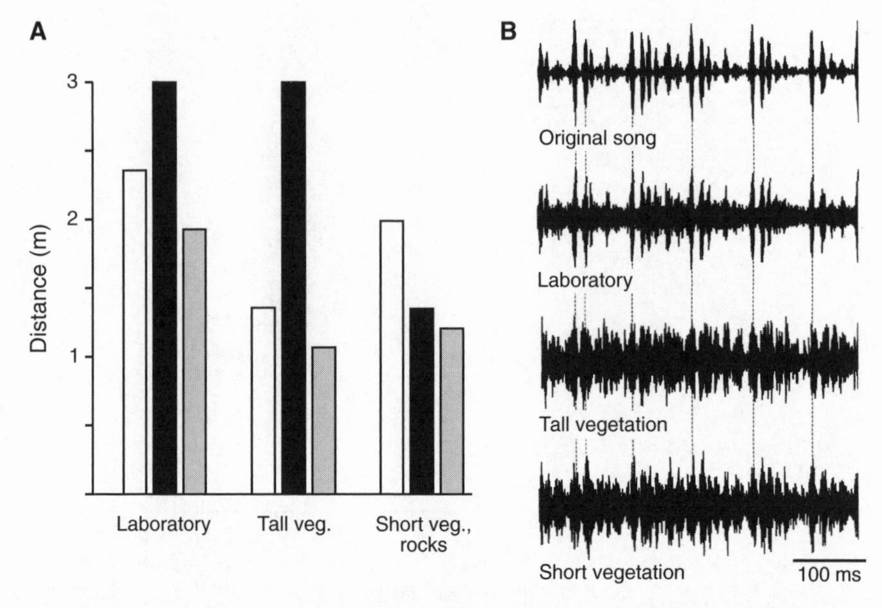

Figure 11.5. Responses of male grasshoppers (*Chorthippus biguttulus*) to attenuated and degraded signals. (A) Estimates of the detection (*open bars*), recognition (*solid bars*), and communication (*hatched bars*) distances of males. Communication distance is less than detection distances presumably because more than just a few of the most sensitive auditory receptors need to be stimulated to elicit the turning response. Recognition distance was determined by playing back and rerecording signals at different distances in the three environments and then playing back these sounds at above-threshold levels. Recognition distances were greater than detection distances in the tall grass and laboratory environments because attenuation and not degradation was the limiting factor in those environments. Recognition distance was less in the normal (short-grass) habitat because the temporal structure of the signal was degraded more severely after propagating for a shorter distance in this habitat than in the other two situations. Data plotted from Lang 2000, table 1. (B) Oscillograms of parts of the song (*top trace*) of a female song recorded at 210 cm from a loudspeaker at 5 cm above the ground in the laboratory and in two natural habitats. From Lang 2000, fig. 9.

that excess (signal) attenuation was clearly the limiting factor in both the laboratory and (especially) in the tall vegetation habitats. By contrast, the limiting factor in the favored habitat was not signal amplitude (detection) but the degradation of the temporal structure of the signal (fig. 11.5B). Lang (2000) demonstrated this by estimating the "recognition" distance in each situation. She did so by playing back and rerecording a female's song, which had been originally recorded at close range, at different distances in the laboratory, tall grass and short grass habitats, thus creating a series of stimuli that differed in the degree of temporal degradation. These sounds were all then played to males at levels above the estimated neural and behavioral thresholds so that males would always be able to detect the stimulus. The recognition distance for sounds that propagated through the laboratory and tall grass environments far exceeded the communication distance (fig. 11.5A), thus indicating that signal attenuation (detection) but not signal integrity limited the communication range. In the short grass habitat, however, the recognition distance was about one-half of the communication distance, indicating that signal integrity and not attenuation was the limiting factor. Because some temporal degradation was evident in all three environments (fig. 11.5B), only the behavioral experiments could directly assess the relative importance of temporal degradation and attenuation.

To our knowledge, no study of anurans has systematically tested the effects of signal degradation on conspecific receivers. Female barking treefrogs (*Hyla gratiosa*) phonotactically responded to playbacks of chorus sounds that had been originally recorded at ground level at 160 m from the chorus, even when the playback level was 6 dB less than the sound pressure level measured at the recording site (chapter 8; Gerhardt and Klump 1988b). At that distance the high-frequency energy in the advertisement calls was severely attenuated and only a few calls of individuals were evident against the background noise. Thus, degraded and partially masked signals were still effective under these conditions, but further experiments, modeled after those of Lang (2000), are needed to assess detection and recognition distances of this and other anurans.

Acoustically Orienting Predators and Parasites

Signaling insects and anurans are subject to predation or parasitism by numerous kinds of organisms that use sound to locate their victims (review: Zuk and Kolluru 1998). One classic example is the túngara frog *Physalemus pustulosus*, in which advertisement calls attract a wide range of predators (Ryan 1985). Some behavioral tactics that seemingly reduce these risks are mentioned in chapters 8–10, and comparative studies potentially provide strong evidence for a direct role of predators and parasites in the evolution of risk-aversive behaviors. That is, species in areas where predation/parasitism risks are high should share signal properties, predator detection mechanisms, and (evasive) behaviors that differ from those of close relatives in areas where such risks are low. Finding such differences within a single, wide-ranging species would provide even stronger evidence.

Bat predation: evolution of risk-aversive behavior in Neotropical katydids? A striking difference in behavior and signal structure between species of Neotropical katydids (family Pseudophyllidae) that live in open areas and those that live in forests seemingly reflects differences in bat predation pressure (reviews: Belwood 1990; Heller 1995). Species living in the forest, where bat predation is high, used vibrational (tremulation) signals as well as acoustic signals, and one species (*Schedocentrus differens*) produced only vibrational signals (Morris et al. 1994). Tremulation was absent in species in open habitats, where bat predation on katydids is much lower. Furthermore, carrier frequency of the song in forests species was very high (> 100 kHz) and call rate was greatly reduced (call-duty cycle was usually less than 3%) in comparison to open-habitat species (Belwood and Morris 1987). These differences should reduce the probability and range of detection by bats.

These general trends are not, however, corroborated by a preliminary study of some forest-living katydids in the Paleotropics. In a study of six species of pseudophyllid katydids in Malaysia, Heller (1995) describes songs and behavior that are similar to those of open-habitat species in the Neotropics (Heller 1995). Tremulation was not detected in the Malaysian species, although no concerted effort was made to discover it. The lack of such signals is, however, consistent with the generally higher rate of acoustic signaling: call-duty cycle was greater than 3% in five of the six species, with a mean of about 10%. Carrier frequencies in the sample of Malaysian species were also much lower (≤ 10 kHz) than those typical of forest-dwelling species in the Neotropics. Heller (1995) speculates that the differences in acoustic behavior reflect a difference in the severity of bat predation in the Neotropics and Paleotropics, although the very low carrier frequencies (< 6 kHz) in some Malaysian species could also reduce predation risks from bats whose hearing sensitivity is poor at these frequencies. Another possibility is that differences in carrier frequency are an incidental consequence of the differences in body and wing shape that provide the camouflage (lateral compression in Neotropical species and dorsoventral flattening in Malaysian species). Obviously, much more sampling of katydids in the other parts of the Old World tropics, as well as studies of their common predators, is needed to test these ideas.

Acoustically orienting parasitoid flies. As discussed in chapters 9 and 10, acoustically orienting parasitoid flies (family Tachinidae) can be a severe source of mortality on some species of field crickets and katydids (e.g., Cade 1974; Walker and Wineriter 1991; Stumpner and Lakes-Harlan 1996; Lehmann and Heller 1998; Allen 1998, 2000). The flies deposit larvae on or near the insect, and the emergence of the pupae kills the host in about a week (in crickets) or two (in the katydid *Sciarasaga quadrata*). One species of acoustically orienting fly in the family Sarcophagidae has also been described as a parasite of the cicada *Okanagana rimosa* (Soper et al. 1976), and the great similarity in the structure of the ears of both kinds of flies appears to be a convincing example of convergent evolution (Lakes-Harlan et al. 1999).

In the katydid *S. quadrata*, the selective pressure for attracting scarce females appears to be so severe that males call persistently during times when the flies search for them, and most individuals are infected by the end of the season (Allen 1998). In field crickets the evolution of the noncalling, satellite tactic and reduced nightly singing duration in some species have been hypothesized to have evolved as responses to the high risks of fly parasitism by

Ormia ochracea (Cade 1975, 1979). Paralleling the situation in the túngara frog and one of its major bat predators (Ryan 1985), playbacks show that females of two species of crickets (*Gryllus lineaticeps* and *G. integer* [= *texensis*]) and their tachinid parasites prefer the same acoustic properties of calling songs (Wagner 1996; Gray and Cade 1999b). Indirect evidence from a third species (*Teleogryllus oceanicus*) is consistent: Males producing songs with acoustic properties favored by females are most likely to be infected (Zuk et al. 1998).

Studies of different populations of the field cricket *Teleogryllus oceanicus* suggest that parasitoids can affect the evolution of some calling behaviors and acoustic signals. This cricket has been studied in Australia and French Polynesia (Mooréa), where parasitoid flies are absent, and in the Hawaiian Islands, where both the cricket and the parasitoid fly *O. ochracea* were introduced by humans (Zuk et al. 1993; Rotenberry et al. 1996; Kolluru 1999). Males from Hilo, Hawaii, where parasitoids are relatively common, showed generally reduced calling compared with males from the two parasitoid-free populations (Zuk et al. 1993; Kolluru 1999). Because, as might be expected (chapter 10), the proportion of time singing is also a reliable predictor of female attraction (Kolluru 1999), the risk of attack appears to have compromised the ability of males to maximize their mating success. Although Hawaiian males tend to start and stop calling more abruptly than do males in Mooréa and Australia (Zuk et al. 1993), they do not shift the calling period to times when the flies are less active (Kolluru 1999).

Comparisons of the structure of the songs of individuals of *T. oceanicus* in parasitized and unparasitized populations provide weak and equivocal support for the hypothesis that the flies have affected the evolution of signal structure. A multivariate statistical analysis showed that a relatively small but statistically significant amount of song variation occurs among populations on three of the Hawaiian Islands (Rotenberry et al. 1996). Males in the population experiencing the least density of parasitoids had songs most like those of males from parasitoid-free areas, and males in the population with the highest density had the most divergent songs. However, most song variation (80–90%) among all populations (Hawaii, Mooréa, and Australia) was not explained by parasitoid density or occurrence, which is consistent with the observation that the songs of the Mooréa population (parasitoid-free) were generally intermediate between those of males from Hawaii and Australia, which is also parasitoid-free (Zuk et al. 1993). Furthermore, Zuk et al (1998) also showed that the probability of infection of Hawaiian males was explained to a significant degree by the duration of the long chirps, suggesting that songs with long chirps are especially attractive to the flies (fig. 11.6). Yet males of *T. oceanicus* from Hilo, Hawaii, where there was a high incidence of parasitoids, produced songs with more pulses per chirp (and hence presumably longer chirps) than did males from parasitoid-free localities (Zuk et al. 1993). Thus, at this level of analysis, these Hawaiian males appear to be producing songs that are more attractive

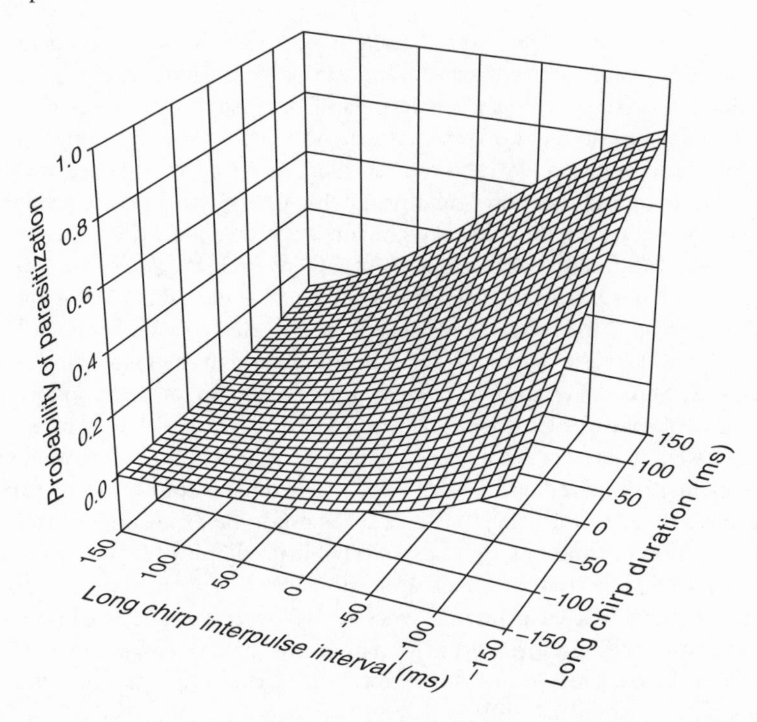

Figure 11.6. Cricket song properties that predict infection by parasitoid flies. Plot showing the probability that a male of *Teleogryllus oceanicus* would harbor larvae of the fly *Ormia ochracea* as a function of two of its calling-song properties. Units are milliseconds, standardized for temperature. From Zuk et al. 1998, fig. 3.

to parasitoids than are the songs of males in parasitoid-free localities. Perhaps an increase in the mating success of males producing such signals compensates for the increased risks. Another possibility is that the overall reduction in singing activity reported by Zuk et al. (1993) could reduce the detection of males by parasitoids.

One other comparison of the behavior of males in Hawaii and parasitoid-free localities fails to support the hypothesis that parasitoids favor the evolution of silent-male satellite behavior (chapter 9). Namely, there is no significant difference in the frequency of occurrence of silent males between Hawaii and the two parasitoid-free localities. Furthermore, recall that infected males of *T. oceanicus* in Hawaii were actually more likely to be silent than were uninfected males (Zuk et al. 1995). This fact underlines the possibility that infection could also affect singing behavior and songs in subtle ways (see also Kolluru 1999). Hence direct tests of the acoustic preferences of the *O. ochracea* with regard to the song properties of *T. oceanicus* are needed to corroborate the indirect evidence so far available (Zuk et al. 1998).

Masking Interference: Environment Noise and Acoustic Interactions between Distantly Related Species

A particularly severe kind of environmental interference is the noise produced by fast-flowing streams and waterfalls. One evolutionary response that appears to have occurred in several anuran species is the production of relatively high, narrow-band signals (Hödl and Amézquita 2001). Even more impressive is the apparently convergent evolution of visual signals in a wide range of taxa (which usually also produce acoustic signals) that live in such habitats (Hödl and Amézquita 2001). Visual signals are also common in diurnal frogs, especially if their bright colors also serve an aposematic function.

Several forms of acoustical interaction take place between species that do not mate with each other. First, there are numerous examples of insects and anurans in which the signaling activity by one species inhibits that of another species (chapter 8). Second, the timing or duration of the signals of one species can be influenced by signaling of another species or (rarely) the two species both change their signals (review of katydid studies: Schatral 1990; Marshall and Gerhardt, unpubl. data for the gray treefrogs *Hyla chrysoscelis* and *H. versicolor*). Third, aggressive interactions, mediated in part by acoustic signals, can occur between species (chapter 9). These interactions might be expected to generate differences between the communication systems of one or both taxa in areas of sympatry (Littlejohn 1981 and below).

Evidence for behavioral shifts caused by interspecific masking comes from removal experiments with katydids. As discussed in chapter 8, singing males of the katydid *Hemisaga denticulata* are inhibited by signaling males of *Mygalopsis marki*, and in both *Hemisaga* and in a Central American katydid (*Neoconocephalus spiza*), the diel periods of singing differ in areas with and without masking taxa (Greenfield 1988; Römer et al. 1989). Because, however, all of the sampled localities were nearby, these studies do not provide direct evidence that interspecific inhibition has promoted population (or species) differences in singing behaviors over evolutionary time. Such shifts might even occur in populations that have never experienced a particular masking species, and playbacks of the songs of non-native species or even white noise might also be effective. Better evidence for an evolutionary effect of interspecific interactions requires systematic comparisons of signals and signaling behaviors in more geographically separated populations with and without masking species.

Two studies of Australian anurans provide circumstantial support for the hypothesis that acoustic interactions could play a role in interspecific exclusions from preferred habitats. Odendaal et al. (1986) speculate that a species with a limited distribution (*Crinia riparia*) might be prevented from expanding its range by another species (*C. signifera*), whose call amplitudes exceed those of *C. riparia* by more than 20 dB SPL. However, even though the intense calls of *C. signifera*, which also form choruses in streamside microhabitats, would probably mask communication by *C. riparia*, the possibility remains that habitat differences alone could explain the distribution of the two species. That is, regardless of whether they occur in the same area, males of *C. riparia* seem to prefer faster-flowing streams than do males of *C. signifera;* such fast-flowing streams are also typical of the areas in which only *C. riparia* occurs. In the second study, which directly assessed interspecific aggression mediated by vocal signals, *C. signifera* was more likely than another sympatric congener *C. parinsignifera* to cease production of advertisement calls during playbacks of calls of the other species (Littlejohn et al. 1985). This asymmetry in response might result in displacement of *C. signifera* from preferred calling sites near the edges of ponds. As stated by the authors, additional fieldwork concerning preferred calling sites and comparisons of calling behaviors in response to heterospecific calls need to be conducted in sympatric and allopatric areas in order to support this hypothesis.

Reproductive Character Displacement

Theoretical and Practical Considerations

Reproductive interactions between closely related species are another potential source of selection for divergence in communication systems. We define the term *reproductive character displacement* as a geographic pattern in which differences in the communication systems of two taxa (or incipient taxa) are greater in sympatry than in allopatry because of selection in sympatry against costly mating mistakes (= reduced viability or fitness of hybrids, wasted gametes, or missed mating opportunities). Sympatric divergence can be accomplished by changes in one or both of the taxa. Note that merely observing the pattern is inconclusive without evidence that selection against mating mistakes has contributed to it. For example, sympatric divergence could arise from interactions between taxa that never mismate with each other (e.g., insect and frog) but whose signals mask one another (see previous section; Littlejohn 1981).

This seemingly simple idea, which was proposed by Dobzhansky (1940), remains a subject of considerable controversy, even with respect to the terminology. The term *reproductive character displacement* is, for example, sometimes restricted to the pattern of geographic variation, whereas a second term, *reinforcement* (sensu Blair 1955), is used to describe the selective process (Howard 1993; Littlejohn 1993). Butlin (1987) further restricts the usage of reinforcement to situations in which there is significant gene exchange between the interacting taxa and uses the term *reproductive character displacement* to describe the process when the interacting taxa rarely, if ever, produce fertile hybrids. Unfortunately, *reinforcement* also refers to a long-established psychological process, thus adding to the semantic confusion surrounding this term (Gerhardt and Schwartz 1995).

On balance, current theoretical opinion suggests that reproductive character displacement is plausible. Howard (1993) reviews various theoretical objections and concludes that none is particularly compelling. Modeling and computer simulations show that reproductive character displacement could occur under a wide variety of conditions (Liou and Price 1994; Kelly and Noor 1996; Kirkpatrick and Servedio 1999; Servedio and Kirkpatrick 1999). Still, we think that the conditions favoring reproductive character displacement are restrictive. Moreover, at least five alternative outcomes of reestablished contact are possible: (1) introgression (fusion); (2) extinction (displacement) of one taxon by the other; (3) tension (stable hybrid) zones; (4) hybrid zones where hybrids are of equal or equivalent fitness to parental types; (5) noninteractive attainment of sympatry (Barton and Hewitt 1985; Butlin 1987; Littlejohn 1993, 1999; Noor 1999). Littlejohn (1993) argues persuasively that the problems of achieving stable sympatry—a prerequisite for reproductive character displacement—are too often ignored. He emphasizes the important role of ecological divergence and the spatial distribution of resources in determining the outcome of interactions between populations that reestablish contact. Differences in the apparent outcomes of interactions (e.g., within the *Pseudacris nigrita* [chorus frog] complex; Fouquette 1975; Gartside 1980; see below) and the heterogeneous nature of some hybrid zones (e.g., between members of two pairs of anuran species—*Geocrinia laevis* and *G. victoriana:* Littlejohn 1988; and *Bombina bombina* and *B. variegata:* Szymura 1993) support Littlejohn's emphasis on ecological factors.

Another controversy concerns the evolutionary significance of reproductive character displacement. Butlin (1987) emphasizes that selection against mating mistakes between taxa that are initially interbreeding could contribute to speciation. Other researchers consider such selection to be incidental to speciation and argue that character displacement is likely to occur only after or near the end of the completion of speciation (e.g., Littlejohn 1993, 1999; Alexander et al. 1997; Marshall and Cooley 2000). The difficulties in resolving this issue are empirical and theoretical. We seldom have good estimates of (1) the frequency and costs of mating mistakes; (2) the extent of genetic incompatibility; (3) the fitness of hybrids; (4) the extent and direction of migration; or (5) the selectivity of receivers in either sympatry or allopatry. Moreover, a perennial debate in evolutionary biology concerns how much gene flow can occur between species. Indeed, definitions that use the extent of interbreeding as a criterion for recognizing species are also subject to considerable debate (e.g., Wheeler and Maier 2000).

The main reason for skepticism about reproductive character displacement is the paucity of good examples (e.g., Walker 1974), about which there is also considerable disagreement. Paterson (1985) denies that any example holds up to scrutiny, whereas Howard (1993) lists numerous examples, most of which are not particularly convincing. Formidable practical problems exist in finding suitable systems to study. First, extensive areas of allopatry are required in order to calibrate differences in sympatric areas (or to compare the steepness of clines from sympatry to adjacent allopatry with that of clines in remote allopatry) (e.g., Grant 1972). Second, it is necessary to identify pairs of interacting species in which signals and receiver selectivity have not diverged so much in allopatry that mistakes in signal identification would be rare when contact is reestablished. These mistakes and their negative consequences are the expected causes of sympatric divergence. Third, there must be evidence that potential mistakes in (acoustic) signal identification are not simply avoided by the use of other cues (visual, chemical). Reproductive character displacement might be more common than generally supposed simply because nearly all studies have focused solely on signal divergence rather than receiver selectivity. If signals diverge significantly in allopatry and receivers remain permissive, then selection in sympatry might operate mainly to sharpen the selectivity of receivers, which often have more to lose from a mating mistake than do senders, without causing further differentiation in signals (e.g., Waage 1979; Gerhardt 1994b; see below). Alexander et al. (1997) suggest that good examples are rare because reproductive character displacement occurs so rapidly that the expected pattern is difficult to detect. Although strong selection can certainly promote rapid change (e.g., Higgie et al. 2000), this hypothesis would seem to depend on the unlikely assumption that the divergence in sympatry will always spread rapidly throughout the rest of the range by gene flow. Following Howard (1993) and Noor (1999), we advocate a case-by-case approach in which the causes of the expected pattern are investigated in a systematic and experimental fashion.

Empirical Evidence: Comparative Studies

Closely related species of insects and anurans that breed synchronously in mixed-species assemblages typically differ significantly in habitat (different calling sites), in peak signaling times within the diel cycle, or in signal structure (e.g., Hödl 1977; Duellman and Pyles 1983; Otte 1989; Helversen and Helversen 1994; Alexander et al. 1997; Garcia-Rutledge and Narins 2001). At the same time, some species in any given community often have breeding ecologies, signals, or both that are very similar to those of species in other (allopatric) communities. At a grosser level of analysis, comparative studies by Coyne and Orr (1989, 1997) of more than a hundred species of *Drosophila* indicate that pairs of sympatric species are more likely to have evolved strong behavioral selectivity that promotes conspecific matings than are pairs of allopatric species. These authors also show that this difference persists even when genetic divergence, which provides an estimate of the time of the original isolation, is taken into account. Without explicitly showing an effect of sympatry, other studies that include estimates of genetic divergence (or phylogenetic relationships) also find that signals and receiver selectivity (or both, as inferred from mating tests) usually evolve more rapidly than do more general traits (including postmating genetic incompatibility), which tend to accumulate changes in a clocklike fashion (e.g., Butlin and Tregenza 1998; Gleason and Ritchie 1998).

Because these studies did not explicitly show an effect of sympatry, these results are sometimes interpreted as evidence that sexual selection often drives speciation in the absence of interspecific interactions (see below). By contrast, divergence in the signals of some allopatrically distributed species or lineages that show significant genetic differentiation is minimal or unimpressive (e.g., insects—*Drosophila pseudoobscura* subspecies: Noor et al. 2000; anurans—lineages of canyon treefrogs *Hyla arenicolor:* Barber 1999a,b; Gerhardt, unpubl. data; species in the *Geocrinia rosea* complex: Roberts and Wardell-Johnson 1995; toads [genus *Bufo*] with isolated distributions in the western United States: Gergus et al. 1997).

The general pattern of species distinctiveness in sympatric assemblages combined with frequent examples of similarities in allopatrically distributed species is often interpreted as evidence that reproductive character displacement might be widespread (Alexander et al. 1997; Coyne and Orr 1989). The underlying assumption is that many of the closely related species that are now sympatric probably interacted reproductively when they first reestablished contact (see Noor 1999 for a critical discussion). Even if this assumption is false, however, we can conclude that, regardless of how different species with different signals achieve syntopy, evolutionary change in the species-identifying elements of the signals of any particular species must, in general, be constrained by the values of those elements in the signals of the other species in the community.

Rare exceptions to the usual pattern of species-specificity of the signals of sympatric species also exist. The ground crickets *Allonemobius fasciatus* and *A. socius* meet in a mosaic zone of contact from New Jersey to Illinois (U.S.A.), where some hybrids are found and where the two species breed in close proximity at the same times during the year (Benedix and Howard 1991). Even though fertilization in cross-species matings is greatly reduced, the ranges of variation in the carrier frequency and pulse rate of songs overlap broadly between all populations of the two species. The only statistically significant difference is in chirp rate, a dynamic property showing considerable within-species variation in all populations, and females of the slower-chirping species from sympatric populations actually prefer the songs representative of the faster-chirping congener (Doherty and Howard 1996). Two species of leaf frogs (*Phyllomedusa distincta* and *P. tetraploidea*) produce indistinguishable advertisement calls (fig. 11.7) despite the fact that they have different chromosome numbers and produce sterile or low-fertility hybrids in areas of sympatry (Haddad et al. 1994). Another possible example involves sibling species of the katydid *Mygalopsis marki* (A and B strains) in southwestern Australia, in which calling songs have diverged only slightly despite one-way genetic incompatibility and the existence of an area of overlap where hybridization is

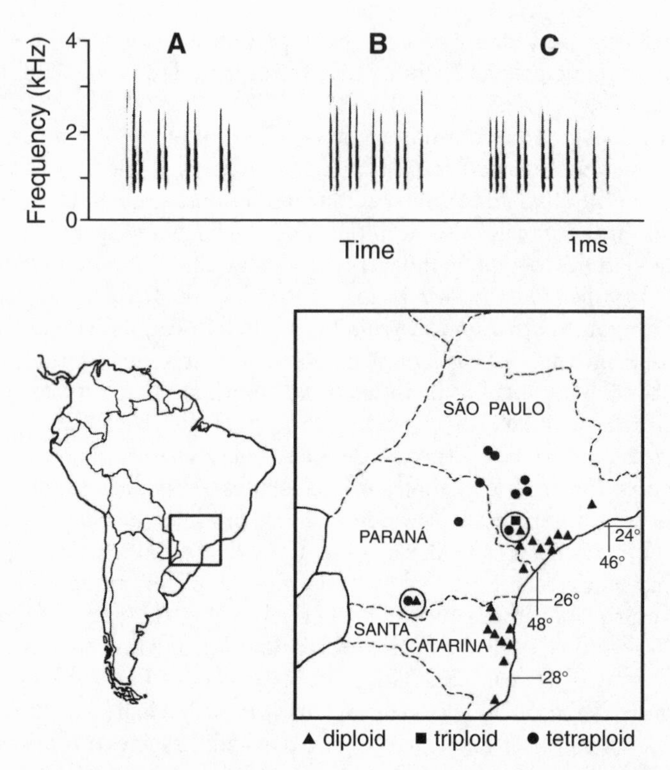

Figure 11.7. Lack of call differentiation in two sympatric, genetically incompatible species of leaf frogs. Sonograms show typical advertisement calls of (A) *Phyllomedusa distincta* (a diploid species); (B) a hybrid (*P. distincta* X *P. tetraploidea*, a triploid); and (C) *P. tetraploidea* (a tetraploid species). These calls vary considerably in duration but overlap extensively between the two species. The map shows the distribution of the two species in southeastern Brazil; circled areas show where both species occur and where triploids were found. From Haddad et al. 1994.

common (Dadour and Johnson 1983; Dadour, pers. comm). Of course, this state of affairs could reflect a recent contact or a stable tension zone. In any event, new molecular techniques might uncover additional examples of cryptic genetic divergence that might occur among sympatrically occurring individuals with similar acoustic signals (Gleason and Ritchie 1998).

Sexual selection is often cited as the most likely source of the initial divergence in allopatry and sympatry (Gleason and Ritchie 1998 and references therein). Otte (1994) provides a contrary viewpoint, and there is certainly evidence that genetic drift and hybridization can also contribute (Driscoll et al. 1994; Shaw 1996b; Roff et al. 1999; review: Butlin and Tregenza 1998). Because female choice is usually either stabilizing or favors increases in the values of signal properties such as duration and call rate (Gerhardt 1991; Ryan and Keddy-Hector 1992; chapter 10), divergence must often depend on strong directional selection (for high values) on one group of isolated populations but not on the other. Day (2000) presents a general model based on the Fisher effect. Divergence could also be achieved if long duration signals were favored in one set of populations and high calling rate in another (see below for a possible example in North American toads).

Empirical Evidence: Geographic Variation in Communication Systems

In this section we examine examples of reproductive character displacement involving congeneric pairs of insect and anuran species. Signals, receiver selectivity, or both show the expected pattern of divergence in sympatry, and the assumption that selection against costly mating mistakes has contributed to this pattern is probable, if not always directly supported by empirical data. In many of these examples, the divergence occurs only, or mainly, in one of the two interacting taxa. Otte (1989) and Littlejohn (1993, 1999) suggest that the taxon showing greater divergence has expanded its range into that of the other taxon.

Crickets. Speciation of endemic crickets on the Hawaiian Islands is impressive. The number of species (about 240), derived from colonization by four general cricket types, is estimated to be about twice that found on the continental United States (Otte 1994). Otte (1989, 1994) provides extensive data on intraspecific and interspecific variation in Hawaiian cricket songs and argues that speciation has mainly occurred within rather than between islands. Based on extensive within-island sampling, Otte (1989) concludes that reproductive character displacement of calling song has frequently occurred. Shaw's (1996b, 1999) phylogenetic analyses, which are based on variation in mitochondrial DNA sequences, support Otte's hypothesis of intra-island speciation in the genus *Laupala*, but suggest different patterns of divergence than those based on morphology and song differences alone (Otte 1989, 1994).

One main difficulty of accepting Otte's conclusions about the widespread occurrence of reproductive character displacement centers on the identity of species with geographically varying song types. For example, three taxa that show expected patterns of sympatric divergence or parallel clines on Oahu—*L. tantalis*, *L. nui*, and *L. pacifica*—can only be distinguished by differences in the songs themselves (Otte 1994). The sympatric divergence of the pulse rate of *L. pacifica* in an area of sympatry with *L. spisa*, from which it differs in song and genitalia, is probably a better example, assuming that all populations presumed to be *L. pacifica* are confirmed as such.

Besides the taxonomic problems, uncertainties exist about the extent to which hybridization has occurred and the role that might be played by shifts in ranges and differential extinctions (e.g., Noor 1995). These problems are compounded by the small spatial scale of the arenas in which interactions may take place between taxa whose arrival times on a given island are also uncertain. In these respects, the situation is less favorable for testing the hypothesis of reproductive character displacement than in continental situations, where we can often be reasonably sure that (incipient) species with partially overlapping distributions were derived from a single taxon whose distribution was fragmented by climate change or some other physical factor. Besides using molecular criteria to characterize the genetic divergence of the putative taxa, further support for Otte's ideas could be provided by evidence that mating mistakes are costly. Behavioral tests are also needed to show that differences in pulse rate typical of sympatric populations reduce the probability of phonotaxis to heterospecific signals in comparison to the differences in pulse rate typical of allopatric populations. The latter proposition should be easy to test because in most of the species of *Laupala* studied so far, females are highly selective with respect to variation in pulse rate and have unimodal pulse rate preference functions (Shaw 2000; Shaw and Herlihy 2000).

Two other studies of field crickets found no evidence of reproductive character displacement. Hill et al. (1972) conducted an extensive study of the field crickets *Teleogryllus commodus* and *T. oceanicus* in Australia. They found no evidence for sympatric divergence in carrier frequency or in the pulse rate of one component of the calling song. Allopatric females were also no more likely to make mating mistakes than sympatric ones, and no interspecific hybrids were detected in the overlap zone. This might be an example in which the degree of allopatric divergence was so great that mating mistakes rarely, if ever, occurred when contact was renewed. Comparable negative results were recently reported for comparisons of allopatric populations of *Gryllus texensis* and populations sympatric with a cryptic congener *G. rubens* in eastern North America (Gray and Cade 2000). Because sympatric divergence requires changes in only one taxon, studies of the signals and preferences of sympatric and allopatric individuals of *G. rubens* are desirable.

Periodical cicadas. Some species of North American periodical cicadas (e.g., 13-year and 17-year species) of the genus *Magicicada* seldom interact because their brief periods of emergence and reproduction only rarely coincide in relatively small areas of geographic overlap. The fact that the morphology and calling songs of some of the pairs of these 13-year and 17-year species are similar, if not indistinguishable, suggests that life-cycle changes contributed to their speciation—a view supported by new genetic evidence (Simon et al. 2000). Within the 13-year and 17-year groups of species, however, synchronously breeding taxa usually have distinctive, species-typical advertisement songs. One exception is a recently described 13-year species (*Magicicada neotredecim*), which shows relatively subtle differences in abdomen color and in the carrier frequency of the song vis-à-vis a sibling species, *M. tredecim* (Marshall and Cooley 2000).

Marshall and Cooley (2000) provide evidence for reproductive character displacement in both carrier frequency and female frequency-preferences in

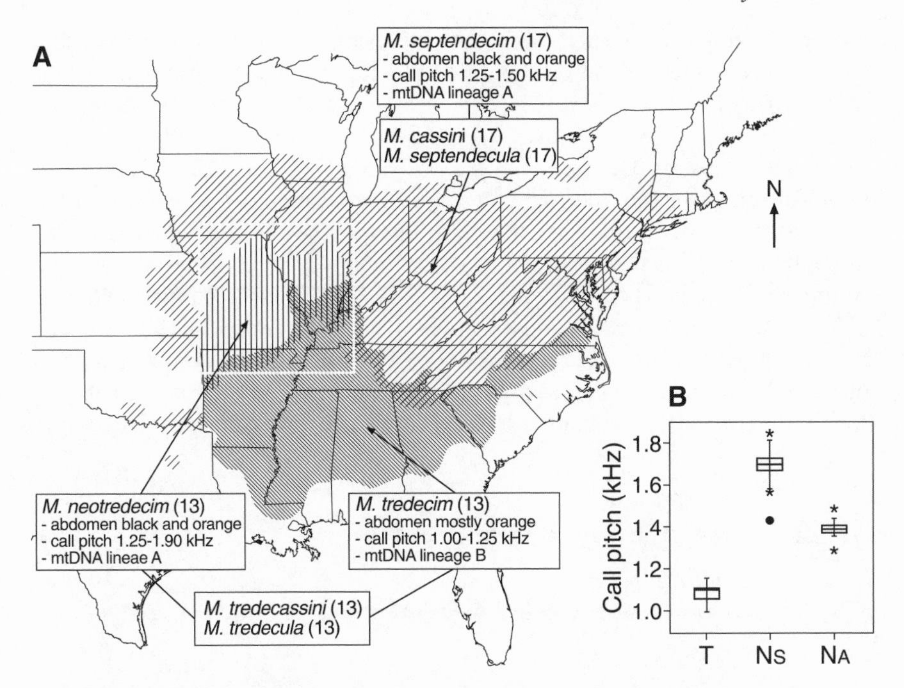

Figure 11.8. Reproductive character displacement of carrier frequency in periodical cicadas. (A) Map showing the distributions of North American species of 13-year and 17-year periodical cicadas (genus *Magicicada*). Some of the criteria used to diagnose these taxa are indicated. The white-bordered box shows the area in which populations of *M. neotredecim* were sampled. (B) Box plots showing median values (*horizontal lines*), first and third quartiles and outliers (*asterisks and dots*) in carrier frequency in *M. tredecim* (T) and *M. neotredecim* in a sympatric population (Ns) and in an allopatric population (NA). Further details in Marshall and Cooley 2000. Differences in frequency were statistically significant in all pairwise comparisons. From Marshall and Cooley 2000, figs. 1, 7.

M. neotredecim. First, the mean carrier frequency of their songs is about 25% higher in areas of sympatry with *M. tredecim* than in areas of allopatry (1.72 kHz vs. 1.52 kHz, respectively). Males of the latter species produce songs with mean carrier frequencies of about 1.12 kHz in both sympatric and allopatric areas (fig. 11.8). Females of *M. neotredecim* from sympatric areas also preferred (as inferred from "wing-flicking" responses) synthetic songs with carrier frequencies that were higher than those of females from allopatric areas (1.72 kHz vs. 1.31 kHz, respectively); females of *M. tredecim* from a sympatric locality had a mean preference of 1.19 kHz. Additional information about the intensity-independence of these frequency preferences is desirable, given the fact that small differences in intensity often negate or even reverse frequency preferences in insects and anurans (chapters 4 and 5). One possibility, which would further support the hypothesis of reproductive character displacement, is that preferences based on the difference in carrier frequency typical of sympatric populations might be resistant to variation in intensity, whereas preferences based on the difference typical of allopatric populations might break down.

Australian treefrogs. A classic example of sympatric divergence in advertisement calls involves the treefrogs (*Litoria verreauxii* and *L. ewingii*) (Littlejohn 1965), which have an extensive (800 km) but narrow (50 km) zone of overlap in southeastern Australia (fig. 11.9). Littlejohn (1999) provides an updated summary. Whereas the pulse rate of the calls was similar in allopatry (means of 73.0 pulses/s in *L. ewingii* and 84.3 pulses/s in *L. verreauxii* at 10C), the difference between the two taxa exceeded 60% in western (deep) sympatry (means of 68 pulses/s in *L. ewingii* and 137.7 pulses/s in *L. verreauxii*) (Littlejohn 1999). Behavioral experiments using synthetic calls indicated that the differ-

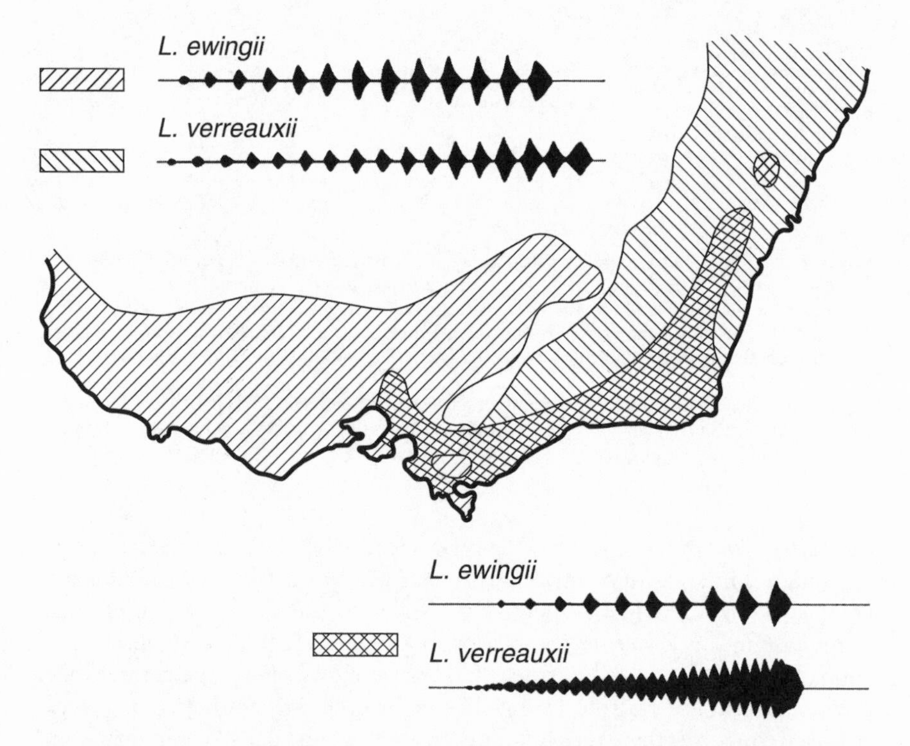

Figure 11.9. Reproductive character displacement in pulse rate and modulation depth in Australian treefrogs. Map showing the distribution of the treefrogs *Litoria ewingii* and *L. verreauxii* in southeastern Australia. Oscillograms at the top of the figure show single notes within advertisement calls of a male of each species from an allopatric population, and oscillograms at the bottom of the figure show single notes of a male of each species from western (deep) sympatry. Notice that the depth of amplitude modulation within the note of the *L. verreauxii* recorded from deep sympatry is greatly reduced compared to the note recorded in allopatry. Modified from Halliday 1980.

ence in pulse rate typical of deep sympatry was adequate for discrimination by females of both species (Loftus-Hills and Littlejohn 1971), which corroborated the results of earlier tests using prerecorded, natural calls (Littlejohn and Loftus-Hills 1968). These earlier experiments had also shown that females of *L. ewingii* from deep sympatry did not prefer the calls of local males to the similar calls of *L. verreauxii* recorded in remote allopatry. Thus, the divergence in pulse rate found in deep sympatry would seem to be adequate to reduce the probability of mismating, which does occur occasionally (Littlejohn 1999). Although the two taxa appear to be genetically compatible as judged by survival to metamorphosis (Watson 1974), nothing is known about the ecological fitness of hybrids. In our view, this remains a good example of reproductive character displacement, which could, however, be improved by additional sampling of the calls of both species in remote allopatry, as recommended by Grant (1972). Experimental studies should also be undertaken to determine the behavioral effectiveness of the smaller call differences found in eastern (shallow) sympatry.

Another change in call structure found in western sympatry might provide a rare chance to learn about qualitative shifts in signals. Namely, within western sympatry there is a significant reduction in the depth of amplitude modulation (pulsing) in the notes of *L. verreauxii* from near 100% in allopatry to a mean of about 50% (Gerhardt and Davis 1988). In many individuals the depth of modulation is less than 10%, suggesting a qualitative shift in temporal coding from a pulsed to an unpulsed call note (Littlejohn 1965; fig. 11.9). Intra-individual variation in the depth of modulation also occurs and appears to be weakly correlated (inversely) with pulse rate (Gerhardt and Davis 1988), a relationship that might account to some extent for the geographic pattern. That is, the shift to very high pulse rates, which might have resulted from reproductive character displacement, could ultimately and indirectly bring about a shift to pulseless notes. It will be essential to test the relative effectiveness of pulsed and weakly modulated (or even unpulsed) notes in phonotactic experiments with females of *L. verreauxii* from western sympatry and other parts of the distribution. In any case, the shift to high pulse rates in western sympatry is sufficient for females of *L. verreauxii* to discriminate in favor of the calls of local males to those of conspecific males from remote allopatry where pulse rate is much lower (Littlejohn and Loftus-Hills 1968).

North American narrow-mouthed toads. The ranges of the eastern (*Gastrophryne carolinensis*) and western (*G. olivacea*) narrow-mouth toads overlap broadly in parts of Texas and Oklahoma. Corroborating the original study of Blair (1955), Loftus-Hills and Littlejohn (1992) demonstrated that the mean difference in carrier frequency is greater between the species in sympatry than in allopatry (728 Hz vs. 291 Hz, respectively); this shift occurs only in *G. carolinensis.* Additional sampling in allopatry and tests of frequency discrimination by females are desirable next steps. Although putative hybrids were identified on the basis of their (intermediate) advertisement calls, the costs of mismating are unknown.

North American chorus frogs. Perhaps the most robust demonstration of the pattern expected of reproductive character displacement involves two species of chorus frogs (*Pseudacris nigrita* and *P. feriarum*) in the southeastern United States (Fouquette 1975). Extensive sampling reveals little geographic variation among allopatric populations in pulse rate, which diverges to a significant degree in the calls of *P. feriarum* in areas of overlap (means of 26.6 pulses/s vs. 16.5 pulses/s in allopatry at 14C) (fig. 11.10). Pulse rate in the advertisement

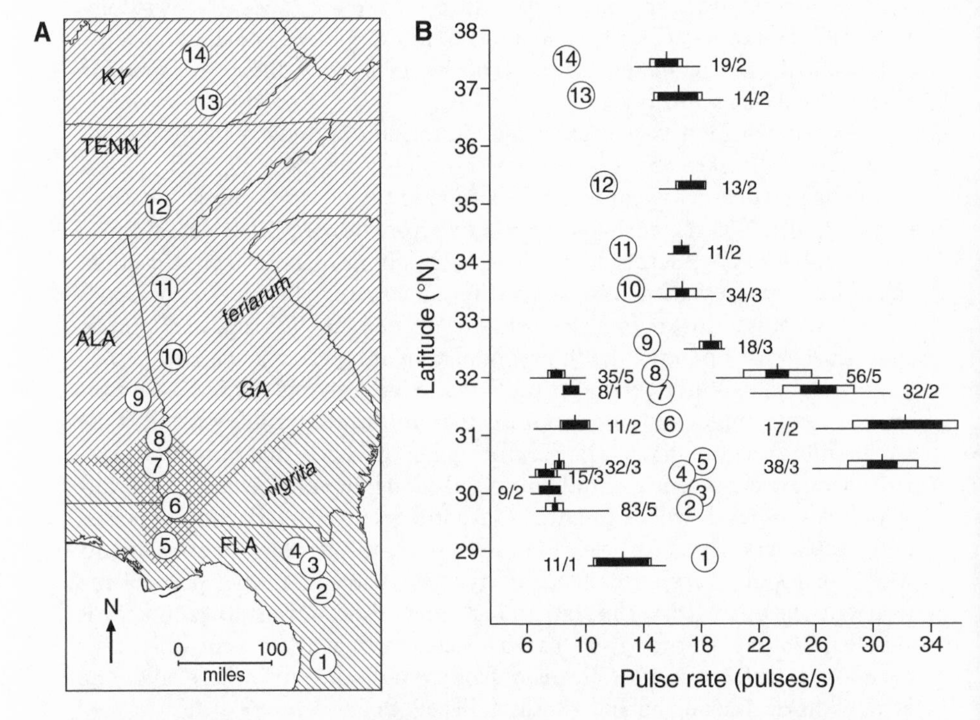

Figure 11.10. Reproductive character displacement of pulse rate in North American chorus frogs. (A) Map showing the distribution and recording sites for chorus frogs *Pseudacris nigrita* and *P. feriarum*, also considered a subspecies of *P. triseriata*. (B) Diagrams graphically show simple statistics (horizontal line = range; open rectangle = standard deviation; filled rectangle = two standard errors; vertical line = mean) for temperature-corrected (to 14°C) pulse rates of the calls of males in sympatric and allopatric populations. Left side of locality numbers = *P. nigrita;* right side of locality numbers = *P. feriarum.* Notice that in the area of overlap (sites 5–8), the pulse rates of *P. feriarum* shift to significantly higher values in comparison with the six allopatric populations to the north (sites 9–14). Whether there is a shift in the calls of *P. nigrita* is uncertain because three well-separated allopatric populations (sites 2–4) have pulse rates that differ more from one allopatric population to the south (site 1) than from sympatric sites (5–8) to the west and northwest. From Fouquette 1975, figs. 1, 2.

calls of *P. nigrita* is nearly the same in sympatry and allopatry (means of 8.2 pulses/s and 8.3 pulses/s, respectively at 14°C). Data are lacking on the degree of pulse-rate selectivity in females and the extent and costs of mating mistakes. One study of *P. triseriata* from another location (Oklahoma, U.S.A.) showed that females were phonotactically selective to prerecorded sounds that differed primarily in temporal properties (Littlejohn 1960).

In another overlap zone between *P. nigrita* and another species in the complex (*P. triseriata*) in Mississippi and Louisiana (U.S.A.), the two taxa form a narrow hybrid zone (estimated to be 9–19 km wide), within which there is evidence for backcrossing (Gartside 1980). Gartside (1980) and Littlejohn (1999) postulate that hybrids are ecologically superior to the parental species in this hybrid zone. The same explanation has also been applied to two hybrid zones involving Australian anurans: (1) the broad zone of overlap with hybridization between the northern and southern call races of the frog *Limnodynastes tasmaniensis* in southeastern Australia (Littlejohn and Roberts 1975); and (2) hybrid zones involving the frogs *Geocrinia laevis* and *G. victoriana* (Littlejohn 1993). However, direct experimental evidence of the ecological superiority of hybrids and recombinants in any of these hybrid zones is still lacking. The existence of these hybrid zones serves to emphasize that the evolution of call differences (and of female phonotactic selectivity as demonstrated in the two species of *Geocrinia;* Littlejohn and Watson 1974) in allopatry does not ensure sympatry, much less reproductive character displacement (Littlejohn 1993, 1999).

North American gray treefrogs. Reproductive character displacement has been studied extensively in the gray treefrog complex. Two well-studied members are *Hyla chrysoscelis*, which is a diploid species composed of two weakly defined lineages that interbreed freely, and *H. versicolor*, which consists of at least three tetraploid species that arose independently from the diploid, probably by autopolyploidization (Ptacek et al. 1994; box 11.1). These frogs occur over most of the eastern one-third of the United States; there are broad areas of allopatry as well as widespread areas where diploids and tetraploids breed synchronously in the same ponds. In some contact zones, the tetraploid lineage is most closely related to the diploid lineage occurring at the same place, but in others, the overlap almost certainly represents a "new" contact between diploids and more distantly related tetraploids. Mismatings, which occur with a frequency as high as 5% in some populations (Gerhardt et al. 1994), result in triploid hybrids with reduced viability in some crosses; hybrids are assumed to be sterile (Johnson 1963). As discussed previously (chapter 4), females of *H. chrysoscelis* are highly selective for signals that differ only in pulse rate, which varies significantly among populations (means of 35–50 pulses/s at 20°C) (Gerhardt 1999).

Given these attributes of the system—especially the occurrence of costly mating mistakes and knowledge that a single call property is necessary and

Box 11.1. Polyploid Speciation and Acoustic Signals in Anurans

Among anurans, autopolyploidy is widespread, occurring in five families and more than a dozen species (review: Bogart 1980). These polyploids usually have been discovered by the observation that morphologically similar (or indistinguishable) males in the same area produce distinctive advertisement calls, such as the gray treefrogs we have discussed numerous times in this monograph.

Allopolyploidy, which results from chromosome duplication in interspecific hybrids, would be expected to produce species with distinctive calls because hybrids inherit genomes from two different species that are likely to have different calls. This appears to be the situation in the polyploid species of clawed frogs (genus *Xenopus*) on the African continent (Tymowska 1992). By contrast, the generation of call differences in autopolyploids is problematic. In the gray treefrog complex, in which tetraploids probably arose at least three times, the calls of these lineages are nearly identical (Ptacek et al. 1994; Gerhardt and Ptacek, unpubl. data). Bogart and Wasserman (1972) noticed that some cell dimensions differed between diploids and tetraploids and speculated that such differences may promote call differences. This hypothesis has some experimental support. Although cell size was not measured, the pulse rate of the calls of artificially produced triploids and tetraploids of the Japanese treefrog *Hyla japonica* were about 13% and 20% slower, respectively, than that in diploid controls, which were individuals that were subjected to the same treatment and rearing conditions as polyploids but did not become polyploids (Ueda 1993). In *H. chrysoscelis*, artificially produced triploid males whose cell sizes were larger than diploid controls (and similar in this respect to triploid hybrids) produced calls that had pulse rates about 13% lower than the diploid controls (fig. 1; Keller and Gerhardt 2000). Several individuals that became triploids but did not have increased cell sizes produced calls with pulse rates that were the same as the diploid controls.

Because the pulse rates of the autotriploids (with larger cell sizes) were not as slow as those of allotriploids (hybrids, which have two sets of chromosomes from *H. versicolor* and one set from *H. chrysoscelis*), we can be rather sure that the effect of polyploidy per se is inadequate to explain the present-day difference (40–60%) in pulse rate between diploids and the tetraploid species, assuming that autotetraploids of *H. chrysoscelis* would show the same percentage drop in pulse rate (20%) as did the autotetraploids of *H. japonica*. This result suggests that selection has also played a role in the divergence of pulse rate in the tetraploids. Of course, selection has almost certainly contributed to the

(box 11.1 continued)

considerable geographical variation in pulse rate that has been documented in *H. chrysoscelis* (see above).

It will be particularly interesting to learn about the preference mechanisms in autotriploid and autotetraploid individuals of *H. chrysoscelis*. Perhaps "pleiotropic" effects of polyploidization might also help to explain the very different acoustic criteria of the wild-type animals (chapter 4).

The origins of other polyploid species are unknown, and sometimes no call differences between presumed diploid and tetraploid species occur (Roberts 1997; Haddad et al. 1994; see fig. 11.7). Phylogenetic analyses using genetic markers will be required to sort out the several possible evolutionary explanations. As we have emphasized throughout this monograph, evolution is opportunistic and universal explanations for the evolution of acoustic communication are unlikely to be found.

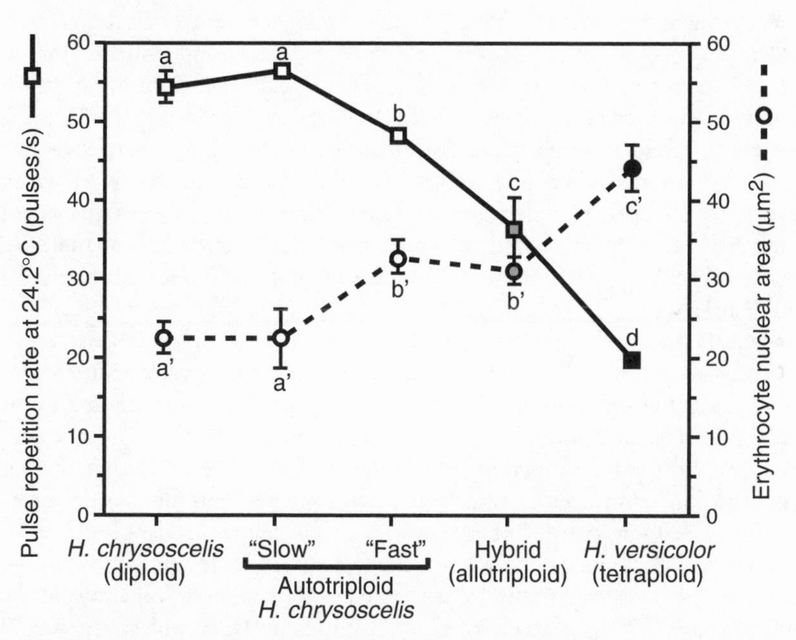

Box 11.1, Figure 1. Plots of temperature-corrected pulse rate (*squares*) and the area of the nucleus of red-cells (*circles*) of diploid controls and autotriploids of *Hyla chrysoscelis*, created by cold-shock of recently fertilized eggs (*open squares and circles*), triploid hybrids between *H. chrysoscelis* and *H. versicolor* (*gray-filled square and circle*), and wild-type *H. versicolor* (*black square and circle*), a tetraploid. Notice that pulse rate is inversely related to nuclear area, but that the pulse rate of autotriploids of *H. chrysoscelis* is significantly higher than that of allotriploids having about the same nuclear area. This result suggests that both cell size and genetic differences affect the pulse rates of the triploid hybrids. Data from Keller and Gerhardt 2001.

sufficient for mate identification in *H. chrysoscelis*—we might expect that the pattern of geographic variation in pulse rate would conform to that expected from reproductive character displacement. That is, pulse rate should be higher in *H. chrysoscelis* calls in areas of overlap with *H. versicolor* (means of 19–24 pulses/s at 20°C) than in allopatric areas. In fact, this expectation is not realized. Pulse rate in *H. chrysoscelis* shows mostly clinal variation (mainly an east-to-west increase), and differences in pulse rate among populations in remote allopatry are greater than any differences between sympatric and adjacent allopatric populations (Gerhardt 1999; Gerhardt et al., in prep.). Because the interspecific differences in pulse rate are relatively large (mean differences of about 47–60% in eastern and western sympatry, respectively) and selection against mating mistakes is strong, the argument could be made that reproductive character displacement occurred rapidly in the past close to the time of speciation (Littlejohn 1993). Moreover, pleiotropic changes in pulse rate in the calls of *H. versicolor* have also contributed to present-day differences between the two species (Keller and Gerhardt 2001; see box 11.1).

Whether or not reproductive character displacement contributed to the present-day species differences in pulse rate, there is good evidence for sympatric divergence in the strength of pulse-rate discrimination by females of *H. chrysoscelis* (Gerhardt 1994b, 1999). As suggested by Waage (1979), if signals are different enough to be distinguished at the time that species interactions commence, then selection might act primarily on receivers without necessarily affecting signal structure. Sympatric enhancement of mate-choice selectivity in receivers, based on visual cues, has been found in males (the choosy sex) of two species of fish (Rundle and Schluter 1998; Gabor and Ryan 2001) and in Darwin's finches (Ratcliffe and Grant 1983).

Demonstrating differences in preference strength in *H. chrysoscelis* between sympatry and allopatry depended, however, on putting females in realistic, worst-case situations (Gerhardt 1994b, 1999). For example, if a female enters a mixed-species chorus and first encounters a heterospecific male, the intensity of his calls will be much greater than those of a distant conspecific male. Call duration also varies considerably within species, and females prefer long to short calls (Gerhardt 1994b; chapter 10); thus, females must often confront a choice of long calls produced by a heterospecific male and short calls produced by a conspecific male. When tested with a difference in call duration (300%) that favored a "heterospecific" stimulus in which pulse rate was 30% less than that of local conspecific males, females from four sympatric populations were much more likely to choose the short signal with the "correct" pulse rate than were females from four remote allopatric populations (fig. 11.11). Comparable results were obtained in comparisons of female selectivity from three sympatric and three allopatric populations in the face of an intensity difference (12 dB) that favored the heterospecific stimulus. Another way of interpreting these results is that females from allopatric populations are less constrained in their choices of calls that differ in pulse rate because no other

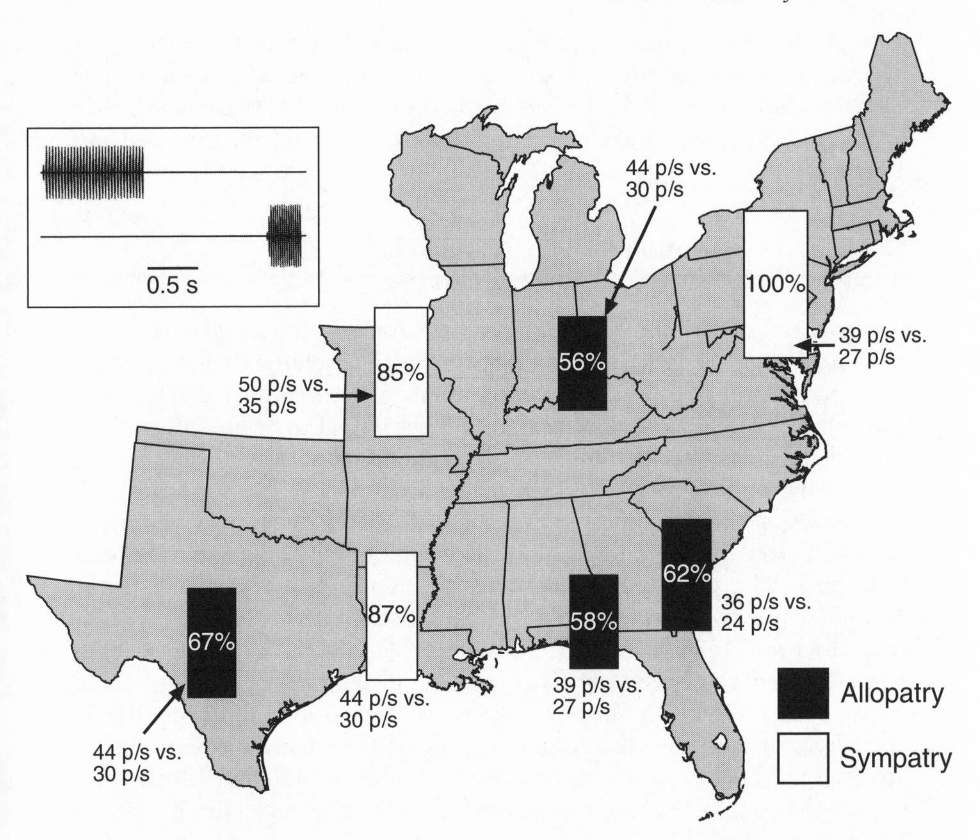

Figure 11.11. Reproductive character displacement of female pulse-rate selectivity in the gray treefrog *Hyla chrysoscelis.* Map showing the percentages of females that chose short synthetic calls (*lower trace in the inset*) with pulse rates close to the mean of that in the calls of males from the same population. The alternative stimulus (*upper trace in the inset*) had a pulse rate that was 30% lower and a call duration that was three times greater than that of the "conspecific" stimulus. White bars are sympatric localities and black bars are allopatric localities. There were two sympatric populations in Missouri, located about 100 km apart, but the results from only one are shown (the proportion of females choosing the short call was about the same). The actual pulse rates of the pairs of alternatives used to test females are also indicated.

species with pulse rates similar to those of conspecific males is present. Comparable data are available from midwife toads *Alytes obstetricans* and *Alytes cisternasii* (Márquez and Bosch 1997b) and spadefoot toads *Spea multiplicata* (Pfennig 2000). In contrast to the usual preferences for high values of gross temporal properties (chapter 10), which were found in allopatric populations of these species, females from sympatric populations did not choose higher values, perhaps because they would thereby run the risk of mating with a heterospecific male. Additional sampling of populations (these last two studies each compared preferences from just one sympatric and one allopatric population) in these species is needed to further support the hypothesis that interspecific interactions are responsible for differences in female selectivity. We emphasize, too, that variation in female preference also occurs among allopatric populations of *H. chrysoscelis*, thus indicating that factors other than interactions with *H. versicolor* can influence receiver selectivity (Gerhardt 1999).

Origins and Evolution of Signals and Selective Acoustic Responses: Comparative Studies

Comparisons of the communication systems of closely related species can generate and test hypotheses about how signals and preferences arise, change, and diverge through evolutionary time. Several general patterns emerge from these studies. First, as we indicated above, behavioral traits are often evolutionarily labile and hence communication systems often diverge more rapidly than do morphological and molecular characteristics. As shown below, this generalization is more robust with respect to quantitatively varying aspects of signals structure (e.g., low vs. high pulse rate) than to qualitative aspects (e.g., pulsed vs. unpulsed signals). Second, as we have repeatedly shown, closely related, sympatric species often use strikingly different acoustic criteria to discriminate the same set of conspecific and heterospecific signals, thus suggesting that receiver mechanisms can be as evolutionarily labile as signals. Third, basic preferences can also be conserved across a set of closely related taxa whose signals show significant differentiation. These results serve to emphasize that different processes are likely to influence different communication systems or even the same system at different times during its evolutionary history.

Informal Analyses

In the tradition of early ethologists, informal hypotheses about patterns of evolution of the communication systems in crickets, grasshoppers, and phaneropterine katydids have been derived from comparisons within broad taxonomic groups (Alexander 1962; Otte 1970, 1992; Heller 1990). These analyses rest on plausible but fallible assumptions. For example, the main criterion for identifying the ancestral state of a trait is its prevalence among species within the group: the ancestral state is assumed to be more likely to be common than rare. A second, even more questionable criterion is that the ancestral pattern is the simplest. Indeed, Lorenz (1941), whose phylogenetic approach to visual displays in ducks (Aves: Anatidae) was surprisingly modern, warned against equating increases in complexity observed in a series of different taxa with evolutionary sequences (see Brooks and McLennan 1991 for a summary of the history of debates about behavioral homology). Despite these caveats, many of the conclusions of these informal analyses reflect a deep comparative understanding of the taxa, and the hypotheses they have generated can, in principle, be tested by the application of modern comparative methods.

In his classic paper, Alexander (1962) argues that because the elaborate stridulatory structures of modern crickets and katydids were unlikely to have evolved full-blown, the first cricket signal was the courtship song, which is usually a low-amplitude, close-range signal. One possibility is that sounds acquiring this function might have occurred inadvertently during lifting and fluttering wing movements that were used to expose the dorsal glands, which disperse pheromones. Indeed, in tree (Oecanthinae) and ground (Nemobiinae) crickets, short-range courtship by chemical cues has been maintained, while the calling song mediates long-range attraction of females (Alexander 1962). With regard to the origins and diversification of long-range calling song, Alexander (1962) proposes that a continuous trill containing pulses of equal amplitude is the ancestral state. In support of this idea, he notes that the trill rates of calling songs often approximate wing-beat frequencies. Although calling-song production might have arisen as a modification of flying, at least some emancipation has occurred in crickets and katydids, whose forewings, which produce sound, may contribute to steering of flight but do not beat at the same frequency as the hindwings that power flight (May et al. 1988). Alexander further argues that a common change consists of breaking these trills into shorter pulse trains, some of which would qualify as chirps. As discussed below, one type of chirp (B1 in Alexander's terminology) is common in European and North American field crickets either because of conservation from a common ancestor or through convergent evolution (Harrison and Bogdanowicz 1995; Huang et al. 2000). Otte (1992) summarizes hypothetical modes of evolutionary changes in pulse intervals and organization in cricket calling songs.

Because the earliest known fossil crickets have stridulatory and auditory organs similar to those of modern species, selection must have strongly favored the evolution of long-range calling songs. Moreover, the similarity of the stridulatory structures and tibial ears of crickets and modern katydids suggests that these structures also occurred in the common ancestor to these two groups (Alexander 1962; Otte 1992). Thus, hearing and sound production must have a long history of selection in the context of intraspecific communication. The subsequent appearance of bats would have then selected for hearing mechanisms that specialized for the detection of these predators. A controversial cladistic analysis based on morphological and behavioral characters (Gwynne 1995) proposes two origins of tibial ears and stridulation, but even here both characters appear very early in the lines leading to modern katydids and crickets.

In grasshoppers, hearing probably evolved before acoustic signaling (Helversen and Helversen 1994). Whereas at least ten forms of sound production have been described (Otte 1970, 1977), the abdominal tympanal ear almost certainly appeared early in the evolution of this group and has been little modified. Recall, too, that even the morphology of ascending auditory interneurons in grasshoppers is similar in species that communicate with sound and those in which sound production has never evolved (see chapters 5 and 6). Thus, the evolution of hearing in grasshoppers might have evolved first as a general mechanism for sensing the environment rather than for intraspecific acoustic communication. Grasshopper ears were probably derived from stretch and vibration detectors (Meier and Reichert 1990), and a demonstration that such pleural receptors can contribute to detection of intraspecific communication signals—and hence be favored during an evolutionary transition in form and function—is available for the bladder grasshopper *Bullacris membracioides* (van Staaden and Römer 1997).

Because most living grasshoppers are diurnal, many acoustic signals probably evolved from sounds made during visual displays and function simultaneously or sequentially with present-day visual signals. Otte (1970) suggests that tactile and chemical signals were the main modes of intraspecific communication in the presumed nocturnal ancestors. Regardless of the order of evolution of acoustic and visual signals, their evolutionary plasticity is remarkable. For example, even within what is considered to be a single species (*Stenobothrus rubicundus*), nearly the same sound pattern is produced by wing stridulation in some populations and by leg stridulation in other populations (Elsner and Wasser 1995). The diagram of figure 11.12 shows examples of how ordinary (leg-wing) stridulation has been combined with various kinds of other signals that may be both visual and acoustic. Otte (1970) argues that the diversification of courtship signals involved several steps: (1) the use of an element in a single context; (2) the use of an element in two or more contexts; (3) the evolutionary elaboration of the elements used in these different contexts; and (4) the subsequent loss of signal function of one or more elements in the original context. In contrast to the bewildering diversity of acoustic and visual elements in courtship signals, some elements of aggressive signals, such as femur jerking, appear to be nearly universal (Otte 1970).

For anurans, acoustic signals might have arisen from sounds produced when air was forcibly ejected through the larynx before or during escape jumping or tactile stimulation (Schmidt 1991), perhaps in the same context as contemporary "release" calls that are produced by a male or nonresponsive female when clasped by a male. As in crickets, these signals would have presumably first been low-intensity, close-range signals. Whereas Alexander (1962) argues on the basis of acoustic similarity that aggressive signaling in crickets arose from modifications of the calling (advertisement) song, aggressive calls in many anurans are more similar to release calls than to advertisement or courtship calls. The best evidence for the evolutionary conservation of aggressive-call structure is the acoustic similarity of these signals in the diploid-tetraploid gray treefrogs (*Hyla versicolor* complex) (chapter 9; Gerhardt 2001). Female anurans of a few species also produce courtship or reciprocation calls (Schlaepfer and Figeroa-Sandí 1998; Emerson and Boyd 1999), which appear to be more similar to male advertisement calls than to release calls, which are given

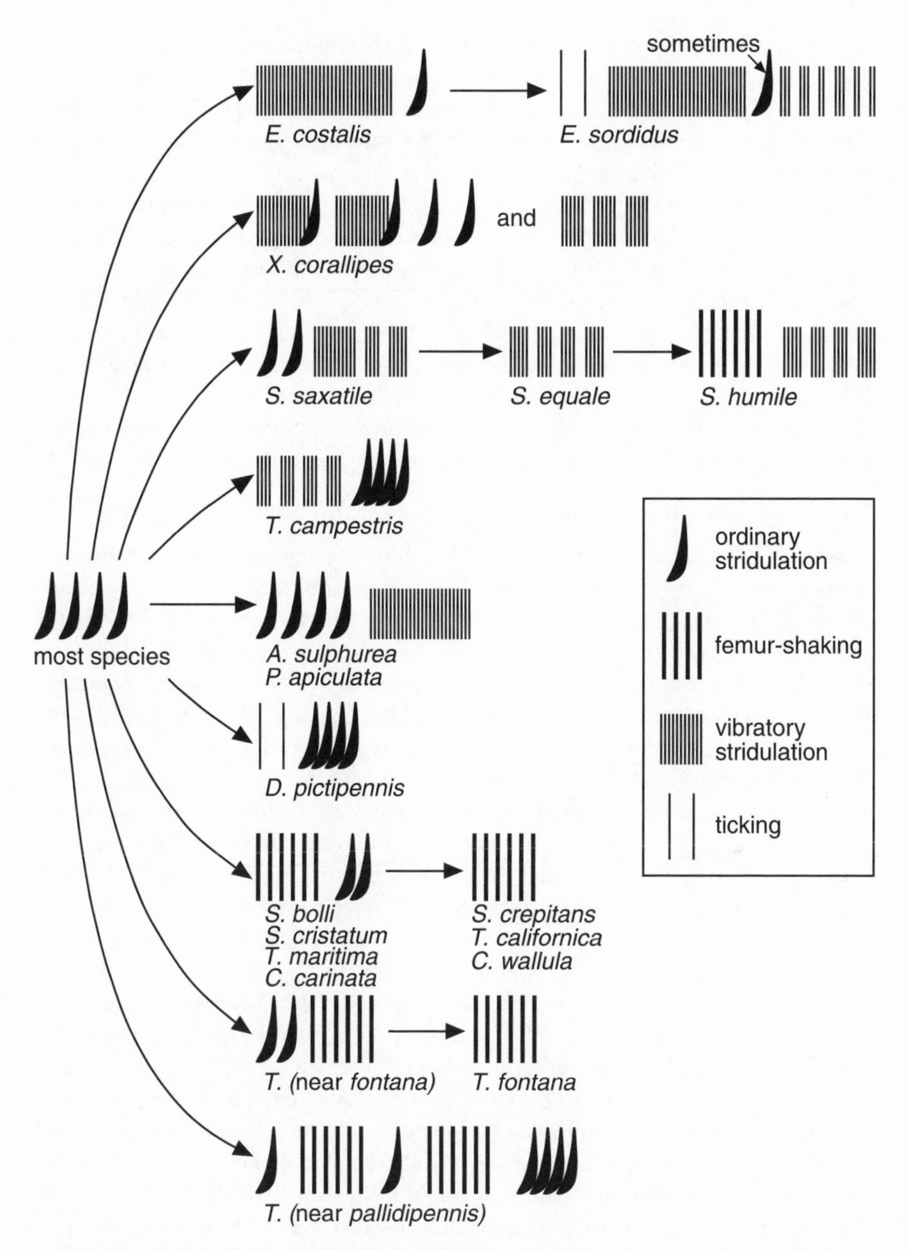

Figure 11.12. Evolutionary scenarios for qualitative changes in courtship signals in grasshoppers (subfamily Oedipodinae). Arrows indicate the probable direction of evolution. On the one hand, the postulated ancestral condition on the left is possessed by a large majority of the species, including close relatives of most of the species on the right. On the other hand, all of the taxa of the first order have complex signals with two or more elements, and this could also be the ancestral condition. Diversification has occurred through both the addition of new elements and the loss of elements. From Otte 1970, fig. 18.

by both sexes. As in female insects that produce sounds in a reproductive context (chapter 2), courtship calls in female anurans appear to have evolved independently on several occasions. The same is true of visual signals that usually augment acoustic signals, especially in situations where background noise levels are high (Hödl and Amézquita 2001).

Repertoires of different signals in both insects and anurans provide two obvious sources of signal diversification among closely related species. First, there can be a change in function (Alexander 1962; Otte 1970). For example, a signal that serves for long-range communication in one species might serve as an aggressive or courtship signal in a close relative. Second, complex signals can arise as combinations of signals originally serving different functions (Alexander 1962; Otte 1970; Helversen and Helversen 1994; Wells 1977). In some species the combination is required for the same function, and in others, each part has different functions (examples and discussion in chapter 9).

Formal Phylogenetic Analyses

In a formal phylogenetic analysis, relationships among extant taxa are inferred from shared derived characteristics and are summarized by trees or cladograms. This kind of information is useful for selecting species for comparative studies and, more controversially, can be used to reconstruct ancestral states (e.g., Cunningham et al. 1998). Despite the difficulties and uncertainties, these procedures can generate and test hypotheses about the broad-scale patterns of evolution of communication systems (e.g., Ryan 1996 and other chapters in Martins 1996).

Modern practitioners often attempt to use data from as many characters as possible ("total evidence"), including proteins, nucleotide sequences, morphology, and behavior. Analyses of behavioral data alone, including acoustic properties of signals, can also result in reasonably well-supported trees, indicating that phylogenetic information is present (e.g., De Queiroz and Wimberger 1993; Martins 1996). However, significant incongruence often exists between such trees and trees based on other characters. A formal analysis of tree incongruence is provided by comparisons of species in the *Physalaemus pustulosus* species group of Neotropical frogs (Canatella et al. 1998). This study compares trees based on total evidence with those derived from subsets of characters such as morphology, mtDNA, and call characters. The tree based on call characters differs from and is not as well supported statistically as the tree based on morphology, which is much more similar to the total-evidence tree (fig. 11.13). These results, and those of another study comparing relationships among different populations of *P. pustulosus* (Ryan et al. 1996), are interpreted as support for the idea that signal traits are subject to more rapid evolution than are other traits—a pattern that we already noted in our discussion of reproductive character displacement.

The patterns that emerge when data on signal structure are superimposed on phylogenetic trees based mainly or exclusively on nonbehavioral characters

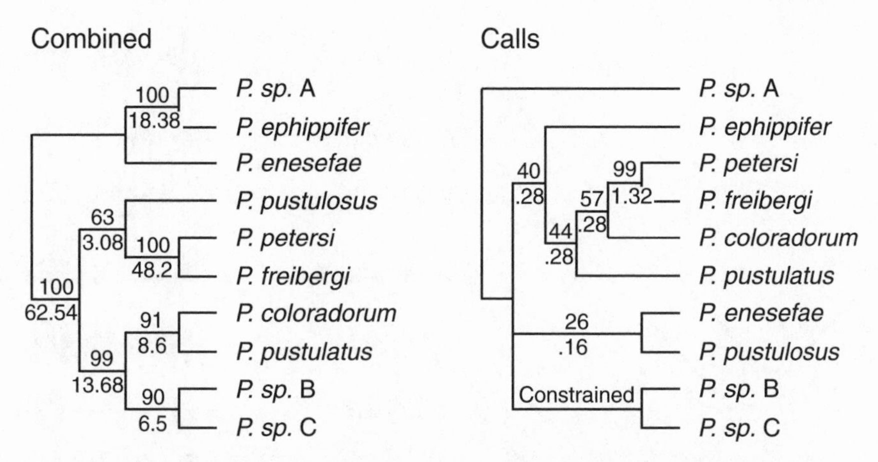

Figure 11.13. Comparison of phylogenetic reconstructions of species in the *Physalaemus pustulosus* species group of Neotropical anurans. The well-supported, combined ("total evidence") tree on the left was constructed by considering variation among species in morphology, mtDNA sequences, allozymes, and calls. The tree on the right used only the call characters. Although at least weakly supported by standard phylogenetic tests, the topology of the tree based on calls alone is highly incongruent compared with other trees based on molecular, biochemical, and morphology data alone, which each resembled the combined tree in most respects. See the text for further details. Numbers above each branch show bootstrap values, and numbers below, decay values. From Canatella et al. 1995, fig. 1.

depend on three factors: (1) the level of signal analysis (qualitative features vs. quantitatively varying features); (2) the scope of the analysis in terms of the taxa considered; and (3) whether the taxa are sympatric or allopatric. Qualitative characters, such as whether signals are pure-tone or broadband, pulsed or unpulsed, and so forth, will often show similarities that reflect phylogenetic relatedness because of shared mechanisms of sound production. But convergent patterns are likely as more distantly related taxa are included simply because different mechanisms can generate similar sound patterns (see chapters 2 and 3). As already documented in our discussion of reproductive character displacement, quantitatively varying properties (e.g., high vs. low pulse rate) usually diverge among closely related species that occur in the same area and can be remarkably similar in species that do not encounter one another regardless of how closely or distantly they are related.

The phylogenetically conservative nature of qualitative properties (basic structure) is illustrated by comparisons of advertisement call structure in two groups of closely related North American anurans (Cocroft and Ryan 1995). Ten species of toads (genus *Bufo*) fall into two distinct clades (fig. 11.14). Two conserved features of all species are the pulsed or trilled nature of the call and the existence of a single, fairly well-defined carrier frequency. The pulse structure is divergent between the clades in that those species clustering with *Bufo cognatus* showed intrapulse modulation (passive AM of Martin 1971; see chapter 2), whereas the pulses of species clustering within the *B. americanus*

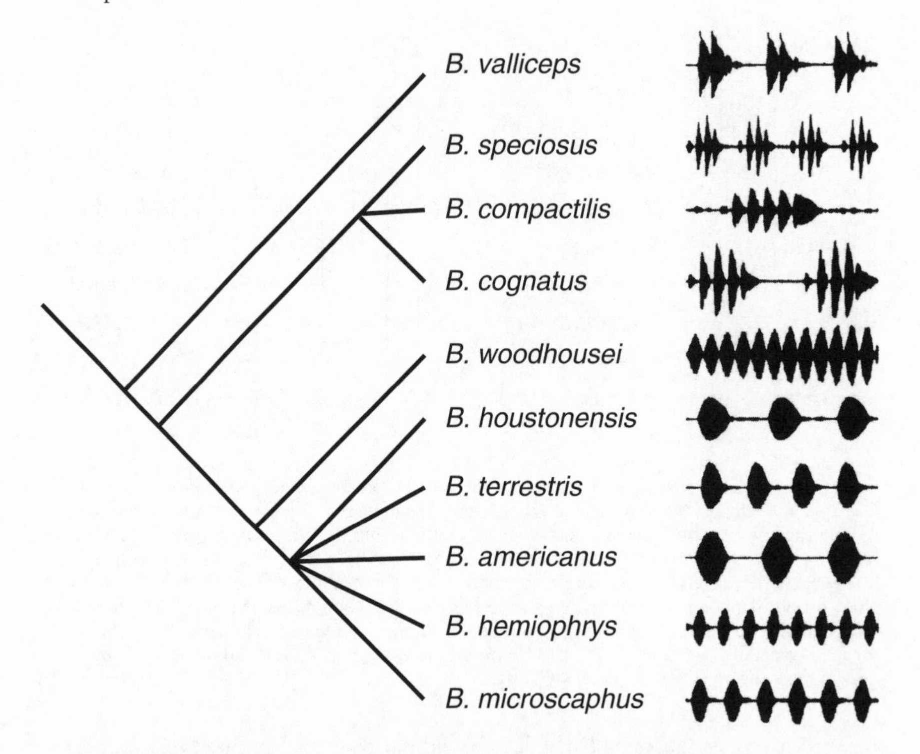

Figure 11.14. Phylogeny of some North American toads (genus *Bufo*) based on immunological distance (Maxson et al. 1981). Oscillograms showing pulse structure are superimposed on the right. Notice that the two clades are characterized by whether or not pulses show intrapulse modulation, which is presumed to be the ancestral state because of its presence in the outgroup species, *B. valliceps*. From Cocroft and Ryan 1995, fig. 6.

clade are unmodulated. In the chorus frogs (genus *Pseudacris*), the carrier frequency is the second harmonic of the spectrum and it increases somewhat from beginning to end (fig. 11.15). Three species in one clade have unpulsed calls, and seven species in the other clade have pulsed calls. A species whose phylogenetic position appears to be intermediate—*Pseudacris ocularis*—produces a two-part call; one part is pulsed, and the other, unpulsed. As in the presumed evolutionary reduction in the depth of modulation (pulses) in the Australian treefrog *Litoria verreauxii* discussed above, further study of *P. ocularis* might provide some insight concerning the processes that affect evolutionary change in qualitative acoustic properties. Clearly, too, the evolutionary loss of either part would result in a different categorical classification (pulsed vs. unpulsed) of its advertisement call. Some of the scenarios (e.g., changes in function of signals with multiple elements; losses of elements) proposed by Otte (1970) and Helversen and Helversen (1994) for the diversification of grasshopper signals might thus apply to these anurans.

Both qualitative and quantitative similarities are found in properties of the

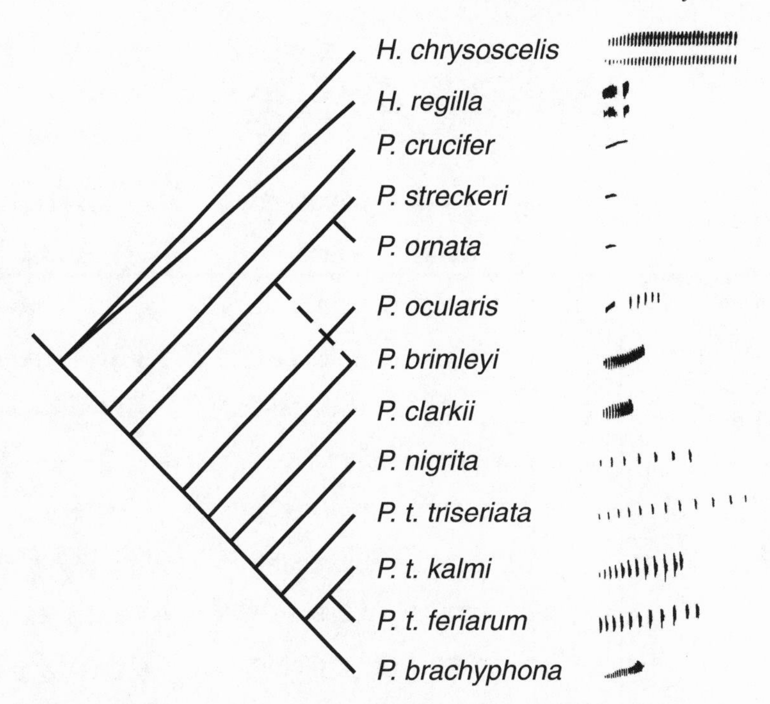

Figure 11.15. Phylogeny of some North American chorus frogs (genus *Pseudacris*) based on a suite of characters, including call characters (Cocroft 1995). The dashed line shows the position of *P. brimleyi* when call characters are omitted from the analysis. Sonograms are superimposed on the right. Notice that in the tree based on all data, the two major clades are characterized by whether the overall call structure is pulsed or not. A pulsed structure is characteristic of the outgroup species, *Hyla chrysoscelis* and *H. regilla.* From Cocroft and Ryan 1995, fig. 7.

vibratory signals of lacewings (Neuroptera: genus *Chrysoperla*) (Henry et al. 1999; fig. 11.16) and of the advertisement calls of microhylid frogs (genus *Cophixalus*) (Olding 1998; fig. 11.17). The lacewing example is grounded in a well-resolved molecular phylogeny and includes some remarkable examples of convergence between European and North American species. Moreover, behavioral testing shows that females of the North American species (*C. adamsi*) do not distinguish between the songs of conspecific and those of their European counterpart (*C.* "*adamsi-K*"). The close convergence of pulse rate in two pairs of microhylid frogs (*C. concinnus* and *C. hosmeri; C. bombiens* and *C. neglectus*) does not pose problems of species identity because the taxa with similar calls are distributed allopatrically. This same general pattern is found among the toads and chorus frogs considered above. The ranges of species having advertisement calls with similar values of quantitatively varying properties (e.g., *Bufo houstonensis* and *B. americanus; Pseudacris ornata* and *P. streckeri; P. brimleyi, P. clarkii,* and *P. brachyphona*) do not overlap.

Another example of qualitative (chirp structure) and quantitative (pulse

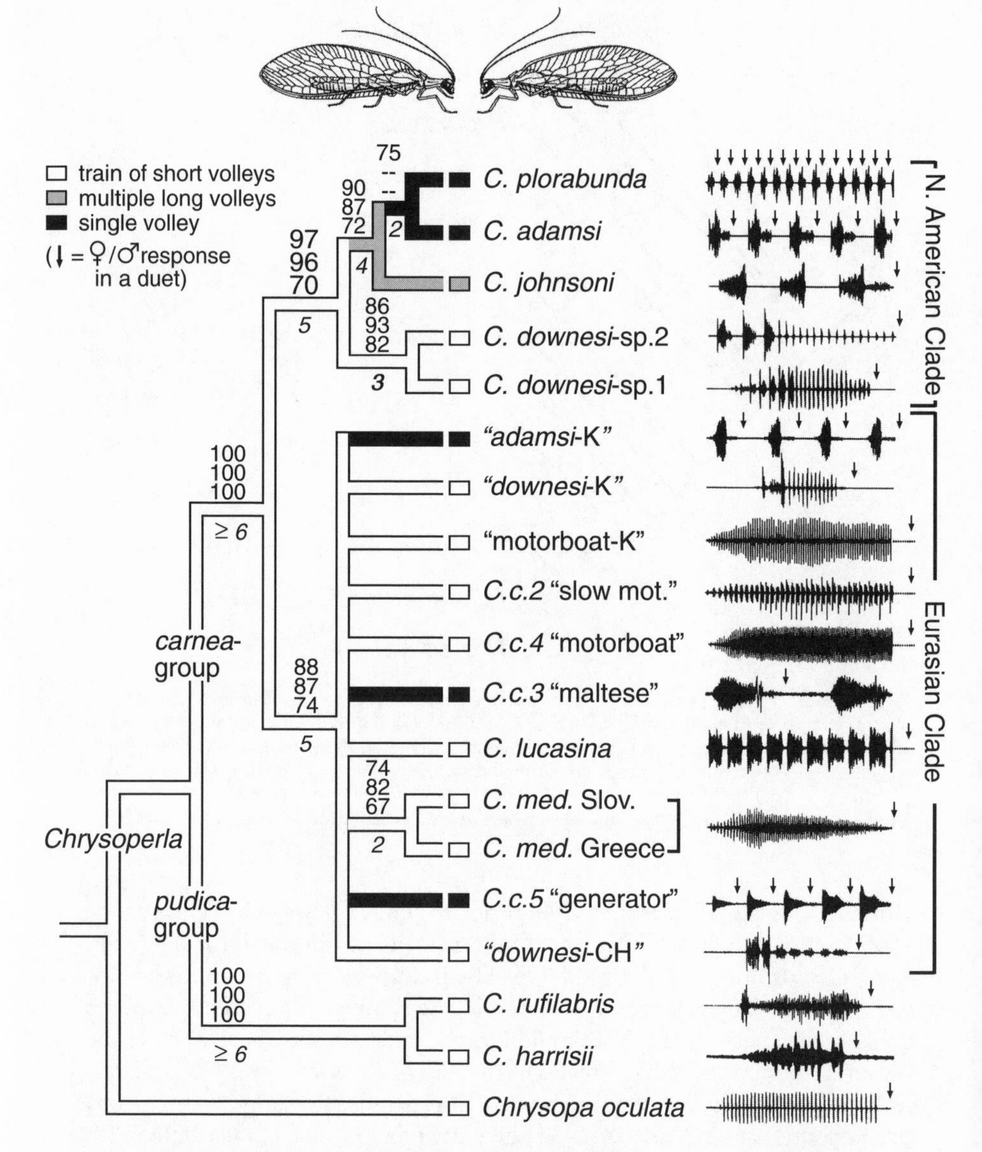

Figure 11.16. Convergence in call (vibratory signals) structure in lacewing. The phylogenetic tree (bootstrap, 50% majority rule) shows relationships among North American and Eurasian green lacewings (Neuroptera) based on analyses of mitochondrial DNA sequences and generated by maximum parsimony (MP), minimum evolution (ME), and maximum likelihood (ML). Numbers above the nodes are bootstrap proportions for each of the three procedures in order from top to bottom (MP, ME, ML). The numbers below each node are decay indices. Oscillograms show 12 s segments of the vibratory signals of males and vertical arrows indicate the timing of female responses in a duet. Production of a train of short volleys is ancestral. Notice the striking similarity of the signals of *Chrysoperla adamsi* from North American and *C. "adamsi-K"* from Europe. These taxa are in highly divergent clades. From Henry et al. 1999, fig. 3.

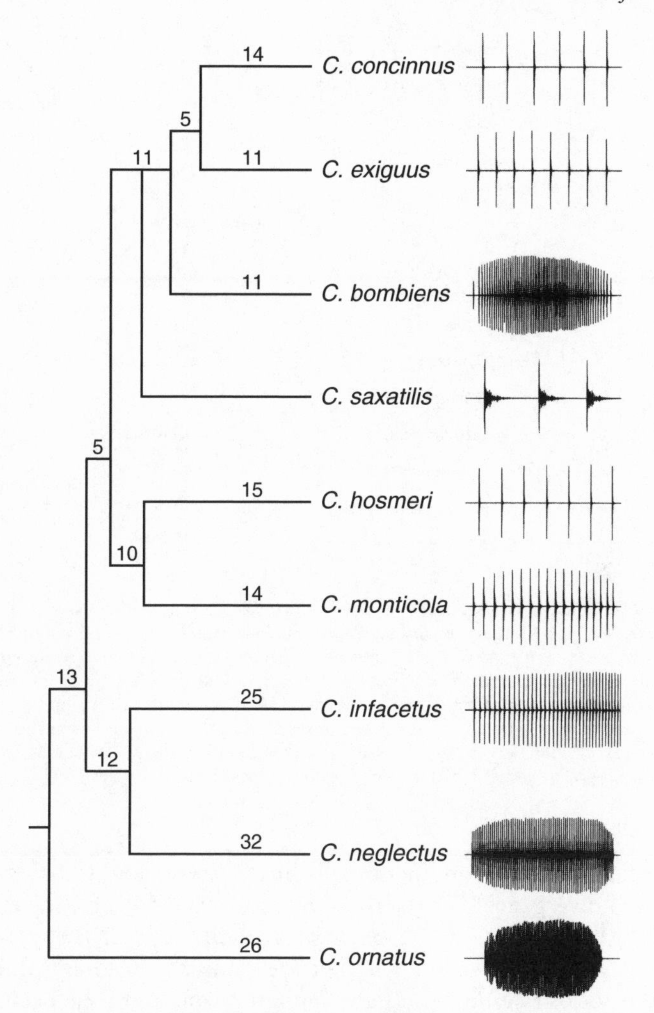

Figure 11.17. Convergence in advertisement calls of allopatric species of microhylid frogs (genus *Cophixalus*) from northeastern Australia. The phylogeny is based on mtDNA sequences (16S), and the numbers represent the number of character changes. Oscillograms of the advertisement calls are shown to the right. From Olding 1998.

rate) similarities in advertisement signals involves the North American field crickets *Gryllus pennsylvanicus* and *G. veletis.* Although found in the same areas, one species breeds in the spring and the other in the fall. Because these species are also very similar in morphology (though genetically incompatible), Alexander and Bigelow (1960) suggested that they speciated *allochronically.* That is, genetic changes that resulted in a shift in the breeding season of some populations could have disrupted gene flow just as effectively as a shift in spatial distribution. Two recent phylogenetic analyses indicate, however, that

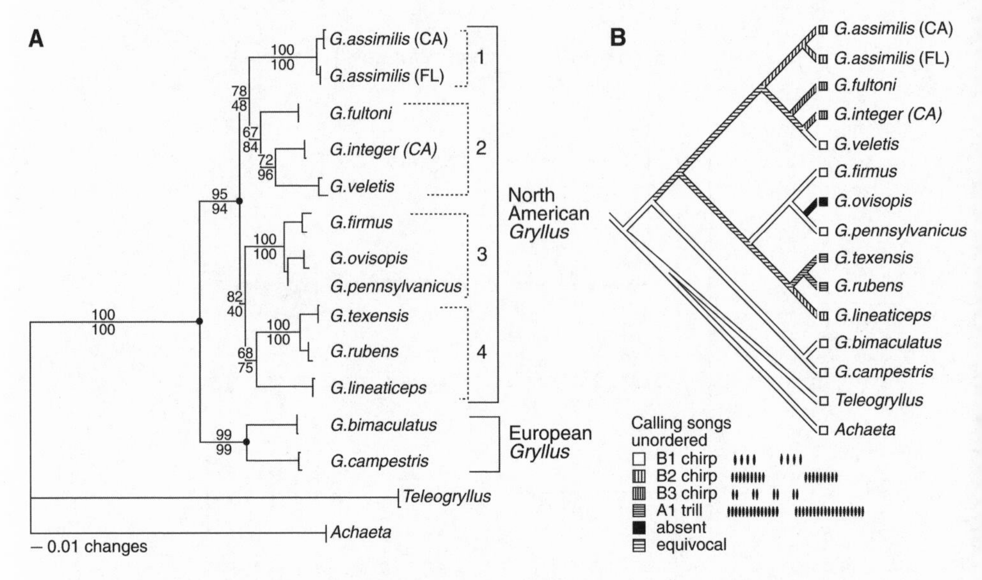

Figure 11.18. Conservation or convergence of chirp structure in field crickets. (A) Minimum evolution tree for North American and European field crickets based on mitochondrial DNA sequences (cytochrome *b* and 16S). Numbers above each branch are bootstrap support values for a weighted parsimony analysis; numbers below each branch are bootstrap values for a maximum likelihood analysis. (B) Simplified tree based on (A) onto which basic calling-song structure has been superimposed. The B1 chirp pattern is assigned to the outgroup (even though the call of *Teleogryllus* is complex) and to the two European *Gryllus*. See the text for further explanation. From Huang et al. 2000, figs. 3, 5. Diagrams of song types from Alexander 1962.

these are not sister taxa, and hence allochronic speciation probably does not explain their divergence (Harrison and Bogdanowicz 1995; Huang et al. 2000; fig. 11.18). Unfortunately, the phylogeny of the field crickets is not well enough resolved to determine whether the similarity in song is due to convergence or conservation. The same limitation applies to the occurrence of the B1 chirp pattern (Alexander 1962) within two clades of North American *Gryllus*.

The analysis of toad vocalizations by Cocroft and Ryan (1995) also considered dynamic call properties such as call duration and call rate. These properties vary within and between clades but show considerable overlap quantitatively. One interesting contrast exists, however, within the *B. cognatus* clade. Call duration is longest (> 20 s) and call rate slowest in *B. cognatus*, whereas call duration is shortest (< 1 s) and call rate highest in *B. speciosus*. As suggested by Cocroft and Ryan (1995), these different patterns might reflect alternative responses to the generally strong directional selection exercised by female anurans on dynamic call properties (see above and chapter 10).

Cocroft and Ryan (1995) also classified acoustic properties according to whether evolutionary changes are likely to be dependent upon or correlated with other attributes of the frogs. For example, carrier frequency and pulse structure in toads are likely to be morphologically constrained by size and special laryngeal structures, respectively (see chapters 2 and 3). Properties such as call rate and call duration are correlated with physiological and behavioral traits such as temperature, energetic costs, and signaling (timing and aggressive) interactions. To the extent that these traits favor or constrain evolutionary change in these acoustic properties, we might expect differences in their relative rates of evolution. Pulse rate and other fine-temporal properties of anuran advertisement calls do not, however, always fit neatly into one of these categories, and this is especially true of insect sound production. For example, fine-scale temporal properties can be influenced by morphology (passive modulation by laryngeal structures in frogs; stridulatory structures in insects) but are also temperature-dependent. Pulse rate is unaffected in nonaggressive (timing) interactions with neighbors but can show graded variation during aggressive encounters (chapter 9). Regardless of how they are categorized, however, the studies reviewed in this chapter show repeatedly that fine-scale temporal properties are almost invariably well differentiated among closely related, sympatric species and often show significant geographical variation (see also Alexander et al. 1997; Ewing 1989). This observation suggests that these traits can change rapidly in response to strong selection despite their relatively low variability within and between males in any given population.

Hidden Preferences and Preexisting Biases

As suggested above, the mutual selective pressures exerted by senders and receivers make it likely that these two components of communication systems coevolve, and results from studies of Hawaiian crickets and European katydids presented in chapter 10 have been interpreted in this way. Other research suggests that mismatches between signals and receivers are common and expected. The hidden-preference hypothesis (Arak and Enquist 1993), which is based on neural network analyses of visual pattern recognition, emphasizes that incidental changes can occur during the evolution of recognition systems that will favor novel signals. This work also predicts that different mechanisms can be used to accomplish the same recognition task, an idea that is consistent with the diversity of acoustic criteria for signal identification and female choice among closely related species that we documented in chapters 4 and 10. Other authors speculate about the behavioral and ecological contexts in which sensory and neural biases originate and their consequences for the evolution of communication systems (review: Endler and Basolo 1998).

Empirical data and verbal models often emphasize that sensory biases can predate signals or even intraspecific communication. At a gross level of analysis, for example, the early origin of the abdominal ear in grasshoppers (Meier and Reichert 1990) and the ultrasonic sensitivity of moths suggest hearing mechanisms evolved in the context of predator detection and have been coopted for intraspecific communication in many kinds of grasshoppers (Otte 1970; Helversen and Helversen 1994) and a few species of moths (review: Greenfield 2001). In general, these models predict that a novel sound is more likely to acquire signal status if it matches the preexisting bias than if it does not. This hypothesis is exemplified by studies of acoustic communication in the *Physalaemus pustulosus* group of anurans (Ryan 1990; Ryan and Rand 1993, 1999), in which novel acoustic elements produced by other species can enhance the attractiveness of conspecific signals that lack these elements. Although such supernormal stimuli had been interpreted earlier as a kind of preadaptation (e.g., Wickler 1967), Ryan and his colleagues were the first to explore signal diversity and sensory biases in a formal phylogenetic context. This approach has inspired numerous studies of preexisting biases in other species and in other sensory modalities (e.g., Sakaluk 2000; other examples in Ryan and Rand 1993; Endler and Basolo 1998).

Much contention and discussion has revolved around two issues: (1) the possibility that preexisting biases can, in principle, result in negative consequences for receivers; and (2) the claim that preexisting bias models provide a strictly alternative explanation to coevolutionary models of sexual selection. Certainly new signals are unlikely to become established if females suffer a *net loss* of fitness by responding to them (Reeve and Sherman 1993), and strong stimulation of the sensory receptors does not always translate into a positive response, as we have seen in examples of correct rejection (see also Gerhardt 1994a; Guilford and Dawkins 1993). However, a preexisting bias can also be viewed as a kind of direct benefit that is maintained by positive selection in its original context (Kirkpatrick and Ryan 1991). Hence preferences for new signals can be considered pleiotropic effects of the bias that can be immediately favored in the new context. Even if responses to new signals have a neutral or negative effect on receivers, the bias can be maintained if selection in the original context is strong enough. For example, a sensory bias that reduced mate-finding costs would outweigh the costs of mating with an inferior male, especially if males did not differ greatly in quality (e.g., Dawkins and Guilford 1996).[1] Because by definition, preexisting biases evolve before the evolutionary appearance of the signal, preferences based on these biases could not have coevolved with signals as predicted by models of indirect selection (Fisher effect and handicap models). The empirical evidence to support this conclusion in a robust way may be more difficult to obtain than has sometimes been supposed. As pointed out by Endler and Basolo (1998), phylogenetic trees that are consistent with preexisting biases still depend on implicit and seldom-tested assumptions about evolutionary rates and speciation events in different

1. Of course, we might then expect that selection would act to favor context-dependent expression of the bias (e.g., Reeve and Sherman 1993).

branches. Moreover, as mentioned above, biases immediately become subject to sexual selection as soon as a new signal becomes established, and thus both are likely to have coevolved to some extent in contemporary populations.

The interpretation that preexisting biases have evolved and been conserved in the *Physalaemus pustulosus* group is based on the existence of female preferences in two species (*P. pustulosus* and *P. coloradorum*) for acoustic characteristics that are not present in the calls of conspecific males but are produced by other species in the group (Ryan and Rand 1993). Whereas in all members of this group (9–10 species) males produce frequency-modulated "whines," acoustic suffixes ("chucks") appear to be produced by only two species (*P. pustulosus* and *P. freibergi*) and in some populations of a third species, *P. petersi* (Canatella et al. 1998; Ryan, pers. comm.). These species occur in a clade found in Central America and South America, east of the Andes Mountains (fig. 11.19). Another species in this clade, *P. pustulatus*, produces an acoustic prefix (a "squawk"). Although no prefixes or chucks are produced in the other clade, found in Ecuador and Peru, west of the Andes, males of *P. coloradorum* sometimes produce double whines. Females of *P. pustulosus* prefer conspecific whines with chucks to whines without chucks, conspecific whines to which squawks of *P. pustulatus* are appended to conspecific whines alone, and double (conspecific) whines to single whines. Females of *P. coloradorum* prefer conspecific whines to which three chucks from the calls of *P. pustulosus* are added. However, females of *P. coloradorum* show only a weak, nonsignificant trend to choose double (conspecific) whines to single whines, even though conspecific males produce such signals. This last result suggests that at a quantitative level of analysis, a general bias for "extra" stimulation has undergone some modification in one or both of these taxa. In our view, further rigorous confirmation that a preference for acoustic additions to single whines is the ancestral state requires information about female preferences from additional species in the group, hence our addition of question marks at the roots of the trees of figure 11.19. Moreover, even if the preference is ancestral, the possibility remains that acoustic appendages evolved at the same time as the preference and were then lost in some lineages (see Quinn and Hews 2000 for an example from visual communication).

Ryan and Rand (1999) have also used phylogenetic methods to estimate and reconstruct the structure of whines at the ancestral nodes using different models of evolution. Synthesizing these signals, they then tested females of *P. pustulosus* in single-stimulus and two-stimulus experiments. Although the various models led to differences in estimated signal structure as well as differences in phonotactic selectivity, the results indicate that the range of effective signal structures exceeds the range of variation in the signals of conspecific calls. In particular, females do not discriminate between conspecific calls and calls with structures estimated by the different models to be representative of the common ancestor of *P. pustulosus* and *P. petersi* (its closest relative, depending on the status of *P. freibergi*, which was formerly lumped with *P. petersi*).

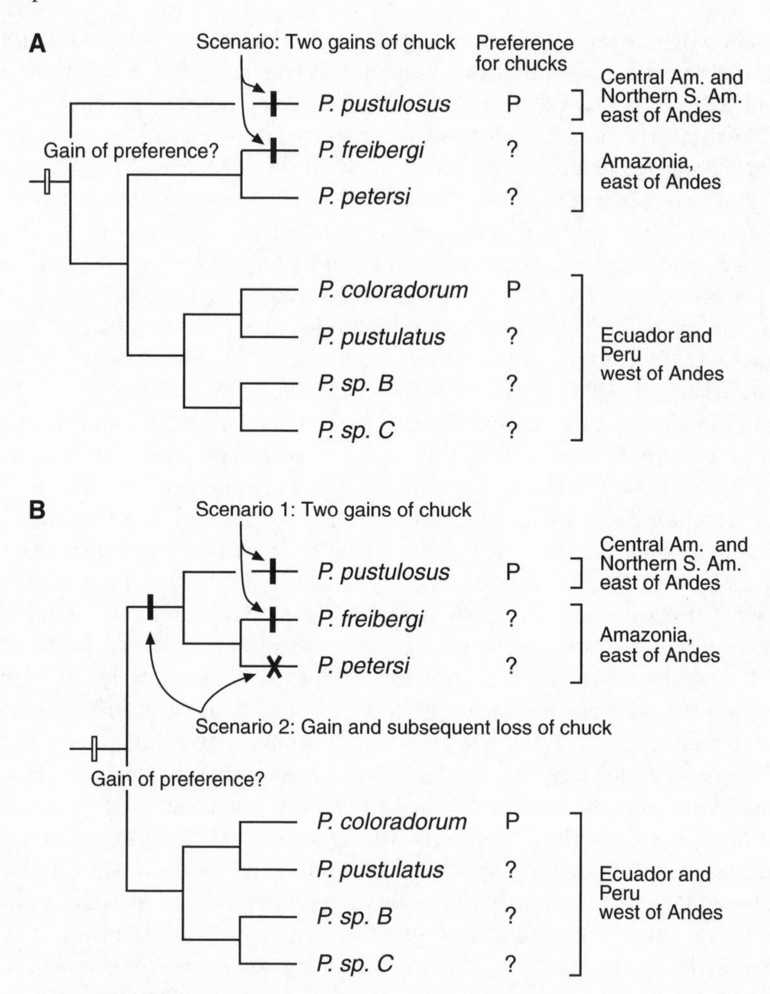

Figure 11.19. Two scenarios for the evolution of the chuck component of calls found in some members of the *Physalaemus pustulosus* species group. The preference for such acoustic appendages has been found in females of the two species in the group that have been tested. Generalized geographical distributions of the species are shown to the right. (A) The tree supported by the original analysis (Ryan et al. 1990). (B) Tree supported by total evidence analysis of Canatella et al. (1995). In the column to the right, P = a preference for acoustic suffixes ("chucks") has been documented. ? = females have not been tested for a preference. We have added a question mark to the ancestral state because we think that the preferences of females of additional species need to be tested in order to be more confident about the ancestral state.

Ryan and Rand (1999) interpret these results as additional support for the preexisting bias model and one of its general corollaries, namely, that signals and preferences do not evolve in a tight coevolutionary way. As these authors suggest, this hypothesis can be further supported by tests of females of *P. petersi* with conspecific and (estimated) calls of the common ancestor. Phelps and

Ryan (2000) have used neural network models to predict patterns of evolution in preferences that have yielded predictions that conform to the way females treat various signal properties. These imaginative approaches exemplify how the comparative method can be used to propose and test various hypotheses about the evolutionary history of communication systems.

Summary and Suggestions for Future Research

At one level of analysis, the intense signals and behaviors (production of burrows, signaling from elevated sites) of some species of insects and anurans can be seen as adaptations to increase communication range. The evolution of intense signals might also be a response to selection for increasing the probability of detection in dense, noisy choruses.

Analysis of patterns of geographical variation and comparative studies provide two ways of identifying environmental factors that might affect long-term evolutionary changes in communication systems. Neither approach has, however, generated much convincing evidence that habitat acoustics (including masking interference by the signals of other species) has been a potent selective force. Indeed, within a particular community, the signals of some species seemingly propagate better in nonpreferred habitats than in preferred habitats. Perhaps the best way to test for environmental effects is to compare the communication system of the same wide-ranging species in different habitats or in localities with and without other species whose signals are sources of masking interference. In one of the few extensive geographical surveys of this kind, small but significant habitat effects on signal structure were demonstrated in cricket frogs (Ryan and Wilczynski 1991). The next step is to assess the behavioral significance of differences in signal degradation that occur after propagation through different habitats as in Lang's (2000) recent study of grasshoppers.

Some predators or parasitoids use the acoustic signals of insects or anurans to locate their prey and thus might be expected to influence the evolution of signal structure and signaling behavior. Some evidence for this hypothesis comes from comparisons of an introduced field cricket (*Teleogryllus oceanicus*) found in areas where parasitoids have also been introduced and in parasitoid-free localities. In areas with parasitoids, overall calling activity appears to be significantly reduced, but some temporal details of the songs of males in areas with parasites might actually attract the flies more effectively than the songs of males from localities without parasitoids.

Although the balance of theoretical opinion now tilts toward the plausibility of reproductive character displacement, the conditions under which it should occur are restrictive, and several other outcomes of contact between formerly isolated groups of populations are perhaps more likely. Comparative studies of insects (primarily *Drosophila*) suggest that reproductive character displacement might be common, but other interpretations are possible.

Indeed, we can only guess at the proportion of species with well-differentiated signals and phonotactic selectivity that have achieved sympatry without interacting, perhaps because they diverged sufficiently in ecology and behavior while still in allopatry. Regardless of how different taxa achieve sympatry, the species-specificity of signals, calling positions, or calling periods suggests that these behaviors are constrained in each species by the presence of other members of the community. A few counterexamples have been discovered, however, and perhaps the expectation that sympatric taxa will always have well-differentiated signals has discouraged searches for cryptic genetic differentiation of taxa with similar signals.

Several examples of the pattern of geographic variation in signals expected from reproductive character displacement have been described in insects (Hawaiian crickets and periodical cicadas) and anurans (Australian treefrogs, North American microhylid and hylid frogs). These examples could all be improved by additional sampling of populations, tests of geographical variation in female selectivity, and estimates of the frequency and costs of mating mistakes. In the diploid gray treefrog *Hyla chrysoscelis*, female selectivity rather than male signals is enhanced in sympatric areas, and this might be a more common result in species in which signals diverge sufficiently during allopatric isolation to be discriminated.

Comparisons of groups of closely related species can reveal patterns of evolution of communication systems that have occurred over relatively long time spans. Comparative studies of insects and anurans in which acoustic communication is prominent provide plausible evolutionary scenarios for signal evolution and diversification. These hypotheses can now be tested by better resolving phylogenetic relationships using strict cladistic criteria and a wide range of characters (morphological, molecular, behavioral). Even though phylogenetic trees based on signal structure can be reasonably well supported by standard comparative methods, these trees are often less robust and disagree with those derived from nonbehavioral characters. This observation might reflect the more rapid evolution of signals and other behaviors in comparison with morphological and molecular traits. At the same time, similarity in signal structure (both in qualitative and quantitative properties) is common, especially in comparisons of allopatrically distributed taxa. In some systems, qualitative similarity seems to reflect phylogenetic relatedness (or conservation of structure), whereas in comparisons of more distantly related taxa, there is good evidence for convergence.

We still know little about the evolution of qualitative, structural differences (e.g., pulsed vs. unpulsed signals, simple vs. compound signals, broadband vs. narrow band signals). Some of the species diversity we now observe must have arisen from changes in function of different signals within a repertoire, or changes or losses of elements of compound signals. Studies of groups of species with both simple and complex signals within a formal phylogenetic context could provide significant insights into signal and receiver evolution.

Indeed, this kind of approach, exemplified by studies of the *Physalaemus pustulosus* species group, has led to the discovery of "hidden" or preexisting sensory biases that can influence the form of stimuli that ultimately achieve the status of signals. Additional, phylogenetically grounded studies of both male signals and female preferences in other groups of frogs and in insects—such as grasshoppers in the genus *Chorthippus*, which have simple and complex signals—are needed in order to learn if this is a general result. Indeed, even the conclusion that preferences for acoustic appendages such as "chucks" have been evolutionarily conserved could be further bolstered by testing preferences of additional species in the *P. pustulosus* group. Nevertheless, the research with this group of frogs serves as a model for a multidisciplinary perspective that attempts to discover broad generalizations by integrating information, obtained by a wide variety of approaches, about mechanisms and phylogeny.

Appendix 1

Theory and Analysis of Communication

Information Theory

In a simple communication system, a *sender* (signaler) produces a signal that propagates through a medium (*channel*) to a *receiver*; a researcher can quantify the production of signals, identify the signaler by nonacoustic criteria, and quantify the behavior of the signaler and receiver after the production of the signal (fig. A1.1). In information theory, *broadcast information* is the reduction in uncertainty about the sender's identity or behavior after production of the signal (Wilson 1975). Production of a particular kind of signal, for example, can indicate that the animal is likely to attack a nearby rival. *Transmitted information* is the reduction in uncertainty about the receiver's behavior after production of the signal (Wilson 1975). After hearing a certain kind of signal, the receiver might approach its source. Changes in the receiver's behavior represent evidence for communication in its everyday usage. Responses of one species to the signals of another species or responses to sounds that occur as an inadvertent result of, say, locomotion would qualify as communication within the context of information theory. In this monograph, however, we concentrate on intraspecific communication, although we also discuss interspecific masking and acoustically orienting predators and parasites of insects and frogs.

Broadcast information depends on statistical patterns of signaling or variation in acoustic properties of signals that can, independently of receivers and external noise, set limits on the communication process. This idea has been applied to assessing the potential for individual recognition by acoustic signals in birds (e.g., Beecher 1989) and the potential for female choice to affect evolutionary change in acoustic signals within populations (Gerhardt et al. 1996). If individuals cannot be distinguished statistically by their signals, then in the long run, there will be no benefit in investing time and energy in discriminating between signals unless signal differences reliably reflect, on average,

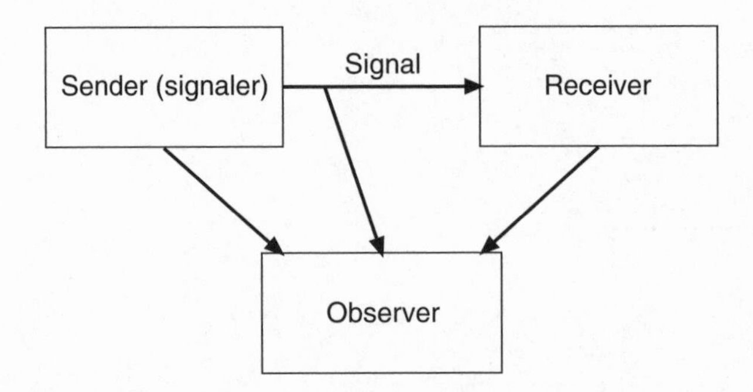

Figure A1.1. Diagram showing basic elements of a communication system: sender (or signaler), receiver, signal, and channel. The channel in the natural world is always imperfect and distorts (degrades) the signal and adds noise. The observer can assess the signal, the channel, and changes in the behavior of the signaler and receiver after a signal has been produced. The observer can also assess, in principle, whether the signaler, receiver, or both benefit from an episode of communication (see fig. A1.2). Modified from Wiley 1983, fig. 1.

differences in the aggressive state or the condition (physical or genetic) of the signaler within the population.

The amount of information (transmitted or broadcast) can be quantified in bits, and information theory has been applied to animal communication systems with mixed success. One problem is that basic assumptions are not fulfilled. One fundamental assumption, for example, is that of constancy of the internal states of signalers and receivers. Yet an animal may not make any overt response to a series of signals, but each signal could indirectly affect the probability of a response to some subsequent signal, say, by stimulating secretion of reproductive hormones. Another complication is that receivers may remember the signals of a particular individual and associate their perception with the results of previous encounters. These problems are compounded by the fact that there might be so much external noise in the communication channel that the receiver only occasionally detects and recognizes signals. Note that in information theory, the lack of motivation of a receiver would also be considered (internal) noise as would errors in encoding made by the signaler. For example, a sender might attack after producing a signal on one occasion and not attack after producing the same signal on another occasion (Wiley 1994).

Another independent assessment concerns the consequences of communication for the signaler and receiver (Wiley 1983). These are usually categorized as shown in figure A1.2. In principle, these consequences can be estimated as fitness losses and gains. In the long run, communication systems evolve to be mutually beneficial to both signalers and receivers of the same species, even though some interactions will favor senders at the expense of receivers and vice versa (see Bradbury and Vehrencamp 1998 for a formal analysis).

		Change in receiver's fitness	
		Increase	Decrease
Change in signaler's fitness	Increase	**Mutuality**	**Deceit (manipulation)**
	Decrease	**Eavesdropping**	**Spite**

Figure A1.2. Categorization of the possible changes in the fitness of signalers and receivers after production of a signal. From Wiley 1983, table 5.1.

Signal Detection and Decision Theory

Because the probability that a receiver will respond appropriately to a signal depends on its internal state as well as the statistical properties of signals and noise in its external environment, biologists and psychologists have sometimes adopted concepts and techniques from signal detection theory, which, like information theory, was originally developed to understand problems of communication in inanimate systems. An important extension of signal detection theory—sometimes called *decision theory*—examines the fitness consequences of responding or not responding, the probabilities of which depend on the response (discrimination) criterion adopted by an individual. Bradbury and Vehrencamp (1998) provide a clear and robust treatment of decision theory and show the mathematical and logical connections between information theory, signal detection theory, and optimality approaches to animal communication. Wiley (1994) discusses many of the same issues and provides examples from acoustic communication in anurans.

The starting point for applying these concepts is the matrix shown in figure A1.3. There are, from the point of animal communication, two desirable outcomes. First, the animal will benefit when it responds appropriately to a (correct) signal (a hit), and, second, when it refrains from responding when a

Signal*	Possible Responses	
Present	Response (HIT)	No Response (MISS)
Absent	Response (FALSE ALARM)	No Response (CORRECT REJECTION)

*Also applies to discrimination of two signals

Figure A1.3. Classification of the responses of a receiver under two conditions. Notice that detection and discrimination problems are equivalent. Modified from Wiley 1994, table 7.1.

signal is absent or incorrect (a correct rejection). Two generally undesirable events are responding when there is no signal or when the signal is incorrect (a false alarm) or failing to respond when a correct signal has been produced (a miss). One inescapable conclusion is that it is impossible for an animal (or system) to set its criterion for responding so that it increases its rate of hits without simultaneously increasing its rate of false alarms. Conversely, changing the criterion to increase the rate of correct rejections will necessarily increase the rate of misses.

For any given criterion, then, the rate of hits can be increased relative to the false-alarm rate only by increasing the signal-to-noise ratio or, in the case of discrimination between two different signals, by increasing the difference in whatever acoustic properties serve to distinguish them (Wiley 1994). These points are illustrated in figure A1.4, which also shows how *receiver operator characteristics* (ROC) are derived. These curves show the ratio of hits to false alarms expected by adopting different threshold criteria for a given overlap in the distributions of some diagnostic property of the two signals to be discriminated. Three examples in the figures show normal distributions, for which the calculations are relatively simple, but the distributions could, in principle, be non-normal and differ between the two signals. In practice, ROCs are unlikely to be generated for many kinds of animals because they require an enormous number of behavioral tests.

Signal detection theory does, however, make explicit some practical procedures that should probably be included in most playback studies and serves to emphasize that a failure to respond can be just as important an indicator of discrimination as a positive response. For example, researchers need to include trials without a signal, or with no difference between two signals, in order to estimate the false-alarm rate of the subject. By doing so, a researcher might discover that certain test conditions, or certain other values of test signals than the one being varied, may affect the probability of response or discrimination.

Most studies of animal communication have considered only positive responses as evidence for recognition or discrimination. In fact, there is good evidence for correct rejection in several species of insects (see chapter 4), where increases in signal levels (within biologically reasonable limits) relative to external noise decrease the probability of response. The best interpretation is that elements of the signals that identify them as inappropriate are masked by external noise at low but not high signal-to-noise levels. If the animal fails to respond at low levels and does so at high levels, the best interpretation is that the signal is marginally attractive (has at least some key properties but perhaps with suboptimal values), the animal's motivation is low, or both. The last two possibilities can be checked by presenting a signal known to be highly effective.

Most importantly, however, signal detection theory provides insights about the evolution of signals and receiver selectivity as well as response variability.

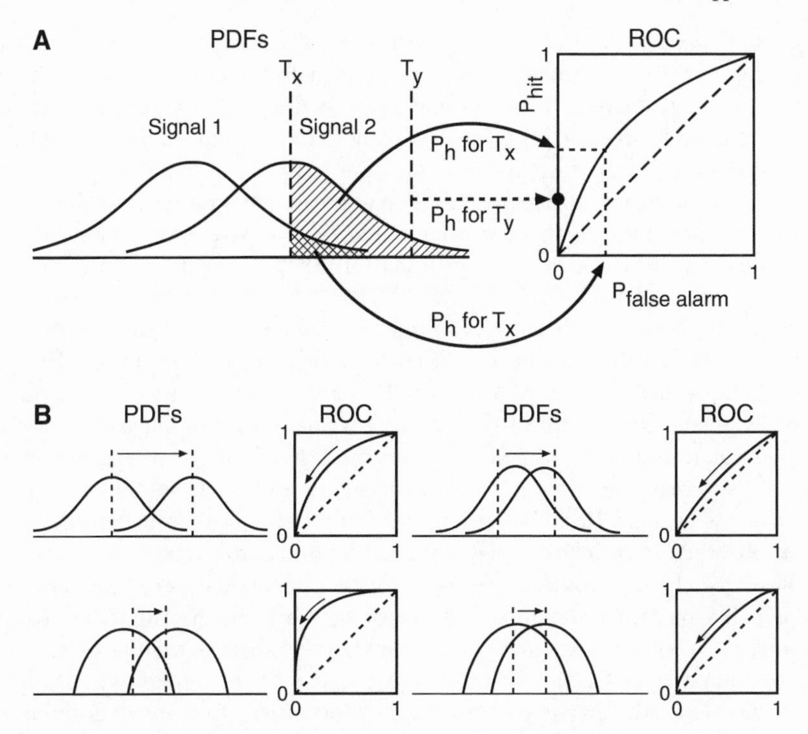

Figure A1.4. (A) The distributions to the left represent the probability density functions (PDFs) of the values of some diagnostic characteristic of two signals, or, equivalently, a signal and noise. Because there is considerable overlap between the distributions, there is only a small part of the distribution (the area to the right of line T_y) where responses will always be correct (hits), assuming that signal 2 is the correct signal; however, the price of this strict criterion is a high probability (the area under the curve of signal 2 to the left of line T_y) that signals will not be detected (misses). If the rate of hits is to be increased, say, by adopting a more lax criterion (line T_x), then the proportion of incorrect responses (false alarms) will increase. The receiver operating characteristic (ROC) curve to the right (*solid curved line*) is generated by varying the threshold criterion over a range encompassing both distributions and then plotting the probability of correct responses (P_{HIT}) as a function of the probability of false alarms (P_{FA}). These probabilities are computed by integrating the areas under the curves to the right of the criterion value. (B) If the difference between signals (or of a signal and background noise) is increased, so that the distributions of values of the diagnostic property overlap to a lesser degree (compare right and left plots), then the same criterion will result in a greater proportion of hits to false alarms, and the greater the convexity of the ROC curve (its distance from the diagonal toward the upper-left corner of the square), the better the performance of the animal. Although ROC curves usually assume Gaussian probability density functions, as in the top two plots, other distributions, which might more realistic model the sensory systems of animals, can also be used, as shown in the bottom pair of plots. Figures modified from Wiley 1994, fig. 7.4.

This link becomes evident when the consequences of a choice of the threshold criterion are considered. The net benefit (or so-called *utility*) of any such choice of criterion will simply be the sum of the fitness gains and losses that result from each of the four possibilities shown in figure A1.3. Gains and losses also depend on the particular environment in which communication occurs. These points are best made with examples. Consider a female that is choosing her mate on the basis of acoustic signals alone. If she is confronted with the signals of two different conspecific males, then the distributions of some diagnostic acoustic property are likely to overlap considerably, and thus the probability of a "mistake" is high. The female's criterion, then, would depend on the relative frequency of good and bad males (and hence the probability of encountering them so they could be compared), the difference in their quality, and the costs of discrimination, that is, not merely picking the closest or loudest male.

Suppose, however, that males of another species are present whose signals also overlap the distribution of conspecific calls in the acoustic dimension used for discrimination. Mating with a heterospecific male will result in a much greater loss of fitness than mating with an inferior conspecific male. In this situation the female should adopt a high threshold for acceptance, even though she may reject some conspecific males and incur somewhat higher costs of assessment. Responses to heterospecific signals that have sometimes been observed in no-choice trials can also be understood from this perspective because the probability of response (or the threshold criterion) is affected, in part, by the frequency of occurrence of signals to be discriminated. If the frequency of conspecific signals is low (or they are absent), then not responding to a heterospecific signal would, in many species, be tantamount to missing a female's only chance to reproduce. Another example, in which females receive direct nutritional benefits from potential mates, highlights the possibility that the benefits of indiscriminate behavior must sometimes outweigh the costs (e.g., Heller et al. 1997).

Appendix 2

Analysis and Description of Acoustic Signals

A pure tone is a sinusoidal waveform and serves as a fundamental illustration of the relationship between time-domain and frequency-domain descriptions (fig. A2.1). That is, the spectrum of a sinusoidal waveform (fig. A2.1A) of infinite duration consists of a single (frequency) component with a frequency that is the reciprocal of the period of the sinusoid (e.g., 1/1 ms = 1000 Hz or 1 kHz; fig. A2.1B). Sound propagates as longitudinal waves, and the distance between successive condensations and rarefactions (in the far field) is the *wavelength* (fig. A2.1C), which depends on both frequency and the velocity of sound, which varies according to the conducting medium. In the example of figure A2.2, the wavelength is approximately 34 cm because the frequency 1000 Hz and the speed of sound in air is about 344 m/s.

The amplitude of a sound in air is usually measured and expressed in decibels (dB) sound pressure level (SPL) relative to a reference sound pressure (P_{ref}: 20 μPa (Pascal) or 2×10^{-4} μbar, which are both equivalent to the average human threshold at 1 kHz; thus, SPL in dB = $20 \log_{10} P/P_{ref}$). Sound intensity is proportional to the pressure squared, hence intensity in dB = $10 \log_{10} I/I_{ref}$, where the I_{ref} is typically 10^{-12} watt/m^2. Note that, regardless of how amplitude was measured, decibels can also be used without a reference to quantify the *relative* amplitude of two signals or different frequency components of the same signal (details in Bradbury and Vehrencamp 1998).

Additional components are introduced in the spectrum whenever the waveform is not exactly sinusoidal or infinite in duration (fig. A2.2). The short pulses common in insect and frog signals have additional energy present in a narrow band of frequencies around the *carrier* (or *dominant*) *frequency*, which is defined as the frequency or frequencies with the greatest amount of acoustic energy. If two or more components have the same (maximum) amplitude, then the signal can be considered to have multiple carrier frequencies. In

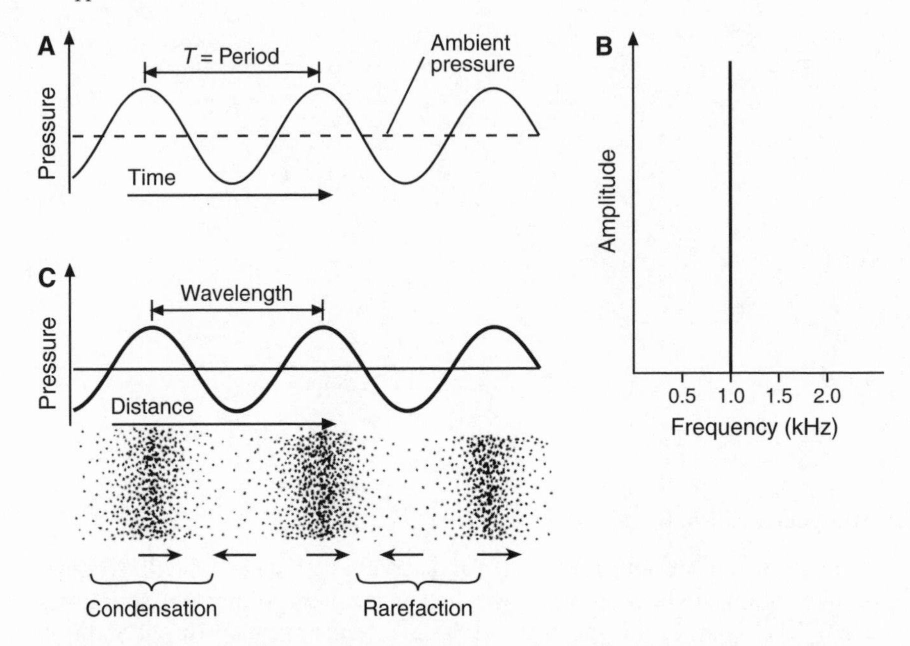

Figure A2.1. (A) Diagram showing the waveform of a sinusoidal signal with a period of 1 ms. (B) If this signal were continuous, then its spectrum would consist of a single component at 1000 Hz or 1 kHz. (C) Diagram showing the waveform of a sinusoidal signal, where the change in pressure is plotted against distance from the source. The distance between successive condensations and rarefactions, shown diagrammatically below the waveform, is the wavelength. Modified from Yost and Nielsen 1985, fig. 2.1.

practice, the amplitude of the most energetic component is assigned a value of 0 decibels (dB), and that of other components are expressed in relative terms, for example, −3 dB, −6 dB, and so on. The carrier frequency and other components could be assigned absolute values in dB SPL if estimates of their bandwidth were available, but these values would be restricted to the distance at which measurements of overall sound pressure level were made in a particular environment.

If a sinusoidal waveform is modulated in frequency (frequency modulation) or amplitude (amplitude modulation), then in addition to the carrier frequency, there are spectral components called *sidebands* (fig. A2.3). Sidebands arise from the multiplication of the carrier wave by a modulating wave; Greenewalt (1968) provides a simple mathematical explanation. If the rate of modulation is rapid relative to the time over which spectral averaging occurs, then sidebands are resolved in the frequency domain, and their separation (frequency interval) from the carrier frequency is exactly equal to the period (rate) of modulation. Sidebands can be confused with harmonics, which, however, are always integral multiples of a fundamental (lowest) frequency and represent different modes of vibration of the structures producing the sound.

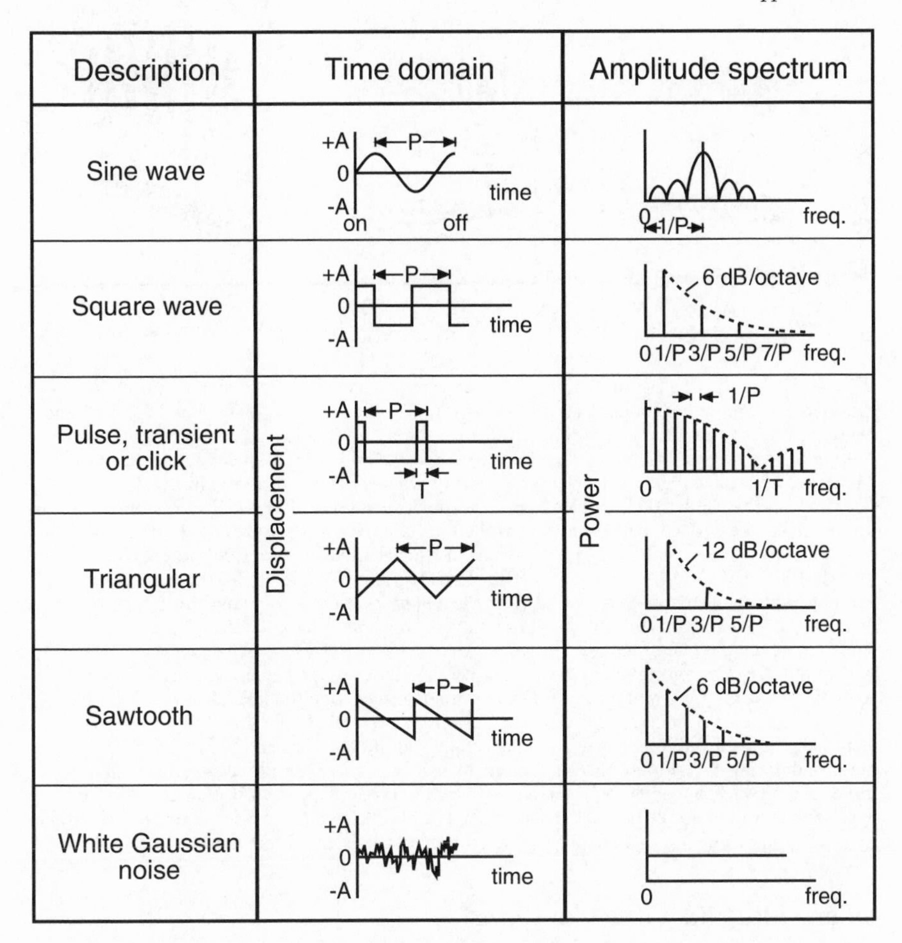

Figure A2.2. Diagrams showing the relationship between time and frequency-domain descriptions of various acoustic waveforms. In principle, any periodic waveform can be decomposed into a series of sinusoidal components, each of which will be an integral multiple of the lowest, or fundamental, frequency (whether or not the fundamental frequency is present in a given signal). A list of the amplitudes and phases of these components comprise the specifications needed to resynthesize the original amplitude-time waveform. Notice that the spectrum shown for the sinusoidal signal does not consist of a single component but rather shows a spectrum representative of a short burst of sinusoidal sound. By definition, white noise contains all frequencies, the relative amplitudes and phase relationships of which randomly vary. P = periodicity of the repeating waveform; 1/P = fundamental frequency. Modified from Yost and Nielsen 1985, fig. 3.2.

Amplitude and frequency modulation can both occur in the production of animal sounds, and the form of these modulations is often not sinusoidal. Bradbury and Vehrencamp (1998) provide an excellent treatment of the relationship between the time-domain and frequency-domain descriptions of signals with complex modulations.

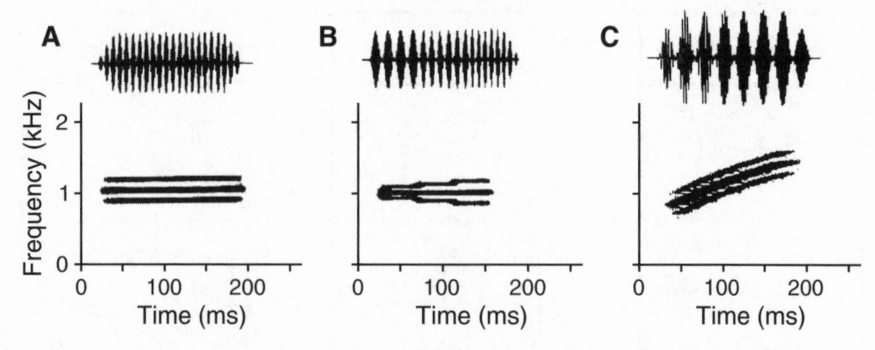

Figure A2.3. (A) Oscillogram (amplitude vs. time) and sonogram (frequency vs. time) of a sinusoidally amplitude-modulated (AM rate of 150/s) carrier tone of 1 kHz. The effective bandwidth of the analyzing filter of the spectrum analyzer is about 50 Hz, and so the sonogram (resulting from a frequency-domain analysis) resolves the modulation frequency as a pair of sidebands, which are the two frequency components (with frequencies of 1150 Hz and 850 Hz) above and below the 1000 Hz carrier frequency (horizontal line at 1 kHz). The three-component spectrum can be derived mathematically as the multiplication of two sinusoids (1 kHz and 0.15 kHz) (see Greenewalt 1968, p. 89), or, in an equivalent fashion, the addition of sinusoids of 850, 1000, and 1150 Hz with appropriate relative amplitudes and phase relationships (as computed by a Fourier analysis) would yield the amplitude-time waveform shown in the oscillogram. (B) Oscillogram and sonogram of a sinusoidally amplitude-modulated carrier tone of 1 kHz in which the rate of modulation increases from 85 to 200 Hz. Notice that the sideband separation (frequency intervals below and above the carrier frequency) increases in a corresponding fashion. (C) Oscillogram and sonogram of a frequency-modulated (0.9–1.2 kHz) carrier tone that is amplitude-modulated at a constant rate of 150 Hz. Because the AM rate was constant, the sideband interval remained constant even though the carrier frequency swept upward in frequency. If these three components had been harmonically related (integral multiples of a missing fundamental frequency whose initial frequency was 150 Hz), then the three components would not be parallel in the display; rather the frequency intervals at the end of the sweep would be much greater than at the beginning of the sweep. From Gerhardt 1998, fig. 6.

Bandwidth and Time Constants

Bandwidth is a fundamental quality that helps to frame questions about signal structure, sound production, and sound reception. In the frequency domain description of a sound, the bandwidth is usually expressed as a *quality factor*, or *Q-value*, a dimensionless quantity that is computed by dividing the frequency range that is within 3 dB of the maximum peak amplitude by the frequency having the maximum amplitude (fig. A2.4). Narrow-band signals, such as nearly pure-tone sounds, will have very high Q-values, and broadband signals, low Q-values. Oscillators or resonators that produce and radiate sounds and tympanic membranes and other structures that vibrate in response to sounds also have Q-values that depend on their physical properties such as mass, stiffness (or its reciprocal, compliance), and tension (Bennet-Clark 1997). Thus, when excited into vibration by short signals—for example, the impulses produced by impacts of stridulatory teeth and files in insect sound production—the Q-value of coupled resonators will largely determine the

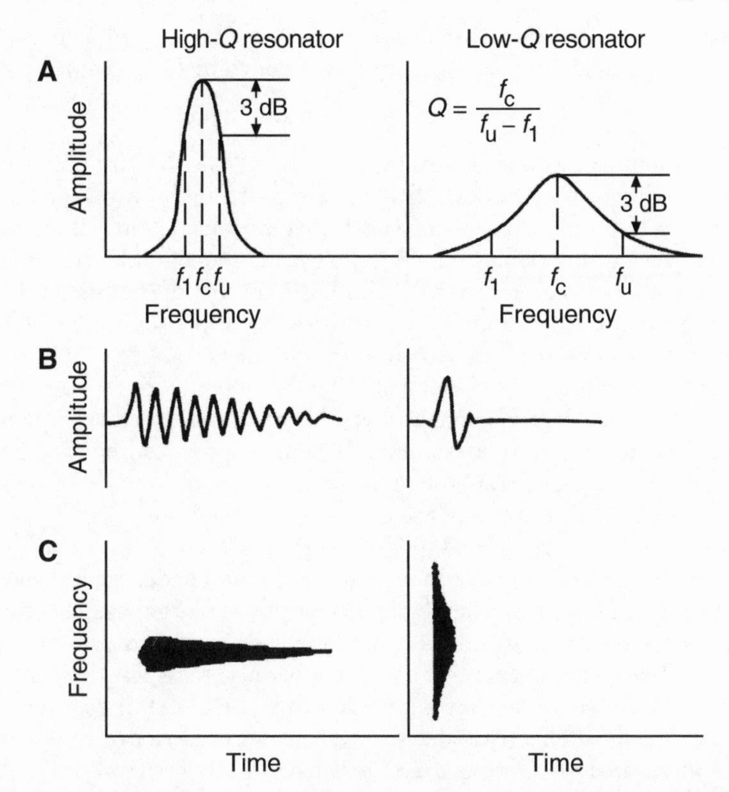

Figure A2.4. Properties of oscillators and filters in the frequency and time domains. (A) The curves at the top of the figure show the amplitude versus frequency responses of sharply tuned (high-Q) and broadly tuned (low-Q) mechanical oscillators, as well as the method for computing Q_{3dB}. (B) The diagrammatic oscillograms show the vibration of high- and low-Q oscillators (or filters) in response to a transient signal, one which has an extremely short duration (microsecond range) and wide-band spectrum. A high-Q oscillator or filter will produce a sustained vibration, with a number of cycles (Q) that have an amplitude within 3 dB of the amplitude of the first cycle. This sustained oscillation is sometimes called ringing, and the vibration is characterized as having a low degree of damping. A low-Q oscillator or filter produces a highly damped vibration that more faithfully approximates the duration of the transient signal. (C) Diagrams depicting sonograms of the vibrational responses of high- and low-Q oscillators (filters). From Bradbury and Vehrencamp 1998.

bandwidth of the signals that are ultimately radiated. As receivers, high-Q mechanical systems will respond effectively only if signals have reasonable amounts of energy at or near their resonant frequency, whereas low-Q systems will respond to signals with frequencies that fall within a broad range. Q-values (= Q_{3dB}) are usually normalized by dividing the bandwidth at 3 dB (on either side of the resonant frequency) by the resonant frequency.

In the time domain, differences in the temporal properties of sounds produced by structures with low and high Q-values are observed when these structures are excited into vibration by very short broadband signals called *impulses.* In the time domain, Q_{3dB} can be computed by the formula

$$Q = \pi/\ln\ [\text{decrement in amplitude}]$$

Low-Q structures respond by producing short signals that are only slightly longer than the exciting impulse; the output is said to be highly damped. High-Q structures respond by producing signals that are much longer than the impulse, and yet another definition of Q_{3dB} is the number of cycles of oscillation that have amplitudes no more than 3 dB less than that of the first cycle produced in response to an impulse. Clearly, low-Q responses are effective for maximizing temporal resolution: two short impulses separated by a brief time period would result in two vibrational events separated in time. High-Q systems, although providing for superior resolution of signals with different frequencies, cannot preserve fine-temporal elements produced by the primary oscillator: the undamped response to the first of two short impulses would obscure the beginning of the response to the second impulse.

These same ideas are applied to filtering properties on the receiver side of an auditory system (chapter 5). Indeed, if the magnitude of a typical auditory neuron's response, as quantified by its firing rate, is plotted against stimulus frequency, the resulting iso-intensity function has a form similar to the curves in figure A2.4A. Tuning curves are inverted because they plot threshold, the frequency to which the neuron is most sensitive, against stimulus frequency, and hence the most effective value (equivalent to the carrier or resonant frequency in an analysis of resonators) will have the lowest threshold. More specifically, the characteristic frequency (CF), the frequency with the lowest threshold, is equivalent to f_c in figure A2.4A, and f_l and f_u correspond to the frequencies that were 3 dB above the lowest threshold on the low- and high-frequency sides of the CF, respectively. Because of a relatively recent tradition, however, neurophysiologists studying vertebrates normally compute Q-values based on the frequency range that is 10 dB above minimum threshold (Q_{10dB}), so it is important to specify the level at which the bandwidth is being measured (Bennet-Clark 1999b).

Trade-offs between the time and frequency responses of low- and high-Q systems illustrate an important idea, that of the *uncertainty principle.* Producing (primary and secondary resonators) or resolving (tympanic membranes, neural filters) narrow-band signals comes at the cost of reductions in producing or resolving fine-temporal detail and vice versa.

Appendix 3

Basics of Environmental Acoustics

Insects and anurans are usually recorded at close range, where the signal-to-noise ratio is favorable and where the effects of environmental degradation on signal structure are minor. Thus, the structures of acoustic signals considered in chapter 2 and elsewhere in this monograph will often be considerably altered at the places where receivers detect and respond to signals in nature. The basic principles of environmental acoustics are well established, and excellent reviews have been available for many years (e.g., Michelsen 1978, 1983; Wiley and Richards 1978, 1982). The text by Bradbury and Vehrencamp (1998) and reviews by Römer (1992, 1993, 1998) provide well-illustrated updates, and more technical treatments of these topics are found in Harris (1991).

Signal Attenuation

The decrease in signal amplitude with distance is nearly always considered with reference to the drop in amplitude expected by spherical (geometrical) spreading, which conforms to the inverse-square (sound-intensity) or inverse-distance (sound pressure) law (fig. A3.1). A useful and often-cited expression is that for every doubling of distance, sound intensity (and sound pressure level) will drop by about 6 dB. Aside from the effects of the environment, the inverse-square law rests on two assumptions: (1) that the amplitude of sound waves is measured in the *acoustic far field;* and (2) that the sound source is an idealized *monopole*, radiating sound equally in all directions in a nonreverberant environment. The first of these assumptions usually holds, at least approximately at the distances at which sound pressure levels of orthopteran insect and anuran signals are usually measured (1–2 wavelengths of the lowest emphasized frequencies) because the effective dimensions of the sound-generating structures are small. The second assumption is seldom met. As illustrated in part 2 of this appendix, insects and anurans signal from the ground,

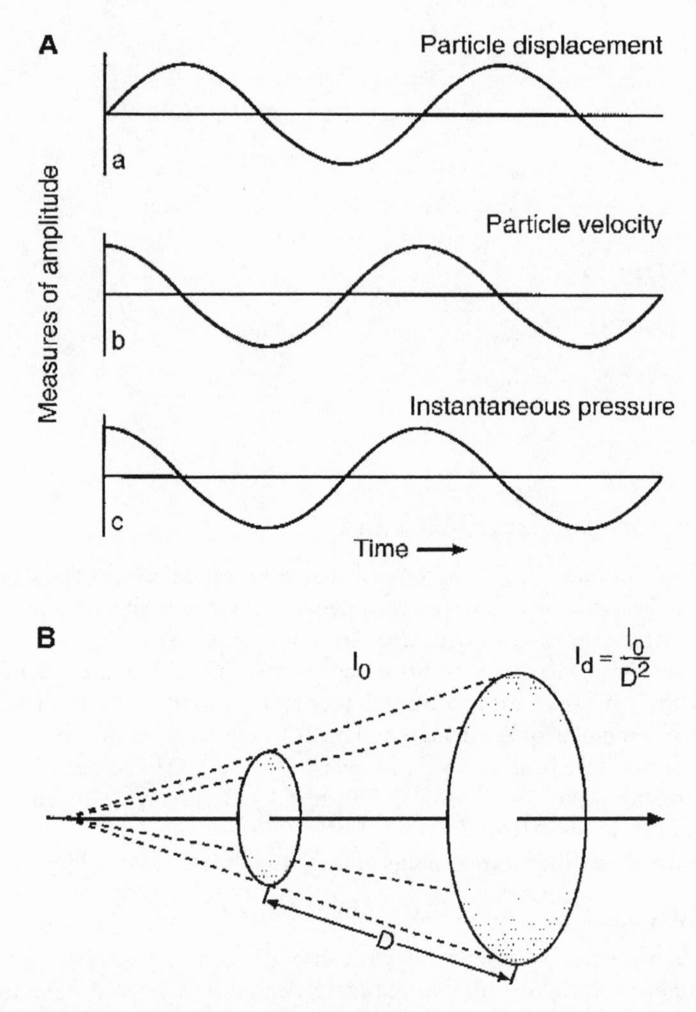

Figure A3.1. (A) Phase relationships between the particle (air molecules, or more generally molecules making up the medium) displacement, particle velocity, and sound pressure of a sinusoidal signal. In the acoustic far field, changes in sound pressure and particle velocity are in phase as shown in this diagram; sound intensity can be calculated from either measure. Close to the sound source, in the near field, where the particle displacement predominates, particle velocity and sound pressure components are not in phase, and the sound field is not only complex but differs depending on the dimensions and behavior of the sound source. Fletcher (1992) provides a clear discussion. (B) Diagram showing the basis for the inverse square law, which holds when measurements are made in the acoustic far field. The acoustic energy per unit area necessarily decreases with distance because of the area through which the sound propagates increases. For each doubling of distance, sound intensity (energy passing through a unit area/unit time) drops by the square of the distance, and sound pressure is inversely related to distance. From Yost and Nielson 1985, fig. 2.2; and Bradbury and Vehrencamp 1998, fig. 2.4, modified.

trees and bushes, or even burrows, which effectively direct sound waves even if the sound-radiating structures approximate monopoles. Moreover, some insect systems approximate dipoles (e.g., Bennet-Clark 1988) or otherwise show directional beaming of sound (e.g., Michelsen and Elsner 1999).

Sound amplitude usually drops by more than expected by spherical spreading because of absorption (a minor effect) by the atmosphere and objects in the environment, scattering, refraction, and reflection (see below). The additional drop in amplitude is called *excess attenuation*, which is usually expressed in decibels (dB) per unit distance (e.g., 100 m). Two general rules usually apply: (1) All things being equal, low frequencies usually suffer less excess attenuation than do high frequencies (fig. A3.2); and (2) excess attenuation is

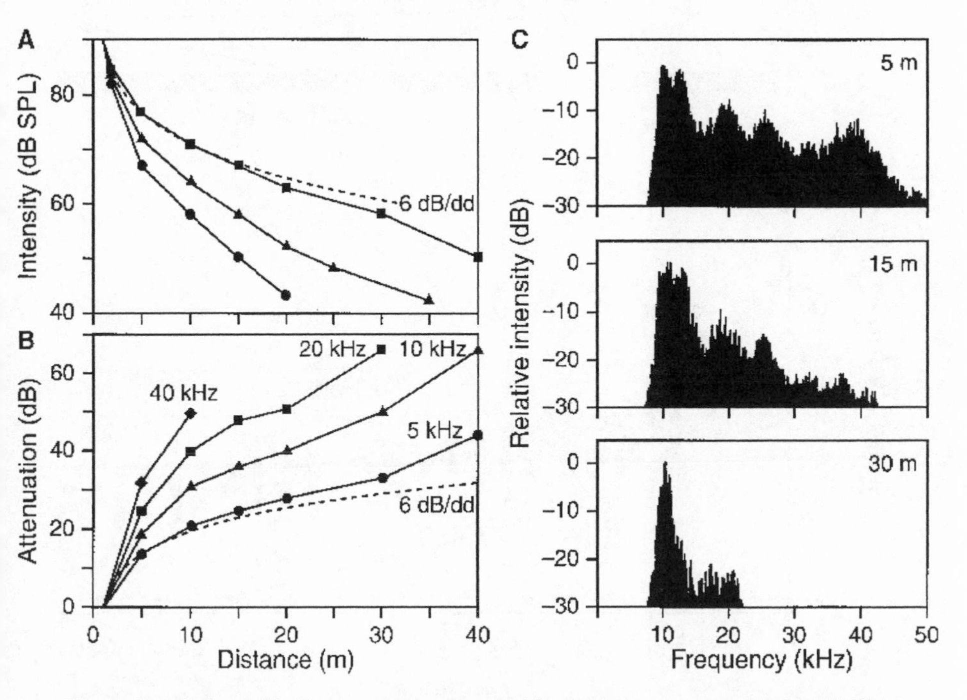

Figure A3.2. Effects of habitat, source elevation, and frequency on excess attenuation. (A) Plot of attenuation of the calling song of a katydid, *Tettigonia viridissima*, as a function of distance from singing males in open grassland and bushland. The microphone was placed at the same height as the singing insect, which was 2 m in the grassland (*squares*) and 1.5 m (*triangles*) and 0.75 m (*circles*) in the bushland. The dotted line shows the attenuation of the song expected from spherical spreading alone. (B) Attenuation of pure tones in a bushland habitat as a function of frequency at a height of 1 m. Instead of using a microphone to estimate attenuation, the change in threshold in an auditory interneuron (omega cell) in a portable preparation is shown using the thresholds determined at a distance of 1 m as a reference. Notice the severe excess attenuation of the 40 kHz signal at 5 and 10 m. (C) Frequency-dependent attenuation of the broadband calling song of the katydid. The series of power spectra show that, congruent with the neurophysiological estimates of (B), the high-frequency components are much more severely attenuated than are the low-frequency components. From Römer 1992, figs. 1, 2; Römer 1998, fig. 3.5.

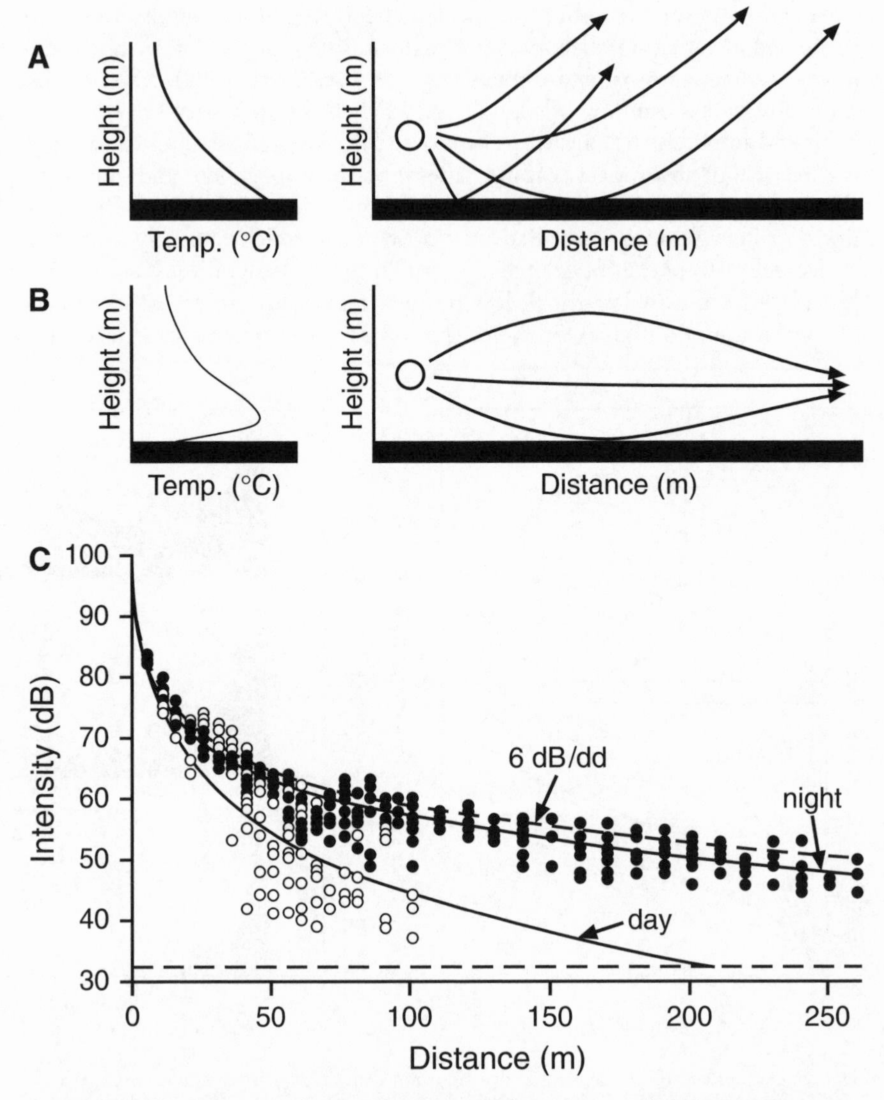

Figure A3.3. Temperature gradients and redirection of sound waves. (A) When temperature decreases with elevation, as it usually does, sound waves are refracted upward toward the medium (cooler air) in which its speed is lower, effectively creating a sound shadow at the same height as the signaler and lower at some distance. (B) At night, temperature inversions, in which a layer of cool air is trapped near the ground, can create an efficient channeling of sound, which will be refracted back to lower heights when it encounters the warmer air mass above. (C) Channels such as the one diagrammed in (B) can result in rates of attenuation that are close to or sometimes less than expected by the inverse distance law (shown by dashed line), as shown for night-time propagation of the calling song of the bladder grasshopper *Bullacris membracioides.* Notice that excess attenuation during the day was much greater. Similar effects were reported for the attenuation of chorus sound in the barking treefrog (Gerhardt and Klump 1988b). From Römer 1998, fig. 3.3.

greatest when signalers are on or close to the substratum, especially if the substratum is porous or vegetated and amplitude is also measured near ground level (figs. A3.2, fig. 3.4B). Elevating either the signaler or the measuring microphone can reduce excess attenuation, especially of high-frequency signals or components of broadband signals, as documented for several species of acoustic insects (fig. 3.4; Arak et al. 1992; Paul and Walker 1979).

Excess attenuation can be increased or reduced by atmospheric effects, such as wind and temperature inversions, which act by refracting or redirecting sound waves. Indeed, the rule that signaling from on or near the ground increases excess attenuation is violated at night when cool air is trapped near the ground, providing a channel that reduces spherical spreading (fig. A3.3; Gerhardt and Klump 1988b; van Staaden and Römer 1997).

Signal Degradation

Scattering and reflection of sound from objects in the environment not only contribute to excess attenuation but also distort the temporal and spectral

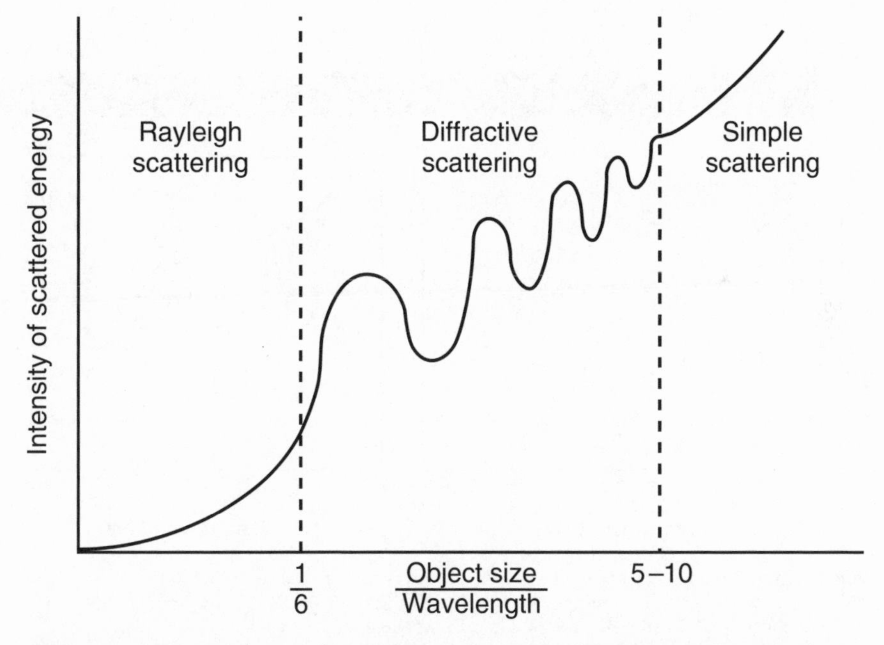

Figure A3.4. Diagram showing the magnitude and patterns of scattering as a function of wavelength (reciprocal of frequency) and the size of objects in the sound field. This plot assumes that wavelength is held constant and object size is varied. Scattering increases as the relative size of the object increases. The same relationship would be seen for Rayleigh (omnidirectional) scattering and diffractive scattering (combination of omnidirectional scattering and bending of waves around the object) if object size were held constant and wavelength varied, but simple scattering (most of the wave is reflected, and little energy diffracts around the object) is considered to be solely a function of object size. From Bradbury and Vehrencamp 1998, fig. 2.6.

structure of signals. First, these phenomena are wavelength-dependent, and scattering takes several forms (fig. A3.4). In general, high-frequency sounds are more affected by scattering and reflection than are low-frequency sounds because their wavelengths are more likely to approximate the size of objects in the environment than are the wavelengths of low-frequency sounds. Second, scattering and reflections result in secondary waves that arrive at the receiver's position via different, longer paths than the direct sound. This means that the secondary waves, which also include various kinds of boundary waves (Bradbury and Vehrencamp 1998), usually have a different phase and amplitude than the direct sound waves. The interaction between direct and indirect waves at the position of the receiver (or measuring microphone) can be constructive or destructive depending on wavelength, distance, and how secondary

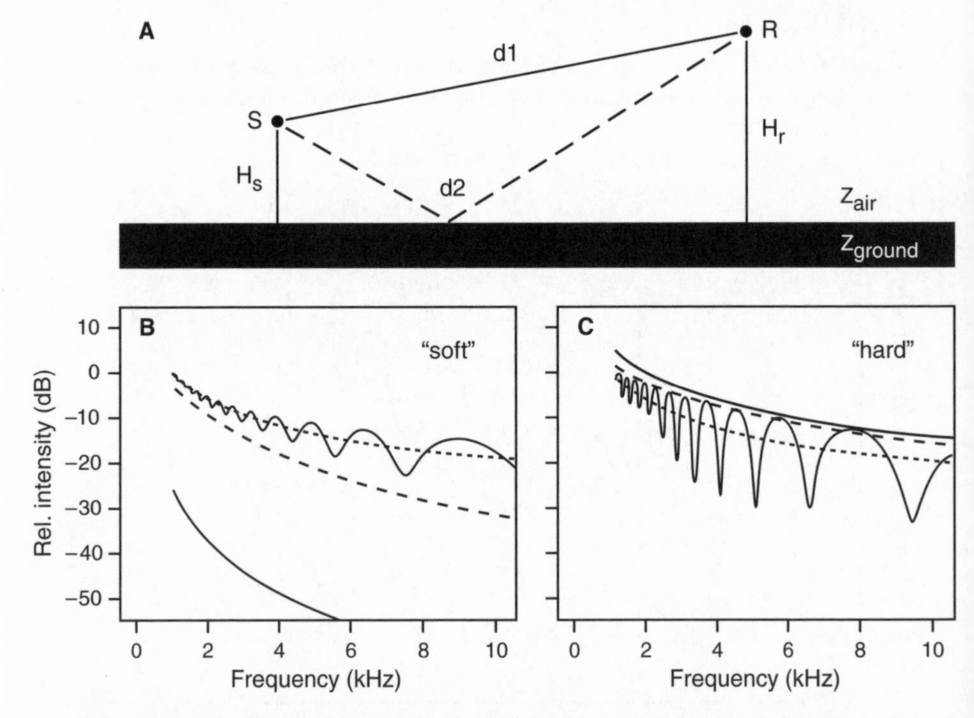

Figure A3.5. Interference of direct and indirect waves that are redirected from the substratum. (A) Diagram showing the geometry of a signaler (S) and receiver (R) that are both elevated (H_s and H_r) above the substratum. The solid line shows the path of direct sound waves that travel between the signaler and receiver. The dashed lines show the path of an indirect wave that is reflected by the substratum to the receiver. The path and phase of the indirect wave will depend on the effective impedance of the ground (Z_{ground}) relative to that of the air (Z_{air}). (B) and (C) Patterns of sound attenuation expected as a function of frequency and signaler-receiver geometry when the substratum is porous ("soft"; B) and hard (C). Dotted lines = attenuation expected from spherical spreading; dashed lines = attenuation expected when the signaler is elevated by 1 m and the receiver is located on the ground; solid wavy lines = attenuation expected if both signaler and receiver are elevated; solid smoothly curving lines = attenuation expected if both signaler and receiver are on the ground. From Römer 1998; as modified from Forrest 1994, fig. 3.4.

waves interact (change phase) with the particular substratum. The interference effects are nonmonotonic when both the signaler and receiver are elevated (fig. A3.5).

Distortion of temporal properties of signals arises from the fact that secondary waves reach the receiver after a delay. These *reverberations* or echoes will obscure the offset of the signal and effectively mask the repetition rate of rapidly pulsed signals (fig. A3.6). *Amplitude fluctuations* are another source of temporal degradation, which is not frequency-dependent and does not involve interactions of secondary waves reflected by objects in the environment. Rather, amplitude fluctuations, which can be caused by wind gusts or rising cells of warm air (especially common during the day in open habitats), disrupt the signal at random times, but usually at rates below about 20 Hz (Richards and Wiley 1980).

These two forms of temporal degradation predict that the temporal structure of animal signals will differ depending on general features of the habitat. Species that signal in dense, forested areas, where many objects will contribute to reverberations, might be expected to produce relatively long, slowly modulated signals. By contrast, animals living in open areas, especially if they signal on warm sunny days, should produce signals with modulation rates in excess of 20 Hz, thereby increasing the chances that temporal structure of

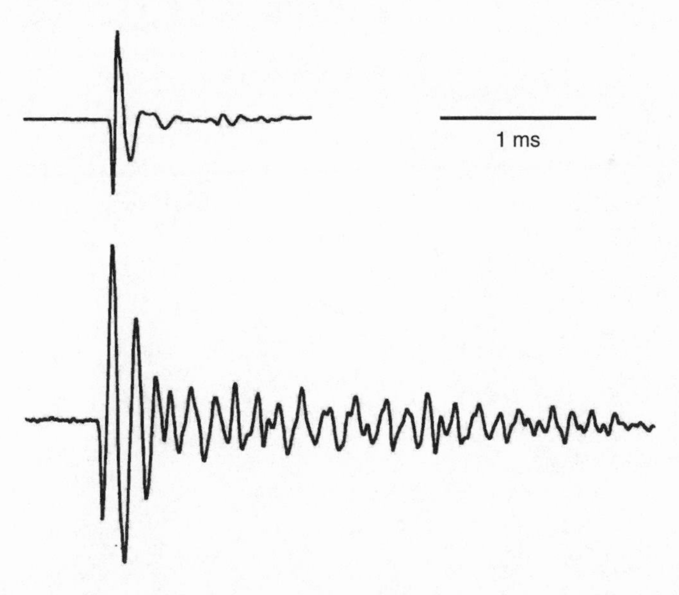

Figure A3.6. Temporal distortion by reverberation. Oscillograms show the direct waveform of a sound impulse recorded in the free field (*top trace*), close to the source (but in the far field) where no scattering objects are present, and after propagating 2 m through natural vegetation (*bottom trace*). If a series of such impulses were repeated at a rapid rate, as in the songs of many insects and frogs, the onsets of successive pulses would also be masked. From Michelsen 1994, fig. 5.2.

some parts of the signal will not reach the signaler without being masked by amplitude fluctuations. Although this prediction is supported to some extent by correlations between habitat and signal design in songbirds (Richards and Wiley 1980), so far the evidence is meager for correlations between any feature of signal design and habitat for insects and anurans (chapter 11).

Appendix 4

Patterns of Female Preference for Acoustic Properties of Long-Range Signals

Taxon	Preference Pattern	Selection Mode	Type of Test	Reference
A. Preferences Based on Carrier Frequency				
Orthoptera: Gryllidae				
Gryllus campestris	MF > HF	ST/WK?	2 sp Choice Ar	Simmons & Ritchie 1996
Laupala cerasina	MF > LF, MF > HF	ST/WK	2 sp Choice Ar	Shaw & Herlihy 2000
Oecanthus nigricornis	LF > HF; LF = MF = HF	ST/WK	2 sp Choice Ar	Brown et al. 1996
Orthoptera: Tettigoniidae				
Kawanaphila nartee	HF > MF > LF	SD (H > M, L)	2 sp Choice Ar	Gwynne & Bailey 1988
Phaneroptera nana	MF > LF; MF > HF	ST/WK	No-choice FAR	Tauber & Pener 2000
Requena verticalis	HF Band: HF > LF	H > L	2 sp Choice Ar	Bailey & Yeoh 1988
Homoptera				
Cystosoma saundersii	LF = MF = HF	NP	No-choice Fl	Doolan & Young 1989
Anura: Bufonidae				
Bufo americanus–Maine	LF = HF	NP?	2 sp Choice Ar	Sullivan 1992
Bufo americanus–Indiana	LF > HF*	L > H	2 sp Choice Ar	Howard & Palmer 1995
Bufo calamita–UK	LF = MF = HF; LF = HF	NP	2 sp Choice Ar	Arak 1988
Bufo calamita–Spain	LF = MF = HF; LF = HF	NP	2 sp Choice Ar	Tejedo 1992
Bufo rangeri	LF = MF = HF; LF = HF	NP	2 sp Choice Ar	Cherry 1993
Bufo valliceps	LF = HF	NP?	2 sp Choice Ar	Wagner & Sullivan 1995
Bufo viridis–Italy 1	LF > HF	L > H	2 sp Choice Ar	Giacoma et al. 1997
Bufo viridis–Italy 2	MF > LF, MF > HF	ST/WK	2 sp Choice Ar	Castellano & Giacoma 1998
Bufo woodhousei	LF = HF	NP?	2 sp Choice Ar	Sullivan 1983
Anura: Discoglossidae				
Alytes cisternasii	MF > LF; MF > HF; LF > HF	ST/WK	7 sp Choice Ar	Márquez & Bosch 1997a
Alytes muletensis	MF > LF; MF > HF; LF = HF	ST/WK	2 sp Choice Ar	Dyson et al. 1997
Alytes obstetricans	MF > LF; MF > HF; LF > HF	ST/WK	7 sp Choice Ar	Márquez & Bosch 1997a
Anura: Hylidae				
Acris crepitans–Indiana	LF > MF > HF	SD (L > M, H)	2 sp Choice Ar	Ryan et al. 1992
Acris crepitans–Texas 1	LF = MF; MF > HF	ST/WK	2 sp Choice Ar	Ryan et al. 1992
Acris crepitans–Texas 2	LF > MF; MF = HF	SD (L > M, H)	2 sp Choice Ar	Ryan et al. 1992
Hyla chrysoscelis–Indiana	LF > HF; LF = MF = HF	ST/WK	2 sp Choice Ar	Morris & Yoon 1989

Appendix 4 continued

Taxon	Preference Pattern	Selection Mode	Type of Test	Reference
Hyla chrysoscelis–Missouri	MF > LF; MF > HF	ST/WK	2 sp Choice Ar	Gerhardt & Tanner, unpubl.
Hyla cinerea	MF > LF; MF > HF	ST/WK	2, 4 sp Choice Ar	Gerhardt 1987, 1991
Hyla ebraccata	LF > MF = HF	SD (L > M, H)	2 sp Choice Ar	Wollerman 1998
Hyla gratiosa	MF > LF; MF > HF	ST/WK	2 sp Choice Ar	Gerhardt 1981b
				Murphy & Gerhardt 2000
Hyla versicolor	MF > LF; MF > HF	ST/WK	2 sp Choice Ar	Gerhardt & Tanner, unpubl.
Pseudacris crucifer	MF > LF; MF = HF	ST/WK	2 sp Choice Ar	Doherty & Gerhardt 1984
Anura: Hyperoliidae				
Hyperolius marmoratus–SA	LF > HF**	L > H	2 sp Choice Ar	Dyson & Passmore 1988a
Hyperolius marmoratus–Zim	MF > LF; MF > HF	ST/WK	2 sp Choice Ar	Grafe 1997a
Anura: Leptodactylidae				
Eleutherodactylus coqui	LF = MF = HF	NP	2 sp Choice Ar	Lopez & Narins 1991
Physalaemus pustulosus	LF > HF	L > H	2sp Choice Ar	Ryan & Rand 1993
Anura: Myobatrachidae				
Uperoleia laevigata	Female-size dependent**	N/A	2 sp Choice Ar	Robertson 1990
Anura: Pelobatidae				
Spea multiplicata	LF > HF or LF = HF	ST/WK?	2 sp Choice Ar	Pfennig 2000
B. Preferences Based on Pulse (= Syllable) Rate				
Orthoptera: Gryllidae				
Acheta domesticus	MPR > LPR; MPR > HPR	ST/WK	No-Choice Ar	Stout et al. 1983
Acheta domesticus	LPR = MPR = HPR	NP	2 sp Choice Ar	Stout and McGhee 1988
Gryllus bimaculatus	MPR >> HPR; MPR > LPR	ST/WK	2 sp Choice LC	Doherty 1985c
Gryllus campestris	MPR >> HPR; MPR > LPR	ST/WK	No-Choice LC	Thorson et al. 1982
Gryllus firmus	MPR >> HPR; MPR > LPR	ST/WK	2 sp Choice LC	Doherty & Storz 1992
Gryllus firmus	MPR > HPR; MPR > LPR	ST/WK	No-Choice FL	Pires & Hoy 1992a
Gryllus integer	LPR > MPR	SD (L > M)	No-Choice LC	Hedrick & Weber 1998
Gryllus rubens	MPR >> HPR; MPR > LPR	ST/WK	2 sp Choice LC	Doherty & Callos 1991
Laupala cerasina	MPR > LPR; MPR > HPR	ST/WK	2 sp Choice Ar	Shaw & Herlihy 2000
Laupala kohalensis	MPR > LPR; MPR > HPR	ST/WK	2 sp. Choice Ar	Shaw 2000
Laupala paraniga	MPR > LPR; MPR > HPR	ST/WK	2 sp. Choice Ar	Shaw 2000
Teleogryllus commodus: chirps	MPR > HPR; MPR = LPR	ST/WK	No-Choice LC	Hennig & Weber 1997
Teleogryllus commodus: trills	MPR >> HPR; MPR > LPR		No-Choice LC	Hennig & Weber 1997
Teleogryllus oceanicus: chirps	MPR > LPR; MPR > HPR	ST/WK	No-Choice LC	Hennig & Weber 1997
Teleogryllus oceanicus: chirps	MPR >> HPR; MPR > LPR		No-Choice FL	Doolan & Pollack 1985

Orthoptera: Acrididae				
Chorthippus biguttulus	MPR >> HPR; MPR > LPR	ST/WK	No-Choice FAR	Helversen & Helversen 1994
Orthoptera: Tettigoniidae				
Phaneroptera nana	MPR > LPR; MPR > HPR	ST/WK	No-Choice FAR	Tauber & Pener 2000
Homoptera				
Cystosoma saundersii	MPR >> LPR; MPR > HPR	ST/WK	No-Choice FPR	Doolan & Young 1989
Anura: Bufonidae				
Bufo calamita	MPR > LPR; MPR > HPR	ST/WK	2 sp Choice Ar	Arak 1988
Bufo viridis	MPR > LPR; MPR > HPR	ST/WK	2 sp Choice Ar	Castellano & Giacoma 1998
Anura: Hylidae				
Hyla chrysoscelis	MPR >> LPR; MPR > HPR	ST/WK	2 sp Choice Ar	Gerhardt & Schul, in prep.
Hyla ebraccata	MPR > HPR; MPR = LPR	ST/WK	2 sp Choice Ar	Wollerman 1998
Hyla microcephala	MPR >> LPR	M > L	2 sp Choice Ar	Schwartz 1987b
Hyla versicolor	MPR >> HPR; MPR > LPR	ST/WK	2 sp Choice Ar	Gerhardt & Doherty 1988
C. Preferences Based on Short Call (Pulse, Phrase) Duration (< 0.5 s)				
Orthoptera: Gryllidae				
Laupala cerasina	LDR > MDR > SDR	SD (H > M > L)	2 sp Choice Ar	Shaw & Herlihy 2000
Oecanthus nigricornis	LDR = MDR = SDR	NP	2 sp Choice Ar	Brown et al. 1996
Orthoptera: Tettigoniidae				
Phaneroptera nana	LDR > MDR > SDR	SD (H > M > L)	No-Choice FAR	Tauber & Pener 2000
Orthoptera: Acrididae				
Omocestus viridulus	MDR > SDR; MDR > LDR	ST/WK	2 sp Choice Ar	Eíriksson 1993
Homoptera				
Cystosoma saundersii	MDR > SDR; MDR > LDR	ST/WK	No-Choice FPR	Doolan & Young 1989
Anura: Bufonidae				
Bufo rangeri	LDR = MDR = SDR	NP	2 sp Choice Ar	Cherry 1993
Anura: Discoglossidae				
Alytes muletensis	LDR = MDR = SDR	NP	2 sp Choice Ar	Dyson et al. 1997
Alytes obstetricans–Spain 1	MDR > SDR; MDR = LDR; LDR > SDR	ST/WK	2 sp Choice Ar	Márquez & Bosch 1997b
Alytes obstetricans–Spain 2	MDR = SDR; MDR > LDR; LDR = SDR	ST/WK	2 sp Choice Ar	Márquez & Bosch 1997b
Anura: Hylidae				
Hyla cinerea	MDR > SDR; MDR = LDR	ST/WK	2 sp Choice Ar	Gerhardt 1987
Hyla meridionalis	MDR > SDR; MDR = LDR	ST/WK	2 sp Choice Ar	Schneider 1982
Pseudacris crucifer	MDR > SDR; MDR > LDR	ST/WK	2 sp Choice Ar	Doherty & Gerhardt 1984
Pseudacris regilla	LDR > MDR	H > L	2 sp Choice Ar	Straughan 1975

Appendix 4 continued

Taxon	Preference Pattern	Selection Mode	Type of Test	Reference
Anura: Hyperoliidae				
Hyperolius marmoratus–ZIM	MDR > LDR***; MDR > SDR	SD (H > M > L)	2 sp Choice Ar	Grafe 1997a
D. Preferences Based on Number of Pulses (Syllables)				
Orthoptera: Gryllidae				
Acheta domesticus	HDR > LDR	H > L	2 sp Choice Ar	Gray 1998
Gryllus bimaculatus	LDR = MDR; MDR > SDR	ST/WK	No-Choice YM	Shuvalov & Popov 1993
Gryllus campestris	LDR > MDR > SDR	SD (H > M > L)	No-Choice YM	Elsner & Popov 1978
Gryllus integer	MSLN > HSLN; MSLN > LSLN	ST/WK	No-Choice LC	Prosser et al. 1997; Hedrick & Weber 1998
Gryllus lineaticeps	HSLN > LSLN	H > L	2 sp Choice LC	Wagner 1996
Gryllus texensis	HSLN > MSL > LSLN	SD (H > M > L)	No-Choice LC	Wagner et al. 1995
Gryllus texensis	MSLN > HSLN; MSLN > LSLN	ST/WK	No-Choice LC	Gray & Cade 1999a,b
Orthoptera: Acrididae				
Chorthippus dorsatus–A Greece	LSLN = MSLN = HSLN	NP	No-Choice FAR	Stumpner & Helversen 1994
Chorthippus dorsatus–A Germ.	MSLN > LSN; MSLN = HSLN	ST/WK	No-Choice FAR	Stumpner & Helversen 1994
Chorthippus dichrous–A	LSLN > MSLN; MSLN = HSLN	SD (L > M, H)	No-Choice FAR	Stumpner & Helversen 1994
Chorthippus loratus–A	HSLN > MSLN > LSLN	SD (H > M > L)	No-Choice FAR	Stumpner & Helversen 1994
Orthoptera: Tettigoniidae				
Ephippiger ephippiger MONO	LSLN > HSLN***	L > H	2 sp Choice Ar	Ritchie 1996
Ephippiger ephippiger POLY	MSLN > LSLN; MSLN > HSLN	ST/WK	2 sp Choice Ar	Ritchie 1996
Phaneroptera nana	LSLN (< 2) = MSLN = HSLN	NP	No-Choice FAR	Tauber & Pener 2000
Phaneroptera nana	HSLN > LSLN	L > H	Choice FAR	Cited in Tauber & Pener 2000
Anura: Hyperoliidae				
Hyperolius tuberilinguis	MPN > LPN; MPN > HPN	ST/WK	2 sp Choice Ar	Pallet & Passmore 1988
E. Preferences Based on Long Call (Phrase) Duration (> 0.5 s)				
Orthoptera: Gryllidae				
Gryllus integer	LDR > SDR****	SD (H > M, L)	No-Choice LC	Hedrick 1986
Orthoptera: Acrididae				
Chorthippus biguttulus	MDR > SDR; MDR = LDR	ST/WK	No-Choice FAR	Helversen & Helversen 1994
Chorthippus dorsatus: B phrase	LDR > MDR > SDR	SD (H > M, L)	No-Choice FAR	Stumpner & Helversen 1994
Chorthippus dichrous: B phrase	LDR > MDR > SDR	SD (H > M, L)	No-Choice FAR	Stumpner & Helversen 1994
Chorthippus loratus: B phrase	SDR > MDR > LDR	SD (L > M, H)	No-Choice FAR	Stumpner & Helversen 1994

Orthoptera: Tettigoniidae				
Scudderia curvicauda (.5–1 s)	LDR > SDR	H > L	No-Choice FAR	Tuckerman et al. 1993
Anura: Bufonidae				
Bufo americanus	LDR > SDR	H > L	2 sp Choice Ar	Sullivan 1992
Bufo valliceps	LDR > MDR > SDR	SD (H > M, L)	2 sp Choice Ar	Wagner & Sullivan 1995
Bufo viridis	LDR > SDR	H > L	2 sp Choice Ar	Castellano & Giacoma 1998 Castellano et al. 2000
Bufo woodhousei	LDR > SDR	H > L	2 sp Choice Ar	Sullivan 1983
Anura: Hylidae				
Hyla chrysoscelis	LDR > MDR > SDR	SD (H > M, L)	2 sp Choice Ar	Gerhardt et al. 1996
Hyla versicolor	LDR > MDR > SDR	SD (H > M, L)	2 sp Choice Ar	Gerhardt 1991
Anura: Pelobatidae				
Spea multiplicata	LDR = SDR	NP?	2 sp Choice Ar	Pfennig 2000
F. Preferences Based on Call (Chirp) Rate				
Orthoptera: Gryllidae				
Acheta domesticus	HCR > LCR	H > L	2 sp Choice Ar	Stout & McGhee 1988
Allonemobius fasciatus	HCR > LCR	H > L	2 sp Choice Ar	Doherty & Howard 1996
Allonemobius socius	HCR > LCR	H > L	2 sp Choice Ar	Doherty & Howard 1996
Gryllus bimaculatus	HCR > LCR	H > L	2 sp Choice LC	Doherty 1985c
Gryllus integer	MCR > LCR; MCR > HCR	ST/WK	No-Choice LC	Hedrick & Weber 1998
Gryllus lineaticeps	HCR > LCR	H > L	2 sp Choice Ar	Wagner 1996
Gryllodinus kerkerinensis	HCR > LCR	H > L	No-Choice YM	Popov & Shuvalov 1977
Teleogryllus oceanicus	HCR > LCR	H > L	No-Choice FL	Pollack & Hoy 1981
Anura: Bufonidae				
Bufo americanus	HCR > LCR (effort)	H > L	2 sp Choice Ar	Sullivan 1992
Bufo calamita	HCR > LCR; MCR = LCR	ST/WK	2 sp Choice Ar	Arak 1988
Bufo rangeri	HCR > LCR	H > L	2 sp Choice Ar	Cherry 1993
Bufo valliceps	HCR > LCR	H > L	2 sp Choice Ar	Wagner & Sullivan 1995
Bufo woodhousei	HCR > LCR	H > L	2 sp Choice Ar	Sullivan 1983
Anura: Discoglossidae				
Alytes cisternasii	HCR > LCR	H > L	2 sp Choice Ar	Bosch & Márquez 1996
Alytes muletensis	HCR > LCR	H > L	2 sp Choice Ar	Dyson et al. 1997
Alytes obstetricans	HCR > LCR	H > L	2 sp Choice Ar	Bosch & Márquez 1996
Anura: Hylidae				
Hyla cadaverina	HCR > LCR	H > L	2 sp Choice Ar	Straughan 1975
Hyla cinerea	HCR > MCR > LCR	SD (H > M, L)	2 sp Choice Ar	Gerhardt 1987

Appendix 4 continued

Taxon	Preference Pattern	Selection Mode	Type of Test	Reference
Hyla chrysoscelis	HCR > LCR	H > L	2 sp Choice Ar	Morris & Yoon 1989
Hyla ebraccata	HCR > LCR	H > L	2 sp Choice Ar	Wells & Bard 1987
Hyla gratiosa	HCR > MCR > LCR	SD (H > M > L)	2 sp Choice Ar	Murphy & Gerhardt 1996, 2000
Hyla meridionalis	HCR > LCR	H > L	2 sp Choice Ar	Schneider 1982
Hyla microcephala	HCR > LCR	H > L	2 sp Choice Ar	Schwartz 1986
Hyla versicolor	HCR > MCR > LCR	SD (H > M > L)	2 sp Choice Ar	Gerhardt 1991
Pseudacris crucifer	HCR > LCR	H > L	2 sp Choice Ar	Forester & Czarnowsky 1985
Pseudacris regilla	HCR > LCR	H > L	2 sp Choice Ar	Straughan 1975
Anura: Hyperoliidae				
Hyla marmoratus–SA	HCR > MCR > LCR	SD (H > M > L)	2 sp Choice Ar	Passmore et al. 1992
Hyla marmoratus–Zim	HCR > MCR > LCR	SD (H > M > L)	2 sp Choice Ar	Grafe 1997a
Anura: Leptodactylidae				
Eleutherodactylus coqui	HCR > MCR > LCR	SD (H > M, L)	2 sp Choice Ar	Lopez & Narins 1991
Physalaemus pustulosus	HCR > MCR > LCR	SD (H > M > L)	2 sp. Choice Ar	Bosch et al. 2000
Anura: Pelobatidae				
Spea multiplicata Pop. 1	HCR > MCR	SD (H > M)	2 sp Choice Ar	Pfennig 2000
Spea multiplicata Pop. 2	MCR > HCR	ST/WK?	2 sp Choice Ar	Pfennig 2000

Key: HF = high frequency; MF = intermediate frequency; LF = low-frequency; HPR = high pulse rate; MPR = intermediate pulse rate; LPR = low pulse rate; LDR = long duration; MDR = intermediate duration; SDR = short duration; HSLN = high syllable (pulse) number; MSLN = intermediate syllable number; LSLN = low syllable number; HCR = high call rate; MCR = intermediate call rate; LCR = low call rate; SA = South Africa; Zim = Zimbabwe; Germ. = Germany; "A" = Syllables produced at the beginning of two-part songs; "B" = second phrase of two-part songs; MONO = "monosyllabic" call race; POLY = "polysyllabic" call race. Values to the left of inequality signs were preferred. Equality sign = no preference. Selection mode refers to the expected pattern of selection on acoustic properties in natural populations. ST/WK (= stabilizing or weakly directional); SD = strongly directional (see operational definitions in chapter 10); NP = no preference. The selection mode or strength of directional selection is indeterminate if only low and high values were tested or if intermediate values were preferred in a test against only low or only high values. Test Types: sp = speaker(s); Ar = arena; FAR = female acoustic response; LC = locomotion compensator; YM = Y-maze; FL = tethered flight assay; FPR = female pheromone response.

*Preferences for LF were abolished if the HF signal led in time.

**Females preferred a carrier frequency that would be produced by a male of about 70% of her weight.

***Values of HSLN stimuli did not occur in these populations.

****HDR = long trill sequence rarely interrupted versus SDR = trill sequence frequently interrupted.

Literature Cited

Adams, E. S., and M. Masterton-Gibbons. 1995. The cost of threat displays and the stability of deceptive communication. *J. Theor. Biol.* 175:405–21.

Aertsen, A. M. H. J., M. S. M. G. Vlaming, J. J. Eggermont, and P. I. M. Johannesma. 1986. Directional hearing in the grassfrog (*Rana temporaria* L.): II Acoustics and modelling of the auditory periphery. *Hearing Res.* 21:17–40.

Aitkin, L. M. 1986. *The auditory midbrain. structure and function in the central auditory pathway*. Clifton, N.J.: Humana Press.

Aitkin, P. G., and R. R. Capranica. 1984. Auditory input of a vocal nucleus in the frog *Rana pipiens*: Hormonal and seasonal effects. *Exp. Brain Res.* 57:33–39.

Alder, T. B., and G. J. Rose. 1998. Long-term temporal integration in the anuran auditory system. *Nature Neuroscience* 1:519–23.

———. 2000. Integration and recovery processes contribute to the temporal selectivity of neurons in the midbrain of the northern leopard frog, *Rana pipiens*. *J. Comp. Physiol. A* 186:923–37.

Alexander, A. J., and M. van Staaden. 1989. Alternative sexual tactics in male bladder grasshoppers (Orthoptera: Pneumoridae). In *Alternative life-history styles of animals*, ed. M. N. Bruton, 261–77. Dordrecht: Kluwer Academic Publishers.

Alexander, R. D. 1960. Sound communication in Orthoptera and Cicadidae. In *Animal sounds and communication*, ed. W. E. Lanyon and W. N. Tavolga, 38–92. Washington, D.C.: Amer. Inst. Biol. Sci. Publ. No. 7.

———. 1961. Aggressiveness, territoriality and sexual behaviour in field crickets (Orthoptera: Gryllidae). *Behaviour* 17:130–223.

———. 1962. Evolutionary change in cricket acoustical communication. *Evolution* 16:443–67.

———. 1975. Natural selection and specialized chorusing behavior in acoustical insects. In *Insects, science and society*, ed. D. Pimentel, 35–77. New York: Academic Press.

Alexander, R. D., and R. S. Bigelow. 1960. Allochronic speciation in field crickets and a new species, *Acheta veletis*. *Evolution* 14:334–46.

Alexander, R. D., D. C. Marshall, and J. R. Cooley. 1997. Evolutionary perspectives on insect mating. In *The evolution of mating systems in insects and arachnids*, ed. J. C. Choe and B. J. Crespi, 4–31. Cambridge: Cambridge University Press.

Alexander, R. D., and T. E. Moore. 1958. Studies on the acoustical behavior of seventeen-year cicadas (Homoptera: Cicadidae: Magicicada). *Ohio J. Sci.* 58:102–7.

———. 1962. The evolutionary relationships of the 17-year and 13-year cicadas, and three new species (Homoptera, Cicadidae, *Magicicada). Misc. Publ. Mus. Zool. Univ. of Michigan* 121:1–59.

Alfs, B., and H. Schneider. 1973. Vergleichend-anatomische Untersuchungen am Labyrinth zentraleuropäischer Froschlurch-Arten (Anura). *Z. Morph. Tiere* 76: 129–43.

Allen, G. R. 1998. Diel calling activity and field survival of the bushcricket, *Sciarasaga quadrata* (Orthoptera: Tettigoniidae): A role for sound-locating parasitic flies? *Ethology* 104:645–60.

———. 2000. Call structure variability and field survival among bushcrickets exposed to phonotactic parasitoids. *Ethology* 106:409–23.

Andersson, M. 1994. *Sexual selection*. Princeton: Princeton University Press.

Arak, A. 1983a. Male-male competition and mate choice in anuran amphibians. In *Mate choice*, ed. P. Bateson, 67–107. Cambridge: Cambridge University Press.

———. 1983b. Vocal interactions, call matching and territoriality in a Sri Lankan treefrog, *Philautus leucorhinus* (Rhacophoridae). *Anim. Behav.* 31:292–302.

———. 1988. Female mate selection in the natterjack toad: Active choice or passive attraction? *Behav. Ecol. Sociobiol.* 22:317–27.

Arak, A., and T. Eíriksson. 1992. Choice of singing sites by male bushcrickets *Tettigonia viridissima* in relation to signal propagation. *Behav. Ecol. Sociobiol.* 30:365–72.

Arak, A., T. Eíriksson, and T. Radesäter. 1990. The adaptive significance of acoustic spacing. *Behav. Ecol. Sociobiol.* 26:1–7.

Arak, A., and M. Enquist. 1993. Hidden preferences and the evolution of signals. *Phil. Trans. R. Soc. Lond. B* 340:207–13.

Arnold, S. J. 1994. Multivariate inheritance and evolution: A review of concepts. In *Quantitative genetic studies*, ed. C. R. B. Boake, 17–48. Chicago: University of Chicago Press.

Arnold, S. J., and D. Duvall. 1994. Animal mating systems: A synthesis based on selection theory. *Am. Natur.* 143:705–19.

Atkins, G., S. Ligman, F. Burghardt, and J. F. Stout. 1984. Changes in phonotaxis by the female cricket *Acheta domesticus* L. after killing identified acoustic interneurons. *J. Comp. Physiol. A* 154:795–804.

Autrum, H. J. 1940. Über Lautäußerungen und Schallwahrnehmung bei Arthropoden. II. Das Richtungshören von Locusta und Versuch einer Hörtheorie für Tympnalorgane vom Locustidentyp. *Z. Vergl. Physiol.* 28:326–52.

Axtell, R. W. 1958. Female reaction to the male call in two anurans (Amphibia). *SW Nat.* 3:70–76.

Ayre, D. J., P. Coster, W. J. Bailey, and J. D. Roberts. 1984. Calling tactics in *Crinia georgiana* (Anura: Myobatrachidae): Alternation and variation in call duration. *Aust. J. Zool.* 32:463–70.

Backwell, P. R. Y. 1988. Functional partitioning in the two-part call of the leaf folding frog *Afrixalus brachycnemis*. *Herpetologica* 44:1–7.

Bailey, W. J. 1970. The mechanics of stridulation in bush crickets (Tettigonioidea, Orthoptera). I. The tegminal generator. *J. Exp. Biol.* 52:495–505.

———. 1990. The ear of the bushcricket. In *The Tettigoniidae: Biology, systematics and evolution*, ed. W. J. Bailey and D. C. R. Rentz, 217–47. Bathurst: Crawford House Press.

———. 1991. *Acoustic behaviour of insects: An evolutionary perspective*. London: Chapman and Hall.

Bailey, W. J., H. C. Bennet-Clark, and N. H. Fletcher. 2001. Acoustics of a small Australian burrowing cricket: The control of low-frequency pure-tone songs. *J. Exp. Biol.* 204:2827–41.

Bailey, W. J., and W. B. Broughton. 1970. The mechanics of stridulation in bush crickets (Tettigoniidae, Orthoptera). II. Conditions for resonance in the tegminal generator. *J. Exp. Biol.* 52:507–17.

Bailey, W. J., R. R. Cunningham, and L. Lebel, 1990. Song power, spectral distribution and female phonotaxis in the bush cricket *Requena verticalis* (Tettigonidae: Orthoptera): Active choice or passive attraction. *Anim. Behav.* 40:33–42.

Bailey, W. J., and G. Field. 2000. Acoustic satellite behaviour in the Australian bush-cricket *Elephantodetta nobilis* (Phaneropterinae; Tettigoniidae; Orthoptera). *Anim. Behav.* 59:361–69.

Bailey, W. J., M. D. Greenfield, and T. E. Shelly. 1993. Transmission and perception of acoustic signals in the desert clicker, *Ligurotettix coquilletti* (Orthoptera: Acrididae). *J. Insect Behav.* 6:141–54.

Bailey, W. J., and J. D. Roberts. 1981. The bioacoustics of the burrowing frog *Heleioperus* (Leptodactylidae). *J. Nat. Hist.* 15:259–88.

Bailey, W. J., and L. S. Simmons. 1991. Male-male behavior and sexual dimorphism of the ear of a Zaprochiline tettigoniid (Orthoptera: Tettigoniidae). *J. Insect Behav.* 4:51–65.

Bailey, W. J., and D. R. Thiele. 1983. Male spacing behavior in the Tettigoniidae: An experimental approach. In *Orthopteran mating systems: sexual competition in a diverse group of insects*, ed. D. T. Gwynne and G. K. Morris, 163–84. Boulder: Westview Press.

Bailey, W. J., P. C. Withers, M. Endersby, and K. Gaull. 1993. The energetic costs of calling in the bushcricket *Requena verticalis* (Orthoptera: Tettigoniidae: Listroscelidinae). *J. Exp. Biol.* 178:21–37.

Bailey, W. J., and P. B. Yeoh. 1988. Female phonotaxis and frequency discrimination in the bushcricket *Requena verticalis*. *Physiol. Entomol.* 13:363–72.

Bakker, T. C. M., and A. Pomiankowski. 1995. The genetic basis of female mate preferences. *J. Evol. Biol.* 8:129–71.

Balakrishnan, R., O. von Helversen, and D. von Helversen. 2001. Song pattern recognition in the grasshopper *Chorthippus biguttulus*: The mechanism of syllable onset and offset detection. *J. Comp. Physiol. A* 187:255–64.

Ball, E. E., and K. G. Hill. 1978. Functional development of the auditory system of the cricket, *Teleogryllus commodus*. *J. Comp. Physiol.* 127:131–38.

Ball, E. E., B. P. Oldfield, and K. Michel-Rudolph. 1989. Auditory organ structure, development and function. In *Cricket behavior and neurobiology*, ed. F. Huber, T. E. Moore, and W. Loher, 391–422. Ithaca: Cornell University Press.

Barber, P. H. 1999a. Patterns of gene flow and population genetic structure in the canyon treefrog, *Hyla arenicolor* (Cope). *Molecular Ecol.* 8:563–76.

———. 1999b. Phylogeography of the canyon treefrog, *Hyla arenicolor* (Cope) based on mitochondrial DNA sequence data. *Molecular Ecol.* 8:547–62.

Barton, N. H., and G. M. Hewitt. 1985. Analysis of hybrid zones. *Ann. Rev. Ecol. Syst.* 16:113–48.

Bass, A. 1992. Dimorphic male brains and alternative reproductive tactics in a vocalizing fish. *Trends Neurosci.* 15:139–45.

Bass, A. H., and R. Baker. 1997. Phenotypic specification of hindbrain rhombomeres and the origins of rhythmic circuits in vertebrates. *Brain, Behav. Evol.* 50, Suppl. 1:3–16.

Bauer, M., and Helversen, O. von 1987. Separate localization of sound recognising and sound producing neural mechanisms in a grasshopper. *J. Comp. Physiol. A* 161:95–101.

Bee, M. A., and H .C. Gerhardt. 2001a. Habituation as a mechanism of reduced aggression between neighboring territorial male bullfrogs, *Rana catesbeiana*. *J. Comp. Psych.* 115:68–82.

———. 2001b. Neighbor-stranger discrimination by territorial male bullfrogs (*Rana catesbeiana*): I. Acoustic basis. *Anim. Behav.* 62:1129–40.

———. 2001c. Neighbor-stranger discrimination by territorial male bullfrogs (*Rana catesbeiana*): II. Perceptual basis. *Anim. Behav.* 62:1141–50.

Bee, M. A., and S. A. Perrill. 1996. Responses to conspecific advertisement calls in the green frog (*Rana clamitans*) and their role in male-male competition. *Behaviour* 133:283–301.

Bee, M. A., S. A. Perrill, and P. C. Owen. 1999. Size assessment in simulated territorial encounters between male green frogs (*Rana clamitans*). *Behav. Ecol. Sociobiol.* 45:177–84.

———. 2000. Male green frogs lower the pitch of acoustic signals in defense of territories: A possible dishonest signal of size? *Behav. Ecol.* 11:169–77.

Beecher, M. D. 1989. Signalling systems for individual recognition: An information theory approach. *Anim. Behav.* 38:248–61.

Bellman, H. 1985. *A field guide to the grasshoppers and crickets of Britain and northern Europe*. London: William Collins and Sons.

Belwood, J. J. 1990. Anti-predator defences and ecology of Neotropical forest katydids, especially the Pseudophyllinae. In *The Tettigoniidae: behaviour, systematics, evolution*, ed. W. J. Bailey and D. C. F. Rentz, 6–26. Bathurst: Crawford House Press.

Belwood, J. J., and G. K. Morris. 1987. Bat predation and its influence on calling behaviour in neotropical katydids. *Science* 238:64–67.

Benedix, J. H., and D. J. Howard. 1991. Calling song displacement in a zone of overlap and hybridization. *Evolution* 45:1751–59.

Benedix, J. H., M. Pedemonte, R. Velluti, and P. N. Narins. 1994. Temperature dependence of two-tone suppression in the northern leopard frog, *Rana pipiens pipiens*. *J. Acoust. Soc. Am.* 96:2738–45.

Bennet-Clark, H. C. 1971. Acoustics of insect song. *Nature (London)* 234:255–59.

———. 1984. Insect hearing: Acoustics and transduction. In *Insect communication*, ed. T. Lewis, 49–82. London: Academic Press.

———. 1987. The tuned singing burrow of mole crickets. *J. Exp. Biol.* 128:383–409.

———. 1989. Cricket songs and the physics of sound production. In *Cricket behavior and neurobiology*, ed. F. Huber, T. E. Moore, and W. Loher, 227–61. Ithaca: Cornell University Press.

———. 1997. Tymbal mechanisms and the control of song frequency in the cicada *Cyclochila australasiae*. *J. Exp. Biol.* 200:1681–94.

———. 1998. Size and scale effects as constraints in insect sound communication. *Philos. Trans. Royal Soc. B* 353:407–19.

———. 1999a. Resonators in insect sound production: How insects produce loud pure-tone songs. *J. Exp. Biol.* 202:3347–57.

———. 1999b. Which Qs to choose: Questions of quality in bioacoustics. *Bioacoustics* 9:351–59.

Bennet-Clark, H. C. and Y. Leroy. 1978. Regularity versus irregularity in specific songs of closely-related drosophilid flies. *Nature (London)* 217:442–44.

Bennet-Clark, H. C., and D. Young. 1992. A model for the mechanism of sound production in cicadas. *J. Exp. Biol.* 173:123–53.

———. 1994. The scaling of song frequency in cicadas. *J. Exp. Biol.* 191:291–94.

———. 1998. Sound radiation by the bladder cicada *Cystosoma saundersii*. *J. Exp. Biol.* 201:701–15.

Bentley, D. R. 1969. Intracellular activity in cricket neurons during generation of song patterns. *Z. vergl. Physiol.* 62:267–83.

———. 1977. Control of cricket song patterns by descending interneurons. *J. Comp. Physiol.* 116:19–38.

Bentley, D. R., and R. R. Hoy. 1970. Postembryonic development of adult motor patterns in crickets: A neural analysis. *Science* 170:1409–11.

Beranek, L. L. 1954. *Acoustics.* New York: McGraw-Hill.

Bergeijk, W. A. van. 1962. Variation on a theme of Bekesy: A model of binaural interaction. *J. Acoust. Soc. Am.* 34:1431–37.

Bertram, S. M. 2000. The influence of age and size on temporal male signaling behaviour. *Anim. Behav.* 60:333–39.

Bibikov, N. G. 1974. Impulse activity of torus semicircularis neurons of the frog *Rana temporaria*. *J. Evol. Biochem. Physiol.* 10:36–41.

Bigelow, R. S. 1960. Interspecific hybrids and speciation in the genus *Acheta* (Orthoptera: Gryllidae). *Can. J. Zool.* 38:509–14.

Blair, W. F. 1955. Mating call and stage of speciation in the *Microhyla oliveacea–M. carolinenesis* complex. *Evolution* 9:469–80.

Blauert, J. 1983. *Spatial hearing. The psychophysics of human sound localization*. 2nd ed. Cambridge: MIT Press.

Boake, C. R. B. 1984a. Male displays and female preferences in the courtship of a gregarious cricket. *Anim. Behav.* 32:690–97.

———. 1984b. Natural history and acoustic behaviour of a gregarious cricket. *Behaviour* 89:241–50.

———. 1989. Repeatability as an indicator of the heritability of and selection on mating behavior. *Evol. Ecol.* 3:173–82.

———. 1991. Coevolution of senders and receivers of sexual signals: genetic coupling and genetic correlations. *Trends Ecol. Evol.* 6:225–27.

Bodenheimer, F. 1928. Materialien zur Geschichte der Entomologie. Band 1, ch. 3: *Das scholastische Mittelalter*, section b: Albertus Magnus, 175. Berlin: W. Junk.

Bodnar, D. A. 1996. The separate and combined effects of harmonic structure, phase, and FM on female preferences in the barking treefrog (*Hyla gratiosa*). *J. Comp. Physiol. A* 178:173–82.

Bodnar, D. A., and R. R. Capranica. 1994. Encoding of phase spectra by the peripheral auditory system of the bullfrog. *J. Comp. Physiol. A* 178:157–71.

Bond, A. B. 1989. Toward a resolution of the paradox of aggressive displays: 1. Optimal deceit in the communication of fighting ability. *Ecology* 81:29–46.

Bosch, J. A., and R. Márquez. 1996. Acoustic competition in male midwife toads *Alytes obstetricans* and *Alytes cisternasii*: Response to neighbor size and calling rate. *Ethology* 102:841–55.

———. 2000. Tympanum fluctuating asymmetry, body size and mate choice in female midwife toads (*Alytes obstetricans*). *Behaviour* 137:1211–22.

Bosch, J., A. S. Rand, and M. J. Ryan. 2000. Signal variation and call preferences in the túngara frog, *Physalaemus pustulosus. Behav. Ecol. Sociobiol.* 49:62–66.

Bourne, G. R. 1992. Lekking behavior in the neotropical frog, *Ololygon rubra. Behav. Ecol. Sociobiol.* 31:173–80.

———. 1993. Proximate costs and benefits of mate acquisition at leks of the frog *Ololygon rubra. Anim. Behav.* 45:1051–59.

Bourne, G. R., A. C. Collins, A. M. Holder, and C. L. McCarthy. 2001. Vocal communication and reproductive behavior of the frog *Colostethus beebei* in Guyana. *J. Herpetology* 35:272–81.

Boyan, G. S. 1981. Two-tone suppression of an identified auditory neurone in the brain of the cricket *Gryllus bimaculatus* (DeGeer). *J. Comp. Physiol.* 144:117–25.

———. 1998. Development of the insect auditory system. In *Comparative hearing: Insects*, ed. R. R. Hoy, A. N. Popper, and R. R. Fay, 97–138. New York: Springer-Verlag.

———. 1999. Presynaptic contributions to response shape in an auditory neuron of the grasshopper. *J. Comp. Physiol. A* 184:279–94.

Boyan, G. S., and J. H. Fullard. 1986. Interneurones responding to sound in the tobacco budworm moth *Heliothis virescens* (Noctuidae): Morphological and physiological characteristics. *J. Comp. Physiol. A* 156:391–404.

Boyd, P. J., and D. B. Lewis. 1983. Peripheral auditory directionality in the cricket (*Gryllus campestris* L., *Teleogryllus oceanicus* Le Guillon). *J. Comp. Physiol.* 153:523–32.

Boyd, P., R. Kühne, S. Silver, and D. B. Lewis. 1984. Two-tone suppression and song coding by ascending neurones in the cricket *Gryllus campestris* L. *J. Comp. Physiol. A* 154:423–30.

Bradbury, J. 1985. Contrasts between insects and vertebrates in the evolution of male display, female choice, and lek mating. In *Experimental behavioral ecology and sociobiology*, ed. B. Hölldobler and M. Lindauer, 271–89. Sunderland, Mass.: Sinauer Associates.

Bradbury, J. W., and S. L. Vehrencamp. 1998. *Principles of animal communication.* Sunderland, Mass.: Sinauer Associates.

Breckow, J., and M. Sippel. 1985. Mechanics of the transduction of sound in the tympanal organ of adults and larvae of locusts. *J. Comp. Physiol.* 157:619–29.

Brenowitz, E. A. 1989. Neighbor call amplitude influences aggressive behaviour and intermale spacing in choruses of the Pacific treefrog (*Hyla regilla*). *Ethology* 83:69–79.

Brenowitz, E. A., G. Rose, and R. R. Capranica. 1985. Neural correlates of temperature coupling in the vocal communication system of the gray treefrog (*Hyla versicolor*). *Brain Res.* 359:364–67.

Brenowitz, E. A., and G. J. Rose. 1997. Plasticity of aggressive thresholds in *Hyla regilla*: Discrete accommodation to encounter calls. *Anim. Behav.* 53:353–61.

———. 1999. Female choice and plasticity of male calling behaviour in the Pacific treefrog. *Anim. Behav.* 57:1337–42.

Brenowitz, E. A., W. Wilczynski, and H. H. Zakon. 1984. Acoustic communication in spring peepers. Environmental and behavioural aspects. *J. Comp. Physiol.* 155: 585–92.

Bridges, A. S., and M. E. Dorcas. 2000. Temporal variation in anuran calling behavior: Implications for surveys and monitoring programs. *Copeia* 2000:587–92.

Brodfuehrer, P. D., and R. R. Hoy. 1989. Integration of ultrasound and flight inputs on descending neurons in the cricket brain. *J. Exp. Biol.* 145:157–71.

Brooks, R., and D. Kemp. 2001. Can older males deliver the good genes? *Trends Ecol. Evol.* 16:308–13.

Brooks, D. R., and D. A. McLennan. 1991. *Phylogeny, ecology, and behavior: A research program in comparative biology.* Chicago: University of Chicago Press.

Broughton, W. B. 1963. Method in bio-acoustic terminology. In *Acoustic behaviour of animals*, ed. R.-G. Bushnel, 3–24. Amsterdam: Elsevier.

Brown, L. E., and J. R. Pierce 1967. Male-male interactions and chorusing intensities of the great plains toad, *Bufo cognatus. Copeia* 1967:149–54.

Brown, W. D. 1999. Mate choice in tree crickets and their kin. *Annu. Rev. Entomol.* 44:371–96.

Brown, W. D., J. Wideman, M. C. B. Andrade, A. C. Mason, and D. T. Gwynne. 1996. Female choice for an indicator of male size in the song of the black-horned cricket, *Oecanthus nigircornis* (Orthoptera: Gryllidae: Oecanthinae). *Evolution* 50:2400–11.

Brzoska, J., W. Walkowiak, and H. Schneider. 1977. Acoustic communication in the grassfrog (*Rana t. temporaria*): Calls, auditory thresholds and behavioral responses. *J. Comp. Physiol.* 118:173–86.

Bucher, T. L., M. J. Ryan, and G. A. Bartholomew. 1982. Oxygen consumption during resting, calling and nest building in the frog *Physalaemus pustulosus. Physiol. Zool.* 55:10–22.

Buck, J. 1988. Synchronous rhythmic flashing in fireflies. II. *Q. Rev. Biol.* 63:265–89.

Burk, T. 1983. Male aggression and female choice in a field cricket (*Teleogryllus oceanicus*): The importance of courtship song. In *Orthopteran mating systems: Sexual competition in a diverse group of insects*, ed. D. T. Gwynne and G. K. Morris, 97–119. Boulder: Westwood.

Burmeister, S., and W. Wilczynski. 2001. Social signals influence hormones independently of calling behavior in the treefrog (*Hyla cinerea*). *Hormones and Behavior* 38: 201–9.

Burmeister, S., J. Konieczka, and W. Wilczynski. 1999a. Agonistic encounters in a cricket frog (*Acris crepitans*) chorus: Behavioral outcomes vary with local competition and within the breeding season. *Ethology* 105:335–47.

Burmeister, S., W. Wilczynski, and M. J. Ryan. 1999b. Temporal call changes and prior experience affect graded signalling in the cricket frog. *Anim. Behav.* 57:611–18.

Bush, S. L. 1997. Vocal behavior of males and females in the Majorcan midwife toad. *J. Herpetol.* 31: 251–57.

Bush, S. L., H. C. Gerhardt, and J. Schul. 2001. Pattern recognition and call preferences in treefrogs (Anura: Hylidae): A quantitative analysis using a no-choice paradigm. *Anim. Behav.* 63:7–14.

Butler, A. B. 1995. The dorsal thalamus of jawed vertebrates: A comparative view. *Brain, Behav. Evol.* 46:209–23.

Butlin, R. K. 1987. Speciation by reinforcement. *Trends Ecol. Evol.* 2:8–13.

———. 1993. The variability of mating signals and preferences in the brown planthopper, *Nilaparva lugens* (Homoptera: Delphacidae). *J. Insect Behav.* 6:125–40.

———. 1994. Genetic variation in mating signals and responses. In *Speciation and the recognition concept: Theory and application*, ed. D. M. Lambert and H. G. Spencer, 327–66. Baltimore: Johns Hopkins University Press.

Butlin, R. K., and G. M. Hewitt. 1988. Genetics of behavioural and morphological differences between parapatric subspecies of *Chlorthippus parallelus*. (Orthoptera: Acrididae). *Biol. J. Linn. Soc.* 33:233–48.

Butlin, R. K., and M. G. Ritchie. 1989. Genetic coupling in mate recognition systems: What is the evidence? *Biol. J. Linn. Soc.* 37:237–46.

Butlin, R. K., and T. Tregenza. 1998. Levels of genetic polymorphism: Marker loci versus quantitative traits. *Phil. Trans. R. Soc. London B* 353:187–98.

Butlin, R. K., G. M. Hewitt, and S. F. Webb. 1985. Sexual selection for an intermediate optimum in *Chorthippus brunneus* (Orthoptera: Acrididae). *Anim. Behav.* 33: 1281–92.

Byrne, P. G., and J. D. Roberts. 1999. Simultaneous mating with multiple males reduces fertilization success in the myobatrachid frog *Crinia georgiana*. *Proc. R. Soc. Lond. B* 266:717–21.

Cade, W. H. 1975. Acoustically orienting parasitoids: Fly phonotaxis to cricket song. *Science* 190:1312–13.

———. 1979. The evolution of alternative male reproductive strategies in field crickets. In *Sexual selection and reproductive competition in insects*, ed. M. S. Blum and N. A. Blum, 343–78. New York: Academic Press.

———. 1981a. Alternative male strategies: Genetic differences in cricket. *Science* 212: 563–64.

———. 1981b. Field cricket spacing, and the phonotaxis of crickets and parasitoid flies to clumped and isolated cricket songs. *Z. Tierpsychol.* 55:365–75.

Cade, W. H., and E. S. Cade. 1992. Male mating success, calling and searching behaviour at high and low densities in the field cricket, *Gryllus integer*. *Anim. Behav.* 43: 49–56.

Cade, W. H., and D. Otte. 1982. Alternation calling and spacing patterns in the field cricket *Acanthogryllus fortipes* (Orthoptera; Gryllidae). *Can. J. Zool.* 60:2916–20.

Cade, W. H., and D. R. Wyatt. 1984. Factors affecting calling behaviour in field crickets, *Teleogryllus* and *Gryllus* (age, weight, density and parasites). *Behaviour* 88:61–75.

Canatella, D. C., D. M. Hillis, P. T. Chippindale, L. Weigt, A. S. Rand, and M. J. Ryan. 1998. Phylogeny of frogs of the *Physalaemus pustulosus* species group, with an examination of data incongruence. *Syst. Biol.* 47:311–35.

Capranica, R. R. 1965. *The evoked vocal response of the bullfrog: A study of communication by sound*. Cambridge: MIT Press.

———. 1992. Untuning of the tuning curve: Is it time? *Sem. Neurosci.* 4:401–8.

Capranica, R. R., and A. J. M. Moffat. 1983. Neurobehavioral correlates of sound communication in anurans. In *Advances in vertebrate neuroethology*, ed. J. P. Ewert, R. R. Capranica, and D. J. Ingle, 701–30. New York: Plenum Press.

Capranica, R. R., G. J. Rose, and E. A. Brenowitz. 1985. Time resolution in the auditory system of anurans. In *Time resolution in auditory systems*, ed. A. Michelsen, 58–73. Heidelberg: Springer-Verlag.

Casseday, J. H., and E. Covey. 1996. A neuroethological theory of the operation of the inferior colliculus. *Brain, Behav. Evol.* 47:311–36.

Castellano, S., and C. Giacoma. 1998. Stabilizing and directional female choice for male calls in the European green toad. *Anim. Behav.* 56:275–87.

Castellano, S., A. Rosso, F. Laoretti, S. Doglio, and C. Giacoma. 2000. Call intensity and female preferences in the European green toad. *Ethology* 106:1129–41.

Charalambous, M., R. K. Butlin, and G. M. Hewitt. 1994. Genetic variation in song and female song preference in the grasshopper *Chorthippus brunneus* (Orthoptera: Acrididae). *Anim. Behav.* 47:399–411.

Cherry, E. C. 1953. Some experiments on the recognition of speech, with one and with two ears. *J. Acoust. Soc. Am.* 25:975–79.

Cherry, M. I. 1993. Sexual selection in the raucous toad, *Bufo rangeri*. *Anim. Behav.* 45: 359–73.

Choe, J. C., B. J. Crespi, eds. 1997. *The evolution of mating systems in insects and arachnids*. Cambridge: Cambridge University Press.

Christensen-Dalsgaard, J., and M. B. Jørgensen. 1996a. One-tone suppression in the frog auditory nerve. *J. Acoust. Soc. Am.* 100:451–57.

———. 1996b. Sound and vibration sensitivity of VIIIth nerve fibers in the grassfrog, *Rana temporaria*. *J. Comp. Physiol. A* 179:437–45.

Chung, S. G., A. Pettigrew, and M. Anson. 1978. Dynamics of the amphibian middle ear. *Nature (London)* 272:142–47.

Clarey, J. C., P. Barone, and T. J. Imig. 1992. Physiology of thalamus and cortex. In *The mammalian auditory pathway: Neurophysiology*, ed. A. N. Popper and R. R. Fay, 232–334. New York: Springer-Verlag.

Clutton-Brock, T. H., and S. D. Albon. 1979. The roaring of red deer and the evolution of honest advertisement. *Behaviour* 69:145–70.

Cocroft, R. B. 1994. A cladistic analysis of chorus frog phylogeny (Hylidae: *Pseudacris*). *Herpetologica* 50:420–37.

Cocroft, R. B., and M. J. Ryan. 1995. Patterns of mating call evolution in toads and chorus frogs. *Anim. Behav.* 49:283–303.

Condon, C. J., S.-H. Chang, and A. S. Feng. 1991. Processing of behaviorally relevant temporal parameters of acoustic stimuli by single neurons in the superior olivary nucleus of the leopard frog. *J. Comp. Physiol. A* 168:709–25.

Coyne, J. A., and H. A. Orr. 1989. Patterns of speciation in *Drosophila*. *Evolution* 43: 362–81.

———. 1997. 'Patterns of speciation in *Drosophila*' revisited. *Evolution* 51:295–303.

Crawford, J. D. 1997. Feature-detecting auditory neurons in the brain of a sound-producing fish. *J. Comp. Physiol. A* 180:439–50.

Cunningham, C. W., K. E. Omland, and T. H. Oakley. 1998. Reconstructing ancestral states: A critical appraisal. *Trends Ecol. Evol.* 13:361–66.

Dadour, I. R. 1989. Temporal pattern changes in the calling song of the katydid *Mygalopsis marki* Bailey in response to conspecific song (Orthoptera: Tettigoniidae). *J. Insect Behav.* 2:199–215.

Dadour, I. R., and W. J. Bailey. 1985. Male agonistic behaviour of the bushcricket *Mygalopsis marki* Bailey in a response to conspecific song (Orthoptera: Tettigoniidae). *Z. Tierpsych.* 70:320–30.

Dadour, I. R., and M. S. Johnson. 1983. Genetic differentiation, hybridization, and reproductive isolation in *Mygalopsis marki* Bailey (Orthoptera: Tettigoniidae). *Aust. J. Zool.* 31:353–60.

Dambach, M. 1989. Vibrational responses. In *Cricket behavior and neurobiology*, ed. F. Huber, T. E. Moore, and W. Loher, 178–97. Ithaca: Cornell University Press.

Dambach, M., and A. Gras. 1995. Bioacoustics of a miniature cricket, *Cycloptiloides canariensis* (Orthoptera: Gryllidae: Mogoplistinae). *J. Exp. Biol.* 198:721–28.

Dambach, M., H.-G. Rausche, and G. Wendler. 1983. Proprioceptive feedback influences the calling song of the field cricket. *Naturwissenschaften* 70:417.

Davis, H. 1965. A model for transducer action in the cochlea. *Cold Spring Harb. Symp. Quant. Biol.* 30:181–90.

Davis, M. S. 1987. Acoustically mediated neighbor recognition in the North American bullfrog, *Rana catesbeiana*. *Behav. Ecol. Sociobiol.* 21:185–90.

Dawkins, M. S., and T. Guilford. 1996. Sensory bias and the adaptiveness of female choice. *Am. Natur.* 148:937–42.

Daws, A. G. 1996. Resonance and frequency selectivity in insect sound communication. Ph.D. diss., University of Melbourne, Australia.

Daws, A. G., H. C. Bennet-Clark, and N. H. Fletcher. 1996. The mechanism of tuning of the mole cricket singing burrow. *Bioacoustics* 7:81–117.

Daws, A. G., and R. M. Hennig. 1996. Tuning of the peripheral auditory system of the cicada, *Cyclochila australasiae*. *Zoology-Analysis of Complex Systems* 99:175–88.

Denton J. S., and T. J. C. Beebee. 1993. Reproductive strategies in a female-biased population of natterjack toads *Bufo calamita*. *Anim. Behav.* 46:1169–75.

De Querioz, A., and P. H. Wimberger. 1993. The usefulness of behavior for phylogeny estimation: Levels of homoplasy in behavioral and morphological characters. *Evolution* 47:46–60.

DeSutter-Grandcolas, L. 1998. Broad-frequency modulation in cricket (Orthoptera, Grylloidea) calling songs: Two convergent cases and a functional hypothesis. *Can. J. Zool.* 76:2148–63.

DeWinter, A. J. 1992. The genetic basis and evolution of acoustic mate recognition signals in a *Ribautodelphax* planthopper (Homoptera, Delphacidae) 1. The female call. *J. Evol. Biol.* 5:249–65.

Diekamp, B. M., and H. C. Gerhardt. 1992. Midbrain auditory sensitivity in the spring peeper (*Pseudacris crucifer*): Correlations with behavioral studies. *J. Comp. Physiol. A* 171:245–50.

———. 1995. Selective phonotaxis to advertisement calls in the gray treefrog *Hyla versicolor:* Behavioral experiments and neurophysiological correlates. *J. Comp. Physiol. A* 177:173–90.

Dobler, S., K.-G. Heller, and O. von Helversen. 1994a. Song pattern recognition and an auditory time window in the female bushcricket *Ancistrura nigrovittata* (Orthoptera: Phaneropteridae). *J. Comp. Physiol. A* 175:67–74.

Dobler, S., A. Stumpner, and K.-G. Heller. 1994b. Sex-specific spectral tuning for the partner's song in the duetting bushcricket *Ancistrura nigrovittata* (Orthoptera: Phaneropteridae). *J. Comp. Physiol. A* 175:303–10.

Dobzhansky, T. 1940. Speciation as a stage in evolutionary divergence. *Am. Natur.* 74:312–21.

Doherty, J. A. 1985a. Phonotaxis in the cricket, *Gryllus bimaculatus* de Geer: Comparisons of choice and no-choice paradigms. *J. Comp. Physiol. A* 157:279–89.

———. 1985b. Temperature coupling and "trade-off" phenomena in the acoustic communication system of the cricket, *Gryllus bimaculatus* De Geer (Gryllidae). *J. Exp. Biol.* 114:17–35.

———. 1985c. Trade-off phenomena in calling song recognition and phonotaxis in the cricket, *Gryllus bimaculatus* (Orthoptera, Gryllidae). *J. Comp. Physiol. A* 156: 787–801.

———. 1991. Song recognition and localization in the phonotaxis behavior of the field cricket, *Gryllus bimaculatus* (Orthoptera: Gryllidae). *J. Comp. Physiol. A* 168:213–22.

Doherty, J. A., and J. D. Callos. 1991. Acoustic communication in the trilling field cricket, *Gryllus rubens* (Orthoptera: Gryllidae). *J. Insect Behav.* 4:67–82.

Doherty, J. A., and H. C. Gerhardt. 1984. Evolutionary and neurobiological implications of selective phonotaxis in the spring peeper (*Hyla crucifer*). *Anim. Behav.* 32: 875–81.

Doherty, J. A., and D. J. Howard. 1996. Lack of preference for conspecific calling songs in female crickets. *Anim. Behav.* 51:981–89.

Doherty, J. A., and R. R. Hoy. 1985. Communication in insects III. The auditory behavior of crickets: Some views of genetic coupling, song recognition, and predator detection. *Quart. Rev. Biol.* 60:457–72.

Doherty, J. A., and M. M. Storz. 1992. Calling song and selective phonotaxis in the field crickets, *Gryllus firmus* and *G. pennsylvanicus* (Orthoptera: Gryllidae). *J. Insect Behav.* 5:555–69.

Doolan, J. M. 1981. Male spacing and the influence of female courtship behaviour in the bladder cicada, *Cystosoma saundersii* Westwood. *Behav. Ecol. Sociobiol.* 9:269–76.

Doolan, J. M., and R. C. MacNally. 1981. Spatial dynamics and breeding ecology in the cicada, *Cystosoma saundersii:* The interaction between distributions of resources and intraspecific behaviour. *J. Anim. Ecol.* 50:925–40.

Doolan, J. M., and G. S. Pollack. 1985. Phonotactic specificity of the cricket *Teleogryllus oceanicus:* Intensity-dependent selectivity for temporal properties of the stimulus. *J. Comp. Physiol. A* 157:223–33.

Doolan, J. M., and D. Young. 1989. Relative importance of song parameters during flight phonotaxis and courtship in the bladder cicada *Cystosoma saudersii*. *J. Exp. Biol.* 141:113–31.

D'Orgeix, C. A., and B. J. Turner. 1995. Multiple paternity in the red-eye treefrog *Agalychnis callidryas* (Cope). *Molecular Ecology* 4:505–8.

Doty, G. V., and A. M. Welch. 2001. Advertisement call duration indicates good genes for offspring feeding rate in gray tree frogs (*Hyla versicolor*). *Behav. Ecol. Sociobiol.* 49:150–56.

Doupe, A. J. 1997. Song- and order-selective neurons in the songbird anterior forebrain and their emergence during vocal development. *J. Neuroscience* 17:1147–67.

Drewry, G. E., and A. S. Rand. 1983. Characteristics of an acoustic community: Puerto Rican frogs of the genus *Eleutherodactylus*. *Copeia* 1983:941–53.

Driscoll, D. A. 1998. Genetic structure of the frogs *Geocrinia lutea* and *Geocrinia rosea* reflects extreme population divergence and range changes, not dispersal barriers. *Evolution* 52:1147–57.

Dudley, R., and A. S. Rand. 1991. Sound production and vocal sac inflation in the túngara frog, *Physalaemus pustulosus* (Leptodactylidae). *Copeia* 1991:460–70.

Duellman, W. E., and R. A. Pyles. 1983. Acoustic resource partitioning in anuran communities. *Copeia* 1983:639–49.

Duellman, W. E., and L. Trueb. 1986. *Biology of amphibians*. New York: McGraw-Hill.

Dumortier, B. 1963a. Ethological and physiological study of sound emissions in Arthropoda. In *Acoustic behaviour of animals*, ed. R.-G. Bushnel, 583–654. Amsterdam: Elsevier.

———. 1963b. Morphology of sound emission in arthropoda. In *Acoustic behaviour of animals*, ed. R.-G. Bushnel, 377–45. Amsterdam: Elsevier.

———. 1963c. The physical characteristics of sound emissions in Arthropoda. In *Acoustic behaviour of animals*, ed. R.-G. Bushnel, 346–73. Amsterdam: Elsevier.

Dunia, R., and P. M. Narins. 1989a. Temporal integration in an anuran auditory nerve. *Hearing Research* 39:287–98.

———. 1989b. Temporal resolution in frog auditory-nerve fibers. *J. Acoust. Soc. Am.* 85:1630–38.

Dyson, M. L., S. L. Bush, and T. R. Halliday. 1997. Phonotaxis by females in the Majorcan midwife toad. *Behaviour* 135:213–30.

Dyson, M. L., S. P. Henzi, and N. I. Passmore. 1994. The effect of changes in the relative timing of signals during female phonotaxis in the reed frog, *Hyperolius marmoratus*. *Anim. Behav.* 48:679–85.

Dyson, M. L., and N. I. Passmore. 1988a. The combined effect of intensity and the temporal relationships of stimuli on phonotaxis in female painted reed frogs *Hyperolius marmoratus*. *Anim. Behav.* 36:1555–56.

———. 1988b. Two-choice phonotaxis in *Hyperolius marmoratus* (Anura: Hyperoliidae): The effect of temporal variation in presented stimuli. *Anim. Behav.* 36:648–52.

———. 1992. Effect of intermale spacing on female frequency preferences in the painted reed frog. *Copeia* 1992:1111–14.

Dyson, M. L., N. I. Passmore, B. J. Bishop, and S. P. Henzi. 1992. Male behavior and correlates of mating success in a natural population of African painted reed frogs (*Hyperolius marmoratus*). *Herpetologica* 48:232–42.

Eberhard, W. G. 1996. *Female control: Sexual selection by cryptic female choice*. Princeton: Princeton University Press.

Eggermont, J. J. 1988. Mechanisms of sound localization in anurans. In *The evolution of the amphibian auditory system*, ed. B. Fritzsch, M. J. Ryan, W. Wilczynski, T. E. Hetherington, and W. Walkowiak, 307–36. New York: John Wiley and Sons.

———. 1990. Temporal modulation transfer functions for single neurons in the dorsal medullary nucleus and torus semicircularis of the grass frog. *Hearing Research* 45:1–14.

Ehret, G., E. Keilwerth, and T. Kamada. 1994. The lung-eardrum pathway in three treefrog and four dendrobatid frog species: Some properties of sound transmission. *J. Exp. Biol.* 195:329–43.

Eíriksson, T. 1993. Female preference for specific pulse duration of male songs in the grasshopper, *Omocestus viridulus*. *Anim. Behav.* 45:471–77.

———. 1994. Song duration and female response behaviour in the grasshopper *Omocestus viridulus*. *Anim. Behav.* 47:707–12.

Elephandt, A., I. Eistettler, A. Fleig, E. Günther, M. Hainich, S. Hepperle, and B. Traub. 2000. Hearing threshold and frequency discrimination in the purely aquatic frog *Xenopus laevis* (Pipidae): Measurement by means of conditioning. *J. Exp. Biol.* 203:3621–29.

Elliott, C .J. H., and U. T. Koch. 1983. Sensory feedback stabilising reliable stridulation in the field cricket *Gryllus campestris L. Anim. Behav.* 31:887–901.

———. 1985. The clockwork cricket. *Naturwissenschaften* 72:150–53.

Elliott, C. J. H., U. T. Koch, K.-H. Schäffner, and F. Huber. 1982. Wing movements during cricket stridulation are affected by mechanosensory input from wing hair plates. *Naturwissenschaften* 69:288–89.

Elsner, N. 1968. Die neuromuskulären Grundlagen des Werbegesangs der roten Keulenheuschrecke *Gomphocerippus rufus L. Z. vergl. Physiol.* 60:308–50.

———. 1974a. Neural economy: Bifunctional muscles and common central pattern elements in leg and wing stridulation of the grasshopper *Stenobothrus rubicundus Germ.* (Orthoptera: Acrididae). *J. Comp. Physiol.* 89:227–36.

———. 1974b. Neuroethology of sound production in gomphocerine grasshoppers. I. Song patterns and stridulatory movements. *J. Comp. Physiol.* 88:72–102.

———. 1975. Neuroethology of sound production in Gomphocerine grasshoppers (Orthoptera: Acrididae). II. Neuromuscular Activity underlying stridulation. *J. Comp. Physiol.* 97:291–322.

———. 1981. Developmental aspects of insect neuroethology. In *Behavioral development*, ed. K. Immelmann, B. G. Barlow, L. Petrinovich, and M. Main, 474–90. Cambridge: Cambridge University Press.

———. 1983a. Insect stridulation and its neurophysiological basis. In *Bioacoustics: A comparative approach*, ed. B. Lewis, 69–92. London: Academic Press.

———. 1983b. A neuroethological approach to the phylogeny of leg stridulation in gomphocerine grasshoppers. In *Neuroethology and behavioral physiology*, ed. F. Huber and H. Markl, 54–68. Heidelberg: Springer-Verlag.

———. 1994. The search for the neural centers of cricket and grasshopper song. In *Neural basis of behavioural adaptations. Fortschr. Zool.* 39:167–93.

Elsner, N., and A. V. Popov. 1978. Neuroethology of acoustic communication. *Adv. Insect Physiol.* 13:229–335.

Elsner, N., and G. Wasser. 1995. The transition from leg to wing stridulation in two geographically distinct populations of the grasshopper *Stenobothrus rubicundus. Naturwissenschaften* 82:384–86.

Emerson, S. B., and S. K. Boyd. 1999. Mating vocalizations of female frogs: Control and evolutionary mechanisms. *Brain, Behav. Evol.* 53:187–97.

Emlen, S. T. 1968. Territoriality in the bullfrog, *Rana catesbeiana. Copeia* 1968:240–43.

Emlen, S. T., and L. W. Oring. 1977. Ecology, sexual selection, and the evolution of mating systems. *Science* 197:185–93.

Endepols, H., and W. Walkowiak. 1999. Influence of descending forebrain projections on processing of acoustic signals and audiomotor integration in the anuran midbrain. *European J. Morphol.* 37:182–84.

———. 2001. Integration of ascending and descending inputs in the auditory midbrain of anurans. *J. Comp. Physiol. A* 186:1119–33.

Endepols, H., W. Walkowiak, and H. Luksch. 2000. Chemoarchitecture of the anuran auditory midbrain. *Brain Res. Revs.* 33:179–98.

Endler, J. A. 1992. Signals, signal conditions, and the direction of evolution. *Am. Natur.* 139:125–53.

———. 1999. Evolutionary implications of the interactions between animal signals and the environment. In *Animal signals*, ed. Y. Espmark, T. Amundsen, and T. Rosenqvist. Trondheim, Norway: Tapir Publishers.

Endler, J. A., and A. L. Basolo. 1998. Sensory ecology, receiver biases and sexual selection. *Trends Ecol. Evol.* 13:415–20.

Enquist, M., and A. Arak. 1993. Selection of exaggerated male traits by female aesthetic senses. *Nature (London)* 361:446–48.

Enquist, M., and O. Leimar. 1983. Evolution of fighting behaviour: Decision rules and assessment of relative strength. *J. Theoret. Biol.* 102:387–410.

Epping, W. J. M., and J. J. Eggermont. 1986. Sensitivity of neurons in the auditory midbrain of the grassfrog to temporal characteristics of sound. II. Stimulation with amplitude-modulated sound. *Hearing Research* 24:55–72.

Esch, H., F. Huber, and D. W. Wohlers. 1980. Primary auditory neurons in crickets: Physiology and central projections. *J. Comp. Physiol.* 137:27–38.

Evans, A. R. 1983. A study of the behaviour of the Australian field cricket *Teleogryllus commodus* (Walker)(Orthoptera: Gryllidae) in the field and in habitat simulations. *Z. Tierpsychol.* 62:269–90.

Ewing, A. W. 1989. *Arthropod bioacoustics: Neurobiology and behavior.* Ithaca: Comstock/Cornell.

Falls, J. B. 1982. Individual recognition by sounds in birds. In *Acoustic communication in birds*, ed. D. E. Kroodsma and E. H. Miller, 2:237–78. New York: Academic Press.

Farris, H. E., T. G. Forrest, and R. R. Hoy. 1997. The effects of calling song spacing and intensity on the attraction of flying crickets (Orthoptera: Gorylliidae: Nemobiinae). *J. Insect Behav.* 10:639–53.

Faulkes, Z., and G. S. Pollack. 2000. Effects of inhibitory timing on contrast enhancement in auditory circuits in crickets (*Teleogryllus oceanicus*). *J. Neurophysiology* 84:1247–55.

Faure, P. A., and R. R. Hoy. 2000a. Auditory symmetry analysis. *J. Exp. Biol.* 203:3209–23.

———. 2000b. Neuroethology of the katydid T-cell. II. Responses to acoustic playback of conspecific and predatory signals. *J. Exp. Biol.* 203:3243–54.

Feaver, M. N. 1983. Pair formation in the katydid *Orchelimum nigripes* (Orthoptera: Tettigonidae). In *Orthopterean mating systems: Sexual selection in a diverse group of insects*, ed. D. T. Gwynne and G. K. Morris, 205–39. Boulder: Westview Press.

Fellers, G. M. 1979. Aggression, territoriality and mating behavior in North American treefrogs. *Anim. Behav.* 27:107–19.

Feng. A. S. 1980. Directional characteristics of the acoustic receiver of the leopard frog (*Rana pipiens*): A study of eighth nerve auditory responses. *J. Acoust. Soc. Am.* 68:1107–14.

———. 1982. Quantitative analysis of intensity-rate and intensity-latency functions in peripheral auditory nerve fibers of northern leopard frogs (*Rana p. pipiens*) *Hearing Research* 6:242–46.

———. 1986a. Afferent and efferent innervation patterns of the cochlear nucleus (dorsal medullary nucleus) of the leopard frog. *Brain Res.* 367:183–91.

———. 1986b. Afferent and efferent innervation patterns of the superior olivary nucleus of the leopard frog. *Brain Res.* 364:167–71.

Feng, A. S., and R. R. Capranica. 1976. Sound localization in anurans: I. Evidence of binaural interaction in the dorsal medullary nucleus of the bullfrog (*Rana catesbeiana*). *J. Neurophysiol.* 39:871–81.

———. 1978. Sound localization in anurans. II. Binaural interaction in superior olivary nucleus of the green tree frog (*Hyla cinerea*). *J. Neurophysiol.* 41:43–54.

Feng, A. S., H. C. Gerhardt, and R. R. Capranica. 1976. Sound localization behavior of the green treefrog (*Hyla cinerea*) and the barking treefrog (*Hyla gratiosa*). *J. Comp. Physiol. A* 107:241–52.

Feng, A. S., J. C. Hall, and D. M. Gooler. 1990. Neural basis of sound pattern recognition in anurans. *Prog. Neurobiol.* 34:313–29.

Feng, A. S., J. C. Hall, and S. Siddique. 1991. Coding of temporal parameters of complex sounds by frog auditory nerve fibers. *J. Neurophysiol.* 65:424–45.

Feng, A. S., and W. Lin. 1991. Differential innervation patterns of three divisions of frog auditory midbrain (torus semicircularis). *J. Comp. Neurol.* 306:613–30.

———. 1996. The neuronal architecture of the dorsal nucleus (cochlear nucleus) of the frog *Rana pipiens pipiens*. *J. Comp. Neurol.* 366:320–34.

Feng, A. S., W. Y. Lin, and L. Sun. 1994. Detection of gaps in sinusoids by frog auditory nerve fibers: Importance in AM coding. *J. Comp. Physiol.* A 175:531–46.

Feng, A. S., and N. A. M. Schellart. 1999. Central auditory processing in fish and amphibians. In *Comparative hearing: Fish and amphibians*, ed. R. R. Fay and A. N. Popper, 218–68. New York: Springer-Verlag.

Field, L. H., and T. Matheson. 1998. Chordotonal organs in insects. *Adv. Insect Physiol.* 27:1–228.

Fischer, F. P., H. Schubert, S. Fenn, and U. Schulz. 1996. Diurnal song activity of Grassland orthoptera. *Acta Oecologica-Int. J. Ecol.* 17:345–64.

Fletcher, N. H. 1992. *Acoustic systems in biology*. Oxford: Oxford University Press.

Fletcher, N. H., and S. Thwaites. 1979a. Acoustic analysis of the auditory system of the cricket *Teleogryllus commodus* (Walker). *J. Acoust. Soc. Am.* 66:350–57.

———. 1979b. Physical models for the analysis of acoustic systems in biology. *Quart. Rev. Biophysics* 12:25–65.

Fonseca, P. J. 1994. *Acoustic communication in cicadas (Homoptera, Cicadidae): Sound production and sound reception.* Ph.D. diss., University of Lisbon, Portugal.

———. 1996. Sound production in cicadas: Timbal muscle activity during calling song and protest song. *Bioacoustics* 7:13–31.

Fonseca, P. J., and R. M. Hennig. 1996. Phasic action of the tensor muscle modulates the calling song in cicadas. *J. Exp. Biol.* 199:1535–44.

Fonseca, P. J., R. M. Hennig, and D. Münch. 2000. How cicadas interpret acoustic signals. *Nature (London)* 405:297–98.

Fonseca, P. J., and A. V. Popov. 1994. Sound radiation in a cicada: The role of different structures. *J. Comp. Physiol.* 175:349–61.

Forester, D. C., and R. Czarnowsky. 1985. Sexual selection in the spring peeper, *Hyla crucifer* (Amphibia, Anura): Role of the advertisement call. *Behaviour* 92:112–28.

Forester, D. C., and D. V. Lykens. 1986. Significance of satellite males in a population of spring peepers (*Hyla crucifer*). *Copeia* 1986:719–24.

Forester, D. C., D. V. Lykens, and W. K. Harrison. 1989. The significance of persistent vocalisation by the spring peeper, *Pseudacris crucifer* (Anura: Hylidae). *Behaviour* 133:197–208.

Forester, D. C., and K. J. Thompson. 1998. Gauntlet behaviour as a male sexual tactic in the American toad (Amphibia: Bufonidae). *Behaviour* 135:99–119.

Forrest, T. G. 1983. Phonotaxis in mole crickets: Its reproductive significance. *Florida Entomol.* 63:45–53.

———. 1991. Power output and efficiency of sound production by crickets. *Behav. Ecol.* 2:327–38.

———. 1994. From sender to receiver: Propagation and environmental effects of acoustic signals. *Amer. Zool.* 34:644–54.

Forrest, T. G., and R. Raspet. 1994. Models of female choice in acoustic communication. *Behav. Ecol.* 5:293–303.

French. B. W., and W. Cade. 1989. Sexual selection at varying population densities in male field crickets, *Gryllus veletis* and *Gryllus pennsylvanicus*. *J. Insect Behav.* 2: 105–22.

Friedl, T. 1992. *Populationsbiologie, Rufverhalten und Fortpflanungsverhalten beim Europäischen Laubfrosch (Hyla arborea).* Diplom Thesis, Technical University, Munich.

Fries, G., and N. Elsner. 1996. Transection of intraganglionic connections causes synchrony of hindleg stridulation in the gomphocerine grasshopper *Stenobothrus lineatus*. *Naturwisssenschaften* 83:284–87.

Frishkopf, L. S., R. R. Capranica, and M. H. Goldstein Jr. 1968. Neural coding in the bullfrog's auditory system—a teleological approach. *Proc. IEEE* 56:968–79.

Fullard, J. H. 1998. The sensory coevolution of moths and bats. In *Comparative hearing: Insects*, ed. R. R. Hoy, A. N. Popper, and R. R. Fay, 279–326. New York: Springer-Verlag.

Fullard, J. H., and J. E. Yack. 1993. The evolutionary biology of insect hearing. *Trends Ecol. Evol.* 8:248–52.

Fuzessery, Z. M. 1988. Frequency tuning in the anuran central auditory system. In *The evolution of the amphibian auditory system*, ed. B. Fritzsch, M. J. Ryan, W. Wilczynski, T. E. Hetherington, and W. Walkowiak, 253–73. New York: John Wiley and Sons.

Fuzessery, Z. M. and A. S. Feng. 1981. Frequency representation in the dorsal medullary nucleus of the leopard frog, *Rana p. pipiens*. *J. Comp. Physiol. A* 143:339–47.

———. 1983a. Frequency selectivity in the anuran auditory medulla: Excitatory and inhibitory tuning properties of single neurons in the dorsal medullary and superior olivary nuclei. *J. Comp. Physiol. A* 150:107–19.

———. 1983b. Mating call selectivity in the thalamus and midbrain of the leopard frog (*Rana p. pipiens*): Single and multiunit analyses. *J. Comp. Physiol. A* 150:333–44.

Gabor, C. R., and M. J. Ryan. 2001. Geographical variation in reproductive character displacement in mate choice by male sailfin mollies. *Proc. Roy. Soc. Lond. B* 268: 1063–70.

Galliart, P. L., and K. C. Shaw. 1991. Effect of intermale distance and female presence on the nature of chorusing by males of the katydid *Amblycorpha parvipennis* (Orthoptera: Tettigoniidae) males. *Florida Entomol.* 74:559–68.

———. 1996. The effect of variation in parameters of the male calling song of the katydid, *Amblycorypha parvipennis* (Orthoptera; Tettigoniidae), on female phonotaxis and phonoresponse. *J. Insect Behav.* 9:841–55.

Gambs, R. D., and M. J. Littlejohn. 1979. Acoustic behavior of males of the Rio Grande leopard frog (*Rana berlandieri*): An experimental analysis through field playback trials. *Copeia* 1979:643–50.

Gans, C. 1973. Sound production in the Salientia: Mechanisms and evolution of the emitter. *Amer. Zool.* 13: 1179–94.

Garcia-Rutledge, E. J., and P. M. Narins. 2001. Shared acoustic resources in an old world frog community. *Herpetologica* 57:104–16.

Garton, J. S., and R. A. Brandon. 1975. Reproductive ecology of the green treefrog, *Hyla cinerea*, in southern Illinois (Anura: Hylidae). *Herpetologica* 31:150–61.

Gartside, D. F. 1980. Analysis of a hybrid zone between chorus frogs of the *Pseudacris nigrita* complex in the southern United States. *Copeia* 1980:56–66.

Gatz, A. J. 1981a. Non-random mating by size in American toads, *Bufo americanus*. *Anim. Behav.* 29:1004–12.

———. 1981b. Size selective mating in *Hyla versicolor* and *Hyla crucifer*. *Herpetologica* 15:114–16.

Gayou, D. C. 1984. Effects of temperature on the mating call of *Hyla versicolor*. *Copeia* 1984:733–38.

Gerhardt, H. C. 1974. The significance of some spectral features in mating call recognition in the green treefrog (*Hyla cinerea*). *J. Exp. Biol.* 61:229–41.

———. 1975. Sound pressure levels and radiation patterns of the vocalizations of some North American frogs and toads. *J. Comp. Physiol. A* 102:1–12.

———. 1976. Significance of two frequency bands in long distance vocal communication in the green treefrog. *Nature (London)* 261:692–94.

———. 1978a. Discrimination of intermediate sounds in a synthetic call continuum by female green tree frogs. *Science* 199:1089–91.

———. 1978b. Mating call recognition in the green treefrog (*Hyla cinerea*): The significance of some fine-temporal properties. *J. Exp. Biol.* 74:59–73.

———. 1978c. Temperature coupling in the vocal communication system of the gray treefrog *Hyla versicolor*. *Science* 199:992–94.

———. 1981a. Mating call recognition in the barking treefrog (*Hyla gratiosa*): Responses to synthetic calls and comparisons with the green treefrog (*Hyla cinerea*). *J. Comp. Physiol.* 144:17–25.

———. 1981b. Mating call recognition in the green treefrog (*Hyla cinerea*): Importance of two frequency bands as a function of sound pressure level. *J. Comp. Physiol.* 144:9–16.

———. 1982. Sound pattern recognition in some North American treefrogs (Anura: Hylidae): Implications for mate choice. *Am. Zool.* 22:581–95.

———. 1983. Acoustic communication in treefrogs. *Verh. Dtsch. Zool. Ges.* 1983: 25–35.

———. 1987. Evolutionary and neurobiological implications of selective phonotaxis in the green treefrog (*Hyla cinerea*). *Anim. Behav.* 35:1479–89.

———. 1988. Acoustic properties used in call recognition by frogs and toads. In *The evolution of the amphibian auditory system*, ed. B. Fritzsch, M. J. Ryan, W. Wilczynski, T. E. Hetherington, and W. Walkowiak, 455–83. New York: John Wiley and Sons.

———. 1991. Female mate choice in treefrogs: Static and dynamic acoustic criteria. *Anim. Behav.* 42:615–35.

———. 1994a. The evolution of vocalization in frogs and toads. *Annu. Rev. Ecol. Syst.* 25:293–324.

———. 1994b. Reproductive character displacement of female mate choice in the grey treefrog *Hyla chrysoscelis*. *Anim. Behav.* 47:959–69.

———. 1998. Acoustic signals of animals: Field measurements, recording, analysis and description. In *Techniques in the study of animal communication by sound*, ed. S. Hopp, M. Owren, and C. Evans, 1–25. New York: Springer-Verlag.

———. 2001. Acoustic communication in two groups of closely related treefrogs. In *Advances in behavioral biology*, ed. P. J. B. Slater, J. S. Rosenblatt, C. T. Snowdon, and T. J. Roper, 99–167. New York: Academic Press.

Gerhardt, H. C., S. Allan, and J. J. Schwartz . 1990. Female green treefrogs (*Hyla cinerea*) do not selectively respond to signals with a harmonic structure in noise. *J. Comp. Physiol. A* 166:791–94.

Gerhardt, H. C., R. E. Daniel, S. A. Perrill, and S. Schramm. 1987. Mating behaviour and male mating success in the green treefrog. *Anim.Behav.* 35:1490–503.

Gerhardt, H. C., and M. D. Davis. 1988. Variation in the coding of species identity in the advertisement calls of *Litoria verreauxi* (Anura: Hylidae). *Evolution* 42:556–65.

Gerhardt, H. C., B. Diekamp, and M. Ptacek. 1989. Inter-male spacing in choruses of the spring peeper, *Pseudacris (Hyla) crucifer*. *Anim. Behav.* 38:1012–24.

Gerhardt, H. C., and J. A. Doherty. 1988. Acoustic communication in the gray treefrog, *Hyla versicolor:* Evolutionary and neurobiological implications. *J. Comp. Physiol. A* 162:261–78.

Gerhardt, H. C., M. L. Dyson, and S. D. Tanner. 1996. Dynamic acoustic properties of the advertisement calls of gray treefrogs: Patterns of variability and female choice. *Behav. Ecol.* 7:7–18.

Gerhardt, H. C., S. I. Guttman, and A. A. Karlin. 1980. Natural hybrids between *Hyla cinerea* and *Hyla gratiosa:* Morphology, vocalization, and electrophoretic analysis. *Copeia* 1980:577–84.

Gerhardt, H. C., and G. M. Klump. 1988a. Masking of acoustic signals by the chorus background noise in the green treefrog: A limitation on mate choice. *Anim. Behav.* 36:1247–49.

———. 1988b. Phonotactic responses and selectivity of barking treefrogs (*Hyla gratiosa*) to chorus sounds. *J. Comp. Physiol. A* 163:795–802.

Gerhardt, H. C., and K. M. Mudry. 1980. Temperature effects on frequency preferences and mating call frequencies in the green treefrog, *Hyla cinerea* (Anura: Hylidae). *J. Comp. Physiol.* 137:1–6.

Gerhardt, H. C., and J. Rheinlaender. 1980. Accuracy of sound localization in a miniature dendrobatid frog. *Naturwissenschaften* 67:362–63.

———. 1982. Localization of an elevated sound source by the green treefrog. *Science* 217:663–64.

Gerhardt, H. C., J. D. Roberts, M. A. Bee, and J. J. Schwartz. 2001. Call matching in the quacking frog (*Crinia georgiana*). *Behav. Ecol. Sociobiol.* 48:243–51.

Gerhardt, H. C., and J. Schul. 1999. A quantitative analysis of behavioral selectivity for pulse-rise time in the gray treefrog, *Hyla versicolor. J. Comp. Physiol. A* 185:33–40.

Gerhardt, H. C., and J. J. Schwartz. 1995. Interspecific interactions in anuran courtship. In *Amphibian biology: Social behaviour*, ed. H. Heatwole and B. K. Sullivan, 603–32. Chipping Norton, NSW: Surrey Beatty and Sons.

———. 2001. Auditory tuning and frequency preferences in anurans. In *Anuran communication*, ed. M. J. Ryan, 73–85. Washington: Smithsonian Institution Press.

Gerhardt, H. C., S. D. Tanner, C. M. Corrigan, and H. C. Walton. 2000. Female preference functions based on call duration in the gray treefrog (*Hyla versicolor*). *Behav Ecol.* 11:663–69.

Gerhardt, H. C., and G. F. Watson. 1995. Within-male variability in call properties and female choice in the grey treefrog. *Anim. Behav.* 50:1187–91.

Giacoma, C., C. Zugolaro, and L. Beani. 1997. The advertisement call of the green toad (*Bufo viridis*): Consistency, variability and role in mate choice. *Herpetologica* 53:454–64.

Gibson, R. M., and J. W. Bradbury. 1985. Sexual selection in lekking sage grouse: Phenotypic correlates of male mating success. *Behav. Ecol. Sociobiol.* 18:117–23.

Girgenrath, M., and R. L. Marsh. 1997. In vivo performance of trunk muscles in tree frogs during calling. *J. Exp. Biol.* 200:3101–8.

Gish, S. L., and E. S. Morton. 1981. Structural adaptations to local habitat acoustics in Carolina wren songs. *Zeit. f. Tierpsychol.* 56:74–84.

Given, M. F. 1987. Vocalizations and acoustic interactions of the carpenter frog, *Rana virgatipes*. *Herpetologica* 43:467–81.

———. 1988. Growth rate and the cost of calling activity in male carpenter frogs, *Rana virgatipes*. *Behav. Ecol. Sociobiol.* 22:153–60.

———. 1996. Intensity modulation of calls in *Bufo woodhouseii fowleri*. *Copeia* 1996: 970–77.

———. 1999. Frequency alteration of the advertisement call in the carpenter frog, *Rana virgatipes*. *Herpetologica*. 55:304–17.

Givois, V., and G. S. Pollack. 2000. Sensory habituation of auditory receptor neurons: Implications for sound localization. *J. Exp. Biol.* 203:2529–37.

Gleason, J. M., and M. G. Ritchie. 1998. Evolution of courtship song and reproductive isolation in the *Drosophila willstoni* species complex: Do sexual signals diverge the most quickly? *Evolution* 52:1493–500.

Godwin, G. J., and S. M. Roble. 1983. Mating success in male treefrogs, *Hyla chrysoscelis* (Anura: Hylidae). *Herpetologica* 39:141–46.

Gogala, M. 1995. Songs of four cicada species from Thailand. *Bioacoustics* 6:101–16.

Gogala, M., and K. Riede. 1995. Time sharing of song activity by cicadas in Temengor Forest Reserve, Hulu Perak, and in Sabah, Malaysia. *Malayan Nature J.* 49:48–54.

Gooler, D. M., C. J. Condon, J.-H. Xu, and A. S. Feng. 1993. Sound direction influences the frequency-tuning characteristics of neurons in the frog inferior colliculus. *J. Neurophysiol.* 69: 1018–30.

Gooler, D. M., and A. S. Feng. 1992. Temporal coding in the frog auditory midbrain: The influence of duration and rise-fall time on the processing of complex amplitude-modulated stimuli. *J. Neurophysiol.* 67:1–22.

Gooler, D. M., J.-H. Xu, and A. S. Feng. 1996. Binaural inhibition is important in shaping the free-field frequency selectivity of single neurons in the inferior colliculus. *J. Neurophysiol.* 76:2580–94.

Grafe, T. U. 1995. Graded aggressive calls in the African painted reed frog *Hyperolius marmoratus* (Hyperoliidae). *Ethology* 101:67–81.

———. 1996a. Energetics of vocalization in the African reed frog (*Hyperolius marmoratus*). *Comp. Biochem.Physiol.* 114:235–43.

———. 1996b. The function of call alternation in the African reed frog (*Hyperolius marmoratus*): Precise call timing prevents auditory masking. *Behav. Ecol. Sociobiol.* 38:149–58.

———. 1997a. Costs and benefits of mate choice in the lek-breeding reed frog, *Hyperolius marmoratus*. *Anim. Behav.* 53:1103–17.

———. 1997b. Use of metabolic substrates in the gray treefrog, *Hyla versicolor:* Implications for calling behavior. *Copeia* 1997:356–62.

———. 1999. A function of synchronous chorusing and a novel female preference shift in an anuran. *Proc. R. Soc. London B* 266:2331–36.

Grafe, T. U., R. Schmuck, and K. E. Linsenmair. 1992. Reproductive energetics of the African reed frogs *Hyperolius viridiflavus* and *Hyperolius marmoratus*. *Physiol. Zool.* 65:153–71.

Grafe, T. U., J. O. Steffen, and C. Stoll. 2000. Vocal repertoire and effect of advertisement call intensity on calling behaviour in the West African tree frog, *Leptopelis viridis*. *Amphibia-Reptilia* 21:13–23.

Grafen, A. 1990. Biological signals as handicaps. *J. Theor. Biol.* 144:517–46.

Gramoll, S. 1988. Activity of metathoracic interneurons during stridulation in the acridid grasshopper *Omocestus viridulus* L. *J. Comp. Physiol. A* 163:813–25.

Gramoll, S., and N. Elsner. 1987. Morphology of local "stridulation" interneurons in the metathoracic ganglion of the acridid grasshopper *Omocestus viridulus* L. *J. Comp. Neurol.* 263:593–606.

Grant, P. R. 1972. Convergent and divergent character displacement. *Biol. J. Linn. Soc.* 4:39–68.

Gray, D. A. 1997. Female house crickets, *Acheta domesticus*, prefer the chirps of large males. *Anim. Behav.* 54:1553–62.

Gray, D. A., and W. H. Cade. 1999a. Quantitative genetics of sexual selection in the field cricket, *Gryllus integer*. *Evolution* 53:848–54.

———. 1999b. Sex, death and genetic variation: Natural and sexual selection on cricket song. *Proc. R. Soc. London B* 266:707–9.

———. 2000. Sexual selection and speciation in field crickets. *Proc. Natl. Acad. Sci. USA* 97:14449–54.

Gray, E. G. 1960. The fine structure of the insect ear. *Phil. Trans. R. Soc. B* 243: 75–94.

Green, A. J. 1990. Determinants of chorus participation and the effects of size, weight and competition on advertisement calling in the túngara frog, *Physalaemus pustulosus* (Leptodactylidae). *Anim. Behav.* 39:620–38.

Greenacre, M., M. G. Ritchie, B. C. Byrne, and C. P. Kyriacou. 1993. Female song preference and the *period* gene of *Drosophila melanogaster*. *Behav. Genet.* 23:85–90.

Greenewalt, C. H. 1968. *Bird song: Acoustics and physiology*. Washington: Smithsonian Institution Press.

Greenfield, M. D. 1983. Unsychronized chorusing in the coneheaded katydid *Neoconocephalus affinis* (Beauvois). *Anim. Behav.* 31:102–12.

———. 1988. Interspecific acoustic interactions among katydids *Neoconocephalus:* Inhibition-induced shifts in diel activity. *Anim. Behav.* 36:684–95.

———. 1993. Inhibition of male calling by heterospecific signals: artifact or chorusing or abstinence during suppression of female phonotaxis? *Naturwissenschaften* 80: 570–73.

———. 1994a. Cooperation and conflict in the evolution of signal interactions. *Annu. Rev. Ecol. Syst.* 25:97–126.

———. 1994b. Synchronous and alternating choruses in insects and anurans: Common mechanisms and diverse functions. *Am. Zool.* 34:605–15.

———. 2002. *Signalers and receivers: Mechanisms and evolution of arthropod communication.* Oxford: Oxford University Press.

Greenfield, M. D., and R. L. Minckley. 1993. Acoustic dueling in tarbush grasshoppers: Settlement of territorial contests via alternation of reliable signals. *Ethology* 95:309–26.

Greenfield, M. D., and A. S. Rand. 2000. Frogs have rules: Selective attention algorithms regulate chorusing in *Physalaemus pustulosus* (Leptodactylidae). *Ethology* 106: 331–47.

Greenfield, M. D., and I. Roizen. 1993. Katydid synchronous chorusing is an evolutionarily stable outcome of female choice. *Nature (London)* 364:618–20.

Greenfield, M. D., and K. C. Shaw. 1983. Adaptive significance of chorusing with special reference to the Orthoptera. In *Orthopteran mating systems: Sexual competition in a diverse group of insects*, ed. D. T. Gwynne and G. K. Morris, 1–27. Boulder: Westview Press.

Greenfield, M. D., and T. E. Shelly. 1985. Alternative mating strategies in a desert grasshopper: Evidence of density-dependence. *Anim. Behav.* 33:1192–210.

———. 1990. Territory-based mating systems in desert grasshoppers: Effects of host plant distribution and variation. In *Biology of grasshoppers*, ed. R. F. Chapman and S. Joern, 315–35. New York: John Wiley and Sons.

Greenfield, M. D., M. K. Tourtellot, and W. A. Snedden. 1997. Precedence effects and the evolution of chorusing. *Proc. R. Soc. Lond. B* 264:1355–61.

Gross, M. R. 1996. Alternative reproductive strategies and tactics: Diversity within sexes. *Trends Ecol. Evol.* 11:92–98.

Gwynne, D. T. 1982. Mate selection by female katydids (Orthoptera: Tettigoniidae, *Conocephalus nigropleurem*). *Anim. Behav.* 30:734–38.

———. 1987. Sex-biased predation and the risky mate-locating behaviour of male tick-tock cicadas (Homptera: Cicadidae). *Anim. Behav.* 35:571–76.

———. 1993. Food quality controls sexual selection in Mormon crickets by altering male mating investment. *Ecology* 74:1406–13.

———. 1995. Phylogeny of the Ensifera (Orthoptera): A hypothesis supporting multiple origins of acoustical signalling, complex spermatophores, and maternal care in crickets, katydids, and weta. *J. Orthoptera Research* 4:203–18.

Gwynne, D. T., and W. J. Bailey. 1988. Mating system, mate choice and ultrasonic calling in a zaprochiline katydid (Othroptera: Tettigonidae). *Behaviour* 105:202–23.

Gwynne, D. T., and L. S. Simmons. 1990. Experimental reversal of courtship roles in an insect. *Nature (London)* 346:172–74.

Hack, M. A. 1997. The energetic costs of fighting in the house cricket, *Acheta domesticus. Behav. Ecol.* 8:28–36.

Haddad, C. F. B. 1991. Satellite behavior in the neotropical treefrog *Hyla minuta*. *J. Herpetol.* 25:226–29.

Haddad, C. F. B., J. P. Pombal, and R. F. Batistic. 1995. Hybridization between diploid and tetraploid species of leaf-frogs, genus *Phyllomedusa* (Amphibia). *J. Herpetol.* 28: 425–30.

Hainfeld C. A., S. L. Boatright-Horowitz, and A. M. Simmons. 1996. Discrimination of phase spectra in complex sounds by the bullfrog (*Rana catesbeiana*). *J. Comp. Physiol A* 179:75–87.

Halfmann, K., and N. Elsner. 1978. Larval stridulation in Acridid grasshoppers. *Naturwissenschaften* 65:265–66.

Hall, J. C. 1994. Central processing of communication sounds in the anuran auditory system. *Amer. Zool.* 34:670–84.

———. 1999. GABAergic inhibition shapes frequency tuning and modifies response properties in the auditory midbrain of the leopard frog. *J. Comp. Physiol. A* 185: 479–91.

Hall, J. C., and Feng, A. S. 1986. Neural analysis of temporally patterned sounds in the frog's thalamus: Processing of pulse duration and pulse repetition rate. *Neuroscience Letters* 63:215–20.

———. 1987. Evidence for parallel processing in the frog's auditory thalamus. *J. Comp. Neurol.* 258:407–19.

———. 1988. Influence of envelope rise time on neural responses in the auditory system of anurans. *Hearing Research* 36:261–76.

———. 1990. Classification of discharge patterns of single neurons in the dorsal medullary nucleus of the northern leopard frog. *J. Neurophysiol.* 64:1460–73.

———. 1991. Temporal processing in the dorsal medullary nucleus of the northern leopard frog (*Rana pipiens pipiens*). *J. Neurophysiol.* 66:955–73.

Halliday, T. R. 1980. *Sexual strategy*. Oxford: Oxford University Press.

Halliday, T. R., and M. Tejedo. 1995. Intrasexual selection and alternative mating behaviour. In *Amphibian biology: Vol. 2: Social behaviour*, ed. H. Heatwole and B. K. Sullivan, 419–68. Chipping Norton, NSW: Surrey Beatty and Sons.

Hamilton, W. D. 1971. Geometry of the selfish herd. *J. Theoret. Biol.* 31:95–311.

Hanson, F. E. 1978. Comparative studies of firefly pacemakers. *Fed. Proc.* 37:2158–64.

Hardt, M. 1988. Zur Phonotaxis von Laubheuschrecken: Eine vergleichende verhaltensphysiologische und neurophysiologisch-anatomische Untersuchung. Ph.D. diss., Universität Bochum.

Harrison. P. A. 1987. Vocal behavior in the south-eastern Australian tree frogs, *Litoria ewingi* and *L. verreauxi* (Anura: Hylidae). M.Sc. thesis, University of Melbourne, Melbourne, Australia.

Harrison, P., and M. J. Littlejohn. 1985. Diphasy in the advertisement calls of *Geocrinia laevis* (Anura: Leptodactylidae): Vocal responses of males during playback. *Behav. Ecol. Sociobiol.* 19:67–73.

Harrison, R. G. 1990. Hybrid zones: Windows on evolutionary processes. In *Oxford surveys in evolutionary biology*, ed. D. Futuyma and J. Antonovics, 7:69–128. Oxford: Oxford University Press.

Harrison, R. G., and S. M. Bogdanowicz. 1995. Mitochondrial DNA phylogeny of North American field crickets: Perspectives on the evolution of life cycles, songs, and habitat associations. *J. Evol. Biol.* 8:209–32.

Hartley, J. C., and D. J. Robinson. 1976. Acoustic behavior of both sexes of the speckled bush cricket *Leptophyes punctatissima*. *Physiol. Entomol.* 1:21–25.

Harvey, P. H., and M. D. Pagel. 1991. *The comparative method in evolutionary biology*. Oxford: Oxford University Press.

Heatwole, H., and B. K. Sullivan, eds. 1995. *Amphibian biology: Vol. 2: Social behaviour*. Chipping Norton, NSW: Surrey Beatty and Sons.

Hedrick, A. V. 1986. Female preferences for male calling bout duration in a field cricket. *Behav. Ecol. Sociobiol.* 19:73–77.

———. 1988. Female choice and the heritability of attractive male traits: An empirical study. *Am. Natur.* 132:267–76.

———. 2000. Crickets with extravagant mating songs compensate for predation risks with extra caution. *Proc. R. Soc. London B* 267:671–75.

Hedrick, A. V., and L. M. Dill. 1993. Mate choice by female crickets is influenced by predation risk. *Anim. Behav.* 46:267–76.

Hedrick, A. V., and T. Weber. 1998. Variance in female responses to the fine structure of male song in the field cricket, *Gryllus integer. Behav. Ecol.* 9:582–91.

Hedwig, B. 1986a. On the role in stridulation of plurisegmental interneurons of the acridid grasshopper *Omocestus viridulus* L. I. Anatomy and physiology of descending cephalothoracic interneurons. *J. Comp. Physiol. A* 158:413–27.

———. 1986b. On the role in stridulation of plurisegmental interneurons of the acridid grasshopper *Omocestus viridulus* L. II. Anatomy and physiology of ascending and T-shaped interneurons. *J. Comp. Physiol. A* 158:429–44.

———. 1988. Activation and modulation of auditory receptors in *Locusta migratoria* by respiratory movements. *J. Comp. Physiol. A* 162:237–46.

———. 1990a. The intersegmental network underlying stridulation in the grasshopper *Omocestus viridulus.* In *Sensory systems and communication in arthropods*, ed. F. G. Gribakin, K. Wiese, and A. V. Popov, 189–92. Basel: Birkhäuser Verlag.

———. 1990b. Modulation of auditory responsiveness in stridulating grasshoppers. *J. Comp. Physiol. A* 167:847–56.

———. 1992a. On the control of stridulation in the acridid grasshopper *Omocestus viridulus* L. I. Interneurons involved in rhythm generation and bilateral coordination. *J. Comp. Physiol. A* 171:117–28.

———. 1992b. On the control of stridulation in the acridid grasshopper *Omocestus viridulus* L. II. Shaping of hindleg movements by spiking and non-spiking interneurons. *J. Comp. Physiol. A* 171:129–40.

———. 1994. A cephalothoracic command system controls stridulation in the acridid grasshopper *Omocestus viridulus* L. *J. Neurophysiol.* 72:2015–25.

———. 1995. Kontrolle des Stridulationsverhaltens von Feldheuschrecken durch deszendierende Hirnneurone. *Verh. Dtsch. Zool. Ges.* 88/2:181–90.

———. 1996. A descending brain neuron elicits stridulation in the cricket *Gryllus bimaculatus (de Geer). Naturwissenschaften* 83:428–29.

———. 2000a. Control of cricket stridulation by a command neuron: Efficacy depends on the behavioral state. *J. Neurophysiol.* 83:712–22.

———. 2000b. Singing and hearing: Neuronal mechanisms of acoustic communication in Orthopterans. *Zoology* 103:140–49.

Hedwig, B., and R. Heinrich. 1997. Identified descending brain neurons control different stridulatory motor patterns in an acridid grasshopper. *J. Comp. Physiol. A* 180:285–94.

Hedwig, B., and J. Meyer. 1994. Auditory information processing in stridulating grasshoppers: Tympanic membrane vibrations and neurophysiology. *J. Comp. Physiol. A* 174:121–31.

Heiligenberg, W. 1991. *Neural nets in electric fish.* Cambridge: MIT Press.

———. 1994. The detection and generation of electric communication signals in gymnotiform fish. *Fortschr. Zool.* 39:13–24.

Heinrich, R. 1995. Mikrochirurgische und pharmakologische Untersuchungen über die neuralen Grundlagen des Heuschreckengesanges. Ph.D. diss. University of Göttingen.

Heinrich, R., and N. Elsner. 1997. Central nervous control of hindleg coordination in stridulating grasshoppers. *J. Comp. Physiol. A* 180:257–69.

Heinrich, R., B. Hedwig, and N. Elsner. 1997. Cholinergic activation of stridulatory behaviour in the grasshopper *Omocestus viridulus* L. *J. Exp. Biol.* 200:1327–37.
Heinrich, R., K. Rozwod, and N. Elsner. 1998. Neuropharmacological evidence for inhibitory cephalic control mechanisms of stridulatory behavior in grasshoppers. *J. Comp. Physiol. A* 183:389–99.
Heinrich, R., B. Wenzel, and N. Elsner. 2001. Pharmacological brain strimulation releases elaborate stridulatory behavior in gomphocerine grasshoppers—conclusions for the organization of central nervous control. *J. Comp. Physiol. A* 187:155–69.
Heller, K.-G. 1988. *Bioakustik der Europäischen Laubheuschrecken.* Weikersheim: Verlag Joseph Margraf.
———. 1990. Evolution of song pattern in east Mediterranean Phaneropterinae: Constraints by the communication system. In *The Tettigoniidae: Biology, systematics and evolution*, ed. W. J. Bailey and D. C. F. Rentz, 130–51. Bathurst: Crawford House Press.
———. 1995. Acoustic signaling in paleotropical bushcrickets (Orthoptera: Tettigoniidae: Pseudophyllidae): Does predation pressure by eavesdropping enemies differ in the Paleo- and Neotropics? *J. Zool. Lond.* 237:469–85.
Heller, K.-G., and D. von Helversen. 1986. Acoustic communication in phaneropterid bushcrickets: Species-specific delay of female stridulatory response and matching male sensory time window. *Behav. Ecol. Sociobiol.* 18:189–98.
Heller, K.-G., and O. von Helversen. 1993. Calling behavior in bushcrickets of the genus *Poecilimon* with differing communication systems (Orthoptera: Tettigoniidae: Phaneropteridae). *J. Insect Behav.* 6:361–77.
Heller, K.-G., O. von Helversen, and M. Sergejeva. 1997. Indiscriminate response behaviour in a female bushcricket: Sex role reversal in selectivity of acoustic mate recognition? *Naturwissenschaften* 84:252–55.
Heller, K.-G., J. Schul, and S. Ingrisch. 1997. Sex-specific differences in song frequency and tuning of the ears in some duetting bushcrickets (Orthoptera: Tettigonioidae: Phaneropteridae). *Zoology* 100:110–18.
Helversen, D. von 1972. Gesang des Männchens und Lautschema des Weibchens bei der Feldheuschrecke *Chorthippus biguttulus* (Orthoptera, Acrididae). *J. Comp. Physiol.* 81:381–422.
———. 1984. Parallel processing in auditory pattern recognition and directional analysis by the grasshopper *Chorthippus biguttulus* L. (Acrididae). *J. Comp. Physiol. A* 154:837–46.
———. 1993. "Absolute steepness" of ramps as an essential cue for auditory pattern recognition by a grasshopper (Orthoptera: Acrididae; *Chorthippus biguttulus* L.). *J. Comp. Physiol. A* 172:633–39.
———. 1997. Acoustic communication and orientation in grasshoppers. In *Orientation and communication in arthropods*, ed. M. Lehrer, 301–41. Basel: Birkhäuser Verlag.
Helversen, D. von, and O. von Helversen. 1981. Korrespondenz zwischen Gesang und auslösendem Schema bei Feldheuschrecken. *Nova acta Leopoldina N.F.* 54, no. 245:449–62.
———. 1983. Species recognition and acoustic localization in acridid grasshoppers: A behavioral approach. In *Neuroethology and behavioral physiology*, ed. F. Huber and H. Markl, 95–107. Berlin: Springer-Verlag.
———. 1987. Innate receiver mechanisms in the acoustic communication of orthop-

teran insects. In *Aims and methods in neuroethology*, ed. D. M. Guthrie, 104–50. Manchester: Manchester University Press.

———. 1995. Acoustic pattern recognition and orientation in orthopteran insects: Parallel or serial processing. *J. Comp. Physiol. A* 177:767–74.

———. 1997. Recognition of sex in the acoustic communication of the grasshopper *Chorthippus biguttulus* (Orthoptera, Acrididae). *J. Comp. Physiol. A* 180: 373–86.

———. 1998. Acoustic pattern recognition in a grasshopper: Processing in the time or frequency domain? *Biol. Cybernetics* 79:467–76.

Helversen, D. von, and J. Rheinlaender. 1988. Interaural intensity and time discrimination in an unrestrained grasshopper: A tentative behavioural approach. *J. Comp. Physiol. A* 162: 333–40.

Helversen, D. von, and G. Wendler. 2000. Coupling of visual to auditory cues during phonotactic approach in the phaneropterine bushcricket *Poecilimon affinis*. *J. Comp. Physiol. A* 186:729–36.

Helversen, O. von, and D. von Helversen. 1994. Forces driving co-evolution of song and song recognition in grasshoppers. *Fortschritte der Zoologie* 39:253–84.

Hennig, R. M. 1988. Ascending auditory interneurons in the cricket *Teleogryllus commodus* (Walker): Comparative physiology and direct connections with afferents. *J. Comp. Physiol. A* 163:135–43.

———. 1990a. Neuronal control of the forewings in two different behaviours: Stridulation and flight in the cricket, *Teleogryllus commodus*. *J. Comp. Physiol. A* 167:617–27.

———. 1990b. Neuronal organization of the flight motor pattern in the cricket, *Teleogryllus commodus*. *J. Comp. Physiol. A* 167:629–39.

———. 1992. Mechanisms of motor pattern switching in crickets: Stridulation and flight. In *Neurobiology of motor programme selection—new approaches to the study of behavioural choice*, ed. J. Kien, C. R. McCorhan, and W. Winlow, 105–22. Oxford: Pergamon Press.

———. 2001. Different filter characteristics for temporal patterns in two closely related cricket species. In *Göttingen neurobiology report 2001*, ed. N. Elsner and G. W. Kreutzberg, 359. Stuttgart: Thieme Verlag.

Hennig, R. M., and D. Otto. 1995/96. Distributed control of song pattern generation revealed by lesions to the thoracic ganglia. *Zoology* 99:268–76.

Hennig, R. M., and T. Weber. 1997. Filtering of temporal parameters of the calling song by cricket females of two closely related species: A behavioral analysis. *J. Comp. Physiol. A* 180:621–30.

Henry, C. S., M. L. Wells, and C. M. Simon. 1999. Convergent evolution of courtship songs among cryptic species of the *Carnea* group of green lacewings (Neuoptera: Chrysopidae: *Chrysoperla*). *Evolution* 53:1165–79.

Henzi, S. P., M. L. Dyson, S. E. Piper, N. E. Passmore, and P. Bishop. 1995. Chorus attendance by male and female Painted Reed Frogs (*Hyperolius marmoratus*): Environmental factors and selection pressures. *Functional Ecol.* 9:485–91.

Hetherington, T. E. 1992a. The effects of body size on the evolution of the amphibian middle ear. In *The evolutionary biology of hearing*, ed. D. B. Webster, R. R. Fay, and A. N. Popper, 421–38. New York: Springer-Verlag.

———. 1992b. The effects of body size on functional properties of middle ear systems of anuran amphibians. *Brain, Behav. Evol.* 39:133–42.

Higgie, M., S. Chenoweth, and M. W. Blows. 2000. Natural selection and the reinforcement of mate recognition. *Science* 290:519–21.

Hill, K. G. 1974. Carrier frequency as a factor in phonotactic behaviour of female crickets (*Teleogryllus commodus*). *J. Comp. Physiol.* 93:7–18.

Hill, K. G., J. J. Loftus-Hills, and M. J. Littlejohn. 1972. Pre-mating isolation between the Australian field crickets *Teleogryllus commodus* and *T. oceanicus* (Orthoptera: Gryllidae). *Aust. J. Zool.* 20:153–63.

Hillery, C. M., and P. N. Narins. 1984. Neurophysiological evidence for a traveling wave in the amphibian inner ear. *Science* 225:1037–39.

Hillman, D. E. 1976. Morphology of the peripheral and central vestibular systems. In *Frog neurobiology*, ed. R. Llinas and W. Precht, 452–80. Berlin: Springer-Verlag.

Hissman, K. 1990. Strategies of mate finding in the European field cricket (*Gryllus campestris*) at different population densities: A field study. *Ecol. Entom.* 15:281–91.

Hoback, W. W., and W. E. Wagner Jr. 1997. The energetic costs of calling in the variable field cricket, *Gryllus lineaticeps*. *Physiol. Entom.* 22:286–90.

Hödl, W. 1977. Call differences and calling site segregation in anuran species from Central Amazonian floating meadows. *Oecologia* 28:351–63.

Hödl, W., and A. Amézquita. 2001. Visual signaling in anuran amphibians. In *Anuran communication*, ed. M. J. Ryan, 121–41. Washington: Smithsonian Institution Press.

Höglund, J., and R. V. Alatalo. 1995. *Leks.* Princeton: Princeton University Press.

Honegger, H. W., and R. Campan. 1989. Vision and visually guided behavior. In *Cricket behavior and neurobiology*, ed. F. Huber, T. E. Moore, and W. Loher, 147–77. Ithaca: Cornell University Press.

Horseman, G., and F. Huber. 1994a. Sound localization in crickets. I. Contralateral inhibition of an ascending interneuron (AN1) in the cricket *Gryllus bimaculatus*. *J. Comp. Physiol. A* 175:389–98.

———. 1994b. Sound localization in crickets. II. Modelling the role of a simple neural network in the prothoracic ganglion. *J. Comp. Physiol. A* 175:399–413.

Howard, D. S. 1993. Reinforcement: Origin, dynamics and fate of an evolutionary hypothesis. In *Hybrid zones and the evolutionary process*, ed. R. G. Harrison, 46–69. Oxford: Oxford University Press.

Howard, D. S., and S. H. Berlocher. 1998. *Endless forms: Species and speciation*. Oxford: Oxford University Press.

Howard, R. D. 1978a. The evolution of mating strategies in bullfrogs, *Rana catesbeiana*. *Evolution* 32:850–71.

———. 1978b. The influence of male-defended oviposition sites on early embryo mortality in bullfrogs. *Ecology* 59:789–98.

———. 1984. Alternative mating behaviors of young male bullfrogs. *Amer. Zool.* 24: 397–406.

Howard, R. D., and J. G. Palmer. 1995. Female choice in *Bufo americanus:* Effects of dominant frequency and call order. *Copeia* 1995:212–17.

Howard, R. D., H. H. Whiteman, and T. I. Schueller. 1994. Sexual selection in American toads: A test of a good-genes hypothesis. *Evolution* 48:1286–300.

Howard, R. D., and J. R. Young. 1998. Individual variation in male vocal traits and female mating preferences in *Bufo americanus*. *Anim. Behav.* 55:1165–79.

Hoy, R. R. 1992.The evolution of hearing in insects as an adaptation to predation from bats. In *The evolutionary biology of hearing*, ed. D. B. Webster, R. R. Fay, and A. N. Popper, 115–29. New York: Springer-Verlag.

———. 1994. Ultrasound acoustic startle in flying crickets: Some neuroethological and comparative aspects. *Fortschr. Zool.* 39:227–41.

———. 1998. Acute as a bug's ear: An informal discussion of hearing in insects. In *Comparative hearing: Insects*, ed. R. R. Hoy, A. N. Popper, and R. R. Fay, 1–17. New York: Springer-Verlag.

Hoy, R. R., G. Casaday, and S. R. Rollins. 1978. Absence of auditory afferents alters the growth pattern of an identified auditory interneuron. *Soc. Neurosci. Abstr.* 4:115.

Hoy, R. R., and R. C. Paul. 1973. Genetic control of song specificity in crickets. *Science* 180:82–83.

Huang, Y., G. Ortí, M. Sutherlin, A. Duhachek, and A. Zera. 2000. Phylogenetic relationships of North American field crickets inferred from mitochondrial DNA data. *Molecular Phylog. and Evol.* 17:48–57.

Huber, F. 1955. Sitz und Bedeutung nervöser Zentren für Instinkthandlungen beim Männchen von *Gryllus campestris* L. *Z. Tierpsychol.* 12:12–48.

———. 1960. Untersuchungen über die Funktion des Zentralnervensystems und insbesondere des Gehirnes bei der Fortbewegung und der Lauterzeugung der Grillen. *Z. vergl. Physiol.* 44:60–132.

———. 1962a. Central nervous control of sound production in crickets and some speculations on its evolution. *Evolution* 16:429–42.

———. 1962b. Lokalisation und Plastizität im Zentralnervensystem der Tiere. *Zool. Anz. Suppl.* 26:200–67.

———. 1975. Principles of motor co-ordination in cyclically recurring behavior in insects. In *"Simple" nervous systems*, ed. P. N. R. Usherwood and D. R. Newth, 381–413. New York: Edward Arnold.

———. 1987. Plasticity in the auditory system of crickets: Phonotaxis with one ear and neuronal reorganization within the auditory pathway. *J. Comp. Physiol. A* 161: 583–604.

———. 1990. Cricket neuroethology: Neuronal basis of intraspecific acoustic communication. *Adv. Study Behav.* 19:299–356.

Huber, F., H.-U. Kleindienst, T. E. Moore, K. Schildberger, and T. Weber. 1990. Acoustic communication in periodical cicadas: Neuronal responses to songs of sympatric species. In *Sensory systems and communication in arthropods*, ed. F. G. Gribakin, K. Wiese, and A. V. Popov, 217–28. Basel: Birkhäuser Verlag.

Huber, F., H.-U. Kleindienst, T. Weber, and J. Thorson. 1984. Auditory behavior of the cricket. III. Tracking of male calling song by surgically and developmentally one-eared females and the curious role of the anterior tympanum. *J. Comp. Physiol. A* 155:725–38.

Huber, F., T. E. Moore, and W. Loher. 1989. *Cricket behavior and neurobiology*. Ithaca: Comstock/Cornell University Press.

Huber, F., and J. Thorson. 1985. Cricket auditory communication. *Sci. Am.* 253: 60–68.

Huber, F., D. W. Wohlers, and T. E. Moore. 1980. Auditory nerve and interneurone responses to natural sounds in several species of cicadas. *Physiol. Entomol.* 5:25–45.

Hubl, L., and H. Schneider. 1979. Temperature and auditory thresholds: Bioacoustic studies of the frogs *Rana r. ridibunda*, *Hyla a. arborea*, and *H. a. savignyi* (Anura: Amphibia). *J. Comp. Physiol. A* 130:17–27.

Hustert, R., E. Lodde, and W. Gnatzy. 1999. Mechanosensory pegs constitute stridulatory files in grasshoppers. *J. Comp. Neurol.* 410:444–56.

Ibáñez, D. R. 1993. Female phonotaxis and call overlap in the Neotropical glassfrog *Centrolenella granulosa*. *Copeia* 1993:846–50.

Imaizumi, K., and G. S. Pollack. 1999. Neural coding of sound frequency by cricket auditory receptors. *J. Neurosci.* 19:1508–16.

Jacobs, K., B. Otte, and R. Lakes-Harlan. 1999. Tympanal receptor cells of *Schistocerca gregaria:* Correlation of soma positions and dendrite attachment sites, central projections and physiologies. *J. Exp. Zool.* 283:270–85.

Jacobson, S. K. 1985. Reproductive behavior and male mating success in two species of glass frogs (Centrolenelidae). *Herpetologica* 41:396–404.

Jaramillo, C., A. S. Rand, R. Ibáñez, and R. Dudley. 1997. Elastic structures in the vocalization apparatus of the Tungara frog *Physalaemus pustulosus* (Leptodactylidae). *J. Morphology* 233:287–95.

Jaslow, A. P., T. E. Hetherington, and R.E. Lombard. 1988. Structure and function of the amphibian middle ear. In *The evolution of the amphibian auditory system*, ed. B. Fritzsch, M. J. Ryan, W. Wilczynski, T. E. Hetherington, and W. Walkowiak, 69–92. New York: John Wiley and Sons.

Jaslow, A. P., and R. E. Lombard. 1996. Hearing in the neotropical frog, *Atelopus chiriquiensis. Copeia* 1996:428–32.

Jatho, M. 1995. Untersuchungen zur Schallproduktion und zum phonotaktischen Verhalten von Laubheuschrecken (Orthoptera: Tettigoniidae). Ph.D. diss., Philipps University, Marburg.

Jatho, M., J. Schul, O. Stiedl, and K. Kalmring. 1994. Specific differences in sound production and pattern recognition in tettigoniids. *Behav. Processes* 31:293–300.

Jehle, R., and A. Arak. 1998. Graded call variation in the Asian cricket frog *Rana nicrobariensis*. *Bioacoustics* 9:35–48.

Jennions, M. D., P. R. Y. Backwell, and N. I. Passmore. 1995. Repeatability of mate choice: The effect of size in the African painted reed frog, *Hyperolius marmoratus*. *Anim. Behav.* 49:181–86.

Jennions, M. D., and N. I. Passmore. 1993. Sperm competition in frogs: Testis size and a "sterile male" experiment on *Chiromantis xerampelina* (Rhacophoridae). *Biol. J. Linnean Soc.* 50:211–20.

Jennions, M. D., and M. Petrie. 1997. Variation in mate choice and mating preferences: A review of causes and consequences. *Biological Reviews* 72:283–327.

Jia, F.-Y., M. D. Greenfield, and R. D. Collins. 2000. Genetic variance of sexually selected traits in waxmoths: Maintenance by genotype X environmental interaction. *Evolution* 54:953–67.

Johnson, C. F. 1959. Genetic incompatibility in the call races of *Hyla versicolor* LeConte in Texas. *Copeia* 1959:327–35.

Johnstone, R. A. 1998. Conspiratorial whispers and conspicuous displays: Games of signal detection. *Evolution* 52:1554–63.

Jones, M. D. R. 1966a. The acoustic behaviour of the bush cricket *Pholidoptera griseoaptera*. I. Alternation, synchronization, and rivalry between males. *J. Exp. Biol.* 45: 15–30.

———. 1966b. The acoustic behaviour of the bush cricket *Pholidoptera griseoaptera*. II. Interaction with artificial sound signals. *J. Exp. Biol.* 45:31–44.

Jørgensen, M. B. 1991. Comparative studies of the biophysics of directional hearing in anurans. *J. Comp. Physiol. A* 169:591–98.

Jørgensen, M. B., and J. Christensen-Dalsgaard. 1997a. Directionality of auditory nerve fiber responses to the pure tone stimuli in the grassfrog, *Rana temporaria*. I. Spike rate responses. *J. Comp. Physiol. A* 180:493–502.

———. 1997b. Directionality of auditory nerve fiber responses to the pure tone stimuli in the grassfrog, *Rana temporaria*. II. Spike timing. *J. Comp. Physiol. A* 180: 503–11.

Jørgensen, M. B., and H. C. Gerhardt. 1991. Directional hearing in gray treefrog *Hyla versicolor:* Eardrum vibrations and phonotaxis. *J. Comp. Physiol. A* 169:177–83.

Jørgensen, M. B., and M. Kanneworff. 1998. Middle ear transmission in the grass frog, *Rana temporaria*. *J. Comp. Physiol. A* 182:59–64.

Jørgensen, M. B., B. Schmitz, and J. Christensen-Dalsgaard. 1991a. Biophysics of directional hearing in the frog *Eleutherodactylus coqui*. *J. Comp. Physiol. A* 169:177–83.

———. 1991b. Directionality of auditory nerve fiber responses to the pure tone stimuli in the grassfrog, *Rana temporaria*. I. Spike rate responses. *J. Comp. Physiol. A* 180: 493–502.

Josephson, R. K. 1984. Contraction dynamics of flight and stridulatory muscles of tettigoniid insects. *J. Exp. Biol.* 108:77–96.

Josephson, R. K., and R. C. Halverson. 1971. High frequency muscles used in sound production by a katydid. I. Organization of the motor system. *Biol. Bull.* 141: 411–33.

Josephson, R. K., and D. Young. 1981. Synchronous and asynchronous muscles in cicadas. *J. Exp. Biol.* 91:219–37.

———. 1985. A synchronous insect muscle with an operating frequency greater than 500 Hz. *J. Exp. Biol.* 118:185–208.

Kalmring, K. 1983. Convergence of auditory and vibratory senses at the neuronal level of the ventral cord in grasshoppers: Its probable importance for behaviour in the habitat. *Fortschr. Zool.* 28:129–41.

Kalmring, K., A. Keuper, and W. Kaiser. 1990. Aspects of acoustic and vibratory communication in seven European bushcrickets. In *The Tettigoniidae: Biology, systematics and evolution*, ed. W. J. Bailey and D. C. F. Rentz, 191–216. Bathhurst: Crawford House Press.

Kalmring, K., R. Kühne, and B. Lewis. 1983. The acoustic behaviour of the bushcricket *Tettigonia cantans*. III. Coprocessing of auditory and vibratory information in the central nervous system. *Behav. Processes* 8:213–28.

Kalmring, K., B. Lewis, and A. Eichendorf. 1978. The physiological characteristics of the primary sensory neurons of the complex tibial organ of *Decticus verrucivorus* L. (Orthoptera, Tettigonioidae). *J. Comp. Physiol.* 127:109–21.

Karban, R. 1982. Increased reproductive success at high densities and predation satiation for periodical cicadas. *Ecology* 63:321–28.

Kavanagh, M. W., and D. Young. 1989. Bilateral symmetry of sound production in the mole cricket, *Gryllotalpa australis*. *J. Comp. Physiol A* 166:43–49.

Keller, M. J., and H. C. Gerhardt. 2001. Polyploidy affects call structure in gray treefrogs. *Proc. R. Soc. London B* 268:341–45.

Kelley, D. B. 1980. Auditory and vocal nuclei of frog brain concentrate sex hormones. *Science* 207:553–55.

———. 1999. Vocal communication in *Xenopus laevis*. In *The design of animal communication*, ed. M. Hauser and M. Konishi, 9–35. Cambridge: MIT Press.

———. 2001. Hormonal regulation of motor output in amphibians; *Xenopus laevis* vocalizations as a model system. *Hormones, Brain and Behavior*, in press.

Kelley, D. B., and D. L. Gorlick. 1990. Sexual selection and the nervous system. *Bioscience* 40:275–83.

Kelly, J. K., and M. A. F. Noor. 1996. Speciation by reinforcement: A model derived from studies of *Drosophila*. *Genetics* 143:1485–97.

Keuper, A., K. Kalmring, A. Schatral, W. Latimer, and W. Kaiser. 1986. Behavioral adaptations of ground living bushcrickets to the properties of sound propagation in low grassland. *Oecologia* 70:414–22.

Keuper, A., and K. Kühne. 1983. The acoustic behavior of the bushcricket, *Tettigonia cantans*. II. Transmission of airborne sound and vibration signals in the biotope. *Behav. Processes* 8:125–46.

Keuper A., S. Weidemann, K. Kalmring, and D. Kaminski. 1988. Sound production and sound emission in seven species of European Tettigoniids. I. The different parameters of the song and their relation to the morphology of the bushcricket. *Bioacoustics* 1:31–48.

Kiflawi, M., and D. A. Gray. 2001. Size-dependent response to conspecific mating calls by male crickets. *Proc. R. Soc. London B* 267:2157–61.

Kime, N. M., A. S. Rand, and M. Kapfer. 1998. Consistency of female choice in the túngara frog: A permissive preference for complex characters. *Anim. Behav.* 55: 641–49.

Kime, N. M., W. R. Turner, and M. J. Ryan. 2000. The transmission of advertisement calls in central American frogs. *Behav. Ecol.* 11:71–83.

Kirkpatrick, M., and M. J. Ryan. 1991. The evolution of mating preferences and the paradox of the lek. *Nature (London)* 350:33–38.

Kirkpatrick, M., and R. M. Servedio. 1999. The reinforcement of mating preferences on an island. *Genetics* 151:865–84.

Kleindienst, H.-U., U. T. Koch, and D. W. Wohlers. 1981. Analysis of the cricket auditory system by acoustic stimulation using a closed sound field. *J. Comp. Physiol.* 141:283–96.

Kleindienst, H.-U., D. W. Wohlers, and O. N. Larsen. 1983. Tympanal membrane motion is necessary for hearing in crickets. *J. Comp. Physiol.* 151:397–400.

Klomberg, K. F., and C. A. Marler. 2000. The neuropeptide arginine vasotocin alters male call characteristics involved in social interactions in the grey treefrog *Hyla versicolor*. *Anim. Behav.* 59:807–12.

Kluge, A. G. 1981. The life history, social organization, and parental behavior of *Hyla rosenbergi* Boulenger, a nest-building gladiator frog. *Misc. Publ. Mus. Zool. Univ. Michigan* 160:1–170.

Klump, G. M., and H. C. Gerhardt. 1987. Use of non-arbitrary acoustic criteria in mate choice by female gray treefrogs. *Nature (London)* 326:286–88.

———. 1989. Sound localization in the barking treefrog. *Naturwissenschaften* 76: 35–37.

———. 1992. Mechanisms and function of call-timing in male-male interactions in frogs. In *Playback and studies of animal communication*, ed. P. K. McGregor, 153–74. New York: Plenum Press.

Knorr, A. 1976. Central control of mating call production and spawning in the tree frog *Hyla arborea savignyi* (Audouin): Results of electrical stimulation of the brain. *Behav. Processes* 1: 295–317.

Koch, U. T., C. J. H. Elliott, K.-H. Schaeffner, and H.-U. Kleindienst. 1988. The mechanics of stridulation of the cricket, *Gryllus campestris*. *J. Comp. Physiol. A* 162: 213–23.

Kogo, N., S. F. Perry, and J. E. Remmers. 1994. Neural organization of the ventilatory activity of the frog, *Rana catesbeiana*. Part I. *J. Neurobiol.* 25:1067–79.

———. 1997. Laryngeal motor control in frogs: Role of vagal and laryngeal feedback. *J. Neurobiol.* 33:213–22.

Kogo, N., and J. E. Remmers. 1994. Neural organization of the ventilatory activity of the frog, *Rana catesbeiana*. Part II. *J. Neurobiol.* 25:1080–94.

Kokko, H., and J. Lindström. 1996. Evolution of female preference for old mates. *Proc. R. Soc. Lond. B* 263:1533–38.

Kokko, H., W. J. Sutherland, J. Lindström, J. D. Reynolds, and A. MacKenzie. 1998. Individual mating success, lek stability, and the neglected limitations of statistical power. *Anim. Behav.* 56:755–62.

Kolluru, G. R. 1999. Variation and repeatability of calling behavior in crickets subject to a phonotactic parasitoid fly. *J. Insect Behav.* 12:611–26.

Konishi, M. 1994. Neural mechanisms of auditory image formation. In *The cognitive neurosciences*, ed. M. S. Gazzaniga, 269–77. Cambridge: MIT Press.

Köppl, C. 1995. Otoacoustic emissions as an indicator for active cochlear mechanics: A primitive property of vertebrate auditory organs. In *Advances in hearing research*, ed. G. A. Manley, G. M. Klump, C. Köppl, H. Fastl, and H. Oeckinghaus, 207–16. Singapore: World Scientific.

Kössl, M., and G. S. Boyan. 1998. Otoacoustic emissions from a nonvertebrate ear. *Naturwissenschaften* 85:124–27.

Kotiaho, J. S. 2000. Testing the assumptions of conditional handicap theory: Costs and condition dependence of a sexually selected trait. *Behav. Ecol. Sociobiol.* 48:188–94.

Krahe, R. 1997. Verarbeitung verhaltensrelevanter Lautattrappen durch die aufsteigenden auditorischen Interneurone von Felheuschrecken. Ph.D. diss., Humboldt University, Berlin.

Krahe, R., and B. Ronacher. 1993. Long rise times of sound pulses in grasshopper songs improve the directionality cues received by the CNS from the auditory receptors. *J. Comp. Physiol. A* 173:425–34.

Krupa, J. J. 1989. Alternative mating tactics in the Great Plains Toad. *Anim. Behav.* 37:1035–43.

Kruse, K. C. 1981. Mating success, fertilization potential, and male body size in the American toad (*Bufo americanus*). *Herpetologica* 37:228–33.

Kupfermann, I., and K. R. Weiss. 1978. The command neuron concept. *Behav. Brain Sci.* 1:3–39.

Kusano, T., M. Toda, and K. Fukuyama. 1991. Testes size and breeding systems in Japanese anurans with special reference to large testes in the treefrog *Rhacophorus arboreus* (Amphibia: Rhacophoridae). *Behav. Ecol. Sociobiol.* 29:27–31.

Kutsch, W. 1969. Neuromuskuläre Aktivität bei verschiedenen Verhaltensweisen von drei Grillenarten. *Z. vergl. Physiol.* 63:335–78.

———. 1989. Formation of the receptor system in the hindlimb of the locust embryo. *Roux's Arch. Dev. Biol.* 198:39–47.

Kutsch, W., and F. Huber. 1970. Zentrale versus periphere Kontrolle des Gesanges von Grillen (*Gryllus campestris*). *Z. vergl. Physiol.* 67:147–59.

———. 1989. Neural basis of song production. In *Cricket behavior and neurobiology*, ed. F. Huber, T. E. Moore, and W. Loher, 262–309. Ithaca: Cornell University Press.

Kutsch, W., and D. Otto. 1972. Evidence for spontaneous song production independent of the head ganglia in *Gryllus campestris L. J. Comp. Physiol.* 81:115–19.

Lakes-Harlan, R., H. Stölting, and A. Stumpner. 1999. Convergent evolution of insect hearing organs from a preadaptive structure. *Proc. R. Soc. London B* 266:1161–67.

Lakes-Harlan, R., A. Stumpner, and G. R. Allen. 1995. Functional adaptations of the auditory system of two parasitoid fly species, *Therobia leonidei* and *Homotrixa spec.* In *Nervous system and behavior*, ed. M. Burrows, T. Matheson, P. I. Newland, and H. Schuppe, 358. Stuttgart: Thieme Verlag.

Lance, S. L., and K. D. Wells. 1993. Are satellite spring peepers physiologically inferior to calling males? *Copeia* 1993:1162–65.

Lande, R. 1981. Models of speciation by sexual selection on polygenic traits. *Proc. Natl. Acad. Sci., USA* 78:3721–25.

Lang, F. 2000. Acoustic communication distances of a gomophocerine grasshopper. *Bioacoustics* 10:233–58.

Lang, F., and N. Elsner. 1994. Leg movement and hearing: biophysics and electrophysiology of the tympanal organ in *Locusta migratoria. J. Comp. Physiol. A* 175: 251–60.

Larsen, O. N., H.-U. Kleindienst, and A. Michelsen. 1989. Biophysical aspects of sound reception. In *Cricket behavior and neurobiology*, ed. F. Huber, T. E. Moore, and W. Loher, 364–90. Ithaca: Cornell University Press.

Larsen, O. N., and A. Michelsen. 1978. Biophysics of the ensiferan ear. III. The cricket ear as a four-input system. *J. Comp. Physiol.* 123:217–27.

Latimer, W., and D. B. Lewis. 1986. Song harmonic content as a parameter determining acoustic orientation behaviour in the cricket *Teleogryllus oceanicus* (Le Guillon). *J. Comp. Physiol. A* 158:583–91.

Lea, J., M. Dyson, and T. Halliday. 2001. Calling by male midwife toads stimulates females to maintain reproductive condition. *Anim. Behav.* 61:373–77.

Lee, H.-J., and W. Loher. 1993. The mating strategy of the short-tailed cricket *Anurogryllus arborerus* de Geer. *Ethology* 95:327–44.

Lehmann, G. U. C., and K. G. Heller. 1998. Bushcricket song structure and predation by the acoustically orienting parasitoid fly *Therobia leonidei* (Diptera: Tachinidae: Ormiini). *Behav. Ecol. Sociobiol.* 43:239–45.

Lemon, R. E., and J. Struger. 1980. Acoustic entrainment to randomly generated calls by the frog, *Hyla crucifer. J. Acoust. Soc. Am.* 67:2090–95.

Lewis D. B. 1974. The physiology of the tettigoniid ear. I. The implications of the anatomy of the ear to its function in sound reception. *J. Exp. Biol.* 60:821–37.

Lewis, E. R. 1978. Comparative studies of anuran auditory papillae. *Scan. Electr. Microsc.* 1978 (II):633–42.

———. 1981. Suggested evolution of tonotopic organization in the frog amphibian papilla. *Neurosci. Lett.* 21:131–36.

———. 1992. Convergence of design in vertebrate acoustic sensors. In *The evolutionary biology of hearing*, ed. D. B. Webster, R. R. Fay, and A. N. Popper, 163–84. New York: Springer-Verlag.

Lewis, E. R., R. A. Baird, E. L. Leverenz, and H. Koyama. 1982. Inner ear: Dye injection reveals peripheral origins of specific sensitivities. *Science* 215:1641–43.

Lewis, E. R., E. L. Leverenz, and W. S. Bialek. 1985. *The vertebrate inner ear*. Boca Raton: CRC Press.

Lewis, E. R., E. L. Leverenz, and H. Koyama. 1982. The tonotopic organization of the bullfrog amphibian papilla, an auditory organ lacking a basilar membrane. *J. Comp. Physiol.* 145:437–45.

Lewis, E. R., and C. W. Li. 1975. Hair cell types and distributions in the otolithic and auditory organs of the bullfrog. *Brain Res.* 83:35–50.

Lewis, E. R., and E. R. Lombard. 1988. The amphibian inner ear. In *The evolution of the amphibian auditory system*, ed. B. Fritzsch, M. J. Ryan, W. Wilczynski, T. E. Hetherington, and W. Walkowiak, 93–124. New York: John Wiley and Sons.

Lewis, E. R., and P. M. Narins. 1999. The acoustic periphery of Amphibians: Anatomy and physiology. In *Comparative hearing: Fish and amphibians*, ed. R. R. Fay and A. N. Popper, 101–54. Springer Handbook of Auditory Research. New York: Springer-Verlag.

Libersat, F., and R. R. Hoy. 1991. Ultrasonic startle behavior in bushcrickets (Orthoptera: Tettigoniidae). *J. Comp. Physiol. A* 169:507–14.

Libersat, F., J. A. Murray, and R .R. Hoy. 1994. Frequency as a releaser in the courtship song of two crickets, *Gryllus bimaculatus* (de Geer) and *Teleogryllus oceanicus*: A neuroethological analysis. *J. Comp. Physiol. A* 174:485–94.

Licht, L. E. 1976. Sexual selection in toads (*Bufo americanus*). *Can. J. Zool.* 54:1277–84.

Lindquist, E. D., and T. E. Hetherington. 1996. Field studies on visual and acoustic signaling in the "earless" Panamanian golden frog, *Atelopus zeteki*. *J. Herpetology* 30: 347–54.

Lindquist, E. D., T. E. Hetherington, and S. F. Volman. 1998. Biomechanical and neurophysiological studies on audition in eared and earless harlequin frogs (*Atelopus*). *J. Comp. Physiol A* 183:265–71.

Liou, L. W., and T. D. Price. 1994. Speciation by reinforcement of premating isolation. *Evolution* 48:1451–59.

Littlejohn, M. J. 1960. Call discrimination and potential reproductive isolation in *Pseudacris triseriata* females from Oklahoma. *Copeia* 1960:370–71.

———. 1965. Premating isolation in the *Hyla ewingi* complex (Anura:Hylidae). *Evolution* 19:234–43.

———. 1977. Long-range acoustic communication in anurans: An integrated and evolutionary approach. In *The reproductive biology of amphibians*, ed. D. H. Taylor and S. I. Guttman, 263–94. New York: Plenum Press.

———. 1981. Reproductive isolation: A critical review. In *Evolution and speciation*, ed. W. R. Atchley and D. S. Woodruff, 298–334. Cambridge: Cambridge University Press.

———. 1993. Homogamy and speciation: a reappraisal. In *Oxford surveys of evolutionary biology*, ed. D. Futuyma and J. Antonovics, 135–64. Oxford: Oxford University Press.

———. 1999. Variation in advertisement calls of anurans across zonal interactions: The evolution and breakdown of homogamy. In *Geographic variation in behavior*, ed. S. A. Foster and J. A. Endler, 209–33. New York: Oxford University Press.

———. 2001. Patterns of differentiation in temporal properties of acoustic signals of animals. In *Anuran communication*, ed. M. J. Ryan, 102–20. Washington: Smithsonian Institution Press.

Littlejohn, M. J., and P. A. Harrison. 1985. The functional significance of the diphasic advertisement call of *Geocrinia victoriana* (Anura: Leptodactylidae). *Behav. Ecol. Sociobiol.* 16:363–73.

Littlejohn, M. J., P. A. Harrison, and R. C. MacNally. 1985. Interspecific acoustic interactions in sympatric populations of *Ranidella signifera* and *R. parinsignifera* (Anura: Leptodactylidae). In *The biology of Australasian frogs and reptiles*, ed. G. Grigg, R. Shine, and H. Ehrmann, 287–96. Chipping Norton, NSW: Surrey Beatty and Sons.

Littlejohn, M. J., and J. J. Loftus-Hills. 1968. An experimental evaluation of premating isolation in the *Hyla ewingi* complex (Anura: Hylidae). *Evolution* 22:659–63.

Littlejohn, M. J., and A. A. Martin. 1969. Acoustic interaction between two species of leptodactylid frogs. *Anim. Behav.* 17:785–91.

Littlejohn, M. J., and J. D. Roberts. 1975. Acoustic analysis of an intergrade zone between two call races of the *Limnodynastes tasmaniensis* complex (Anura: Leptodactylidae) in south-eastern Australia. *Aust. J. Zool.* 23:113–22.

Littlejohn, M. J., and G. F. Watson. 1974. Mating call discrimination and phonotaxis by females of the *Crinia laevis* complex (Anura: Leptodactylidae). *Copeia* 1974: 171–75.

Littlejohn, M. J., G. F. Watson, and J. R. Wright. 1993. Structure of advertisement call of *Litoria ewingi* (Anura: Hylidae) introduced into New Zealand from Tasmania. *Copeia* 1993:60–67.

Lloyd, J. E. 1977. Bioluminescence and communication. In *How animals communicate*, ed. T. A. Sebeok, 164–83. Bloomington: Indiana University Press.

Loftus-Hills, J. J. 1974. Analysis of an acoustic pacemaker in Strecker's chorus frog, *Pseudacris streckeri* (Anura:Hylidae). *J. Comp. Physiol.* 90:75–87.

Loftus-Hills, J. J., and B. M. Johnstone. 1970. Auditory function, communication, and the brain-evoked response in anuran amphibians. *J. Acoust. Soc. AM.* 47:1131–38.

Loftus-Hills, J. J., and M. J. Littlejohn. 1971. Mating call sound intensities of anuran amphibians. *J. Acoust. Soc. Am.* 49:1327–29.

———. 1992. Reinforcement and reproductive character displacement in *Gastrophyrne carolinensis* and *G. olivacea* (Anura: Microhylidae): A reexamination. *Evolution* 46: 896–906.

Löhe, G., and H.-U. Kleindienst. 1994. The role of the medial septum in the acoustic trachea of the cricket *Gryllus bimaculatus.* II. Influence on directionality of the auditory system. *J. Comp. Physiol. A* 174:601–6.

Loher, W. 1989. Temporal organization of reproductive behavior. In *Cricket behavior and neurobiology*, ed. F. Huber, T. E. Moore, and W. Loher, 83–113. Ithaca: Cornell University Press.

Loher, W., and M. Dambach. 1989. Reproductive behavior. In *Cricket behavior and neurobiology*, ed. F. Huber, T. E. Moore, and W. Loher, 43–82. Ithaca: Cornell University Press.

Loher, W., and F. Huber 1965. Nervous and endocrine control of sexual behaviour in a grasshopper (*Gomphocerus rufus L.*). *Symp. Soc. Exp. Biol.* 20:483–90.

Loher, W., T. Weber, and F. Huber. 1993. The effect of mating on phonotactic behaviour in *Gryllus bimaculatus* DeGeer. *Physiol. Entomol.* 18:57–66.

Lombard, E. R., and I. R. Straughan. 1974. Functional aspects of anuran middle ear structures. *J. Exp. Biol.* 61:57–71.

Lopez, P. T., and P. M. Narins. 1991. Mate choice in the Neotropical frog, *Eleutherodactylus coqui*. *Anim. Behav.* 41:757–72.

Lopez, P. T., P. M. Narins, E. R. Lewis, and S. W. Moore. 1988. Acoustically-induced call modification in the white-lipped frog, *Leptodactylus albilabris*. *Anim. Behav.* 36: 1295–308.

Lorenz, K. 1941. Vergleinchende Bewungsstudien an Anatinen. *Suppl. J. Orthin.* 89: 194–294.

Lucas, J. R., R. D. Howard, and J. G. Palmer. 1996. Callers and satellites: Chorus behaviour in anurans as a stochastic dynamic game. *Anim. Behav.* 51: 501–18.

Luksch, H., and W. Walkowiak. 1998. Morphology and axonal projection patterns of auditory neurons in the midbrain of the painted frog, *Discoglossus pictus*. *Hearing Res.* 122:1–17.

Luksch, H., W. Walkowiak, A. Muno, and H. J. tenDonkelaar. 1996. The use of in vitro preparations of the isolated amphibian central nervous system in neuroanatomy and electrophysiology. *J. Neurosci. Methods* 70:91–102.

Machens, C. K., M. B. Stemmler, P. Prinz, R. Krahe, B. Ronacher, and A. V. M. Herz. 2001. Representation of acoustic communication signals by insect auditory receptor neurons. *J. Neuroscience* 21:3215–27.

Mac Nally, R. C. 1981a. On the reproductive energetics of chorusing males: Costs and patterns of call production in two sympatric species of *Ranidella* (Anura). *Oikos* 42:82–91.

———. 1981b. On the reproductive energetics of chorusing males: Energy depletion profiles, restoration and growth in two sympatric species of *Ranidella* (Anura). *Oecologica* 51:181–88.

Marler, C. A., J. Chu, and W. Wilczynski. 1995. Arginine vasotocin injection increases probability of calling in cricket frogs but causes call changes characteristic of less aggressive males. *Hormones and Behavior* 29:554–70.

Marler, C. A., and M. J. Ryan. 1996. Energetic constraints and steroid hormone correlates of male calling behaviour in the túngara frog. *J. Zool. London* 240:397–409.

Márquez, R. 1993. Male reproductive success in two midwife toads, *Alytes obsteticans* and *A. cisternasii*. *Behav. Ecol. Sociobiol.* 32:283–91.

———. 1995. Female choice in the midwife toads (*Alytes obsteticans* and *A. cisternasii*). *Behaviour* 132:151–61.

Márquez, R., and J. Bosch. 1997a. Female preference in complex acoustical environments in the midwife toads *Alytes obstetricans* and *Alytes cisternasii*. *Behav. Ecol.* 8: 588–94.

———. 1997b. Male advertisement call and female preference in sympatric and allopatric midwife toads. *Anim. Behav.* 54:1333–45.

Márquez, R., and M. Tejedo. 1990. Size-based mating pattern in the tree-frog *Hyla arborea*. *Herpetologica* 46:172–290.

Marsh, R. L. 1999. Contractile properties of muscles used in sound production and locomotion in two species of gray tree frog. *J. Exp. Biol.* 202:3215–23.

Marshall, D. C., and J. R. Cooley. 2000. Reproductive character displacement and speciation in periodical cicadas, with description of a new species, 13-year *Magicicada neotredecim*. *Evolution* 54:1313–25.

Martin, S. D., D. A. Gray, and W. H. Cade. 2000. Fine-scale temperature effects on cricket calling song. *Can. J. Zool.* 78:706–12.

Martin, W. F. 1967. The mechanism and evolution of sound production in the toad genus *Bufo*. M.A. thesis, University of Texas, Austin.

———. 1971. Mechanics of sound production in toads of the genus *Bufo*: Passive elements. *J. Exp. Zool.* 176:273–94.

———. 1972. Evolution of vocalization in the genus *Bufo*. In *Evolution in the genus Bufo*, ed. W. F. Blair, 279–309. Austin: University of Texas Press.

Martin, W. F., and C. Gans. 1972. Muscular control of the vocal tract during release signaling in the toad *Bufo valliceps*. *J. Morph.* 137:1–28.

Martins, E. P., ed. 1996. *Phylogenies and the comparative method in animal behavior*. Oxford: Oxford University Press.

Martins, M., J. P. Pombal, and C. F. B. Haddad. 1998. Escalated aggressive behaviour and facultative parental care in the nest building gladiator frog, *Hyla faber*. *Amphibia-Reptilia* 19:65–73.

Martof, B. F. 1961. Vocalization as an isolating mechanism in frogs. *Amer. Midl. Nat.* 65:118–26

Mason, A. C. 1991. Hearing in a primitive ensiferan: The auditory system of *Cyphoderris monstrosa* (Orthoptera: Haglidae). *J. Comp. Physiol. A* 168:351–63.

Mason, A. C., G. K. Morris, and R. R. Hoy. 1999. Peripheral frequency mismatch in the primitive ensiferan *Cyphoderris monstrosa* (Orthoptera: Haglidae). *J. Com. Physiol. A* 184:543–51.

Mason, A. C., and K. Schildberger. 1993. Auditory interneurons in *Cyphoderris monstrosa* (Orthoptera: Haglidae). *J. Comp. Physiol. A* 171:749–57.

May, M. L., P. D. Brodfuehrer, and R. R. Hoy. 1988. Kinematic and aerodynamic aspects of ultrasound-induced negative phonotaxis in flying Australian field crickets *Teleogryllus oceanicus*. *J. Comp. Physiol. A* 164:243–50.

Maynard Smith, J. 1982. *Evolution and the theory of games*. Cambridge: Cambridge University Press.

Maynard Smith, J., and G. A. Parker. 1976. The logic of asymmetric contests. *Anim. Behav.* 24:159–75.

McClelland, B. E., W. Wilczynski, and M. J. Ryan. 1996. Correlations between call characteristics and morphology in male cricket frogs (*Acris crepitans*). *J. Exp. Biol.* 199:1907–19.

McLean, H. A., N. Kimura, N. Kogo, S. F. Perry, and J. E. Remmers. 1995. Fictive respiratory rhythm in the isolated brainstem of frogs. *J. Comp. Physiol. A* 176:703–13.

McLean, H. A., and J. E. Remmers. 1997. Characterization of respiratory-related neurons in the isolated brainstem of the frog. *J. Comp. Physiol. A* 181:153–59.

McLister, J. D. 2001. Physical factors affecting the cost and efficiency of sound production in the treefrog *Hyla versicolor*. *J. Exp. Biol.* 204:69–80.

McLister, J. D., E. D. Stevens, and J. P. Bogart. 1995. Comparative contractile dynamics of calling and locomotor muscles in three hylid frogs. *J. Exp. Biol.* 198:1527–38.

Megela, A. L. 1984. Diversity of adaptation patterns in responses of eighth nerve fibers in the bullfrog, *Rana catesbeiana*. *J. Acoust. Soc. Am.* 75:1155–62.

Megela, A. L., and R. R. Capranica. 1981. Response patterns to tone bursts in peripheral auditory system of anurans *J. Neurophysiol.* 46:465–78.

Meier, T., and H. Reichert. 1990. Embryonic development and evolutionary origin of the orthopteran auditory organs. *J. Neurobiol.* 21:592–610.

Meixner, A. J., and K. C. Shaw. 1986. Acoustic and associated behavior of the coneheaded katydid *Neoconocephalus nebrascensis* (Orthoptera Tettigoniidae). *Ann. Entomol. Soc. Am.* 79:554–65.

Metzner, W. 1993. The jamming avoidance response in *Eigenmannia* is controlled by two separate motor pathways. *J. Neuroscience* 13:1862–78.

Metzner, W., and S. Viete. 1996. The neuronal basis of communication and orientation in the weakly electric fish, *Eigenmannia. Naturwissenschaften* 83:71–77.

Meyer, J., and N. Elsner. 1996. How well are frequency sensitivities of grasshopper ears tuned to species-specific song spectra? *J. Exp. Biol.* 199:1631–42.

———. 1997. Can spectral cues contribute to species separation in closely related grasshoppers? *J. Comp. Physiol. A* 180:171–80.

Michel, K. 1974. Das Tympanalorgan von *Gryllus bimaculatus* DeGeer (Saltatoria, Gryllidae). *Z. Morph. Tiere* 77:285–315.

———. 1975. Das Tympanalorgan von *Cicada orni* L (Cicadina, Homoptera). Eine Licht-und elektronen-mikroskopische Untersuchung. *Zoomorphologie* 82:63–78.

Michelsen, A. 1966. Pitch discrimination in the locust ear: Observations on single sense cells. *J. Insect Physiol.* 12:1119–31.

———. 1968. Frequency discrimination in the locust ear by means of four groups of receptor cells. *Nature (London)* 220:585–86.

———. 1971a. The physiology of the locust ear. I. Frequency sensitivity of single cells in the isolated ear. *Z. vergl. Physiol.* 71:49–62.

———. 1971b. The physiology of the locust ear. II. Frequency discrimination based upon resonances in the tympanum. *Z. vergl. Physiol.* 71:63–101.

———. 1978. Sound reception in different environments. In *Perspectives in sensory ecology*, ed. A. B. Ali, 345–73. New York: Plenum Press.

———. 1983. Biophysical basis of sound communication. In *Bioacoustics: A comparative approach*, ed. B. Lewis, 3–38. London: Academic Press.

———. 1992. Hearing and sound communication in small animals: Evolutionary adaptations to the laws of physics. In *The evolutionary biology of hearing*, ed. D. B. Webster, R. R. Fay, and A.N. Popper, 61–77. New York: Springer-Verlag.

———. 1994. Directional hearing in crickets and other small animals. *Fortschr. Zoologie* 39:195–207.

———. 1998a. Biophysical basis of sound localization in insects. In *Comparative hearing: Insects*, ed. R. R. Hoy, A. N. Popper, and R. R. Fay, 18–62. New York: Springer-Verlag.

———. 1998b. The tuned cricket. *News Physiol. Sci.* 13:31–38.

Michelsen, A., and N. Elsner. 1999. Sound emission and the acoustic far field of a singing acridid grasshopper (*Omocestus viridulus* L.). *J. Exp. Biol.* 202:1571–77.

Michelsen, A., and P. Fonseca. 2000. Spherical sound radiation patterns of singing grass cicadas *Tympanistalua gastrica*. *J. Comp. Physiol. A* 186:163–68.

Michelsen, A., M. B. Jørgensen, J. Christensen-Dalsgaard, and R. R. Capranica. 1986. Directional hearing in awake, unrestrained treefrogs. *Naturwissenschaften* 73:682–83.

Michelsen, A., and O. N. Larsen . 1978. Biophysics of the ensiferan ear. I. Tympanal vibrations in bushcrickets (Tettigoniidae) studied with laser vibrometry. *J. Comp. Physiol.* 123:193–203.

———. 1983. Strategies for acoustic communication in complex environments. In *Neuroethology and behavioral physiology*, ed. F. Huber and H. Markl, 321–31. Berlin: Springer-Verlag.

———. 1985. Hearing and sound. In *Comprehensive insect physiology, biochemistry and pharmacology*, ed. G. A. Kerkut and L. I. Gilbert, 495–556. Oxford: Pergamon Press.

Michelsen A., O. N. Larsen, and A. Surlykke. 1985. Auditory processing of temporal cues in insect songs: Frequency or time domain? In *Time resolution in auditory systems*, ed. A. Michelsen, 3–27. Heidelberg: Springer-Verlag.

Michelsen, A., and G. Löhe. 1995.Tuned directionality in cricket ears. *Nature (London)* 375:369.

Michelsen, A., and H. Nocke. 1974. Biophysical aspects of sound communication in insects. *Adv. Insect Physiol.* 10:247–96.

Michelsen, A., A. V. Popov, and B. Lewis. 1994. Physics of directional hearing in the cricket *Gryllus bimaculatus*. *J. Comp. Physiol. A* 175:153–64.

Michelsen, A., and K. Rohrseitz. 1995. Directional sound processing and interaural sound transmission in a small and a large grasshopper. *J. Exp. Biol.* 198:1817–27.

———. 1997. Sound localisation in a habitat: An analytical approach to quantifying the degradation of directional cues. *Bioacoustics* 7:291–313.

Michelsen, A., ed. 1985. *Time resolution in auditory systems.* Heidelberg: Springer-Verlag.

Miller, J. P., and A. I. Selverston. 1979. Rapid killing of single neurons by irradiation of intracellular injected dye. *Science* 206:702–4.

Minckley, R. L., and M. D. Greenfield. 1995. Psychoacoustics of female phonotaxis and the evolution of male signal interactions in Orthoptera. *Ethology, Ecol. and Evol.* 7:235–43.

Minckley, R. L., M. D. Greenfield, and M. K. Tourtellot. 1995. Chorus structure in tarbush grasshoppers: Inhibition, selective phonoresponse, and signal competition. *Anim. Behav.* 50:579–94.

Mitchell, N. J. 2001. Males call from wetter nests: Effects of substrate water potential on reproductive behaviours of terrestrial toadlets. *Proc. Royal Soc. London B* 268: 87–93.

Miyamoto, M. M., and J. H. Cane. 1980. Behavioral observations on noncalling males in Costa Rican *Hyla ebraccata*. *Biotropica* 12:225–27.

Mohneke, R., and H. Schneider. 1979. Effect of temperature upon auditory thresholds in two anuran species, *Bombina v. variegata* and *Alytes o. obstetricans* (Amphibia: Discoglossidae). *J. Comp. Physiol.* 130:9–16.

Moiseff, A., and R. R. Hoy. 1983. Sensitivity to ultrasound in a identified auditory interneuron in the cricket: A possible neural link to phonotactic behavior. *J. Comp. Physiol.* 152:155–67.

Moiseff, A., G. S. Pollack, and R. R. Hoy. 1978. Steering responses of flying crickets to sound and ultrasound: Mate attraction and predator avoidance. *Proc. Natl. Acad. Sci.* USA 75:4052–56.

Moller, A. P. 2001. Female preference for symmetric calls in a grasshopper. *Ethology, Ecology and Evolution* 13:261–72.

Moore, S. W., E. R. Lewis, P. M. Narins, and P. T. Lopez. 1989. The call-timing algorithm of the white-lipped frog, *Leptodactylus albilabris*. *J. Comp. Physiol. A* 164: 309–19.

Mörchen, A., J. Rheinlaender, and J. Schwartzkopff. 1978. Latency shift in insect auditory nerve fibers. *Naturwissenschaften* 65:656.

Morris, G. K. 1970. Sound analysis of *Metrioptera sphagnorum* (Orthoptera: Tettigoniidae). *Canadian Entomol.* 102:363–68.

Morris, G. K., and J. H. Fullard. 1983. Random noise and congeneric discrimination in *Conocephalus* (Orthoptera: Tettigoniidae). In *Orthopteran mating systems: Sexual competition in a diverse group of insects*, ed. D. T. Gwynne and G. K. Morris, 73–96. Boulder: Westview Press.

Morris, G. K., A. C. Mason, and P. Wall. 1994. High ultrasonic and tremulation signals in neotropical katydids (Orthoptera: Tettigoniidae). *J. Zool. London* 233:129–63.

Morris, M. R. 1989. Female choice of large males in the treefrog *Hyla chrysoscelis*: The importance of identifying the scale of choice. *Behav. Ecol. Sociobiol.* 25:275–81.

———. 1991. Female choice of large males in the treefrog *Hyla ebraccata*. *J. Zool. London* 223:371–78.

Morris, M. R., and S. L. Yoon. 1989. A mechanism for female choice of large males in the treefrog *Hyla chrysoscelis*. *Behav. Ecol. Sociobiol.* 25:65–71.

Morrison, C., J. M. Hero, and W. P. Smith. 2001. Mate selection in *Litoria chloris* and *Litoria xanthomera*: Females prefer smaller males. *Austral Ecol.* 26:223–32.

Möss, D. 1971. Sinnesorgane im Bereich des Flügels der Feldgrille (*Gryllus campestris L.*) und ihre Bedeutung für die Kontrolle der Singbewegung und die Einstellung der Flügellage. *Z. vergl. Physiol.* 73:53–83.

Mudry, K. M., and R. R. Capranica. 1987a. Correlation between auditory thalamic areas evoked responses and species-specific call characteristics. I. *Rana catesbeiana* (Anura: Ranidae). *J. Comp. Physiol. A* 160:477–89.

———. 1987b. Correlation between auditory thalamic area evoked potentials and species-specific call characteristics II. *Hyla cinerea* (Anura: Hylidae). *J. Comp. Physiol. A* 161:407–16.

Muller, K. L. 1998. The role of conspecifics in habitat settlement in a territorial grasshopper. *Anim. Behav.* 56:479–85.

Münch, D. 1999. Frequenz- und Zeitverarbeitung durch thorakale auditorische Interneurone bei Zikaden (*Tettigetta josei*). Diploma thesis, Humboldt Universität, Berlin.

Murphey, R. K., and M. D. Zaretsky. 1972. Orientation to calling song by female crickets, *Scapsipedus marginatus* (Gryllidae). *J. Exp. Biol.* 56:335–52.

Murphy, C. G. 1994a. Chorus tenure of male barking treefrogs, *Hyla gratiosa*. *Anim. Behav.* 48:763–77.

———. 1994b. Determinants of chorus tenure in barking treefrogs (*Hyla gratiosa*). *Behav. Ecol. Sociobiol.* 34:285–94.

———. 1998. Interaction-independent sexual selection and the mechanism of sexual selection. *Evolution* 52:8–18.

———. 1999. Nightly timing of chorusing by male barking treefrogs (*Hyla gratiosa*): The influence of female arrival and energy. *Copeia* 1999:333–47.

Murphy, C. G., and H. C. Gerhardt. 1996. Evaluating experimental designs for determining mate choice: The effect of amplexus on mate choice by barking treefrogs. *Anim. Behav.* 51:881–90.

———. 2000. Preference functions of individual female barking treefrogs, *Hyla gratiosa*. *Evolution* 54:660–69.

———. 2002. Mate-sampling by female barking treefrogs (*Hyla gratiosa*). *Behav. Ecol.*, in press.

Naguib, M. 1996. Ranging by song in Carolina wrens *Thryothorus ludovicianus*: Effects of environmental acoustics and strength of sound degradation. *Behaviour* 133: 541–59.

Narins, P. M. 1982. Behavioral refractory period in neotropical treefrogs. *J. Comp. Physiol.* 148:337–44.

———. 1987. Coding of signals in noise by amphibian auditory nerve fibers. *Hearing Res.* 26:145–54.

———. 1992a. Biological constraints on anuran acoustic communication: Auditory capabilities of naturally behaving animals. In *The evolutionary biology of hearing*, ed. D. B. Webster, R. R. Fay, and A. N. Popper, 439–54. New York: Springer-Verlag.

———. 1992b. Reduction of tympanic membrane displacement during vocalization of the arboreal frog, *Eleutherodactylus coqui*. *J. Acoust. Soc. Am.* 91:3551–57.

Narins, P. M., and R. R. Capranica. 1978. Communicative significance of the two-note call of the treefrog *Eleutherodactylus coqui*. *J. Comp. Physiol. A* 127:1–9.

———. 1980. Neural adaptation for processing the two-note call of the Puerto Rican treefrog, *Eleutherodactylus coqui*. *Brain, Behav. Evol.* 17:48–66.

Narins, P. M, G. Ehret, and J. Tautz. 1988. Accessory pathway for sound transfer in a neotropical frog. *Proc. Natl. Acad. Sci. USA* 85:1508–12.

Narins, P. M., and C. M. Hillery. 1983. Frequency coding in the inner ear of anuran amphibians. In *Hearing—physiological bases and psychophysics*, ed. R. Klinke and R. Hartmann, 70–75. Berlin: Springer-Verlag.

Narins, P. M., and E. R. Lewis. 1984. The vertebrate ear as an exquisite seismic sensor. *J. Acoust. Soc. Am.* 76:1384–87.

———. 1996. Extended call repertoire of a Madagascar frog. *Biogeogr. de Madagasgar* 1996:403–10.

Narins, P. M., E. R. Lewis, and B. E. McClelland. 2000. Hyperextended call note repertoire of the endemic Madagascar treefrog *Boophis madagascariensis* (Rhacophoridae). *J. Zool., London* 250:283–98.

Narins, P. M., E. R. Lewis, A. P. Purgue, P. J. Bishop, L. R. Minter, and D. P. Lawson. 2001. Functional consequences of a novel middle ear adaptation in the central African frog *Petropedetes parkeri* (Ranidae). *J. Exp. Biol.* 204:1223–32.

Narins, P. M., and S. L. Smith. 1986. Clinal variation in anuran advertisement calls: Basis for acoustic isolation? *Behav. Ecol. Sociobiol.* 19:135–41.

Narins, P. M., and R. Zelick. 1988. The effects of noise on auditory processing and behavior in amphibians. In *The evolution of the amphibian auditory system*, ed. B. Fritzsch, M. J. Ryan, W. Wilczynski, T. E. Hetherington, and W. Walkowiak, 511–36. New York: John Wiley and Sons.

Nelson, C. M., and T. G. Nolen 1997. Courtship song, male agonistic encounters, and female mate choice in the house cricket, *Acheta domesticus* (Orthoptera: Gryllidae). *J. Insect Behav.* 10:557–70.

Nelson, D. A., and P. Marler. 1990. The perception of birdsong and an ecological concept of signal space. In *Comparative perception, Vol. II: Complex signals*, ed. W. C. Stebbins and M. A. Berkley, 443–78. New York: John Wiley and Sons.

Neuweiler, G. 1999. Neuroethology, vertebrates. In *Encyclopedia of neuroscience*, 2nd ed., ed. G. Adelman and B. H. Smith, 1345–52. Amsterdam: Elesvier Science.

Nocke, H. 1971. Biophysik der Schallerzeugung durch die Vorderflügel der Grillen. *Z. vergl. Physiol.* 74:272–314.

———. 1972. Physiological aspects of sound communication in crickets (*Gryllus campestris*). *J. Comp. Physiol.* 80:141–62.

Nolen, T. G., and R. R. Hoy. 1984. Initiation of behavior by single neurons: The role of behavioral context. *Science* 226:992–94.

———. 1986a. Phonotaxis in flying crickets I. Attraction to the calling song and avoidance of bat-like ultrasound. *J. Comp. Physiol. A* 159:423–39.

———. 1986b. Phonotaxis in flying crickets II. Physiological correlates of two-tone suppression of the high-frequency avoidance steering behavior by the calling song. *J. Comp. Physiol. A* 159:441–56.

Noor, M. A. 1995. Speciation driven by natural selection in *Drosophila*. *Nature (London)* 375:674–75.

———. 1999. Reinforcement and other consequences of sympatry. *Heredity* 83:503–8.

Noor, M. A. F., M. A. Williams, D. Alvarez, D. and M. Ruiz-Garcia. 2000. Lack of evolutionary divergence in courtship songs of *Drosophila pseudoobscura* subspecies. *J. Insect Behav.* 13:255–62.

Ocker, W.-G., B. Hedwig, and N. Elsner. 1995. Application of putative neurotransmitters elicits and modulates stridulation in two species of acridid grasshoppers. *J. Exp. Biol.* 198:1701–10.

Odendaal, F. J., C. M. Bull, and S. R. Telford. 1986. Influence of the acoustic environment on the distribution of the frog *Ranidella riparia*. *Anim. Behav.* 34:1836–43.

Oldfield, B. P. 1980. Accuracy of orientation in female crickets, *Teleogryllus oceanicus* (Gryllidae): Dependence on the song spectrum. *J. Comp. Physiol.* 141:93–99.

———. 1982. Tonotopic organization of auditory receptors in Tettigoniidae (Orthoptera: Ensifera). *J. Comp. Physiol.* 147:461–69.

———. 1985. The tuning of auditory receptors in bushcrickets. *Hearing Res.* 17:27–35.

———. 1988. The effect of temperature on the tuning and physiology of insect auditory receptors. *Hearing Res.* 35:151–58.

Oldfield, B. P., and K. G. Hill. 1983. The physiology of ascending auditory interneurons in the tettigoniid *Caedicia simplex* (Orthoptera: Ensifera): Response properties and a model of integration in the afferent pathway. *J. Comp. Physiol. A* 152:495–508.

———. 1986. Functional organization of insect auditory sensilla. *J. Comp. Physiol.* 158:27–34.

Oldfield, B. P., H.-U. Kleindienst, and F. Huber. 1986. Physiology and tonotopic organization of auditory receptors in the cricket *Gryllus bimaculatus* DeGeer. *J. Comp. Physiol. A* 159:457–64.

Oldham, R. S., and H. C. Gerhardt. 1975. Behavioral isolation of the treefrogs *Hyla cinerea* and *Hyla gratiosa*. *Copeia* 1975:223–31.

Olding, P. 1998. The diversity of advertisement call structure found in the Microhylidae of Australia. Ph.D. diss., University of Oxford, Oxford.

Ossiannilsson, F. 1949. Insect drummers. *Opuscula Entomologica, Suppl.* 10:1–145.

Otte, D. 1970. A comparative study of communicative behavior in grasshoppers. *Misc. Publ. Mus. Zool. Univ. Michigan* 141:1–167.

———. 1974. *The American grasshoppers. Vol. 2.* Cambridge: Harvard University Press.

———. 1977. Communication in Orthoptera. In *How animals communicate*, ed. T. A. Sebeok, 334–61. Bloomington: Indiana University Press.

———. 1989. Speciation in Hawaiian crickets. In *Speciation and its consequences*, ed. D. Otte and J. A. Endler, 482–526. Sunderland, Mass.: Sinauer Associates.

———. 1992. Evolution of cricket songs. *J. Orthopteran Res.* 1:24–46.

———. 1994. *The crickets of Hawaii: Origin, systematics and evolution.* Philadelphia: The Orthoperists' Society at the Academy of Natural Sciences of Philadelphia.

Otto, D. 1967. Untersuchungen zur nervösen Kontrolle des Grillengesanges. *Zool. Anz. Suppl.* 31:585–92.

———. 1971.Untersuchungen zur zentralnervösen Kontrolle der Lauterzeugung von Grillen. *Z. vergl. Physiol.* 74:227–71.

———. 1978. Änderungen von Gesangsparametern bei der Grille (*Gryllus campestris L.*) nach Injektion von Pharmaka ins Gehirn. *Verh. Dtsch. Zool. Ges.* 1978:245.

Otto, D., and T. Weber. 1985. Plurisegmental neurones of the cricket *Gryllus campestris* L., that discharge in the rhythm of its own song. *J. Insect Physiol.* 31:537–48.

Páez, V. P., B. C. Bock, and A. S. Rand. 1993. Inhibition of evoked calling of *Dendrobates pumilio* due to acoustic interference from cicada calling. *Biotropica* 25:242–45.

Pallett, J. R., and N. I. Passmore. 1988. The significance of multi-note advertisement calls in a reed frog, *Hyperolius tuberilinguis*. *Bioacoustics* 1:13–23.

Parker, G. A. 1974. Assessment strategy and the evolution of fighting behaviour. *J. Theor. Biol.* 47:223–43.

———. 1982. Phenotype-limited evolutionary stable strategies. In *Current problems in sociobiology*, ed. King's college sociobiology group, 173–201. Cambridge: Cambridge University Press.

Passmore, N. I., P. J. Bishop, and N. Caithness. 1992. Calling behaviour influences mating success in male painted reed frogs, *Hyperolius marmoratus*. *Ethology* 92:227–41.

Passmore, N. I., R. R. Capranica, R. S. Telford, and P. J. Bishop. 1984. Phonotaxis in the painted reed frog (*Hyperolius marmoratus*): The localization of elevated sound sources. *J. Comp. Physiol. A* 154:189–97.

Paterson, H. E. H. 1985. The recognition concept of species. In *Species and speciation*, ed. E. S. Vrba, 21–29. Pretoria: Transvaal Museum Monograph No. 4.

Paul, R. C., and T. J. Walker. 1979. Arboreal singing in a burrowing cricket, *Anurogryllus arboreus*. *J. Comp. Physiol.* 132:217–23.

Payne, R. J. H., and M. Pagel. 1996. When is false modesty a false economy? An optimality model of escalating signals. *Proc. R. Soc. London B* 263:1545–50.

Pearl, C. A., M. Cervantes, M. Chan, U. Ho, R. Shoji, and E. O. Thomas. 2000. Evidence for a mate-attracting chemosignal in the dwarf African clawed frog *Hymenochris*. *Hormones and Behavior* 38:67–74.

Pearson, K. G. 1993. Common principles of motor control in vertebrates and invertebrates. *Annu. Rev. Neurosci.* 16:265–97.

Penna, M., W.-Y. Lin, and A .S. Feng. 1997. Temporal selectivity for complex signals by single neurons in the torus semicircularis of *Pleurodema thaul* (Amphibia: Leptodactylidae). *J. Comp. Physiol. A* 180:313–28.

Penna, M., and R. Solís. 1996. Influence of burrow acoustics on sound reception by frogs *Eupsophus* (Leptodactylidae). *Anim. Behav.* 51:255–63.

———. 1998. Frog call intensities and sound propagation in the South American temperate forest region. *Behav. Ecol. Sociobiol.* 42:371–81.

Perrill, S. A. 1984. Male mating behavior in *Hyla regilla*. *Copeia* 1984:727–32.

Perrill, S. A., and R. E. Daniel. 1983. Multiple eggs clutches in *Hyla regilla*, *H. cinerea*, and *H. gratiosa*. *Copeia* 1983:513–16.

Perrill, S. A., H. C. Gerhardt, and R. Daniel. 1978. Sexual parasitism in the green treefrog (*Hyla cinerea*). *Science* 200:1179–80.

———. 1982. Mating strategy shifts in male green treefrogs (*Hyla cinerea*): An experimental study. *Anim. Behav.* 30:43–48.

Peters, B. H., H. Römer, and V. Marquart. 1986. Spatial segregation of synaptic inputs and outputs in a locust *Locusta-migratoria* auditory interneuron. *J. Comp. Neurol.* 254:34–50.

Petrie, M. 1988. Intraspecific variation in structures that display competitive ability: Large animals invest relatively more. *Anim. Behav.* 36:1174–79.

Pfennig, K. S. 2000. Female spadefoot toads compromise on mate quality to ensure conspecific matings. *Behav. Ecol.* 11:220–27.

Pfennig, K .S., K. Rapa, and R. McNair. 2000. Evolution of male mating behavior: Male spadefoot toads preferentially associate with conspecific males. *Behav. Ecol. Sociobiol.* 48:69–74.

Phelps, S. M., and M. J. Ryan. 2000. History influences signal recognition: Neural network models of túngara frogs. *Proc. Royal S. London B* 267:1633–39.

Pinder, A. C., and A. R. Palmer. 1983. Mechanical properties of the frog ear: Vibration measurements under free and closed-field acoustic conditions. *Proc. R. Soc. London B* 219:371–96.

Pires, A., and R. R. Hoy. 1992a. Temperature coupling in cricket acoustic communication I. Field and laboratory studies of temperature effects on calling song production and recognition in *Gryllus firmus*. *J. Comp. Physiol. A* 171:69–78.

———. 1992b. Temperature coupling in cricket acoustic communication II. Localization of temperature effects on song production and recognition networks in *Gryllus firmus*. *J. Comp. Physiol. A* 171:79–92.

Plewka. 1993. Zur Erkennung zeitlicher Gesangsstrukturen bei Laubheuschreceken: Eine vergleichende Untersuchung der Arten *Tettigonia cantans* und *Leptophytes laticauda*. Ph.D diss., University of Frankfurt.

Pollack, G. S. 1982. Sexual differences in calling song recognition. *J. Comp. Physiol.* 146:217–22.

———. 1986. Discrimination of calling song models by the cricket, *Teleogryllus oceanicus*: The influence of sound direction on neural coding of the stimulus temporal pattern and on phonotactic behavior. *J. Comp. Physiol.* 158:549–61.

———. 1988. Selective attention in an insect auditory neuron. *J. Neurosci.* 8:2635–39.

———. 1994. Synaptic inputs to the omega neuron in the cricket *Teleogryllus oceanicus*: Differences in EPSP waveforms evoked by low and high frequencies. *J. Comp. Physiol. A* 174:83–89.

———. 1998. Neural processing of acoustic signals. In *Comparative hearing: Insects*, ed. R. R. Hoy, A. N. Popper, and R. R. Fay, 139–96, New York: Springer-Verlag.

———. 2001. Neural representation of sound amplitude by functionally different auditory receptors in crickets. *J. Acoust. Soc. Am.* 109:1247–60.

Pollack, G. S., and E. El-Feghaly. 1993. Calling song recognition in the cricket *Teleogryllus oceanicus*: Comparison of the effects of stimulus intensity and sound spectrum on selectivity for temporal pattern. *J. Comp. Physiol. A* 171:759–65.

Pollack, G. S., and R. R. Hoy. 1981. Phonotaxis to individual rhythmic components of a complex cricket *Teleogryllus oceanicus* calling song. *J. Comp. Physiol.* 144:367–74.

———. 1989. Evasive acoustic behavior and its neurobiological basis. In *Cricket behavior and neurobiology*, ed. F. Huber, T. E. Moore, and W. Loher, 340–63. Ithaca: Cornell University Press.

Pollack, G. S., F. Huber, and T. Weber. 1984. Frequency and temporal pattern-dependent phonotaxis of crickets (*Teleogryllus oceanicus*) during tethered flight and compensated walking. *J. Comp. Physiol. A* 154:13–26.

Pollack, G. S., and N. Plourde 1982. Directionality of acoustic orientation in flying crickets. *J. Comp. Physiol.* 146:207–15.

Pomiankowski, A. 1988. The evolution of female mating preferences for male genetic quality. *Oxford Surveys in Evolutionary Biology* 5:136–84.

Pomiankowski, A., and L. Sheridan. 1994. Female choice and genetic correlations. *Trends Ecol. Evol.* 9:343.

Popov, A. V., and V. F. Shuvalov. 1977. Phonotactic behaviour of crickets. *J. Comp. Physiol. A* 119:111–26.

Popov, A. V., V. F. Shuvalov, I. D. Svetlogorskaya, and A. M. Markovich. 1974. Acoustic behavior and the auditory system in insects. In *Mechanoreception*, ed. W. Schwartzkopff, 53:281–306. Bochum: Abh. RWA Wiss. Symp.

Porter, K. R. 1965. Intraspecific variation in mating call of *Bufo coccifer* Cope. *Amer. Midl. Nat.* 74:250–56.

———. 1968. Evolutionary status of a relict population of *Bufo hemiophrys* Cope. *Evolution* 22:583–94.

Pough, F. H., W. E. Magnusson, M. J. Ryan, K. D. Wells, and T. L. Taigen. 1992. Behavioral energetics. In *Environmental physiology of the amphibians*, ed. M. E. Feder and W. W. Burggren, 395–436. Chicago: University of Chicago Press.

Prestwich, K. N. 1994. The energetics of acoustic signaling in anurans and insects. *Amer. Zool.* 34:625–43.

Prestwich, K. N., K. E. Brugger, and M. Topping. 1989. Energy and communication in three species of hylid frogs: Power output and efficiency. *J. Exp. Biol.* 144: 53–80.

Prestwich, K. N., K. M. Lenihan, and D. M. Martin. 2000. The control of carrier frequency in cricket calls: A refutation of the subalar-tegminal resonance/auditory feedback model. *J. Exp. Biol.* 203:585–96.

Prestwich, K. N., and T. J. Walker. 1981. Energetics of singing in crickets: Effect of temperature in three trilling species (Orthoptera: Gryllidae). *J. Comp. Physiol.* 143: 199–212.

Pröhl, H., and W. Hödl. 1999. Parental investment, potential reproductive rates, and mating system in the strawberry dart-poison frog, *Dendrobates pumilio*. *Behav. Ecol. Sociobiol.* 46:215–20.

Prosser, M. R., A. M. Murray, and W. H. Cade. 1993. The influence of female age on phonotaxis during single and multiple song presentations in the field cricket, *Gryllus integer* (Orthoptera: Gryllidae). *J. Insect Behav.* 10:437–49.

Ptacek, M. B., H. C. Gerhardt, and R. D. Sage. 1994. Speciation by polyploidy in treefrogs: Multiple origins of the tetraploid, *Hyla versicolor*. *Evolution* 48:898–908.

Purgue, A. P. 1995. The sound broadcasting system of the bullfrog (*Rana catesbeiana*). Ph.D. diss., University of Utah.

———. 1997. Tympanic sound radiation in the bullfrog *Rana catesbeiana*. *J. Comp. Physiol. A* 181:438–45.

Purgue, A. P., and P. M. Narins. 2000a. Mechanics of the inner ear of the bullfrog (*Rana catesbeiana*): The contact membranes and the periodic canal. *J. Comp. Physiol. A* 186: 481–88.

———. 2000b. A model for energy flow in the inner ear of the bullfrog (*Rana catesbeiana*). *J. Comp. Physiol. A* 186:489–95.

Quinn, V. S., and D. K. Hews. 2000. Signals and behavioural responses are not coupled in males: Aggression affected by replacement of an evolutionarily lost colour signal. *Proc. R. Soc. London B* 267:755–58.

Rabb, G. B. 1960. On the unique sound production of the Surinam toad, *Pipa pipa*. *Copeia* 1960:368–69.

Ragge, D. R., and W. J. Reynolds. 1998. *The songs of the grasshoppers and crickets of western Europe*. Colchester: Harley Books.

Ralin, D. B. 1977. Evolutionary aspects of mating call variation in a diploid-tetraploid species complex of treefrogs (Anura). *Evolution* 31:721–36.

Rand, A. S., and M. J. Ryan. 1981. The adaptive significance of a complex repertoire in a neotropical frog (*Physalaemus pustulosus*). *Zeit. f. Tierpsychol.* 57:209–14.

Ratcliffe, L. M., and P. R. Grant 1983. Species recognition in Darwin's finches (*Geospiza*, Gould). II. Geographic variation in mate preference. *Anim. Behav.* 31:1154–65.

Regen, J. 1913. Über die Orientierung des Weibchens von *Liogryllus campestris* L. nach dem Stradulationsschal des Männchens. *Akad. Wiss. Math. Nat. Kl. Abt. I (Wien)* 132:81–88.

Reinhold, K. 1999. Energetically costly behaviour and the evolution of resting metabolic rate in insects. *Functional Ecol.* 13:217–24.

Reinhold, K, M. D. Greenfield, and Y. Jang. 1998. Energetic cost of sexual attractiveness: Ultrasonic advertisement in wax moths. *Anim. Behav.* 55:905–13.

Reis, I. 1996. Die Entwicklung der Stridulationsmuster bei verschiedenen Feldheuschrecken der Gattung *Chorthippus*. Ph.D. diss., Universität Göttingen.

Reiss, R. F. 1964. *Neural theory and modeling*. Stanford: Stanford University Press.

Reyer, H.-U., G. Frei, and C. Som. 1999. Cryptic female choice: Frogs reduce clutch size when amplexed by undesirable males. *Proc. R. Soc. London B* 266:2101–7.

Rheinlaender, J. 1984. Das akustische Orientierungsverhalten von Heuschrecken, Grillen und Fröschen: Eine vergleichende neuro- und verhaltensphysiologische Untersuchung. Habil-Thesis, Universität Bochum.

Rheinlaender, J., and G. Blätgen. 1982. The precision of auditory lateralization in the cricket, *Gryllus bimaculatus*. *Physiol. Entomol.* 7:209–18.

Rheinlaender, J., H. C. Gerhardt, D. D. Yager, and R. R. Capranica. 1979. Accuracy of phonotaxis by the green treefrog (*Hyla cinerea*). *J. Comp. Physiol.* 133:247–55.

Rheinlaender, J., K. Kalmring, A. V. Popov, and H. G. Rehbein. 1976. Brain projections and information processing of biologically significant sounds by two large ventral-cord neurons of *Gryllus bimaculatus* DeGeer (Orthoptera, Gryllidae). *J. Comp. Physiol.* 110:251–69.

Rheinlaender, J., and A. Mörchen 1979. "Time-intensity trading" in locust auditory interneurones. *Nature (London)* 281:672–74.

Rheinlaender, J., and H. Römer. 1986. Insect hearing in the field I. The use of identified nerve cells as "biological microphones." *J. Comp. Physiol. A* 158:647–51.

———. 1990. Acoustic cues for sound localization and spacing in Orthopteran insect. In *The Tettigoniidae, biology, systematics and evolution*, ed. W. J. Bailey and D. C. F. Rentz, 248–64. Bathurst: Crawford House Press.

Rheinlaender, J., W. Walkowiak, and H. C. Gerhardt. 1981. Directional hearing in the green treefrog: A variable mechanism? *Naturwissenschaften* 68:430–31.

Richards, D. G., and R. H. Wiley. 1980. Reverberations and amplitude fluctuations in the propagation of sound in a forest: Implications for animal communication. *Am. Nat.* 115:381–99.

Ritchie, M. G. 1996. The shape of female mating preferences. *Proc. Natl. Acad. Sci. U.S.A.* 93:14628–31.

———. 2000. The inheritance of female preference functions in a mate recognition system. *Proc. R. Soc. London B* 267:327–32.

Ritchie, M. G., I. D. Couzin, and W. A. Snedden. 1995. What's in a song? Female bushcrickets discriminate against the song of older males. *Proc. R. Soc. London B* 262:21–27.

Ritchie, M. G., and J. M. Gleason. 1995. Rapid evolution of courtship song pattern in *Drosophila willstoni* sibling species. *J. Evol. Biol.* 8:463–79.

Ritchie, M. G., and C. P. Kyriacou. 1996. Artificial selection for courtship signal in *Drosophila melanogaster. Anim. Behav.* 52:603–11.

Ritchie, M. G., and S. D. F. Phillips. 1998. The genetics of sexual isolation. In *Endless forms: Species and speciation*, ed. D. J. Howard and S. H. Berlocher, 291–308. Oxford: Oxford University Press.

Ritchie, M. G., D. Sunter, and L. R. Hockham. 1998. Behavioral components of sex role reversal in the tettigoniid bushcricket *Ephippiger ephippiger*. *J. Insect Behav.* 11: 481–91.

Ritke, M. E., and R. D. Semlitsch. 1991. Mating behavior and determinants of male mating success in the gray treefrog, *Hyla chrysoscelis. Can. J. Zool.* 69:246–50.

Roberts, J. D. 1997. Call evolution in *Neobatrachus* (Anura: Myobatrachidae): Speculations on tetraploid origins. *Copeia* 1997:791–801.

Roberts, J. D., R. J. Standish, P. G. Byrne, and P. Doughty. 2000. Synchronous polyandry and multiple paternity in the frog *Crinia georgiana* (Anura: Myobatrachidae). *Anim. Behav.* 57:721–26.

Roberts, J. D., and G. Wardell-Johnson. 1995. Call differences between peripheral isolates of the *Geocrinia rosea* complex (Anura: Myobatrachidae) in southwestern Australia. *Copeia* 1995:899–906.

Robertson, J. G. M. 1984. Acoustic spacing by breeding males of *Uperoleia rugosa* (Anura: Leptodactylidae). *Z. Tierpsychol.* 64:283–93.

———. 1986a. Male territoriality, fighting and assessment of fighting ability in the Australian frog *Uperoleia rugosa. Anim. Behav.* 34:763–72.

———. 1986b. Female choice, male strategies and the role of vocalisations in the Australian frog *Uperoleia rugosa. Anim. Behav.* 34:773–84.

———. 1990. Female choice increases fertilisation success in the Australian frog, *Uperoleia laevigata. Anim. Behav.* 39:639–45.

Robinson, D. 1990. Acoustic communication between the sexes in bushcrickets. In *The Tettigoniidae: Behaviour, systematics, evolution*, ed. W. J. Bailey and D. C. F. Rentz, 110–29. Bathurst: Crawford House Press.

Robinson, D., J. Rheinlaender, and J. C. Hartley. 1986. Temporal parameters of male-female sound communication in *Leptophyes punctatissima. Physiol. Entomol.* 11:317–23.

Roble, S. M. 1985. Observations on satellites males in *Hyla chrysoscelis, Hyla picta*, and *Pseudacris triseriata. J. Herpetol.* 19:432–36.

Roeder, K. D. 1967. *Nerve cells and insect behavior*. Cambridge: Harvard University Press.

Römer, H. 1985. Anatomical representation of frequency and intensity in the auditory system of Orthoptera. In *Acoustic and vibrational communication in insects*, ed. K. Kalmring and N. Elsner, 25–32. Berlin: Parey Verlag.

———. 1987. Representation of auditory distance with a central neuropil of the bushcricket *Mygalopsis marki. J. Comp. Physiol. A* 161:33–42.

———. 1992. Ecological constraints for the evolution of hearing and sound communication in insects. In *The evolutionary biology of hearing*, ed. D. B. Webster, R. R. Fay, and A. N. Popper, 79–94. New York: Springer-Verlag.
———. 1993. Environmental and biological constraints for the evolution of long-range signalling and hearing in acoustic insects. *Phil. Trans. Royal Soc. London B* 340: 179–85.
———. 1998. The sensory ecology of acoustic communication in insects. In *Comparative hearing: Insects*, ed. R. R. Hoy, A. N. Popper, and R. R. Fay, 63–96. New York: Springer-Verlag.
Römer, H., and W. J. Bailey. 1986. Insect hearing in the field II. Male spacing behaviour and correlated acoustic cues in the bushcricket *Mygalopsis marki*. *J. Comp. Physiol. A* 159:627–38.
———. 1998. Strategies for hearing in noise: Peripheral control over auditory sensitivity in the bushcricket *Sciarasaga quadrata* (Austrosaginae: Tettigoniidae). *J. Exp. Biol.* 201:1023–33.
Römer, H., W. Bailey, and I. Dadour. 1989. Insect hearing in the field. III. Masking by noise. *J. Comp. Physiol. A* 164:609–20.
Römer, H., B. Hedwig, and S. R. Ott. 1997. Proximate mechanism of female preference for the leader male in synchronizing bushcrickets (*Mecopoda elongata*). In *From membrane to mind*, ed. N. Elsner and H. Wässle, 322. New York: Thieme Verlag.
Römer, H., and M. Krusch. 2000. A gain-control mechanism for processing of chorus sounds in the afferent auditory pathway of the bushcricket *Tettigonia viridissima* (Orthoptera: Tettigoniidae). *J. Comp. Physiol. A* 186:181–91.
Römer, H., and J. Lewald. 1992. High-frequency sound transmission in natural habitats: Implications for the evolution of insect acoustic communication. *Behav. Ecol. Sociobiol.* 157:631–42.
Römer, H., and V. Marquart. 1984. Morphology and physiology of auditory interneurons in the metathoracic ganglion of the locust. *J. Comp. Physiol. A* 155:249–62.
Römer, H., V. Marquardt, and M. Hardt. 1988. Organization of a sensory neuropile in the auditory pathway of two groups of Orthoptera. *J. Comp. Neurol.* 275:201–15.
Römer, H., and J. Rheinlaender. 1989. Hearing in insects and its adaptation to environmental constraints. In *Biological signal processing*, ed. H. C. Lüttgau and R. Necker, 146–62. Weinheim: VCH Verlag.
Römer, H., J. Rheinlaender, and R. Dronse. 1981. Intracellular studies of auditory processing in the metathoracic ganglion of the locust. *J. Comp. Physiol.* 144:305–12.
Römer, H., and U. Seikowski. 1985. Responses to model songs of auditory neurons in the thoracic ganglia and brain of the locust. *J. Comp. Physiol. A* 156:845–60.
Römer, H., M. Spickermann, and W. Bailey. 1998. Sensory bias for sound intensity discrimination in the bushcricket *Requena verticalis* (Tettigoniidae, Orthoptera). *J. Comp. Physiol. A* 182:595–607.
Römer, H., and J. Tautz. 1992. Invertebrate auditory receptors. In *Advances in comparative and environmental physiology 10*, 185–212. Heidelberg: Springer-Verlag.
Ronacher, B. 1989. Stridulation of acridid grasshoppers after hemisection of thoracic ganglia: Evidence for hemiganglionic oscillators. *J. Comp. Physiol. A* 164:723–36.
———. 1990. Contribution of brain and thoracic ganglia to the generation of the stridulation pattern in *Chorthippus dorsatus*. In *Sensory systems and communication in arthropods*, ed. F. G. Gribakin, K. Wiese, and A. V. Popov, 317–23. Basel: Birkhäuser Verlag.

———. 1991. Contribution of abdominal commissures in the bilateral coordination of the hindlegs during stridulation in the grasshopper *Chorthippus dorsatus*. *J. Comp. Physiol. A* 169:191–200.

Ronacher, B., D. von Helversen, and O. von Helversen. 1986. Routes and stations in the processing of auditory directional information in the CNS of a grasshopper, as revealed by surgical experiments. *J. Comp. Physiol. A* 158:363–74.

Ronacher, B., and R. Krahe. 1998. Song recognition in the grasshopper *Chorthippus biguttulus* is not impaired by shortening song signals: Implications for neuronal coding. *J. Comp. Physiol. A* 183:729–35.

———. 2000. Temporal integration vs. parallel processing: Coping with the variability of neuronal messages in directional hearing of insects. *European J. Neurosci.* 12: 2147–56.

Ronacher, B., and H. Römer. 1985. Spike synchronization of tympanic receptor fibres in a grasshopper (*Chorthippus biguttulus* L., Acrididae). *J. Comp. Physiol. A* 157:631.

Ronacher, B., and A. Stumpner. 1988. Filtering of behaviourally relevant temporal parameters of a grasshopper's song by an auditory interneuron. *J. Comp. Physiol. A* 163:517–23.

Ronacher, B., A. Stumpner, T. Sokoliuk, and B. Herrmann. 1993. Acoustic communication of grasshopper males after lesions in the thoracic connectives: Correlation with the ascending projections of identified auditory neurons. *Zool. Jb.Physiol.* 97: 199–214.

Rose, G. J., and E. A. Brenowitz. 1991. Aggressive thresholds of male Pacific treefrogs for advertisement calls vary with amplitude of neighbors' calls. *Ethology* 89:244–52.

Rose, G. J., E. A. Brenowitz, and R. R. Capranica. 1985. Species specificity and temperature dependency of temporal processing by the auditory midbrain of two species of treefrogs. *J. Comp. Physiol. A* 157:763–69.

Rose, G. J., and R. R. Capranica. 1984. Processing amplitude modulated sounds by the auditory midbrain of two species of toads: matched temporal filters. *J. Comp. Physiol. A* 154:211–19.

———. 1985. Sensitivity to amplitude modulated sounds in the anuran auditory system. *J. Neurophysiol.* 53:446–65.

Rösel von Rosenhof, A. J. [1749] 1978. *Insectenbelustigung 2.* Theil. reprint. Dortmund: Harenburg Kommuinikation.

Rössler, W. 1992a. Functional morphology and development of tibial organs in the legs I, II, and III of the bushcricket *Ephippiger ephippiger* (Insecta, Ensifera). *Zoomorphology* 112:181–88.

———. 1992b. Postembryonic development of the complex tibial organ in the foreleg of the bushcricket *Ephippiger ephippiger* (Orthoptera, Tettigoniidae). *Cell Tissue Res.* 269:505–14.

Rössler, W., A. Hubschen, J. Schul, and K. Kalmring. 1994. Functional morphology of bushcricket ears: Comparison between two species belonging to the Phaneropterinae and Decticinae (Insects, Ensifera). *Zoomorphologie* 114:39–46.

Rotenberry, J. T., M. Zuk, L. W. Simmons, and C. Hayes. 1996. Phonotactic parasitoids and cricket song structure: An evaluation of alternative hypotheses. *Evol. Ecol.* 10:233–43.

Rundle, H. D., and D. Schluter. 1998. Reinforcement of stickleback mate preferences: Sympatry breeds contempt. *Evolution* 52:200–8.

Runkle L. S., K. D. Wells, C. C. Robb, and S. L. Lance. 1994. Individual, nightly and seasonal variation in calling behavior of the gray treefrog, *Hyla versicolor*: Implications for energy expenditure. *Behav. Ecol.* 5:318–25.

Ryan, M. J. 1980. The reproductive behavior of the bullfrog (*Rana catesbeiana*). *Copeia* 1980:108–14.

———. 1985. *The Túngara frog: A study in sexual selection and communication.* Chicago: University of Chicago Press.

———. 1986. Synchronized calling in a treefrog (*Smilisca sila*). *Brain Behav. Evol.* 29: 196–206.

———. 1988. Constraints and patterns in the evolution of anuran acoustic communication. In *The evolution of the amphibian auditory system*, ed. B. Fritzsch, M. J. Ryan, W. Wilczynski, T. E. Hetherington, and W. Walkowiak, 637–77. New York: John Wiley and Sons.

———. 1990. Sensory systems, sexual selection, and sensory exploitation. *Oxford Surveys in Evolutionary Biology* 7:157–95.

———. 1996. Phylogenetics in behavior: Some cautions and expectations. In *Phylogenies and the comparative method in animal behavior*, ed. E. P. Martins, 1–21. Oxford: Oxford University Press.

Ryan, M. J., R. B. Cocroft, and W. Wilczynski. 1990. The role of environmental selection in intraspecific divergence of mate recognition signals in the cricket frog, *Acris crepitans*. *Evolution* 44:1869–72.

Ryan, M. J., and R. C. Drewes. 1990. Vocal morphology of the *Physalaemus pustulosus* species group (Leptodactylidae): Morphological response to sexual selection for complex calls. *Biol. J. Linnean Soc.* 40:37–52.

Ryan, M. J., J. H. Fox, W. Wilczynski, and A. S. Rand. 1990. Sexual selection for sensory exploitation in the frog *Physalaemus pustulosus*. *Nature (London)* 343:66–67.

Ryan, M. J., and A. Keddy-Hector. 1992. Directional patterns of female mate choice and the role of sensory biases. *Am. Natur.* 139:S4–S35.

Ryan, M. J., S. A. Perrill, and W. Wilczynski. 1992. Auditory tuning and call frequency predict population based mating preferences in the cricket frog, *Acris crepitans*. *Am. Natur.* 139:1370–83.

Ryan, M. J., and A. S. Rand. 1990. The sensory basis of sexual selection for complex calls in the túngara frog, *Physalaemus pustulosus* (sexual selection for sensory exploitation). *Evolution* 44:305–14.

———. 1993. Sexual selection and signal evolution: The ghost of biases past. *Phil. Trans. R. Soc. Lond. B* 340:187–95.

———. 1999. Phylogenetic influence on mating call preferences in female túngara frogs, *Physalaemus pustulosus*. *Anim. Behav.* 57:945–56.

Ryan, M. J., A. S. Rand, and L. A. Weigt. 1996. Allozyme and advertisement call variation in the túngara frog, *Physalaemus pustulosus*. *Evolution* 50:2435–53.

Ryan, M. J., and B. K. Sullivan. 1989. Transmission effects on temporal structure in the advertisement calls of two toads, *Bufo woodhousii* and *Bufo valliceps*. *Ethology* 80: 182–89.

Ryan, M. J., M. D. Tuttle, and L. K. Taft. 1981. The costs and benefits of frog chorusing behavior. *Behav. Ecol. Sociobiol.* 8:273–78.

Ryan, M. J., and W. Wilczynski. 1991. Evolution of intraspecific variation in the advertisement call of a cricket frog (*Acris crepitans*, Hylidae). *Biol. J. Linnean Soc.* 44: 249–71.

Ryder, J. J., and M. T. Siva-Jothy. 2000. Male calling song provides a reliable signal of immune function in a cricket. *Proc. R. Soc. Lond. B* 267:1171–75.

———. 2001. Quantitative genetics of immune function and body size in the house cricket, *Acheta domesticus. J. Evol. Biol.* 14:646–53.

Sakaluk, S. K. 1987. Reproductive behaviour of the decorated cricket, *Gryllodes supplicans* (Orthoptera: Gryllidae): Calling schedules, spatial distribution and mating. *Behaviour* 100:202–25.

———. 2000. Sensory exploitation as an evolutionary origin to nuptial food gifts in insects. *Proc. R. Soc. London B* 267:339–43.

Sakaluk, S. K. and A.-K. Eggett. 1996. Female control of sperm transfer intraspecific variation in sperm precedence: Antecedents to the evolution of a courtship food gift. *Evolution* 50:694–703.

Sales, G., and D. Pye. 1974. *Ultrasonic communication by animals.* New York: John Wiley and Sons.

Sanborn, A. F. 2001. Timbal muscle physiology in the endothermic cicada *Tibicen winnemanna* (Homoptera: Cicadidae). *Comp. Biochem. Physiol. A* 130:9–19.

Sanborn, A. F., and P. K. Phillips. 1999. Analysis of acoustic signals produced by the cicada *Platypedia putnami* variety *lutea* (Homoptera: Tibicinidae). *Ann. Entomol. Soc. Am.* 92:451–55.

Sanderson, N., J. M. Szymura, and N. H. Barton. 1992. Variation in mating call across the hybrid zone between the fire-bellied toads *Bombina bombina* and *B. variegata. Evolution* 46:595–607.

Schäffner, K.-H. 1984. Mechanorezeptoren auf den Vorderflügeln der Grillenmännchen und ihre Bedeutung bei der Stridulation. Thesis, Univ. München.

Schäffner, K.-H., and U. T. Koch. 1987a. Effects of wing campaniform sensilla lesions on stridulation in crickets. *J. Exp. Biol.* 129:25–40.

———. 1987b. A new field of wing campaniform sensilla essential for the production of the attractive calling song in crickets. *J. Exp. Biol.* 129:1–23.

Schatral, A. 1990. Interspecific acoustic behaviour among bushcrickets. In *The Tettigoniidae: Behaviour, systematics, evolution*, ed. W. J. Bailey and D. C. F. Rentz, 150–65. Bathurst: Crawford House Press.

Schatral, A., and W. J. Bailey. 1991a. Decisions during phonotaxis in the bushcricket *Requena vericalis* (Orthoptera: Tettigoniidae): Do females change direction to alternative male calls? *Ethology* 88:320–30.

———. 1991b. Song variability and the response to conspecific song and to song models of different frequency contents in males of the bushcricket *Requena verticalis* (Orthoptera: Tettigoniidae). *Behaviour* 116:163–79.

Schildberger, K. 1984. Temporal selectivity of identified auditory neurons in the cricket brain. *J. Comp. Physiol. A* 155:171–85.

———. 1994. The auditory pathway of crickets: Adaptations for intraspecific communication. *Fortschr. Zool.* 39:209–25.

Schildberger, K., and M. Hörner. 1988. The function of auditory neurons in cricket phonotaxis. I. Influence of hyperpolarization of identified neurons on sound localization. *J. Comp. Physiol. A* 163:621–31.

Schildberger, K., F. Huber, and D.W. Wohlers. 1989. Central auditory pathway: Neuronal correlates of phonotactic behavior. In *Cricket behavior and neurobiology*, ed. F. Huber, T. E. Moore, and W. Loher, 423–58. Ithaca: Cornell University Press.

Schildberger, K., and H.-U. Kleindienst. 1989. Sound localization in intact and one-eared crickets: Comparison of neuronal properties with open-loop and closed-loop behavior. *J. Comp. Physiol. A* 165:615–26.

Schildberger, K., J. J. Milde, and M. Hoerner. 1988. The function of auditory neurons in cricket phonotaxis. II. Modulation of auditory responses during locomotion. *J. Comp. Physiol. A* 163:633–40.

Schlaepfer, M. A., and R. Figeroa-Sandí. 1998. Female reciprocal calling in a Costa Rican leaf-litter frog, *Eleutherodactylus podiciferus. Copeia* 1998:1076–80.

Schmidt, R. S. 1973. Central mechanisms of frog calling. *Amer. Zool.* 13:1169–77.

———. 1974a. Neural correlates of frog calling: Independence from peripheral feedback. *J. Comp. Physiol.* 88:321–33.

———. 1974b. Neural correlates of frog calling: Trigeminal tegmentum. *J. Comp. Physiol.* 92:229–54.

———. 1976. Neural correlates of frog calling: Isolated brainstem. *J. Comp. Physiol.* 108:99–113.

———. 1978. Neural correlates of frog calling: Circum-metamorphic "calling" in leopard frog. *J. Comp. Physiol.* 126:49–56.

———. 1988. Mating call phonotaxis in female American toads: Lesions of central auditory system. *Brain Behav. Evol.* 32:119–28.

———. 1992. Neural correlates of frog calling: Production by two semi-independent generators. *Behav. Brain Res.* 50:17–30.

Schmitz, B., H. Scharstein, and G. Wendler. 1982. Phonotaxis in *Gryllus campestris* L. (Orthoptera Gryllidae): I. Mechanism of acoustic orientation in intact females. *J. Comp. Physiol.* 148:431–44.

Schmitz, B., T. D. White, and P. M. Narins. 1992. Directionality of phase locking in auditory nerve fibers of the leopard frog *Rana pipiens pipiens. J. Comp. Physiol. A* 170:589–604.

Schneider, H. 1977. Acoustic behavior and physiology of vocalization in the European tree frog, *Hyla arborea* (L.). In *The reproductive biology of amphibians*, ed. D. H. Taylor and S. I. Guttman, 295–336. New York: Plenum Press.

———. 1982. Phonotaxis bei Weibchen des Kanarischen Laubfroches, *Hyla meridionalis. Zool. Anz., Jena* 208:161–74.

———. 1988. Peripheral and central mechanisms of vocalization. In *The evolution of the amphibian auditory system*, ed. B. Fritzsch, M. J. Ryan, W. Wilczynski, T. E. Hetherington, and W. Walkowiak, 537–58. New York: John Wiley and Sons.

Schul, J. 1994. Untersuchungen zur akustischen Kommunikation bei drei Arten der Gattung *Tettigonia* (Orthoptera, Tettigoniidae). Ph.D. diss., Phillips-University Marburg.

———. 1997. Neuronal basis of phonotactic behaviour in *Tettigonia viridissima*: Processing of behaviourally relevant signals by auditory afferents and thoracic interneuons. *J. Comp. Physiol. A* 180:573–83.

———. 1998. Song recognition by temporal cues in a group of closely related bushcricket species (genus *Tettigonia*). *J. Comp. Physiol. A* 183:401–10.

———. 1999. Neuronal basis for spectral song discrimination in the bushcricket *Tettigonia cantans. J. Comp. Physiol. A* 184:457–61.

Schul, J., and S. L. Bush. 2000. Rate recognition in *Tettigonia cantans*: Evidence against high- and low pass filters. In *Göttingen neurobioloy report*, ed. N. Elsner and G. W. Kreutzberg, Stuttgart: Thieme Verlag.

Schul, J., and M. Fritsch. 1999. Sound intensity discrimination in the absence of directional cues: A behavioural test in the katydid *Tettigonia cantans*. In *Göttingen neurobiology report*, ed. N. Elsner and Eysel, 71. Stuttgart: Thieme Verlag.

Schul, J., D. von Helversen, and T. Weber. 1998. Selective phonotaxis in *Tettigonia cantans* and *T. viridissima* in song recognition and discrimination. *J. Comp. Physiol. A* 182:687–94.

Schul, J., M. Holderied, D. von Helversen, and O. von Helversen. 1999. Directional hearing in grasshoppers: Neurophysiological testing of a biological model. *J. Exp. Biol.* 202:121–33.

Schul, J., and W. Schulze. 2001. Phonotaxis during walking and flying: Are differences in selectivity due to predation pressure? *Naturwissenschaften* 88:438–42.

Schuster, S. M., and M. J. Wade. 1991. Equal mating success among male reproductive strategies in a marine isopod. *Nature (London)* 350:608–10.

Schwartz, J. J. 1986. Male call behavior and female choice in a neotropical frog. *Ethology* 73:116–27.

———. 1987a. The function of call alternation in anuran amphibians: A test of three hypotheses. *Evolution* 41:461–71.

———. 1987b. The importance of spectral and temporal properties in species and call recognition in a neotropical treefrog with a complex vocal repertoire. *Anim. Behav.* 35:340–47.

———. 1989. Graded aggressive calls of the spring peeper, *Pseudacris crucifer*. *Herpetologica* 45:172–81.

———. 1991. Why stop calling? A study of unison bout singing in a neotropical treefrog. *Anim. Behav.* 42:565–78.

———. 1993. Male calling behavior, female discrimination and acoustic interference in the Neotropical treefrog, *Hyla microcephala* under realistic conditions. *Behav. Ecol. Sociobiol.* 32:401–14.

———. 1994. Male advertisement and female choice in frogs: Recent findings and new approaches to the study of communication in a dynamic acoustic environment. *Am. Zool.* 34:616–24.

———. 2001. Call monitoring and interactive playback systems in the study of acoustic interactions among male anurans. In *Anuran communication*, ed. M. J. Ryan, 183–204. Washington: Smithsonian Institution Press.

Schwartz, J. J., B. W. Buchanan, and H. C. Gerhardt. 2001. Female mate choice in the gray treefrog (*Hyla versicolor*) in three experimental environments. *Behav. Ecol. Sociobiol.* 49:443–55.

Schwartz, J. J., and H. C. Gerhardt. 1989. Spatially mediated release from auditory masking in an anuran amphibian. *J. Comp. Physiol. A* 166:37–41.

———. 1995. Directionality of the auditory system and call pattern recognition during acoustic interference in the gray treefrog, *Hyla versicolor*. *Auditory Neurosci.* 1:195–206.

———. 1998. The neuroethology of frequency preferences in the spring peeper. *Anim. Behav.* 56:55–69.

Schwartz, J. J., S. Ressel, and C. Bevier. 1995. Carbohydrate and calling: Depletion of muscle glycogen and the chorusing dynamics of the neotropical treefrog *Hyla microcephala*. *Behav. Ecol. Sociobiol.* 37:125–35.

Schwartz, J. J., and A.M. Simmons. 1990. Encoding of a spectrally-complex communication sound in the bullfrog's auditory nerve. *J. Comp. Physiol. A* 166:489–99.

Schwartz, J. J., and K. D. Wells. 1983a. An experimental study of acoustic interference between two species of neotropical treefrogs. *Anim. Behav.* 31:181–90.

———. 1983b. The influence of background noise on the behavior of a neotropical treefrog, *Hyla ebraccata. Herpetologica* 39:121–29.

———. 1984a. Interspecific acoustic interactions of the neotropical treefrog *Hyla ebraccata. Behav. Ecol. Sociobiol.* 14:211–24.

———. 1984b. Vocal behavior of the neotropical treefrog *Hyla phlebodes. Herpetologica* 40:452–63.

———. 1985. Intra- and interspecific vocal behavior of the neotropical treefrog *Hyla microcephala. Copeia* 1985:27–38.

Selverston, A. I., H.-U. Kleindienst, and F. Huber. 1985. Synaptic connectivity between cricket auditory interneurons as studied by selective photoinactivation. *J. Neurosci.* 5:1283–92.

Semlitsch, R. D. 1994. Evolutionary consequences of non-random mating: Do large males increase offspring fitness in the anuran *Bufo bufo*? *Behav. Ecol. Sociobiol.* 34: 19–24.

Semsar, K., K. F. Klomberg, and C. A. Marler. 1998. Arginine vasotocin increases call–site acquisiton by nonresident male grey treefrogs. *Anim. Behav.* 56:983–87.

Servedio, M. R., and M. Kirkpatrick. 1999. The effects of gene flow on reinforcement. *Evolution* 51:1764–72.

Shaw, K. L. 1996a. Polygenic inheritance of a behavioral phenotype: Interspecific genetics of song in the Hawaiian cricket genus *Laupala. Evolution* 50:256–66.

———. 1996b. Sequential radiations and patterns of speciation in the Hawaiian cricket genus *Laupala* inferred from DNA sequences. *Evolution* 50:37–255.

———. 1999. A nested analysis of song groups and species boundaries in the Hawaiian cricket genus *Laupala. Mol. Phyl. Evol.* 11:332–41.

———. 2000. Interspecific genetics of mate recognition: inheritance of female acoustic preference in Hawaiian crickets. *Evolution* 54:1303–12.

Shaw, K. L., and D. Herlihy. 2000. Acoustic preference functions and song variability in the Hawaiian cricket *Laupala cerasina. Proc. R. Soc. Lond. B* 267:577–84.

Shaw, S. R. 1994. Detection of airborne sound by a cockroach 'vibration detector': A possible missing link in insect auditory evolution. *J. Exp. Biol.* 193:13–47.

Shelly, T. E., and M. D. Greenfield. 1985. Alternative mating strategies in a desert grasshopper: A transitional analysis. *Anim. Behav.* 33:1211–22.

———. 1989. Satellites and transients: Ecological constraints on alternative mating tactics in male grasshoppers. *Behaviour* 109:200–21.

———. 1991. Dominions and desert clickers (Orthoptera: Acrididae): Influences of resources and male signaling on female settlement patterns. *Behav. Ecol. Sociobiol.* 28:133–40.

Shuvalov, V. F., and A. V. Popov. 1993. Mechanisms of chirp duration evaluation in crickets *Gryllus bimaculatus*: A behavioral approach. In *Sensory systems of arthropods*, ed. K. Wiese, 328–35. Basel: Birkhäuser.

Simmons, A. M., R. C. Buxbaum, and M. P. Mirin. 1993a. Perception of complex sounds by the green treefrog, *Hyla cinerea*: Envelope and fine structure cues. *J. Comp. Physiol. A* 173:321–27.

Simmons, A. M., G. Reese, and M. Ferragamo. 1993b. Periodicity extraction in the anuran auditory nerve. II. Phase and temporal fine–structure. *J. Acoust. Soc. Am.* 93: 3374– 3389.

Simmons, D. D., C. Bertolotto, and P. M. Narins. 1992. Innervation of the amphibian and basilar papillae in the leopard frog: reconstruction of single labeled fibers. *J. Comp. Neurol.* 322:191–200.

Simmons, J. A., E. G. Wever and J. M. Pylka. 1971. Periodical cicada: Sound production and hearing. *Science* 171:212–213.

Simmons, L. W. 1987. Female choice contributes to offspring fitness in the field cricket, *Gryllus bimaculatus* (De Geer). *Behav. Ecol. Sociobiol.* 21:313–21.

———. 1988. The calling song of the field cricket *Gryllus bimaculatus* (DeGeer): Constraints on transmission and its role in intermale competition and female choice. *Anim. Behav.* 36:380–94.

———. 1995. Correlates of male quality in the field cricket, *Gryllus campestris* L: Age, size, and symmetry determine pairing success in field populations. *Behav. Ecol.* 6: 376–81.

Simmons, L. W., and W. J. Bailey. 1991. Agonistic communication between males of a zaprochiline katydid (Orthoptera: Tettigoniidae). *Behav. Ecol.* 4:364–68.

Simmons, L. W., and M. G. Ritchie. 1996. Symmetry in the songs of crickets. *Proc. R. Soc. London B* 263:305–11.

Simmons, L. W., R. J. Teale, M. Maier, R. J. Standish, W. J. Bailey, and P. C. Withers. 1992. Some costs of reproduction for male bushcrickets, *Requena verticalis* (Orthoptera: Tettigoniidae): Allocating resources to mate attraction and nuptial feeding. *Behav. Ecol. Sociobiol.* 31:57–62.

Simmons, P., and D. Young. 1978. The tymbal mechanism and song patterns of the bladder cicada, *Cystosoma saundersii*. *J. Exp. Biol.* 76:27–45.

Simon, C., J. M. Tang, S. Dalwadi, G. Staley, J. Deniega, and T. R. Unnasch. 2000. Genetic evidence for assortative mating between 13-year cicadas and sympatric "17-year cicadas with 13-year life cycles" provides support for allochronic speciation. *Evolution* 54:1326–36.

Sismondo, E. 1979. Stridulation and tegminal resonance in the tree cricket *Oecanthus nigricornis* (Orthoptera: Gryllidae: Oecanthinae). *J. Comp. Physiol.* 129:269–79.

———. 1990. Synchronous, alternating, and phase-locked stridulation by a tropical katydid. *Science* 263:823–26.

Smotherman, M. S., and P. M. Narins. 2001. Hair cells, hearing and hopping: A field guide to hair cell physiology in the frog. *J. Exp. Biol.* 203:2237–46.

Snedden, W. A. 1996. Lifetime mating success in male sagebrush crickets: Sexual selection constrained by a virgin male mating advantage. *Anim. Behav.* 51:1119–25.

Snedden, W. A., and M. D. Greenfield. 1998. Females prefer leading males: Relative call timing and sexual selection in katydid choruses. *Anim. Behav.* 56:1091–98.

Soper, R. S., G. E. Shewell, and D. Tyrrell. 1976. *Colcondamyia auditrix* nov. sp. (Diptera: Sarcophagidae); a parasite which is attracted by the mating song of its host, *Okanaga rimosa* (Homoptera: Cicadidae). *Can. Entomol.* 108:61–68.

Souroukis, K., and W. H. Cade. 1993. Reproductive competition and selection on male traits at varying sex ratios in the field cricket, *Gryllus pennsylvanicus*. *Behaviour* 126:45–62.

Stabel, J., G. Wendler, and H. Scharstein. 1989. Cricket phonotaxis depends on recognition of the calling song pattern. *J. Comp. Physiol. A* 165:165–77.

Staudacher, E., and K. Schildberger. 1993. Characteristics of some descending neurons in the walking cricket, *Gryllus bimaculatus*. In *Gene-brain-behaviour*, ed. N. Elsner and M. Heisenberg, 134, Proc. 21th Göttingen Neurobiology Conference. New York: Thieme Verlag.

———. 1998. Gating of sensory responses of descending brain neurons during walking in crickets. *J. Exp. Biol.* 201:559–72.

Stearns, S. C. 1992. *The evolution of life histories.* Oxford: Oxford University Press.

Stephen, R. O., and H. C. Bennet-Clark. 1982. The anatomical and mechanical basis of stimulation and frequency analysis in the locust ear. *J. Exp. Biol.* 99:279–314.

Stephen, R. O., and J. C. Hartley. 1995. Sound production in crickets. *J. Exp. Biol.* 198:2139–52.

Stewart, M. M., and P. J. Bishop. 1994. Effects of increased sound level of advertisement calls on calling male frogs, *Eleutherodactylus coqui. J. Herpetology* 28:46–53.

Stewart, M. M., and A. S. Rand. 1991. Vocalizations and the defense of retreat sites by male and female frogs, *Eleutherodactylus coqui. Copeia* 1991:1013–24.

Stiebler, I. B., and P. M. Narins. 1990. Temperature-dependence of auditory nerve response properties in the frog. *Hearing Res.* 46:63–82.

Stiedl, O., U. Bickmeyer, and K. Kalmring. 1991. Tooth rate impact alteration in the song of males of *Ephippiger ephippiger* Fiebig (Orthoptera: Tettigoniidae) and its consequences for phonotactic behaviour of females. *Bioacoustics* 3:1–16.

Stoddard, P. K. 1996. Vocal recognition of neighbors by territorial passerines. In *Ecology and evolution of acoustic communication in birds*, ed. D. E. Kroodsma and E. H. Miller, 356–74. Ithaca: Cornell University Press.

Stoelting, H., and A. Stumpner. 1998. Tonotopic organization of auditory receptors of the bushcricket *Pholidoptera griseoaptera* (Tettigoniidae, Decticinae). *Cell Tissue Res.* 294:377–86.

Stout, J. F., C. H. deHaan, and R. W. McGhee. 1983. Attractiveness of the male *Acheta domesticus* calling song to females. I. Dependence on each of the calling song features. *J. Comp. Physiol.* 153:509–21.

Stout, J. F., and F. Huber. 1981. Responses to features of the calling song by ascending auditory interneurons in the cricket *Gryllus campestris. Physiol. Entomol.* 6:199–212.

Stout, J. F., and R. W. McGhee. 1988. Attractiveness of the male *Acheta domesticus* calling song to females. II. The relative importance of syllable period, intensity and chirp rate. *J. Comp. Physiol. A* 164:277–87.

Strake, J., H. Luksch, and W. Walkowiak. 1994. Audio-motor interface in anurans. *Europ. J. Morphology* 32:122–26.

Straughan, I. R. 1975. An analysis of the mechanisms of mating call dicrimination in the frogs *Hyla regilla* and *Hyla cadaverina. Copeia* 1975:415–24.

Stumpner, A. 1988. Auditorische thorakale Interneurone von *Chorthippus biguttulus* L.: Morphologische und physiologische Charakterisierung und Darstellung ihrer Filtereigenschaften für verhaltensrelevante Lautattrappen. Thesis, Universität, Erlangen.

———. 1996. Tonotopic organization of the hearing organ in a bushcricket: Physiological characterization and complete staining of auditory receptor cells. *Naturwissenschaften* 83:81–84.

———. 1997. An auditory interneurone tuned to the male song frequency in the duetting bushcricket *Ancistrura nigrovittata* (Orthoptera, Phaneropteridae). *J. Exp. Biol.* 200:1089–101.

———. 1999a. Comparison of morphology and physiology of two plurisegmental sound-activated interneurones in a bushcricket. *J. Comp. Physiol. A* 185:199–205.

———. 1999b. An interneurone of unusual morphology in tuned to the female song frequency in the bushcricket *Ancistrura nigrovittata* (Orthoptera, Phaneropteridae). *J. Exp. Biol.* 202:2071–81.

Stumpner, A., G. Atkins, and J. F. Stout. 1995. Processing of unilateral and bilateral auditory input by the ON1 and L1 interneurons of the cricket *Acheta domesticus* and comparison to other cricket species. *J. Comp. Physiol. A* 177:379–88.

Stumpner, A., and D. von Helversen. 2001. Evolution and function of auditory systems in insects. *Naturwissenschaften* 88:159–70.

Stumpner, A., and O. von Helversen. 1992. Recognition of a two-element song in the grasshopper *Chorthippus dorsatus* (Orthoptera: Gomphocerinae). *J. Comp. Physiol. A* 171:405–12.

———. 1994. Song production and song recognition in a group of sibling grasshopper species (*Chorthippus dorsatus, Ch. dichrous* and *Ch. loratus*: Orthoptera: Acrididae). *Bioacoustics* 6:1–27.

Stumpner, A., and R. Lakes-Harlan. 1996. Auditory interneurones in a hearing fly (*Theorobia leonidei*, Ormiini, Tachnidaem Diptera). *J. Comp. Physiol. A* 178:227–33.

Stumpner, A., and S. Meyer. 2001. Songs and the function of song elements in four duetting bushcricket species (Ensifera, Phaneropteridae, Barbitistes). *J. Insect Behav.* 14:511–34.

Stumpner, A., and B. Ronacher. 1991a. Auditory interneurones in the metathoracic ganglion of the grasshopper *Chorthippus biguttulus.* I. morphological and physiological characterization. *J. Exp. Biol.* 158:391–410.

———. 1991b. Auditory interneurones in the metathoracic ganglion of the grasshopper *Chorthippus biguttulus.* II. Processing of temporal patterns of the song of the male. *J. Exp. Biol.* 158:411–30.

———. 1994. Neurophysiological aspects of song pattern recognition and sound localization in grasshoppers. *Amer. Zool.* 34:696–705.

Stumpner, A., B. Ronacher, and O. von Helversen. 1991. Auditory interneurones in the metathoracic ganglion of the grasshopper *Chorthippus biguttulus.* II. Processing of temporal patterns of the song of the male. *J. Exp. Biol.* 158:411–30.

Suga, N. 1966. Ultrasonic production and its reception in some neotropical Tettigoniidae. *J. Insect Physiol.* 12:1039–50.

———. 1989. Principles of auditory information-processing derived from neuroethology. *J. Exp. Biol.* 146:277–86.

———. 1990. Cortical computational maps for auditory imaging. *Neural Networks* 3:3–21.

Sullivan, B. K. 1982a. Male mating behaviour in the Great Plains toad (*Bufo cognatus*). *Anim. Behav.* 30:939–40.

———. 1982b. Sexual selection in Woodhouse's toad (*Bufo woodhousei*). I. Chorus organization. *Anim. Behav.* 30:680–96.

———. 1983. Sexual selection in Woodhouse's toad (*Bufo woodhousei*). II. Female choice. *Anim. Behav.* 31:1011–17.

———. 1985. Sexual selection and mating system variation in anuran amphibians of the Arizona-Sonora desert. *Great Basin Natur.* 45:688–96.

———. 1992. Sexual selection and calling behavior in the American toad *Bufo americanus. Copeia* 1992:1–7.

Sullivan, B. K., and S. H. Hinshaw. 1990. Variation in advertisement calls and male mating behaviour in the spring peeper (*Pseudacris crucifer*). *Copeia* 1990:1146–50.

———. 1992. Female choice and selection on male calling behaviour in the grey treefrog *Hyla versicolor*. *Anim. Behav.* 44:733–44.

Sullivan, B. K., and M. R. Leek. 1986. Acoustic communication in Woodhouse's toad (*Bufo woodhousei*). I. Response of calling males to variation in spectral and temporal components of advertisement calls. *Behaviour* 98:305–19.

Sullivan, B. K., M. J. Ryan, and P. A. Verrell. 1995. Female choice and mating system structure. In *Amphibian biology, Vol. 2, Social behaviour*, ed. H. Heatwole and B. K. Sullivan, 469–517. Chipping Norton, NSW: Surrey Beatty and Sons.

Sullivan, B. K., and W. E. Wagner Jr. 1988. Variation in advertisement and release calls and social influences on calling behavior in the Gulf Coast toad *Bufo valliceps*. *Copeia* 1998:1014–20.

Summer, K., R. Symula, M. Clough, and T. Cronin. 1999. Visual mate choice in poison frogs. *Proc. R. Soc. London B* 266:2141–45.

Sun, L., W. Wilczynski, A. S. Rand, and M. J. Ryan. 2000. Trade-off in short- and long-distance communication in túngara (*Physalaemus pustuolosus*) and cricket (*Acris crepitans*) frogs. *Behav. Ecol.* 11:102–9.

Szymura, J. M. 1993. Analysis of hybrid zones within *Bombina*. In *Hybrid zones and the evolutionary process*, ed. R. G. Harrison, 261–89. New York: Oxford University Press.

Tachon, G., A.-M. Murray, D. A. Gray, and W. H. Cade. 1999. Agonistic displays and the benefits of fighting in the field cricket, *Gryllus bimaculatus*. *J. Insect Behav.* 12: 533–43.

Taigen, T. L., and K. D. Wells. 1985. Energetics of vocalization by an anuran amphibian (*Hyla versicolor*). *J. Comp. Physiol. B* 155:163–70.

Taigen, T. L., K. D. Wells, and R. L. Marsh. 1985. The enzymatic basis of high metabolic rates in calling frogs. *Physiol. Zool.* 58:719–26.

Tauber, E., D. Cohen, M. D. Greenfield, and M. P. Pener. 2001. Duet singing and female choice in the bushcricket *Phaneroptera nana*. *Behaviour* 138:411–30.

Tauber, E., and M. P. Pener. 2000. Song recognition in female bushcrickets *Phaneroptera nana*. *J. Exp. Biol.* 203:597–603.

Tejedo, M. 1992. Large male mating advantage in natterjack toads, *Bufo calamita*: Sexual competition or energetic constraints? *Anim. Behav.* 44:733–44.

———. 1993. Do male natterjack toads join larger choruses for increased mating success? *Copeia* 1993:75–80.

Telford, S. R. 1985. Mechanisms and evolution of intermale spacing in painted reed frogs (*Hyperolius marmoratus*). *Anim. Behav.* 33:1351–61.

Telford, S. R., M. L. Dyson, and N. I. Passmore. 1989. Mate choice only occurs in small choruses of painted reed frogs (*Hyperolius marmoratus*). *Bioacoustics* 2:47–53.

Telford, S. R., and J. Van Sickle. 1989. Sexual selection in an African toad (*Bufo gutteralis*): The roles of morphology, amplexus displacement and chorus participation. *Behaviour* 110:62–75.

Thompson, N. S., K. LeDoux, and K. Moody. 1994. A system for describing bird song units. *Bioacoustics* 5:267–79.

Thornhill, R., and J. Alcock. 1983. *The evolution of insect mating systems.* Cambridge: Harvard University Press.

Thorson, J., T. Weber, and F. Huber. 1982. Auditory behaviour of the cricket. II.

Simplicity of calling-song recognition in *Gryllus* and anomalous phonotaxis at abnormal carrier frequencies. *J. Comp. Physiol.* 146:361–78.

Tobias, M. L., and D. B. Kelley. 1995. Sexual differentiation and hormonal regulation of the laryngeal synapse in *Xenopus laevis. J. Neurobiol.* 28:515–26.

Tougaard, J. 1996. Energy detection and temporal integration in the noctuid A1 auditory receptor. *J. Comp. Physiol. A* 178:669–77.

Townsend, D. S., and M. M. Stewart. 1994. Reproductive ecology of the Puerto Rican frog *Eleutherodactylus coqui. J. Herpetology* 28:34–40.

Tschuch, G. 1976. Der Einfluß synthetischer Gesänge auf die Weibchen von *Gryllus bimaculatus* de Geer. *Zool. Jb. Physiol.* 80:383–88.

Tuckerman, J. F., D. T. Gwynne, and G. K. Morris. 1993. Reliable acoustic cues for female mate preference in a katydid (*Scudderia curvicauda*, Orthoptera: Tettigoniidae). *Behav. Ecol.* 4:106–13.

Turner, G. F., and T. J. Pitcher. 1986. Attack abatement: A model for group protection by combined avoidance and dilution. *Am. Natur.* 128:228–40.

Tuttle, M. D., and M. J. Ryan. 1982. The role of synchronized calling, ambient light, and ambient noise, in anti-bat-predator behavior of a treefrog. *Behav. Ecol. Sociobiol.* 11:125–31.

Tuttle, M. D., L. K. Taft, and M. J. Ryan. 1982. Evasive behaviour of a frog (*Physalaemus pustulosus*) in response to bat (*Trachops cirrhosus*) predation. *Anim. Behav.* 30: 393–97.

Ulagaraji, S. M., and T. J. Walker. 1975. Response of flying mole crickets to three parameters of synthetic songs broadcast outdoors. *Nature (London)* 253:361–78.

Van Buskirk, J. 1997. Independent evolution of song structure and note structure in American wood warblers. *Proc. R. Soc. London B 264*:755–61.

Van Dijk, P., and H. P. Wit. 1995. Speculations on the relation between emission generation and hearing mechanisms in frogs. In *Advances in hearing research*, ed. G. A. Manley, G. M. Klump, C. Köppl, H. Faslt, and H. Oeckinghaus, 97–104. Singapore: World Scientific Publishing.

Van Gelder, J. J., and H. C. M. Hoedemaekers. 1971. Sound activity and migration during the breeding period of *Rana temporaria* L., *R. arvalis* Nilsson, *Pelobates temporaria* L., and *Rana esculenta* L. *J. Anim. Ecol.* 40:559–68.

Van Staaden, M. J., and H. Römer. 1997. Sexual signalling in bladder grasshopppers: Tactical design for maximizing call range. *J. Exp. Biol.* 200:2597–608.

Waage, J. K. 1979. Reproductive character displacement in Calopteryx (Odonata: Calopterygidae). *Evolution* 33:104–16.

Wada, M., and A. Gorbman. 1977. Mate calling induced by electrical stimulation in freely moving leopard frogs, *Rana pipiens. Hormones and Behavior* 9:41–149.

Wadepuhl, M. 1983. Control of grasshopper singing behavior by the brain: Responses to electrical stimulation. *Z. Tierpsychol.* 63:173–200.

Wagner, W. E., Jr. 1989a. Fighting, assessment, and frequency alternation in Blanchard's cricket frog. *Behav. Ecol. Sociobiol.* 25:429–36.

———. 1989b. Graded aggressive signals in Blanchard's cricket frog: Vocal responses to opponent proximity and size. *Anim. Behav.* 38:1025–38.

———. 1989c. Social correlates of variation in male calling behaviour in Blanchard's cricket frog. *Ethology* 82:27–45.

———. 1992. Deceptive or honest signalling of fighting ability? A test of alternative

hypotheses for the function of changes in call dominant frequency by male cricket frogs. *Anim. Behav.* 44:449–62.
———. 1996. Convergent song preferences between female field crickets and acoustically orienting parasitoid flies. *Behav. Ecol.* 7:279–85.
———. 1998. Measuring female mating preferences. *Anim. Behav.* 55:1029–42.
Wagner, W. E., Jr., and W. W. Hoback. 1999. Nutritional effects on male calling behaviour in the variable field cricket. *Anim. Behav.* 57:89–95.
Wagner, W. E., Jr., R. J. Kelley, K. R. Tucker, and C. J. Harper. 2001. Females receive a lifespan benefit from male ejaculates in a field cricket. *Evolution* 55:994–1001.
Wagner, W. E., Jr., A.-M. Murray, and W. H. Cade. 1995. Phenotypic variation in the mating preferences of female field crickets, *Gryllus integer*. *Anim. Behav.* 49: 1269–81.
Wagner, W. E., Jr., and M. G. Reiser. 2000. The importance of calling song and courtship song in female mate choice in the variable field cricket. *Anim. Behav.* 59: 1219–26.
Wagner, W. E., Jr., and B. K. Sullivan. 1992. Chorus organization in the Gulf Coast Toad (*Bufo valliceps*): Male and female behavior, and the opportunity for sexual selection. *Copeia* 1992:647–58.
———. 1995. Sexual selection in the Gulf Coast Toad *Bufo valliceps*: Female choice based on variable characters. *Anim. Behav.* 49:305–19.
Walker, T. J. 1957. Specificity in the response of female tree crickets (Orthoptera, Gryllidae, Oecanthinae) to calling songs of the males. *Ann. Entomol. Soc. Am.* 50: 626–36.
———. 1962. Factors responsible for intraspecific variation in the calling songs of crickets. *Evolution* 16:407–28.
———. 1969. Acoustic synchrony: Two mechanisms in the snowy tree cricket. *Science* 166:891–94.
———. 1974. Character displacement and acoustic insects. *Am. Zool.* 14:1137–50.
———. 1975a. Effects of temperature, humidity, and age on stridulatory rates in *Atlanticus* spp. (Orthoptera: Tettiogoniidae: Decticinae). *Ann. Entomol. Soc. Am.* 68: 607–11.
———. 1975b. Effects of temperature on rates in poikilothermous nervous systems: Evidence from the calling songs of meadow katydids (Orthoptera: Tettigoniidae: Orcheliumum) and reanalysis of published data. *J. Comp. Physiol.* 101:57–69.
———. 1983a. Diel patterns of calling in nocturnal Orthoptera. In *Orthopteran mating systems: Sexual competition in a diverse group of insects*, ed. D. T. Gwynne and G. K. Morris, 45–72. Boulder: Westview.
———. 1983b. Mating modes and female choice in short-tailed crickets (*Anurogryllus arboreus*). In *Orthopteran mating systems: Sexual competition in a diverse group of insects*, ed. D. T. Gwynne and G. K. Morris, 240–67. Boulder: Westview.
———. 1993. Phonotaxis in female *Ormia ochracea* (Diptera: Tachninidae), a parasitoid of field crickets. *J. Insect Behav.* 6:389–410.
Walker, T. J., and T. G. Forrest. 1989. Mole cricket phonotaxis: Effects of intensity of synthetic calling song (Orthoptera: Gryllotalpidae: *Scapteriscus acletus*). *Florida Entomol.* 72:655–59.
Walker, T. J., and S. A. Wineriter. 1991. Hosts of a phonotactic parasitoid and levels of parasitism (Diptera: Tachinidae: *Ormia ochracea*). *Florida Entomol.* 74:554–59.

Walkowiak, W. 1984. Neuronal correlates of the recognition of pulsed sound signals in the grass frog. *J. Comp. Physiol. A* 155:57–66.

———. 1988. Neuroethology of anuran call recognition. In *The evolution of the amphibian auditory system*, ed. B. Fritszch, M. J. Ryan, W. Wilczynski, T. E. Hetherington, and W. Walkowiak, 485–509. New York: John Wiley and Sons.

Walkowiak, W., M. Berlinger, J. Schul, and H. C. Gerhardt. 1999. Significance of forebrain structures in acoustically guided behavior in anurans. *Europ. J. Morphol.* 37: 177–81.

Walkowiak, W., and H. Luksch. 1994. Sensory motor interfacing in acoustic behavior of anurans. *Am. Zool.* 34:685–95.

Wallach, H., E. B. Newman, and M. R. Rosenzweig. 1949. The precedence effect in sound localization. *Am. J. Psychol.* 62:315–36.

Walton, B. M. 1988. Relationships among metabolic, locomotory, and field measures of organismal performance in the Fowler's toad (*Bufo woodhousei fowleri*). *Physiol. Zool.* 61:107–18.

Wang, G., M. D. Greenfield, and T. E. Shelly. 1990. Inter-male competition for high quality host plants: The evolution of protandry in a territorial grasshopper. *Behav. Ecol. Sociobiol.* 27:191–98.

Wang, J., T. A. Ludwig, and P. M. Narins. 1996. Spatial and spectral dependence of the auditory periphery in the northern leopard frog. *J. Comp. Physiol. A* 178:159–72.

Wasserman, A. O. 1970. Polyploidy in the common tree toad *Hyla versicolor* Le Conte. *Science* 167:385–86.

Watkins, W. A., E. R. Baylor, and A. T. Bowen. 1970. The call of *Eleutherodactylus johnstonei*, the whistling frog of Bermuda. *Copeia* 1970:559–61.

Weber, T., and J. Thorson. 1989. Phonotactic behavior of walking crickets. In *Cricket behavior and neurobiology*, ed. F. Huber, T. E. Moore, and W. Loher, 310–39. Ithaca: Cornell University Press.

Weber, T., J. Thorson, and F. Huber. 1981. Auditory behavior of the cricket. I. Dynamics of compensated walking and discrimination paradigms on the Kramer treadmill. *J. Comp. Physiol.* 141:215–32.

Welch, A. M., R. D. Semlitsch, and H. C. Gerhardt. 1998. Call duration as an indicator of genetic quality in male gray tree frogs. *Science* 280:1928–30.

Wells, K. D. 1977. The social behaviour of anuran amphibians. *Anim. Behav.* 25: 666–93.

———. 1978. Territoriality in the green frog (*Rana clamitans*): Vocalizations and agonistic behaviour. *Anim. Behav.* 26:1051–63.

———. 1988. The effects of social interactions on anuran vocal behavior. In *The evolution of the amphibian auditory system*, ed. B. Fritszch, M. J. Ryan, W. Wilczynski, T. E. Hetherington, and W. Walkowiak, 433–54. New York: John Wiley and Sons.

———. 1989. Vocal communication in a neotropical treefrog, *Hyla ebraccata*: Responses of males to graded aggressive calls. *Copeia* 1989:461–66.

———. 2001. The energetics of calling in frogs. In *Anuran communication*, ed. M. J. Ryan, 45–60. Washington: Smithsonian Institution Press.

Wells, K. D., and K. M. Bard. 1987. Vocal communication in a neotropical treefrog, *Hyla ebraccata*: Responses of females to advertisement and aggressive calls. *Behaviour* 101:200–10.

Wells, K. D., and C. R. Bevier. 1997. Contrasting patterns of energy substrate use in two species of frogs that breed in cold weather. *Herpetologica* 53:70–80.

Wells, K. D., and J. J. Schwartz. 1982. The effect of vegetation on the propagation of calls in the neotropical frog *Centrolenella fleischmanni. Herpetologica* 38:449–55.

———. 1984a. Vocal communication in a neotropical treefrog, *Hyla ebraccata*: Advertisement calls. *Anim. Behav.* 32:405–20.

———. 1984b. Vocal communication in a neotropical treefrog, *Hyla ebraccata*: Aggressive calls. *Behaviour* 91:128–45.

Wells, K. D., and T. L. Taigen. 1986. The effect of social interactions on calling energetics in the gray treefrog (*Hyla versicolor*). *Behav. Ecol. Sociobiol.* 19:9–18.

———. 1989. Calling energetics of a neotropical treefrog, *Hyla microcephala*. *Behav. Ecol. Sociobiol.* 25:13–22.

Wells, K. D., T. L. Taigen, S. W. Rusch, and C. C. Robb. 1995. Seasonal and nightly variation in glycogen reserves of calling gray treefrogs (*Hyla versicolor*). *Herpetologica* 51:359–68.

Wendler, G. 1989. Acoustic orientation of crickets (*Gryllus campestris*) in the presence of two sound sources. *Naturwissenschaften* 76:128–29.

———. 1990. Pattern recognition and localization in cricket phonotaxis. In *Sensory systems and communication in arthropods*, ed. F. G. Gribakin, K. Wiese, and A. V. Popov, 387–94. Basel: Birkhäuser Verlag.

Wendler, G., and G. Löhe. 1993. The role of the medial septum in the acoustic trachea of the cricket *Gryllus bimaculatus.* I. Importance for efficient phonotaxis. *J. Comp. Physiol. A* 173:557–64.

Wenzel, B., N. Elsner, and B. Hedwig. 1998. Microinjection of neuroactive substances into brain neuropil controls stridulation in the cricket *Gryllus bimaculatus* (DeGeer). *Naturwissenschaften* 85:452–54.

Wenzel, B., and B. Hedwig. 1999. Neurochemical control of cricket stridulation revealed by pharmacological microinjections into the brain. *J. Exp. Biol.* 202:2203–16.

Wetzel, D. M., U. L. Haerter, and D. B. Kelley. 1985. A proposed neural pathway for vocalization in South African clawed frogs, *Xenopus laevis. J. Comp. Physiol. A* 157:749–61.

Wever, E. G. 1985. *The amphibian ear.* Princeton: Princeton University Press.

Wheeler, Q. D., and R. Meier, eds. 2000. *Species concepts and phylogenetic theory: A debate.* New York: Columbia University Press.

Whitney, C. L., and J. R. Krebs. 1975. Mate selection in Pacific treefrogs. *Nature (London)* 255:325–26.

Wickler, W., and U. Seibt. 1974. Rufen and Antworten bei *Kassina senegalensis, Bufo regularis* und anderen Anuren. *Z. Tierpsych.* 34:524–37.

Wiegmann, D. D. 1999. Search behaviour and mate choice by female field crickets, *Gryllus integer. Anim. Behav.* 58:1293–98.

Wiese, K., and K. Eilts. 1985. Evidence for matched frequency dependence of bilateral inhibition in the auditory pathway of *Gryllus bimaculatus. Zool. Jb. Physiol.* 89:181–201.

Wiese, K., and K. Eilts-Grimm. 1985. Functional potential of recurrent lateral inhibition in cricket audition. In *Acoustic and vibrational communication in insects*, ed. K. Kalmring and N. Elsner, 33–40. Hamburg: Paul Parey.

Wilbur, H. M., D. I. Rubenstein, and L. Fairchild. 1978. Sexual selection in toads: The roles of female choice and male body size. *Evolution* 32:264–70.

Wilczynski, W., J. D. Allison, and C. A. Marler. 1993. Sensory pathways linking social and environmental cues to endocrine control regions of amphibian forebrains. *Brain, Behav. Evol.* 42:252–64.

Wilczynski, W., and E. A. Brenowitz. 1988. Acoustic cues mediate inter-male spacing in a neotropical frog. *Anim. Behav.* 36:1054–63.

Wilczynski, W., and R. R. Capranica. 1984. The auditory system of anuran amphibians. *Progress Neurobiol.* 22:138.

Wilczynski, W., B. E. McClelland, and A. S. Rand. 1993. Acoustic, auditory, and morphological divergence in three species of neotropical frog. *J. Comp. Physiol. A* 172: 425–38.

Wilczynski, W., A. S. Rand, and M. J. Ryan. 1995. The processing of spectral cues by the call analysis system of the túngara frog, *Physalaemus pustulosus. Anim. Behav.* 49: 911–29.

Wilczynski, W., C. Reisler, and R. R. Capranica. 1987. Tympanic and extratympanic sound transmission in the leopard frog. *J. Comp. Physiol. A* 161:659–69.

Wilczynski, W., and M. J. Ryan. 1999. Geographic variation in animal communication systems. In *Geographic variation in behavior*, ed. S. A. Foster and J. A. Endler, 234–61. New York: Oxford University Press.

Wiley, R. H. 1983. The evolution of communication: information and manipulation. In *Animal behaviour, Vol. 2: Communication*, ed. T. J. Halliday and P. J. B. Slater, 156–89. Oxford: Blackwell Publishers.

———. 1991. Associations of song properties with habitats for terrestrial oscine birds of eastern North America. *Am. Natur.* 138:973–93.

———. 1994. Errors, exaggeration, and deception in animal communication. In *Behavioral mechanisms in evolutionary ecology*, ed. L. A. Real, 157–89. Chicago: University of Chicago Press.

Wiley, R. H., and D. G. Richards. 1978. Physical constraints on acoustic communication in the atmosphere: implications for the evolution of animal vocalizations. *Behav. Ecol. Sociobiol.* 3:69–94.

———. 1982. Adaptations for acoustic communication in birds: Sound transmission and signal detection. In *Acoustic communication in birds, vol. 1*, ed. D. E. Kroodsma, E. H. Miller, and H. Ouellet, 132–81. New York: Academic Press.

Williams, K. S., and C. Simon. 1995. The ecology, behavior and evolution of periodical cicadas. *Ann. Rev. Entomol.* 40:269–95.

Williams, K. S., and K. G. Smith. 1991. Dynamics of periodical cicada chorus centers (Homoptera: Cicadidae). *J. Insect Behav.* 4:275–91.

Wilson, E. O. 1975. *Sociobiology: The new synthesis.* Cambridge: Belknap/Harvard University Press.

Wohlers, D. W., and F. Huber. 1982. Processing of sound signals by six types of neurons in the prothoracic ganglion of the cricket, *Gryllus campestris* L. *J. Comp. Physiol.* 146:161–73.

Wohlers, D. W., J. D. L. Williams, F. Huber, and T. E. Moore. 1979. Central projections of fibers in the auditory and tensor nerves of cicadas (Homoptera: Cicadidae). *Cell Tissue Res.* 203:35–51.

Wolf, H., and O. von Helversen. 1986. "Switching-off" of an auditory interneuron during stridulation in the acridid grasshopper *Chorthippus biguttulus* L. *J. Comp. Physiol. A* 158:861–71.

Wollerman, L. 1998. Stabilizing and directional preferences of female *Hyla ebraccata* for calls differing in static properties. *Anim. Behav.* 55:1619–30.

Wollerman, L., and R. H. Wiley. 2002. Background noise from a natural chorus alters female discrimination of male calls in a Neotropical frog. *Anim. Behav.* 63:15–22.

———. 1999. Acoustic interference limits call detection in a Neotropical frog *Hyla ebraccata*. *Anim. Behav.* 57:529–36.

Woodward, B. D. 1986. Paternal effects on juvenile growth in *Scaphiopus multiplicatus* (the New Mexico spadefoot toad). *Am. Natur.* 128:58–65.

———. 1987. Paternal effects on offspring traits in *Scaphiopus couchi* (Anura: Pelobatidae). *Oecologica* 73:626–29.

Woodward, B. D., and J. Travis. 1991. Parental effects on juvenile growth and survival in spring peepers (*Hyla crucifer*). *Evol. Ecol.* 5:40–51.

Wu, K. H., M. L. Tobias, and D. B. Kelley. 2001. Estrogen and laryngeal strength in *Xenopus laevis*: Opposite effects of acute and chronic exposure. *Neuroendocrinology* 74:22–32.

Wyttenbach, R. A., and R. R. Hoy. 1997. Spatial acuity of ultrasonic hearing in flying crickets. *J. Exp. Biol.* 200:1999–2006.

Wyttenbach, R., M. L. May, and R. R. Hoy. 1996. Categorical perception of sound frequency by crickets. *Science* 273:1542–44.

Xu, J., D. M. Gooler, and A. S. Feng. 1994. Effects of sound direction on the processing of amplitude-modulated signals in the frog inferior colliculus. *J. Comp. Physiol. A* 178:435–45.

———. 1996. Single neurons in the frog inferior colliculus exhibit direction-dependent frequency selectivity to isointensity tone bursts. *J. Acoust. Soc. Am.* 95:2160–70.

Yager, D. D. 1992. A unique sound production mechanism in the pipid anuran *Xenopus borealis*. *Zool. J. Linnean Soc.* 104:351–75.

———. 1999. Structure, development and evolution of insect auditory systems. *Microsc. Res. and Techn.* 47:380–400.

Yamaguchi, A., and D. B. Kelley. 2000. Generating sexually differentiated vocal patterns: Laryngeal nerve and EMG recordings from vocalizing male and female African clawed frogs (*Xenopus laevis*). *J. Neurosci.* 20:1559–67.

Yost, W. A., and D. W. Nielsen. 1985. *Fundamentals of hearing: An introduction*. 2nd ed. New York: Holt, Rinehart & Winston.

Young, A. J. 1971. Studies on the acoustic behaviour of certain orthoptera. *Anim. Behav.* 19:727–43.

Young, D. 1990. Do cicadas radiate sound through their eardrums? *J. Exp. Biol.* 151:41–56.

Young, D., and H. C. Bennet-Clark. 1995. The role of the tymbal in cicada sound production. *J. Exp. Biol.* 198:1001–19.

Young, D., and R. K. Josephson. 1983a. Mechanisms of sound-production and muscle contraction kinetics in cicadas. *J. Comp. Physiol.* 152:183–95.

———. 1983b. Pure-tone songs in cicadas with special reference to the genus *Magicicada*. *J. Comp. Physiol.* 152:197–207.

Zakon, H. H., and W. Wilczynski. 1988. The physiology of the anuran eighth nerve. In *The evolution of the amphibian auditory system*, ed. B. Fritszch, M. J. Ryan, W. Wilczynski, T. E. Hetherington, and W. Walkowiak, 125–55. New York: John Wiley and Sons.

Zaretsky, M. D. 1972. Specificity of the calling song and short term changes in the phonotactic response by female crickets, *Scapsipedus marginatus* (Gryllidae). *J. Comp. Physiol.* 79:153–72.

Zelick, R. D., and P. M. Narins. 1985. Characterization of the advertisement call oscillator in the frog *Eleutherodactylus coqui. J. Comp. Physiol.* 156:223–29.

Zimmerman, B. L. 1983. A comparison of structural features of calls of open and forest habitat frog species in the central Amazon. *Herpetologica* 39:235–46.

Zimmitti, S. J. 1999. Individual variation in morphological, physiological, and biochemical features associated with calling in spring peepers (*Pseudacris crucifer*). *Physiol. Biochem. Zool.* 72:666–67.

Zuk, M. 1988. Parasite load, body size, and age of wild-caught male field crickets (Orthoptera: Gryllidae): Effects on sexual selection. *Evolution* 42:969–76.

Zuk, M., and G. R. Kolluru. 1998. Exploitation of sexual signals by predators and parasitoids. *Quart. Rev. Biol.* 73:415–38.

Zuk, M., J. T. Rotenberry, and L. W. Simmons. 1998. Calling songs of field crickets (*Teleogryllus oceanicus*) with and without phonotactic parasitoid infection. *Evolution* 52:166–71.

Zuk, M., and L. W. Simmons. 1997. Reproductive strategies of the crickets (Orthoptera: Gryllidae). In *The evolution of mating systems in insects and arachnids*, ed. J. C. Choe and B. J. Crespi, 89–109. Cambridge: Cambridge University Press.

Zuk, M., L. W. Simmons, and L. Cupp. 1993. Calling characteristics of parasitized and unparasitized populations of the field cricket *Teleogryllus oceanicus. Behav. Ecol. Sociobiol.* 33:339–43.

Zuk, M., L. W. Simmons, and J. T. Rotenberry. 1995. Acoustically-orienting parasitoids in calling and silent males of the field cricket *Teleogryllus oceanicus. Ecol. Entomol.* 20:380–83.

Zurek, P. M. 1987. The precedence effect. In *Directional hearing*, ed. W. A. Yost and G. Gourevitch, 85–105. Berlin: Springer-Verlag.

Zweifel, R. G. 1968. Effects of temperature, body size, and hybridization on mating calls of toads, *Bufo a. americanus* and *Bufo woodhousei fowleri. Copeia* 1968:269–85.

Index

Page numbers followed by an f indicate a figure; those followed by a t indicate a table.